T0237280

Frontiers in the History of Science

Frontiers in the History of Science is designed for publications of up-to-date research results encompassing all areas of history of science, primarily with a focus on the history of mathematics, physics, and their applications. Graduates and post-graduates as well as scientists will benefit from the selected and thoroughly peer-reviewed publications at the research frontiers of history of sciences and at interdisciplinary "frontiers": history of science crossing into neighboring fields such as history of epistemology, history of art, or history of culture. The series is curated by the Series Editor with the support of an international group of Associate Editors.

Series Editor:

Vincenzo De Risi, Université Paris-Diderot — CNRS, PARIS CEDEX 13, Paris, France

Associate Editors:

Karine Chemla, Paris, France
Sven Dupré, Utrecht, The Netherlands
Moritz Epple, Frankfurt, Germany
Orna Harari, Tel Aviv, Israel
Dana Jalobeanu, Bucharest, Romania
Henrique Leitão, Lisboa, Portugal
David Marshal Miller, Ames, Iowa, USA
Aurélien Robert, Tours, France
Eric Schliesser, Amsterdam, The Netherlands

Eduardo Noble

The Rise and Fall of the German Combinatorial Analysis

 Birkhäuser

Eduardo Noble
Mexico City, Mexico

ISSN 2662-2564 ISSN 2662-2572 (electronic)
Frontiers in the History of Science
ISBN 978-3-030-93819-2 ISBN 978-3-030-93820-8 (eBook)
https://doi.org/10.1007/978-3-030-93820-8

Mathematics Subject Classification: 01-02, 01A50, 01A55

This book is published under the imprint Birkhäuser, www.birkhauser-science.com, by the registered company Springer Nature Switzerland AG
The registered company address is: Gewerbestrasse 11, 6330 Cham, Switzerland

Acknowledgements

I would like to express my deep gratitude to Marco Panza, who shared with me his knowledge of the history of mathematics and encouraged me to undertake this research. His constant advice and indefatigable support helped me in many ways, more than I could enumerate here. He also encouraged me to rework the first French draft of what would become this book and persuaded me to publish it. This book would not have existed without him.

I am also indebted to Carlos Álvarez, Maarten Bullynck, Christian Gilain, Hans Niels Jahnke, and David Rabouin who carefully read my first French draft and made helpful suggestions for improving it.

I want to thank the whole SPHERE research team at CNRS for the opportunity to work in its group when writing the French draft that was the first version of this book. Special thanks go to Karine Chemla, director of SPHERE at that time, for her cordial reception in the team.

Special thanks are given to the anonymous reviewers of the manuscript. Their comments helped me to identify some errors and obscurities, whose correction led, I hope, to improve the argument of the book.

Many thanks also go to Vincenzo De Risi for his interest in including my work in this series, and for his support through the publishing process. His advice and suggestions also enabled me to enhance the quality of the book.

Contents

Introduction

At the end of the eighteenth century, the mathematical landscapes of German-speaking territories in Europe were dominated by the ideas of a particular group of mathematicians known as the "German combinatorial school". This book deals with those ideas. But before moving on to them, it is necessary to consider the existing historiography on the topic, around which a specific historical interpretation of the group and of its scientific work has taken decisive form. Getting acquainted with this traditional interpretation, which concerns the factors involved in the formation of the combinatorial school, its longevity in the context of German mathematics, the distinctive features of its scientific theory, and the relation of this latter to other scientific theories, opens the possibility of better understanding the motivations and goals of this research subject.

1.1 Historiography on the Combinatorial Analysis

Despite the limited number of studies devoted to the combinatorial school, it is not that easy finding your way around their historical narratives. A sort of intellectual struggle begins as soon as our curiosity on the subject leads us to the natural question of just what the combinatorial school is. Carl Immanuel Gerhardt was one of the first historians of mathematics to answer this question (Gerhardt 1877, 201–206). In 1877, he affirms that this school was founded by Carl Friedrich Hindenburg (1739–1808) and adds that Johann Friedrich Pfaff (1765–1825) enjoyed a prominent position in this academic organization. In 2003, Laurence Brockliss writes, in a volume centered on the history of science during the eighteenth century and published by the University of Cambridge, that Pfaff and Hindenburg were the founders of the combinatorial school (Brockliss 2003, 57). Although the vast majority of studies agrees with Gerhardt's opinion, Brockliss's remark affords an instance of the basic inaccuracies that still prevail in our knowledge of this school.

This kind of inaccuracies does not only concern the question of who founded the combinatorial school, but also the question of who joined this group as an active member. Gerhardt claims that, besides Pfaff and the founder Hindenburg, Georg Simon Klügel (1739–1812), Hieronymus Christoph Wilhelm Eschenbach (1764–1797), and Heinrich August Rothe (1773–1842) were members of the combinatorial school. In his paper on combinatorics, which is about twenty pages long and which remains the most detailed and the most comprehensive study on our research subject to date, Eugen Netto adds another three names to the list established by Gerhardt: Heinrich August Töpfer (1758–1833), Christian Kramp (1760–1826), and Johann Karl Burckhardt (1773–1825) (Netto 1908, 201–221). However, while Gerhardt and Netto included these names in their respective lists of members of the combinatorial school on the basis of their partial analysis of some texts of these authors, other historians arbitrarily inflate the list by adding names without justifying their choices, often not even a brief description of any writing of the chosen author is provided. In Bell's account of the combinatorial analysis, one can read that Josef Hoëné Wronski (1776–1853) was linked to this group, but this claim is not supported by any kind of primary source analysis (Bell 1940, 290–291). As time passes, the list of the members of the combinatorial school increases and differs from one historian to another according to the needs of their respective narratives. For instance, Martin suggests that Heinrich Bürmann (?–1817) was an active collaborator in the construction of the combinatorial school's theory, but he does not discuss any of Bürmann's ideas, merely quoting the title of one of Bürmann's works (Martin 1996b, 83–84). Pradier analyzes the work of Johann Nicolaus Tetens (1736–1807) but he does not justify the assertion that Tetens was a member of this school (Pradier 2003, 146). Ferraro includes Bürmann, Tetens, and Moritz von Prasse (1769–1814) in the group (Ferraro 2007, 479–480), and instead of elaborating on his reasons for doing so he directs the reader to (Jahnke 1993) and (Panza 1992, vol. 2, 651–659) for further information, but Jahnke does not address the works of any of those authors in his paper and Panza deals with Prasse's thought exclusively. Admittedly, it is just not reasonable to expect historians to draw up a single and definitive inventory of the mathematicians belonging to the combinatorial school, for the discovery of new historical data is a constituent part of historical research. On the contrary, since the inclusion of a given mathematician in the combinatorial school must be based on methodological decisions as far as historical narratives are concerned, it is reasonable to ask what criteria they used to make those decisions. Given the recurrent lack of primary source analysis in many studies, as those referred to above, it must be recognized that, in most cases, those criteria become indiscernible and hard to assess.

Gerhardt suggests that the combinatorial school has headed the mathematical research carried out in Germany during the "last decades" (*den letzten Jahrzehnten*) (Gerhardt 1877, 201). Considering that his book was published in 1877 and that he thought that Hindenburg was the founder of the combinatorial school, his remark implies that the existence of this school extends over a period of seventy or eighty years. Georg Faber and Alfred Pringsheim seem to agree with Gerhardt's estimation when they say that the influence of the combinatorial school can be perceived in German textbooks until the middle of the nineteenth century (Pringsheim and Faber 1909, 3–4). In his French version of Faber's and Pringsheim's work, Jules Molk goes beyond a simple translation and transforms one statement of Faber and Pringsheim into an entire paragraph about the combinatorial school, in which Molk claims that the combinatorial school continues to be active between 1825 and 1850, and even later (Molk et al. 1911, 3). Similarly, Wilhelm Lorey states that Hindenburg founded the combinatorial school and that university professors of mathematics emerged from this school up to the mid-nineteenth century (Lorey 1916, 27, 29). However, views on this matter are far from unanimous. According to Ferraro, it is possible to put a precise date to the emergence and dissolution of the combinatorial school, which was formed in 1780 and dissolved in 1810 (Ferraro 2007, 479–480). Therefore, its existence corresponds to a shorter period of just thirty years. The foundation year of 1780 comes again and again in the writings of different historians. This date is used by Schubring too, though he is not as categorical as Ferraro, nuancing that this school was "in vigour in Germany since about 1780" (Schubring 2007, 110). He disagrees with respect to the second date. For him, the textbooks related to the combinatorial school's ideas that appeared between 1800 and 1825 prove that this school did not cease its activities before 1825 (Schubring 2005, 562). In fact, in his lecture presented at a Satellite Symposium of the first European Congress of Mathematics held in Paris from April 3 to April 6, 1992, Schubring dated the beginning of the combinatorial school around 1780 and its end around 1820 (Schubring 1996a, 370). Thus, for him, the existence of this school varies between forty and forty-five years. Jahnke, for his part, proposes to distinguish two periods in the history of the combinatorial school (Jahnke 1990, 171). The first took place from 1780 to 1808, where the starting year was probably chosen because of the publication of Hindenburg's book entitled *Infinitinomii dignitatum exponentis indeterminati historia leges ac formulae*, appeared in 1779, and the period ends when Hindenburg dies. The second took place from 1808 to 1840, and it is characterized by the combinatorial school's loss of hegemony in the German academic world, as well as by the lack of innovative mathematical research, which was replaced by the production of textbooks. Thus the temporal situation of the combinatorial school fluctuates within relatively wide limits, which range from thirty to eighty years, with an average of fifty years. There is what seems to be a tacit agreement on the year of the combinatorial school's emergence or foundation, located around 1780. However, a historical reconstruction based on the assumption that this school lasted thirty years necessarily bears little relation to another that attributes a lifetime of fifty or eighty years to the group. The difficulty is clear: given the deep differences between the historical and cultural contexts of a period of three decades and a period of almost a century, the composition, nature, and characteristics of the combinatorial school cannot be the same in both cases. In other words, each of these historical reconstructions deals with a different object of study, ambiguously referred to by the same term of "combinatorial school", and this ambiguity necessarily brings confusion to our comprehension of the subject.

   The theory elaborated by the combinatorial school goes under the name of "combinatorial analysis". In this case, there is absolute unanimity among scholars as to the nature of this theory, which focuses on studying the elementary operations of permutations and combinations in order to then apply them to solve problems belonging to the field of mathematical analysis. Three different terms are currently used to make reference to the combinatorial school's theory in the literature: "combinatorial analysis" (*kominatorische Analysis*), "theory of combinations" (*Kombinationslehre*), and "combinatorics" (*Kombinatorik*). Gerhardt and Netto employ all of them indistinctly (Gerhardt 1877; Netto 1908). When German is not the original language of the historical study, there is a predominant use of the first two terms. For instance, Eric Temple Bell adopts the term "combinatorial analysis", while Schubring and Heuser prefer "theory of combinations" or "combination theory" (Bell 1940, 290–291; Schubring 1996a, 370; Heuser 1996, 48–49). Schubring also uses "combinatorial analysis" when he writes in French (Schubring 1990, 98). Indeed, it seems clear that these terms are taken as synonyms in the relevant literature. Let's now examine in more detail what is involved in the combinatorial analysis according to the relevant literature.

According to a consensus reached by most of the scholars, the combinatorial school advanced the idea that the multinomial theorem is the ultimate principle of mathematical analysis. The most common explanation for this conviction of the combinatorial school is the increasing importance of the binomial theorem in mathematics during the eighteenth century (Gerhardt 1877; Netto 1908). In this sense, the conviction of the combinatorial school is regarded as an attempt to reach more general mathematical results by enlarging the domain of objects subject to the theorem from binomials to any polynomial. However, it is possible to find more adventurous explanations. For Séguin, there is no mathematical motivation behind the combinatorial school's choice of postulating the multinomial theorem as the foundation of analysis, but a philosophical and cultural contextualized motivation (Séguin 2005). Séguin claims that the cultural atmosphere of the time dominated by German philosophers that began to build systems of knowledge based on a single, ultimate principle influenced Hindenburg and lured him to believe that a similar principle should exist within the realm of mathematics. Séguin even dares to affirm without any kind of historical evidence that it was Fichte's philosophy that directly influenced Hindenburg.

However, there have been some dissenting voices, too, about which mathematical proposition played the role of the ultimate principle of analysis for the combinatorial school. At the end of the nineteenth century, Karl Fink seems to suggest that the members of the combinatorial school, except for Hindenburg, considered the binomial theorem as the most important mathematical proposition in analysis, but it is true that Fink's text remains somehow ambiguous on this question (Fink 1890, 115). On the contrary, Sebestik explicitly says that, for the combinatorial school, the "key element" of the entire field of analysis was Taylor's theorem (Sebestik 1992, 84–85). Although Sebestik does not elaborate on this assertion at all, his idea could be argued in two different ways. One could say that Taylor's theorem was considered as the most fundamental theorem in analysis by the combinatorial school because of its interest in problems of reversing series, in which Taylor's theorem plays a central role. Or one could simply say that this assertion was explicitly formulated by Bürmann. Anyway, what is really important at this point is to emphasize the fact that one can find incompatible historical interpretations about one of the crucial issues that deeply impinges on the understanding and explanation of what the German combinatorial analysis was.

An elementary theory of combinations, its application to mathematical analysis, and a fundamental principle represent the constitutive elements of the German combinatorial analysis developed by the combinatorial school, according to the existing historical reconstructions. This general picture, shared, as far as we know, by all researchers that work on the subject, sets the stage for a more general interpretation about the intellectual commitments made by the members of the combinatoiral school regarding the way in which they understood mathematical knowledge. The general conception of mathematics attributed to this school has been associated with some kind of "formalism", as already noted by Gerhardt and Netto (Gerhardt 1877, 205; Netto 1908, 201). But it is possible to distinguish between two different interpretative variants of what this formalism means. The first rests on Gerhardt's and Netto's works and affirms that the combinatorial school conceived mathematics as a general science of formulae, in which every problem can be reduce to the problem of transforming a given formula into another by formal manipulations. In this sense, mathematical analysis, which depends on the theory of series for the members of the combinatorial school, can be seen as a theory focused on the formal transformations of finite and infinite series by means of combinations, as pointed out by Ferreirós Domínguez 1993, 37–38, 1999, 11–13. On the other hand, the second interpretative variant does not originate from a historical narration, either that of Gerhardt and Netto or any other one. It rests instead on some texts written during the first half of the nineteenth century in which the topics discussed and their treatment are close to the combinatorial school's ideas, but in which the theory of combinations is presented as a sort of syntax. From this point of view, the combinatorial analysis can be regarded as a formal syntax that provides the rules for manipulating the symbols of mathematical theories. In this second interpretation, the formalism of the combinatorial school points to the formulation of a common symbolic grammar for mathematical analysis, while in the first interpretation there is no common grammar, but local manipulations of symbols according to algorithms formulated for a given specific case. The syntactic interpretation can be found in (Martin 1996b, 83–84, 1996a, 201; Schubring 1996b, 66). In particular, Schubring claims that, under the influence of the combinatorial school, Jacob Friedrich Fries (1773–1843) called "*Syntaktik*" the most fundamental branch of mathematics. Later, Pulte reproduces Schubring's interpretation about Fries and adds that Fries's conception also results from the influence of Kant's philosophy (Pulte 2006, 117).

In the literature, the combinatorial school's theory has been associated with two different scientific programs, one dating from the seventeenth century and the other from the eighteenth century. This latter program is now known under the name of "algebraic analysis", and the idea that there exists a theoretical relation between this program and the combinatorial school's work had not been formulated until recently. No historical reconstruction carried out in the nineteenth century or in the first half of the twentieth century proposes such an interpretation of the combinatorial school's work. It is worth saying that this school has always been associated with Joseph-Louis Lagrange (1736–1813), who is considered to be the first proponent of the algebraic analysis, but this association was traditionally based on the belief that certain aspects of the combinatorial school's manipulations of infinite series have their roots in Lagrange's thought, in particular those concerned with reversion of series, as can be seen for example in (Reiff 1889, 148–149). On the contrary, during the two last decades of the twentieth century, some historians have advanced the hypothesis that the most general objective of the German combinatorial analysis coincides with that of the algebraic analysis (see, for example, Friedelmeyer 1997, 17; Jahnke 1990, 169–232; Laugwitz 1996, 57–63; Schubring 2005, 562). In these historical narratives, the algebraic analysis is characterized as the Lagrangian project of founding differential and integral calculus on algebraic bases, thus excluding all problematic metaphysical assumptions inherent in questionable concepts, as the concept of "infinitesimal". In this order of ideas, under the influence of the Lagrangian project, the German combinatorial school pursued, according to these narratives, the goal of founding calculus on algebra, with the difference being that algebra was supposed to obey combinatorial laws.

On the other hand, the second program is that of the "universal characteristic" or the "*ars combinatoria*" suggested by Gottfried Wilhelm Leibniz (1646–1716) during the second half of the seventeenth century. In this case, practically every single historian that has worked on the subject is convinced that the German combinatorial analysis is closely related to Leibniz's program (see, for example, Friedelmeyer 1997, 17; Gerhardt 1877, 203; Netto 1908, 201; Peckhaus 1997, 238–240; Schubring 1990, 98; Séguin 2005; Thiel 1995, 20). According to these historical reconstructions, the Leibnizian project consisted in developing a symbolic calculus on the basis of the theory of combinations and whose scope extends to all human knowledge. The influence exerted by Leibniz's thought on Hindenburg and his colleagues was so important that, according to these historians, the German combinatorial analysis can be seen essentially as an attempt to finally bring Leibniz's project into existence. Nevertheless, these scholars recognize that the plan of the German combinatorial analysis is less ambitious than that of the Leibnizian universal characteristic inasmuch as its scope is restricted to mathematics.

In fact, the German combinatorial analysis has also been related to a major cultural movement of the time by one of the scholars mentioned above. Séguin ventures the hypothesis that the early German Romanticism impacted the shape and composition of Hindenburg's mathematical work, principally through Fichte's philosophical thought (Séguin 2006, 64–72). Thus, the nature and characteristics of Hindenburg's mathematics were determined by the nascent romantic ideals, and, at the same time, his mathematics transmitted these ideals to other exponents of the German culture, notably to Friedrich von Hardenberg, Novalis (1772–1801). However, although Novalis's fascination with the German combinatorial analysis is well established (Dyck 1960; Hannah 1981; Jahnke 1991; Margantin 1999; Schefer 2005), Séguin's hypothesis does not seem to be supported by any kind of confirming evidence. It is well known that Novalis wrote some lines about Hindenburg's theory (Novalis 1993), but it cannot be inferred from that that Hindenburg was a Romantic. In fact, it is well known that Novalis was fascinated with new scientific discoveries in general, but this does not mean that such discoveries belong to the Romantic movement. Besides the fact that Séguin does not provide any textual documentation supporting his hypothesis about Hindenburg's Romanticism, it should be noted that this hypothesis depends on another questionable hypothesis, namely, that about Fichte's influence on Hindenburg's work which also rests on nothing more than vague assumptions. In sum, these hypotheses should be considered with caution.

Besides these general interpretations about the combinatorial school and its theory, the historians of the end of the nineteenth century and beginning of the twentieth expressed negative opinions of this period in the history of German mathematics. Gerhardt characterized it as "stagnant" (*stehend*) and Netto as "strange" (*merkwürdig*) (Gerhardt 1877, 201; Netto 1908, 201). Elaborating on Gerhardt's remark, Gino Loria described the mathematical works of the combinatorial school's members as "soporific" (*soporiferi lavori*) and placed on them the entire responsibility for the "lethargy" that had supposedly paralyzed German mathematics for half a century (Loria 1888, 335). It should be noted, however, that Loria provided a rather neutral description of this school in his capital work on the history of mathematics (Loria 1933, 349–351). In 1916, Lorey describes the theory of the combinatorial school as "an old combinatorics" (*alte Kombinatorik*) forgotten because of its "sterile subtleties" (*unfruchtbaren Spitzfindigkeiten*) that distorted the spirit of the "original and very interesting and important combinatorial problems" (*ursprünglich mathematisch gewiß sehr interessanten und wichtigen kombinatorischen Probleme*) of Leibniz, Moivre, Bernoulli, and Euler (Lorey 1916, 29–30). And he talks about Hindenburg as a "narrow-minded" (*Engherzigkeit*) man blinded by his obsession with combinatorics, and thus unable to see beyond his own mathematical horizon (Lorey 1916, 28, footnote 5). According to Bell, this historical episode was nothing but a "ridiculous interlude" in which the "patriotic disciples" of Leibniz (i.e., the members of the combinatorial school), lacking the talent for mathematical sciences, failed to understand the deep insights of the Leibnizian universal characteristic (Bell 1940, 290–291). Florian Cajori was convinced that the combinatorial school produced nothing worthy of esteem and History rewards it with deserved oblivion (Cajori 1919a, 231, 373, 1892, 1). He went as far as to mock their manipulations of divergent series as though the members of the combinatorial school had had the obligation to know beforehand Cauchy's results (Cajori 1890, 362–363). In a more moderated tone, Felix Klein refused to devote any part of his acclaimed lectures on the state of mathematics in the nineteenth century to the combinatorial school under the pretext that its ideas can rather be seen as a "ramification of old scientific trends" than the "beginning of a new scientific development" (Klein 1967, 113). Perhaps this recurrent stigmatization has played a dissuasive role as a deterring factor for scholars to pursue more detailed studies of this historical period. In any case, it was not until the end of the twentieth century that a fresh wave of historical studies appeared, in which the combinatorial school's ideas have been reevaluated by placing them in the mathematical contexts to which they belong. For instance, Hans Niels Jahnke analyzes the impact that this school had on the design and planning of education in the Germany of Humboldt's day (Jahnke 1990); Marco Panza analyzes it in the general framework of the development of mathematical analysis during the eighteenth century (Panza 1992, vol. 2, 651–659); Maarten Bullynck proposes a new approach centered on the components of Hindenburg's work that are related to the history of number theory (Bullynck 2006, 235–271); Gert Schubring shows that it is even possible to objectively address the topic of the strong opposition to the combinatorial school's ideas that emerged in the nineteenth century, without disqualifying them on the basis of personal opinions (Schubring 2007).

In sum, the existing reconstructions of the history of the combinatorial school do not always correspond to a coherent and consistent narrative. Some tensions and disagreements can be detected among historians on elementary matters concerning, for instance, the question of who was the true founder of the combinatorial school, or the more relevant question of what was the fundamental mathematical proposition on which the combinatorial analysis should be based. Nevertheless, there are common traits linking these historical interpretations. It seems clear that they explain the formation of the combinatorial school as the result of a conscious, voluntary decision made by several mathematicians with the aim of creating a scientific team. The research team is focused on developing the theory of combinations, which can also be called "combinatorics" or "combinatorial analysis", since all of these terms are synonyms. The theory of combinations developed by this school is characterized by its generality, its formalism, and its syntactic symbolism. Furthermore, there is no doubt that the general conception of this theory comes from the Leibnizian project of the universal characteristic, and its aims are determined by the aims of Leibniz's project, though Lagrange's algebraic analysis has exerted some influence on these goals. These common traits constitute what will be called here the "traditional interpretation" of our subject matter.

1.2   An Alternative Historical Interpretation of the German Combinatorial Analysis

As our research advanced, we were forced gradually to abandon the traditional historical interpretation of the combinatorial school, and its work, summarized above. This traditional interpretation was in fact our starting point, and our goal consisted in consolidating and completing it in a single volume. However, we realized that the main ideas on which it rests do not stand up to close scrutiny in the light of the available primary sources. Because of this, we could not rely on these ideas and we were compelled to rethink our goals and rewrite the history of the combinatorial analysis from a new viewpoint. This new viewpoint comes from our analysis of primary sources, which allowed us to identify what one could call working assumptions, even if they are rather the result of our research than a general framework externally imposed upon our historical narrative. In what follows, we will present the assumptions on which our historical reconstruction has been built, assumptions discussed in detail throughout this book.

The subject matter of our historical reconstruction is the German combinatorial analysis at the end of the eighteenth century. The general aim of this study is to explain, from a historical point of view, what this mathematical theory was according to its creators. There is no doubt that the history of the combinatorial analysis walks hand in hand with that of the combinatorial school, but the historical reconstruction of the first does not presuppose the historical reconstruction of the second. The members of the German combinatorial school were related to specific academic environments, worked in particular academic institutions, each with their own defining features, they communicated their ideas by different means and interacted with various scientific communities, and each of them had his own intellectual interests and incentives, other than the combinatorial analysis. Each of these elements should be carefully analyzed in a study devoted to the German combinatorial school. In this book, since the combinatorial school is not the object of our historical reconstruction, these elements will be only discussed when necessary, that is, when their discussion is pertinent to provide further clarification around our main subject: the German combinatorial analysis. Thus, these pages tell the story of the emergence of a scientific theory, of its gradual conformation, and of its death.

In our research, we have identified two main historical facts concerning the history of the combinatorial analysis. For the purposes of this introductory section, we will express these historical facts in terms of working assumptions, in particular because their status of "historical facts" has not yet been established. They are incompatible with the traditional interpretation, but we hope that we will be able to show their historical adequacy. The first assumption concerns the historiographical category of "combinatorial school", and the second one that of "combinatorial analysis". They can be formulated as follows:

1. The historiographical category of "combinatorial school" makes reference to a group of mathematicians that worked together to achieve a specific common objective, like a scientific research team. This intellectual association was formed around 1794 and progressively disintegrated from 1800.
2. The historiographical category of "German combinatorial analysis", or simply "combinatorial analysis", makes reference to the mathematical system that the combinatorial school wanted to develop. This system was never completed, but a scientific research program was written by Hindenburg in which the main elements of the system are described. Its first formulation in 1781 appears before the formation of the group. In 1781, this research program is a personal project of Hindenburg and it is composed of four objectives: (a) Organizing mathematical knowledge as a scientific system, (b) Improving and developing combinatorial methods, (c) Creating new mathematical notations in order to express the results of the system, (d) Applying combinatorial methods to the solution of problems belonging to the theory of series. Hindenburg's personal project becomes the scientific research program of the combinatorial school.

According to the first assumption, the period of activity of the combinatorial school has been considerably reduced with respect to the temporal limits proposed by other historians.

This apparent discrepancy can be easily explained. The term "combinatorial school" has been inconsistently and ambiguously used in the historical literature. It has been used to designate, at the same time, both a scientific team and a mathematical current of thought. This confusion has led historians to multiply the temporal limits attributed to the existence of the combinatorial school and to propose excessive and incompatible historical datings, ranging from three decades to almost a century. However, the existence of a current of thought does not necessarily rely on the existence of a particular scientific team. In our view, it is possible to identify the existence of a "combinatorial current of thought" after 1800 in German mathematics, but at this time the combinatorial school no longer exists as a team. Thus, we propose to exclusively use the term "combinatorial school" to designate the group of mathematicians to which point (1) makes reference. To decide who was a member of this school, we propose the criterion that a scholar belonged to the combinatorial school if he contributed to the achievement of the scientific program described in point (2).

Assumption (2) allows us to pave the way for properly characterizing the German combinatorial analysis. With the benefit of hindsight, we know that the German combinatorial analysis should be considered as a project in progress or as an unfinished theory, despite the effort and energy of its creators to complete it and despite the partial goals achieved. This peculiarity of the German combinatorial analysis makes it harder to tackle the problem of its historical reconstruction, especially when it comes to unravel the results or the writings that agree with this program from those that were presented under the name of "combinatorial analysis" but that pursued different goals. Naturally, this question will be carefully discussed throughout the book. But, even before discussing this point in detail, if our historical interpretation is right, it is impossible to reduce the German combinatorial analysis to the theory of combinations. Certainly, the improvement of this theory is listed in point (b) of the combinatorial school's program, but this is only one point of the program. Thus, the term "German combinatorial analysis" is not synonymous with "theory of combinations". In this book, the terms "combinatorics" and "theory of combinations" will be considered synonymous, and, on the other hand, the expression "the combinatorial analysis" always refers to the German combinatorial analysis, unless otherwise indicated.

Other historical facts were discovered during this research, which did not conform to the traditional picture of the combinatorial school. Here again, their status of "historical facts" will be established later, when supporting historical data will be presented and discussed. For the purposes of this section, they are formulated here as subsidiary assumptions, and they complement our two main assumptions. Concerning assumption (1), we will defend the idea that the formation of the combinatorial school was not the result of a conscious, voluntary decision made by a group of scientists, and therefore there is no founder of this school. As remarked in assumption (2), Hindenburg's personal research project became the research program of the combinatorial school, but this does not imply that Hindenburg founded a scientific team. Similarly, it will be seen that nobody took a conscious decision to disband the research team.

Concerning assumption (2), we have identify three subsidiary assumptions. The first one contradicts the only point of agreement reached by almost all historians in their historical reconstructions, and thus this subsidiary assumption is probably the most radical one in our historical narrative. We will argue, on the basis of textual evidence, that the combinatorial analysis does not have its roots in Leibniz's universal characteristic or *ars combinatoria*. In our research, we have observed, moreover, not only that the combinatorial analysis does not find its origins in Leibniz's project, but also that Hindenburg's project cannot be considered as an attempt to accomplish it since, in fact, it bears no relationship to Leibniz's ideas. The German combinatorial analysis was originally related to the Newtonian tradition of the theory of series, and was later explicitly founded by Hindenburg on the work of Abraham de Moivre (1667–1754). It is worth noting that a few studies have reconstructed this part of history without involving Leibniz; for instance, Panza stresses the importance of Moivre's work for the combinatorial school and leaves Leibniz aside (Panza 1992, vol. 2, 651–652).

We have also found that the combinatorial analysis was conceived as a foundational program of mathematical analysis. Indeed, it is possible to find explicit assertions in this regard in Hindenburg's works. It is more difficult to decide whether this foundation concerns the analysis of the finite or the analysis of the infinite, i.e. whether Hindenburg's project can be seen as an attempt to found differential calculus on less controversial bases. In our view, Hindenburg aimed to rebuild the theory of series on a firmer foundation, and thus he intended to reformulate the analysis of the finite by means of combinatorial concepts and methods, which should then be used to reconstruct the theory of series. In this sense, his program aimed to provide mathematical analysis with an alternative that may be less controversial that differential calculus. Hindenburg thought that the multinomial theorem was the link between the combinatorial reconstruction of the analysis of the finite and the theory of series, and as a consequence this theorem should be considered as the mathematical fundamental proposition on which the mathematical system mentioned in point (2a) rests.

Our third subsidiary assumption regarding point (2) states that the combinatorial analysis vanished around 1800. Needless to say, this does not mean that the influence of the combinatorial school's work on German mathematics stopped all of the sudden in 1800. On the contrary, its influence can be perceived far beyond the first half of the nineteenth century, and this is what can be called a combinatorial current of thought in German mathematics during the nineteenth century. Since the accomplishment of the project objectives that characterized the German combinatorial analysis cannot be identified as a goal to be pursued by the mathematicians belonging to this combinatorial current of thought, it can be affirmed that the work of those mathematicians did not contribute to the construction of the German combinatorial analysis, which, hence, disappeared at the turn of the eighteenth and nineteenth centuries.

The German combinatorial analysis is an example of the different programs appeared during the second half of the eighteenth century in order to better explain the relatively new field of mathematics called "analysis". It was an attempt to understand the concepts and methods of this new, mysterious field in terms of the mathematical methods and concepts available at the time. Mocking this program, or any other of the period, on the basis of personal preferences just shows incapacity to discern how science evolves and how historical processes are constituted. In this book, we have tried to understand and explain a historical period of the German scientific thought, as well as the mechanisms involved in the creation of a mathematical theory.

1.3   Content of the Book

This book is composed of five chapters, without counting the introduction and the epilogue. Chapter 2 intends to give a picture of the conceptual genealogy of the German combinatorial analysis, which is of paramount importance to understand its genesis and further development. This conceptual genealogy is not so much a sort of conceptual background lying behind the combinatorial school's work as it is a kind of conceptual branching structure impacting on the evolution of the combinatorial analysis at specific times of its history. Although all the events included in Chap. 2 are chronologically prior to the combinatorial school's works, they are incorporated into the formation of the combinatorial analysis's identity in different time periods, and some of them keep coming back modifying the course of history and the theoretical orientation of the program. These events constitute, certainly, the past of the combinatorial analysis, but they cannot be taken as a unified background which remains homogeneous and unproblematic no matter what happens in the future. The different branches of this conceptual genealogy disrupt the present of the combinatorial analysis at definite moments, causing sensible inflections in its configuration. To better capture these inflections, it would have been convenient if these disruptive events had been incorporated in our narrative in accordance with the moments in which they were taken into account by the members of the combinatorial school. However, from a methodological point of view, it is a better choice to bringing them all together in a single chapter in order to avoid frequent and long digressions in our discourse. Thus, these events form the subject matter of the second chapter.

More precisely, Chap. 2 first presents a historical reconstruction of the binomial theorem from its first formulation by Isaac Newton (1642–1727) up to 1781, that is to say, up until the first draft of Hindenburg's project appeared. Special attention will be paid to the nature of the methods used to demonstrate this theorem, emphasizing its central place in the general conception of mathematical analysis at the time. In particular, it will be seen that the enormous amount of proofs that were carried out during this period reflects the vivid interest of mathematicians from all Europe in understanding what would be the best method to correctly justify the binomial theorem, and at the same time it reflects the great difficulty of finding a correct answer. This also means that this subject had remained in force for a century and it was still a topical theme in the 1770s, as Hindenburg turned his attention to it. Later in the chapter, a similar historical reconstruction of the multinomial theorem is presented for the same period of time. Particular emphasis is placed on the way the problem of raising a multinomial to a given power was stated and on the different methods used to prove it. It is of particular interest for us to understand that the research on the multinomial theorem was considered in general as a secondary topic, whose importance and relevance for mathematics depended on the binomial theorem and it was seen, at best, as a way of fully analyzing the consequences of the binomial theorem. Nevertheless, the research on the multinomial theorem gave rise to a new topic that was profusely discussed during the eighteenth century, namely, that of reversing series. Although no historical reconstruction of the theorem on reversion of series has been included in this chapter, we will examine some elements of its history that are related to the combinatorial analysis.

Chapter 3 is organized around the idea, completely neglected and unexplored to date, that the origin of the German combinatorial analysis is found in Hindenburg's early mathematical works and in his mathematical training. Concerning this latter point, it will be argued that Borz, one of his close teachers, had intended to show that differential methods could be successfully replaced by algebraic ones, at least in certain domains of mathematical inquiry. Hindenburg collaborated with his teacher on this work around 1769, and it is possible that this collaboration planted in him the belief that methods based on the analysis of the finite could be developed as an alternative to differential calculus. This could explain why Hindenburg never used differential methods in his writings and why he strove his entire life to create new mathematical methods similar to algebraic ones. In our view, this is a better explanation of his "algebraic analysis" than the traditional hypothesis that Lagrange's algebraic analysis influenced him at an unknown time by unknown means, since our explanation rests, at least, on existing textual documents.

Concerning Hindenburg's early mathematical work, it will be argued that Hindenburg's interest in mathematical tables during the 1770s convinced him that, on the one hand, mathematical tables were an essential tool to drive progress in mathematics, and, on the other hand, they also convinced him that there is a close relationship between arithmetic and combinatorics. These two beliefs inspired him to approach, for the first time, the study of a particular problem of the theory of series from a new perspective, namely, that of finding a non-recursive formula for calculating the general coefficient of the power series expansion of a multinomial raised to a rational power. Later, he generalized his innovative method to deal with a relatively long list of problems belonging to the theory of series. It will be seen that, through this generalization, the multinomial theorem was overshadowed by the combinatorial method itself. Then, it will be analyzed how Hindenburg wrote the first draft of his research program on the basis of his new combinatorial method, and how this draft can be considered as the first exposition of what is known as the German combinatorial analysis.

Chapter 4 deals with the growing period of the combinatorial analysis from around 1792 to 1798. A main aim of this chapter is to carefully describe the process that led to the formation of the German combinatorial school. It will be discussed the idea that three main factors are involved in this process. First, Hindenburg forged a new generation of students that adopted his mathematical techniques and his general conception of mathematics. Second, he had at his disposal a scientific journal, founded by himself, that served as an optimal vehicle to transmit his combinatorial views on mathematics. Third, Hindenburg's scientific program was plagiarized in 1792, which triggered an intellectual fightback against the plagiarist, coordinated by Hindenburg's former students. The German combinatorial school raised from this intellectual conflict, in which it is possible to identify the characteristic organization of a scientific team. In particular, a combinatorial non-recursive formula for reversing series, established and demonstrated by Hindenburg's former students, was stolen by the plagiarist. Later, some members of the combinatorial school proved that this formula was equivalent to a formula established by Lagrange on the basis of differential calculus. It will be argued that this equivalence could be interpreted by the combinatorial school as a corroborative evidence that differential calculus could be replaced by combinatorial methods.

Another main objective of Chap. 4 is to describe the process of consolidation of the combinatorial school's scientific research program. It will be seen that the consolidation of the German combinatorial analysis depends on two factors. On the one hand, Hindenburg improved his combinatorial methods by modifying the mathematical tables on which they were based. On the other hand, this improvement led Hindenburg to reevaluate Moivre's work in the light of his new mathematical tables. As a result, Hindenburg claimed that Moivre had already implicitly used those combinatorial methods when proving the multinomial theorem. It will be shown that this reevaluation of Moivre's thought was the key to consolidate the German combinatorial analysis inasmuch as it led to the formulation of the theoretical bases on which the combinatorial school's mathematical system of science should rest. There are two theoretical bases. First, the multinomial theorem becomes the fundamental proposition in the field of the theory of series, and therefore in mathematical analysis. Second, the entire system should be based on a new theory of combinations that can be seen, from today's perspective, as a sort of abstract algebra. It was Hindenburg who sketched the ideas behind this new theory of combinations, but no member of the combinatorial school, including Hindenburg, ever undertook to further develop this theory.

In Chap. 5, it will be discussed the idea that the German combinatorial analysis disappeared as a result of four main factors. First, no member of the combinatorial school was ever able to show that the multinomial theorem was effectively the most fundamental proposition in mathematical analysis, particularly because no one was ever able to prove this theorem independently of the binomial theorem and using exclusively the new theory of combinations of the school. This fact cast doubt on whether the multinomial theorem could be regarded as a legitimate foundation of a system. Second, the binomial theorem gained importance in the texts related to the combinatorial analysis that appeared at the beginning of the nineteenth century, and thus these texts did not convey anymore the image of the combinatorial analysis as a well-structured system of science. Third, it is possible to identify the emergence of rival theories at the end of the eighteenth and the beginning of the nineteenth century, which were systematically confused with the German combinatorial analysis, but their goals and general conception of mathematics were significantly different from those of the combinatorial school's project. Fourth, some mathematicians disagreed with the idea of the combinatorial nature of mathematical analysis, but they praised Hindenburg's mathematical results and tried to eliminate their combinatorial elements.

In Chap. 6, the historical consequences of Chap. 5 are discussed in detail. It will be seen that the dissolution of the combinatorial school in the first decade of the nineteenth century led to the production of isolated research, instead of the collaborative research that characterized the work of this school. Particular attention will be paid to the reconstruction of what has been called above the combinatorial current of thought of the nineteenth century inspired by the ideas of the combinatorial school. It will be argued that this current is divided into four main branches. Each branch is characterized by a particular position about the role that the theory of combinations plays in mathematical analysis, but this does not mean that the authors of a given branch hold the same conception of mathematics. A brief final section of Chap. 6 deals with the German combinatorial analysis in non-German-speaking countries.

Given the nature of the subject matter, the discussion of mathematical symbols occupies an important place throughout the book. It would be necessary to distinguish between use and mention of a symbol by means of quotation marks. However, in an effort to lighten the text, quotation marks will be omitted when a mathematical symbol is mentioned, the context being enough to avoid any possible confusion.

As a final remark, we note that all the translations of the primary sources quoted in this book are ours, and we have decided to complement them by including the corresponding original texts in footnotes.

A History of the Binomial and Multinomial Theorems

2

2.1 The Binomial Theorem

Some historians hold that the history of the binomial theorem goes back as far as the *Elements* of Euclid (fl.300 BC) by interpreting proposition 4 of Book II as a particular case of squaring $(a + b)$, where a and b are segments (Coolidge 1949, 147; Heath 1908, 379). In the same vein, there are historical interpretations of the work of Heron of Alexandria according to which Heron used this theorem in the context of extracting cube roots, and something similar applies to the work of Theon of Alexandria (ca. 335–ca. 405) (Bourbaki 1984, 95 note; Deslauriers and Dubuc 1996; Heron of Alexandria 1903, 176–179; Rome 1936, 473; Nikolantonakis and Yao-Yong 2011, 172–175).

This kind of interpretations is not restricted to the Greco-Latin tradition. According to some scholars, in the Chinese tradition a square root algorithm similar to Heron's can be found in the *Nine Chapters on the Mathematical Art*, and it was later reformulated by Liu Hui, in the third century AD, in order to calculate both square and cube roots (Deslauriers and Dubuc 1996, 178–180; Martzloff 1988, 215–216). In a related topic, it has been also argued that Jia Xian invented the arithmetical triangle in the eleventh century to solve certain root extraction problems (Chemla 1994, 210–231; Lay-Yong 1980, 416–423). In India, some applications of an early version of the binomial theorem have been attributed to Āryabhaṭa (476–550 AD) (Ayyangar 1926, 172–173; Datta and Singh 1962, 175; Prakash 1968, 167–168; Raju 2007, 128–130). It has been argued that, in Arabic mathematics, al-Karajī (953–1029) invented a method for calculating any positive integer power of a binomial expression, which eventually led him to the discovery of the arithmetical triangle, and other Arabic mathematicians, such as al-Khayyām (1048–1131), al-Zanjānī (fl. 1257), al-Tūsī (1201–1274), and Ibn al-Bannā (1256–1321), consolidated a vigorous field of study around the work of al-Karajī (Djebbar and Rashed 1981; Djebbar 2005, 86 ff.; Rashed 1972, 1984; Jushkevich 1976, 80; Yadegari 1980).

© The Author(s), under exclusive license to Springer Nature Switzerland AG 2022   17
E. Noble, *The Rise and Fall of the German Combinatorial Analysis*, Frontiers in the History of Science, https://doi.org/10.1007/978-3-030-93820-8_2

In the European tradition, no systematic study about the arithmetical triangle was carried out until the second half of the sixteenth century. Although Jordanus de Nemore (fl. 1225) discussed the procedure for generating the triangular pattern in his *De arithmetica*, the earliest printed reproduction of the pattern itself appeared only in 1527, decorating the frontispiece of Petrus Apian's arithmetical treatise (Apian 1527; Hughes 1989). And here too, as in the other mathematical traditions, the main application of the arithmetical triangle concerned the solution of root extraction problems. That is the case for Michael Stifel (1487–1567) and Johann Scheubel (1494–1570) in Germany, and for Jacques Peletier (1517–1583) in France (Peletier 1549; Scheubel 1545; Stifel 1544, 44 verso). But in Italy Niccolò Tartaglia (1500–1557) and Gerolamo Cardano (1501–1576) examined the properties of the arithmetical triangle in the context of the theory of combinations (Cardano 1570, 131, 185; Tartaglia 1556, vol. 1: 69 verso). There is no doubt that the long history of advances in this field culminated in the conception of the *Traité du triangle arithmétique*, printed by Pascal in 1654 but published until 1665, in which Pascal exposed the relation between triangular numbers, combinatorial numbers, and binomial coefficients (Boyer 1950; Edwards 1987; Pascal 1665). Indeed, his treatise is reputed to contain the first proof of the binomial theorem for positive integral exponents.

Although the algebraic interpretation of ancient mathematics on which depends the ancient history of the binomial theorem can be deemed anachronistic and has been questioned by several scholars, this is not the place to try to decide that question. Suffice it to say that it is a plausible way to interpret them, but that is not the perspective assumed in this chapter. For the purposes of this section, the history of the binomial theorem begins in the seventeenth century when Newton invented it and it was named after him in the eighteenth century. Moreover, the principal aim of this section is not to give an account of Newton's theorem for its own sake, but rather for its historical significance in determining the way in which the binomial theorem was conceived during the eighteenth century. Particular attention will be paid to the reception of Newton's theorem in eighteenth century Germany.

2.1.1 Newton's Theorem

The binomial theorem was discovered by Isaac Newton (1642–1727) as a result of his early mathematical research on the method of quadrature developed by John Wallis (1616–1703) and on the method of tangents of René Descartes (1596–1650). His early mathematical research has been analyzed in excellent monographs and critical editions of his work, which can be consulted with respect to the genesis of Newton's theorem (Panza 1995, 2005; Whiteside 1961, 1967). The German combinatorial school had no access to Newton's early research, and, for the members of this school, the history of the binomial theorem begins with the *Epistola prior* in which Newton states his theorem. This then will be the starting point for the story we are interested in here.

On 12 May 1676, Gottfried Wilhelm Leibniz (1646–1716) addressed a letter to Henry Oldenburg (1618–1677) in which he asks for information regarding the power series expansions of the sine and arcsine functions (Leibniz 1676a). Leibniz found out about these expansions from Georg Mohr (1640–1697) and he was curious about how they have been constructed. The expansions had been calculated by Newton. For instance, the power series expansion of arcsine can be found in one of Newton's papers collected by Whiteside (Newton 1665a, 110). Oldenburg informed Newton about Leibniz's interest and strongly advised him to make public his method, otherwise he might lose priority for his idea. Shortly afterwards, on 13 June 1676, Newton addressed to Oldenburg the so-called *Epistola prior*, which was transmitted to Leibniz on 26 June 1676 and which contains the first statement of the binomial theorem for rational exponents. The theorem is enunciated as follows (Newton 1676b, 49, or Turnbull 1960, 21–54):

It is much easier to extract a root by this *theorem*:

$$\overline{P + PQ}|^{\frac{m}{n}} = P^{\frac{m}{n}} + \frac{m}{n}AQ + \frac{m-n}{2n}BQ + \frac{m-2n}{3n}CQ + \frac{m-3n}{4n}DQ + \&c.,$$

where $P + PQ$ is the quantity whose root, or even any power, or the root of a [given] power, is to be found. P is the first term of this quantity, and Q stands for the other terms divided by the first. And $\frac{m}{n}$ is the numerical index of the power $P + PQ$, whether it is an integer power, or whether it is, so to speak, a fractional one, or whether it is a positive or negative one.[1]

The alphabetical notation consisting in capital letters is a Newtonian innovation, and means: $A = P^{\frac{m}{n}}$, $B = \frac{m}{n}AQ$, $C = \frac{m-n}{2n}BQ$, and so on. This notation works as an abbreviation for the more complex expression of binomial coefficients: 1, $\frac{m}{n}$, $\frac{m}{n} \times \frac{m-n}{2n}$, $\frac{m}{n} \times \frac{m-n}{2n} \times \frac{m-2n}{3n}$, and so on.

To show what kind of problems can be solved with the help of his theorem, Newton provides a lot of examples the aim of which is, more precisely, to illustrate the theoretical and practical importance of his discovery. We will consider here only one example of each kind of problem.

Not surprisingly, the first application regards the problem of extracting roots. In order to calculate the fifth root of the trinomial $c^5 + c^4x - x^5$, Newton proposes two solutions depending on whether the terms of the trinomial are grouped from right to left so as to

---

[1] Sed Extractiones Radicum multum abbreviantur per hoc *Theorema*.

$$\overline{P + PQ}|^{\frac{m}{n}} = P^{\frac{m}{n}} + \frac{m}{n}AQ + \frac{m-n}{2n}BQ + \frac{m-2n}{3n}CQ + \frac{m-3n}{4n}DQ + \&c.$$

Ubi $P + PQ$ significat Quantitatem cujus Radix, vel etiam Dimensio quaevis, vel Radix Dimensionis, investiganda est. P, Primum Terminum quantitatis ejus; Q, reliquos terminos divisos per primum. Et $\frac{m}{n}$, numeralem Indicem dimensionis ipsius $P + PQ$: sive dimensio illa Integra sit, sive (ut ita loquar) Fracta; sive Affirmativa, sive Negativa.

form a binomial, or they are grouped from left to right (Newton 1676b, 50; or Turnbull 1960, 22). In the first case, the terms $c^4x - x^5$ are taken together, and one obtains:

$$\sqrt{5 : c^5 + c^4x - x^5} : = \overline{c^5 + c^4x - x^5}|^{\frac{1}{5}} = c + \frac{c^4x - x^5}{5c^4} + \frac{-2c^8xx + 4c^4x^6 - 2x^{10}}{25c^9} + \&c.,$$

for $m = 1$, $n = 5$, $P = c^5$, and $Q = \frac{c^4x - x^5}{c^5}$. In the second case one gets:

$$\sqrt{5 : c^5 + c^4x - x^5} : = \overline{c^5 + c^4x - x^5}|^{\frac{1}{5}} = -x + \frac{c^4x - c^5}{5x^4} + \frac{2c^8xx + 4c^9x + c^{10}}{25x^9} + \&c.,$$

for $m = 1$, $n = 5$, $P = -x^5$, and $Q = \frac{c^4x + c^5}{-x^5}$. Newton shows in this way that the binomial theorem furnishes the basis upon which it is possible to generalize the series expansions from a binomial to any polynomial. As will be seen in this chapter, this idea was decisive in the eighteenth century.

The second application is related to the method of long division. Let be $m = -1$, $n = 1$, $P = d$ and $Q = \frac{e}{d}$, then:

$$\frac{1}{d + e} = \overline{d + e}|^{-1} = \overline{d + e}|^{-\frac{1}{1}} = \frac{1}{d} - \frac{e}{dd} + \frac{ee}{d^3} - \frac{e^3}{d^4} + \&c.$$

Newton attached great importance to this example because Nicolaus Mercator (1620?–1687) used, in 1668, the method of long division to expand in series the function $\frac{1}{1+x}$. Then, Mercator obtained the expansion of $\ln(1 + x)$ by means of that of $\frac{1}{1+x}$ (Mercator 1668, 28–34; Naux 1971, vol. 2: 57–67). Mercator's procedure was one of the first to be published concerning the manipulation of infinite series. In 1665, Newton worked on the quadrature of the hyperbola $y = \frac{1}{1+x}$, but he did not explicitly relate his research on this subject to the expansion of $\ln(1 + x)$ (Newton 1665b, 113; Whiteside 1961, 255–259). In 1669, Isaac Barrow (1630–1677) transmitted to him a copy of Mercator's *Logarithmotechnia*, and, apparently, this book was a source of incentive for Newton to go deeper into the study of the theory of series. It seems that this example of application suggests that Newton considered the binomial theorem to be the fundamental basis of Mercator's procedure.

The third example concerns the so-called Newton–Raphson method that Newton applied to approximate the roots of algebraic equations. The method was exposed for the first time in the *De analysi*, and it was simplified by Joseph Raphson (1648?–1715?) in 1690 (Cajori 1911; Kollerstrom 1992; Newton 1711, 8–14; Raphson 1690; Ypma 1995). In the *Epistola prior*, the example consists in finding an approximation for the root of the equation $y^3 - 2y - 5 = 0$. One can propose $y = 2$ as an approximate solution for it, and take $y = 2 + p$ as being the exact solution. By substituting $y = 2 + p$ into the original equation and by applying the binomial theorem, one obtains:

$$y^3 = 8 + 12p + 6pp + p^3,$$
$$-2y = -4 - 2p,$$
$$-5 = -5,$$

that is: $-1 + 10p + 6pp + p^3 = 0$. In order to find an approximate solution for this new equation, one can pick $10p - 1 = 0$, that is $p = 0.1$, as the approximation, so that the exact solution will be $p = 0.1 + q$. By substituting $p = 0.1 + q$ into the second equation and by applying the binomial theorem, one obtains another equation: $0.061 + 11.23q + 6.3qq + q^3 = 0$. The exact value of $q = -0.0054 + r$ follows from $11.63q + 0.061 = 0$. Likewise, one can calculate the approximate value of $r = -0.00004852$. Then, an approximate solution for the original equation is given by $y = 2 + 0.1 - 0.0054 - 0.00004852 = 2.09455148$.

A last kind of problem treated in the *Epistola prior* involves the power series expansion of some functions, such as the trigonometric functions arcsine, sine, cosine, and the exponential function. This latter is, however, expanded in a power series by means of a method called, in the eighteenth and nineteenth centuries, "the reversion of series", which was also known as "the inversion of series".

Newton's new method intrigued Leibniz, who asked Newton to comment more thoroughly on how it could be used to obtain Newton's results (Leibniz 1676b, 63). The response to Leibniz's letter is contained in the *Epistola posterior*, addressed to Oldenburg on 24 October 1676, but which was not transmitted to Leibniz until June 1677 (Newton 1676a, 67–70; or Turnbull 1960, 110–129). In the *Epistola posterior*, the first statement of the theorem on reversion of series was formulated in the following two propositions, where $a, b, c, \ldots$ are known coefficients (Newton 1676a, 85):

Proposition 2.1 *Let be $z = ay + byy + cy^3 + dy^4 + ey^5 + \&c.$, then:*

$$y = \frac{z}{a} - \frac{b}{a^3}z^2 + \frac{2bb - ac}{a^5}z^3 + \frac{5abc - 5b^3 - aad}{a^7}z^4$$
$$+ \frac{3aacc - 21abbc + 6aabd + 14b^4 - a^3e}{a^7}z^5 + \&c.$$

Proposition 2.2 *Let be $z = ay + by^3 + cy^5 + dy^7 + ey^9 + \&c.$, then:*

$$y = \frac{z}{a} - \frac{b}{a^4}z^3 + \frac{3bb - ac}{a^7}z^5 + \frac{8abc - aad - 12b^3}{a^{10}}z^7$$
$$+ \frac{55b^4 - 55abbc + 10aabd + 5aacc - a^3e}{a^{13}}z^9 + \&c.$$

Newton did not give any proof at all for these propositions, but showed how to calculate theirs coefficients in two particular cases. Concerning Proposition 2.1, he assumed the following series:

$$z = x + \frac{1}{2}x^2 + \frac{1}{3}x^3 + \frac{1}{4}x^4 + \&c., \tag{2.1}$$

which corresponds to the power series expansion of $\ln\left(\frac{1}{1-x}\right)$, with the aim of expressing x as function of z. As seen above, one of Newton's examples suggests that the binomial theorem can be considered as the basis of the series expansions of powers of polynomials by virtue of the technique of grouping terms. Thus, in particular, the binomial theorem is the theoretical justification of the following powers of z, even if they were certainly calculated by simple multiplication in practice:

$$z^2 = x^2 + x^3 + \frac{11}{12}x^4 + \frac{5}{6}x^5 + \&c.,$$

$$z^3 = x^3 + \frac{3}{2}x^4 + \frac{7}{4}x^5 + \&c.,$$

$$z^4 = x^4 + 2x^5 + \&c.,$$

$$\vdots$$

Then he eliminated the terms involving the higher powers of x from the right-hand side of Eq. 2.1. To eliminate x^2, one subtract $\frac{1}{2}z^2$ from both sides of Eq. 2.1:

$$z - \frac{1}{2}z^2 = x - \frac{1}{6}x^3 - \frac{5}{24}x^4 + \frac{13}{60}x^5 + \&c.$$

By adding $\frac{1}{6}z^3$ to both sides of this latter equation, one gets:

$$z - \frac{1}{2}z^2 + \frac{1}{6}z^3 = x + \frac{1}{24}x^4 + \frac{3}{40}x^5 + \&c.$$

By subtracting $\frac{1}{24}z^4$ from both sides of this latter equation, one gets:

$$z - \frac{1}{2}z^2 + \frac{1}{6}z^3 - \frac{1}{24}z^4 = x + \&c.,$$

and so on. In this way, Newton derived the power series expansion of $1-e^{-z}$. In the second case, corresponding to Proposition 2.2, Newton obtained the power series expansion of the arctangent function using a procedure similar to that just described (Newton 1676a, 84–85).

Even though Newton provides no rigorous justification of any of these propositions, his examples insinuate that the justification of the theorem on the reversion of series rests on the binomial theorem. But on the other hand, such examples do not suffice to establish the law of formation of coefficients in Propositions 2.1 and 2.2. The systematic search for this law was a main task to which considerable efforts of many mathematicians were devoted, as we shall see, during the eighteenth century.

In short, the binomial theorem has proved to be, in the epistolary exchange between Newton and Leibniz, a versatile mathematical tool that can be put to many uses. And from a theoretical point of view, it appears to be the foundation of:

1. The method of root extraction.
2. Mercator's method.
3. The Newton-Raphson method of root finding.
4. The method of raising a polynomial to an arbitrary rational power.
5. The procedure for finding the power series expansion of a given function.
6. The theorem on reversion of series.

Probably only in this perspective can one understand that the binomial theorem for rational exponents became the theorem of Newton in the eighteenth century; that is to say, it was Newton's theorem in the sense that it was conceptualized as the basis of Newton's method of infinite series.

2.1.2 Newton's Theorem and the Method of Fluxions

The first printed version of Newton's binomial theorem appeared in Wallis's *Treatise of algebra* in 1685, which was translated into Latin 8 years later to ensure widespread dissemination (Wallis 1685, 318–320, 330–333; 1693, 357–359, 368–371). However, the systematic reproduction of Newton's theorem in most of eighteenth century manuals on mathematical analysis and algebra cannot be explained solely by reference to the publication of Wallis's treatise. In some measure, the prominent place that the binomial theorem occupied in these mathematical disciplines was due to its relation to Newton's method of fluxions. And this is true even of mathematical practices developed outside of the English mathematical community, like those developed by German mathematicians in the eighteenth century. The purpose of this section is to analyze the relation between the binomial theorem and the method of fluxions in the first half of the eighteenth century.

At about the same time that Newton discovered the binomial theorem, he also discovered the rudiments of his new method of fluxions. In October 1666, Newton organized his discoveries about the method of fluxions into some principles in a tract that was never published in its day, but some of his results were included in the *De analysi* (Newton 1666; Panza 2005, 433–514; Whiteside 1966). Some 5 years later, in 1671, Newton prepared the *Tractatus de methodis serierum et fluxionum* for publication, though it was not published until 1736 by John Colson (1680–1760) in an English edition (Newton 1736; or Whiteside 1969, 32–353). Ultimately Newton revised his theory in the three successive versions of the *Tractatus de quadratura curvarum*, which were written respectively in November 1691, at the beginning of the winter of 1691, and before the spring of 1694 (Whiteside 1976, 24–48, 48–129, 508–561). This latter version finally appeared in 1704 in the second appendix to Newton's *Opticks* (Newton 1704).

In the *De analysi*, the specific terminology of Newton's method of fluxions has not yet been fixed, whereas the very idea of the method is formulated in the following three rules (Newton 1711, 1–5):

Proposition 2.3 *Let a curve be given by* $y = ax^{\frac{m}{n}}$, *then the area under the curve* y *is given by* $\frac{n}{m+n} ax^{\frac{m+n}{n}}$, *where m and n are (positive) integers.*

Proposition 2.4 *If the curve* y *is represented as a sum whose terms are given in the form* $ax^{\frac{m}{n}}$, *then the area under the curve* y *is given by the sum of the areas corresponding to each term.*

Proposition 2.5 *If the curve* y *or any of the terms in the sum that represents it are not given in the form* $ax^{\frac{m}{n}}$, *they must be reduced to that form. Then one applies the preceding rules.*

Thus, Propositions 2.4 and 2.5 depend on the first rule, which can be interpreted as an incipient formulation of the fundamental theorem of calculus.

Newton proved Proposition 2.3 as follows (Newton 1711, 19–20). He supposed that $z = \frac{n}{m+n} ax^{\frac{m+n}{n}}$ is the area under a given curve y. By assigning to the abscissa x an infinitely small increment o, the area itself will take on an infinitely small increment:

$$z + yo = \frac{n}{m+n} a \, (x+o)^{\frac{m+n}{n}} .$$

Then he raised both sides of this equation to the integral power n:

$$(z + yo)^n = c^n \, (x+o)^p ,$$

where $c = \frac{an}{m+n}$ and $p = m + n$. And by applying the binomial theorem he got:

$$z^n + noyz^{n-1} + \&c. = c^n \left(x^p + pox^{p-1} + \&c. \right).$$

Given that $z^n = c^n x^p$, he subtracted this term from both sides of the latter equality, and then, by dividing both sides by o and by eliminating the terms still involving o after division, he got:

$$c^n p x^{p-1} = n y z^{n-1} = \frac{n y z^n}{z} = \frac{n y c^n x^p}{c x^{\frac{p}{n}}}.$$

He divided this expression by $c^n x^p$, and obtained after reduction:

$$p c x^{\frac{p-n}{n}} = n y.$$

By substitution of the values of c and p into this equation, he got the expression of the curve:

$$y = a x^{\frac{m}{n}}.$$

Similarly, if the curve is given by $y = a x^{\frac{m}{n}}$, Newton remarks that one can derive from it the expression of the area $z = \frac{n}{m+n} a x^{\frac{m+n}{n}}$.

Nevertheless, there is no justification for eliminating the terms involving the infinitely small increment o. In fact, it is not clear at all what an "infinitely small increment" would be. Some scholars argue that the symbol "o" would denote an infinitely small time interval, so that one would plot time on the abscissa and the velocity of the increasing area on the ordinate (Boyer 1959, 193; Guicciardini 1999, 20–22). So the product of the ordinate by an infinitely small interval of time gives the area of an infinitely small part, which might be considered as a "moment" of the total area under the curve. Therefore, this latter is the sum of its moments. To find these moments, Newton's argument lays considerable emphasis on the application of the binomial theorem.

While reading of *De analysi* in terms of "moments", "time", and "velocity" remains an interpretation, Newton actually made use of these concepts in his *De methodis*, as we can see it in the way in which the same problem was posed (Newton 1736, 19):

1. The Length of the Space described being continually (that is, at all Times) given; to find the Velocity of the Motion at any Time proposed.
2. The Velocity of the Motion being continually given; to find the Length of the Space described at any Time proposed.

Later, Newton's method will be called "direct method of fluxions" when it deals with the first problem, and "inverse method of fluxions" when it deals with the second one.

In addition to the conceptual framework, in the *De methodis* Newton introduced a new terminology and a new notational system. The new terminology was associated with a kinematical conception of geometrical objects according to which a curve would be generated by the motion of a given point. It is possible that he inherited this idea from Barrow, though it can be found in the work of such figures as René Descartes and others (Guicciardini 2009, 169–179; Westfall 1980, 131). Anyway, Newton's kinematical conception of geometrical objects began to take shape in his tract of October 1666 (Panza 2005, 434 ff.), and in the *De methodis* the term "fluxion" refers to the rate of change or, in other words, it refers to the velocity of the motion which generates the curve. On the other hand, the term "fluent" refers to the quantity generated by motion, that is, it refers to the distance described by the moving point. Regarding notations, the last letters of the alphabet v, x, y, z were employed to denote the fluents, and the middle letters of the alphabet l, m, n, r were employed to denote the corresponding fluxions. But in 1691, Newton replaced this notation of fluxions by the standard notation: $\dot{v}$, $\dot{x}$, $\dot{y}$, $\dot{z}$. To avoid any possible confusion, we will use here this latter. Finally, the symbol "o" is explicitly used to denote an infinitely small interval of time.

In full accordance with a didactic style, the working of the algorithm for the direct method of fluxions is exemplified by calculating the fluxions of the equation $x^3 - ax^2 + axy - y^3 = 0$ (Newton 1736, 24–25). But it is possible to put forward a series of more general steps that could have accounted for how the algorithm works. For instance, given the equation $y = x^n$, we can take $\dot{x}o$ and $\dot{y}o$ as the moments corresponding to the fluents x and y after an infinitely small interval of time has passed. Thus, we get $y + \dot{y}o = (x + \dot{x}o)^n$ and, by applying the binomial theorem, we obtain:

$$y + \dot{y}o = x^n + n\dot{x}ox^{n-1} + \&c.$$

By dividing both sides of this equation by o and by canceling the terms involving o after division, we get after reduction:

$$\dot{y} = n\dot{x}x^{n-1}.$$

In this case, the step consisting of eliminating the terms involving o is justified since—says Newton in reference to his example—"o is supposed to be infinitely little", and then "the Terms that are multiply'd by it will be nothing in respect of the rest" (Newton 1736, 25).

Indeed, a similar argument was made by Newton in the Introduction to the *De quadratura* (Newton 1704, 168–169):

> *Given a quantity x that flows uniformly, find the fluxion of the quantity x^n.* At the same time that the quantity x becomes by flowing $x + o$, the quantity x^n will become $\overline{x + o}|^n$, that is, by the method of infinite series, [it will be] $x^n + nox^{n-1} + \frac{nn-n}{2}oox^{n-2} + \&c$. And the increments o and $nox^{n-1} + \frac{nn-n}{2}oox^{n-2} + \&c$. are to each other as 1 is to $nx^{n-1} + \frac{nn-n}{2}ox^{n-2} + \&c$. Now let these increments vanish, and their last ratio will be 1 to nx^{n-1};

and in consequence, the fluxion of the quantity x is to the fluxion of the quantity x^n as 1 is to nx^{n-1}.[2]

In this argument, the binomial theorem plays the same central role as the one he plays in both the *De analysi* and the *De methodis*, but the step consisting in eliminating the terms involving o does not rest anymore on the concept of "infinitely small quantity", whatever this concept means. In fact, whereas Newton's kinematical conception has been preserved in the *De quadratura*, the concept of "infinitely small quantity" has been completely excluded (Newton 1704, 169):

> To establish analysis on [the basis of] finite quantities and investigating the first or the last ratios of finite nascent or evanescent quantities is in harmony with Ancient geometry. And I wanted to show that, as regards the method of fluxions, it is not needed to introduce infinitely small figures into geometry.[3]

Now the notion of "moment" is to be referred to finite quantities at the very instant in which they appear or disappear by means of continuous motion. Owing to their finite dimension, nascent and evanescent quantities might be compared in a ratio, as any other finite quantity. However, it is as difficult to say precisely what a nascent or evanescent quantity is as it is to say what an infinitely small increment is.

Newton studied carefully the geometry of the ancients from the late 1670s, paying particular attention to the works of Pappus of Alexandria (fourth century AD) and Apollonius of Perga (ca. 240–ca. 170 BC). His studies led him to modify the method of fluxions on at least two occasions. The second one corresponds to the modification of the conceptual framework in *De quadratura* by replacing, as we have seen above, the concept of "infinitely small quantity" with that of "nascent or evanescent quantity". But before writing the *De quadratura*, Newton carried out a most significant modification of his method in the *Principia mathematica* (Newton 1687, 250–254). The modification consisted in eliminating the use of the method of infinite series, and was so radical for Newton that he referred to his "new" method as "the method of first and last ratios" in contrast to the direct method of fluxions. This latter was conceived as an analytical method, while the method of first and last ratios was a synthetic or geometrical one and, in this sense, was perceived as being more rigorous (Guicciardini 1999, 39–98; 2009, 219–

---

[2] *Fluat quantitas x uniformiter & invenienda sit fluxio quantitas x^n*. Quo tempore quantitas x fluendo evadit $x + o$, quantitas x^n evadet $\overline{x + o}|^n$, id est per methodum serierum infinitarum, $x^n + nox^{n-1} + \frac{nn-n}{2}oox^{n-2} + \&c$. Et augmenta o & $nox^{n-1} + \frac{nn-n}{2}oox^{n-2} + \&c$. sunt ad invicem ut 1 & $nx^{n-1} + \frac{nn-n}{2}ox^{n-2} + \&c$. Evanescant jam augmenta illa, & eorum ratio ultima erit 1 ad nx^{n-1}: ideoq; fluxio quantitatis x est ad fluxionem quantitatis x^n ut 1 ad nx^{n-1}.

[3] In finitis autem quantitatibus Analysin sic instituere, & finitarum nascentium vel evanescentium rationes primas vel ultimas investigare, consonum est Geometriae Veterum: & volui ostendere quod in Methodo Fluxionum non opus sit figuras infinite parvas in Geometriam introducere.

229; Kitcher 1973, 46–49). The rigor, however, was not exempt from the problems related to the notion of "infinity". In the fist edition of the *Principia mathematica*, a moment is an infinitely small entity (Newton 1687, 251):

> Moments, when they come to be finite magnitudes, cease to be moments. To be finite is, in some measure, incompatible with their continuous increment or decrement.[4]

On the contrary, in the second edition, moments are said to be the cause of generation for finite quantities, but it is not clear whether they are finite or infinite entities (Newton 1713, 224). Some authors think that Newton abandoned the idea of "infinitesimal" after *De quadratura*'s publication in 1704, and some others think that this idea was a cornerstone of the Newtonian ontology (De Morgan 1852; Lai 1975). Without pretending to resolve this problematic issue, we will limit ourselves to the assertion that Newton never stopped to reflect on the subject, which seems to be corroborated by the second edition of the *Principia mathematica*. In any case, among the different approaches of Newton, the method of first and last ratios was the first to be published, since it appeared in 1687.

In this method, the operational rules for finding the "fluxion" of products, powers, and quotients are justified as follows. Let a and b be the respective moments of the quantities (or geometrical figures) A and B which are generated by continuous motion. Then, the rules are given by:

1. Product rule: the fluxion of the product AB is $aB + bA$. Newton took A and B, and subtracted a half-moment from each one, that is: $A - \frac{1}{2}a$ and $B - \frac{1}{2}b$. Then he got:

$$\left(A - \frac{1}{2}a\right)\left(B - \frac{1}{2}b\right) = AB - \frac{1}{2}aB - \frac{1}{2}bA + \frac{1}{4}ab.$$

Likewise he got:

$$\left(A + \frac{1}{2}a\right)\left(B + \frac{1}{2}b\right) = AB + \frac{1}{2}aB + \frac{1}{2}bA + \frac{1}{4}ab.$$

Then:

$$\left(AB + \frac{1}{2}aB + \frac{1}{2}bA + \frac{1}{4}ab\right) - \left(AB - \frac{1}{2}aB - \frac{1}{2}bA + \frac{1}{4}ab\right) = aB + bA.$$

2. Power rule: the fluxion of A^n is $na A^{n-1}$, for a positive integer power n. Newton showed through the examples $A = B$ and $A = B = C$ that the power rule follows from the product rule.

---

[4] Momenta, quam primum finitae sunt magnitudinis, desinunt esse momenta. Finiri enim repugnat aliquatenus perpetuo eorum incremento vel decremento.

Later Newton generalized the power rule to rational powers. He put $A^{\frac{m}{n}} = B$, and then $A^m = B^n$. By the power rule for integral powers, the fluxion of $A^m = B^n$ is $ma A^{m-1} = nbB^{n-1}$. It follows that $b = \frac{ma A^{m-1}}{n B^{n-1}}$. And then he got $b = \frac{m}{n} a A^{\frac{m-n}{n}}$, which is the fluxion of $B = A^{\frac{m}{n}}$.

3. Reciprocal rule: the fluxion of $\frac{1}{A^n} = A^{-n}$ is $-\frac{na}{A^{n+1}}$. It follows immediately from the power rule.

Thus, in 1687 Newton replaced the method of infinite series by elementary algebraic operations, attributing certain properties of numbers to the infinite small quantities a and b, as the property of divisibility by 2.

Probably, one of the main reasons Newton decided to divulge his method in the first edition of the *Principia mathematica* was that Leibniz exposed the elementary rules of differential calculus in 1684 (Leibniz 1684; Rigaud 1838, 18–24). In particular, Leibniz formulated the same three rules that will appear in Newton's *Principia*, though they are expressed using a differential operator instead of fluxional notation:

1. Product rule: If $y = xv$, then $dy = xdv + vdx$.
2. Power rule: $dx^n = nx^{n-1}dx$, where n is a positive integer.
3. Reciprocal rule: $d\frac{1}{x^n} = -\frac{ndx}{x^{n+1}}$.

Leibniz presented also the quotient rule, underlying the operational power of differential calculus and staying away from the discussion around the validity of his new theory. It was only in 1702 that he faced this question, attempting to show that differential calculus with infinitesimals can be founded on ordinary algebra (Leibniz 1702).

In England, the Scottish mathematician John Craig (1660?–1731) was maybe the first to recognize openly the importance of Leibniz's differential calculus. Furthermore, in 1685 he had the privilege of reading long passages of the *De analysi*, the *De methodis*, the *Epistola prior*, and the *Epistola posterior* in Cambridge, where he met Newton. Craig was working on a book devoted to the quadrature of curves, which was completed in the same year of 1685 taking advantage of Newton's manuscripts (Craig 1685; Whiteside 1976, 3–14; Westfall 1980, 400, 437, 513–515). Craig's book has the double merit of showing how Newton's method of infinite series works, and of introducing Leibniz's differential calculus into England. For instance, he transcribed the Newtonian quadrature of the circle, and on the other hand he used the differential notation and, implicitly, the differential rules of Leibniz (Craig 1685, 14–15, 28–30). Craig's book is indeed the first publication in which Leibniz's calculus cohabits with Newton's method of infinite series. Involuntarily this fact became an obstacle to the transmission of Newton's method of fluxions, since at this period none was able to relate it to the method of first and last ratios presented in the *Principia mathematica*.

In order to ensure the transmission of the method of fluxions, it was of capital importance that some passages of the *De quadratura* appeared in the Latin edition of

Wallis's *Algebra* (Cajori 1919b; Guicciardini 1989; Wallis 1693, 392–396). By this means, John Harris (1666?–1719) became familiar with both the algorithm for the direct method of fluxions and fluxional terminology and notation, gathering this material in the first manual on the theory of fluxions to be published in 1702 (Harris 1702, 115–136). Moreover, Harris established a theoretical hierarchy between the direct method of fluxions (*De quadratura*) and the method of first and last ratios (*Principia*), subordinating this latter to the first (Harris 1702, 119). However, despite the fact that Harris tackled Newton's theory, he did not hesitate to reconstruct it on the basis of the Leibnizian differential calculus. He gave, for instance, the quotient rule, which does not occur in Newton's texts, and referred the reader who wished to go deeply into the theory of fluxions to Leibniz's *Nova methodus* (Harris 1702, 121–122, 131). In 1704, Charles Hayes (1678–1760) published other manual on the theory of fluxions, which was based as well on Leibniz's works (Hayes 1704, 1–15). It is convenient to underline that, even though Nicolas Fatio de Duillier (1664–1753) accused Leibniz of plagiarism in 1699, in a sort of preamble to the quarrel between Newton and Leibniz, the results of differential calculus were commonly used to improve the reconstruction of the theory of fluxions, as attested by Harris's and Hayes's manuals (Fatio de Duillier 1699, 18).

This practice was continued even after the integral publication of Newton's *De quadratura* in 1704. Humphry Ditton (1675–1715) was not unwilling to admit it in the Preface to his *Institution of fluxions* printed in 1706, and in the same year William Jones implicitly used the rules of differential calculus, as Leibniz's quotient rule which was translated into fluxional notation (Ditton 1706; Jones 1706, 229). On the other hand, both of them reproduced the argument that Newton invoked in the *De quadratura* to justify the algorithm of the direct method of fluxions, which is cited above (Ditton 1706, 41–43; Jones 1706, 226). The reproduction of this argument in the few manuals dedicated to Newton's theory will help to fix the idea that the binomial theorem plays an essential role in justifying the algorithm of the direct method of fluxions.

During the decade of the 1730s, this idea will be reviewed in more detail, as we shall see below, in the new manuals on the theory of fluxions that will appear in response to the *Analyst* of George Berkeley (1685–1753), where the Bishop of Cloyne evinces a skeptical attitude toward the correctness of the results of the new calculus (Berkeley 1734; Blay 1986, 243–253; Jesseph 1993, 178–230). The aim of the *Analyst* is to demonstrate that if freethinkers reject religion because of the inherent irrationality of the mysteries of faith, then for the sake of consistency they must reject as well Newton's and Leibniz's new calculus, whose concepts and demonstrative methods are just as irrational as religious beliefs. We will discuss here two objections Berkeley brings against the direct method of fluxions.

The first objection maintains that Newton's proof of the product rule in the *Principia* is unsatisfactory. As seen above, Newton argues that it is possible to operate with moments as if these were ordinary numbers, so as to obtain the fluxion of a given product AB by subtracting a half-moment from both A and B, and then by adding a half-moment to both A and B. But Berkeley points out that there is no good reason to think moments are divisible by 2, and at best it may be possible to operate with a given "entire" moment. Thus the correct procedure to follow in order to find the fluxion of a given product AB is to add the whole of the moment a to A and the whole of the moment b to B. By multiplying these sums, one gets:

$$(A + a)(B + b) = AB + aB + bA + ab.$$

If AB is a rectangle, Berkeley insists that its real increment is given by:

$$aB + bA + ab.$$

The infinitely small quantity ab cannot be neglected when determining the product rule; otherwise, Newton's methodological maxim according to which "the minutest errors are not to be neglected in mathematics"[5] would be violated (Berkeley 1734, 16; Newton 1704, 167). Berkeley's objection appears to be so reasonable and circumspect that, even in the middle of the nineteenth century, some mathematicians like William Rowan Hamilton (1805–1865) and Hermann Weissenborn (1830–1896) echoed it (Cajori 1919b, 91–92).

The second objection concerns Newton's argument for justifying the algorithm of the direct method of fluxions in the *De quadratura* (Berkeley 1734, 21–26). Berkeley holds that the symbol "o" means two contradictory things in Newton's work. For him, when Newton considers the ratio:

$$o : \left(nox^{n-1} + \frac{nn - n}{2} oox^{n-2} + \&c. \right)$$

in order to deduce from it the ratio:

$$1 : \left(nx^{n-1} + \frac{nn - n}{2} ox^{n-2} + \&c. \right),$$

the symbol "o" means "something" —says Berkeley—; or, in other words, it denotes an object whose properties are similar to those of strictly positive numbers. Thus the fraction $\frac{o}{o}$ is equal to 1 and the quotient $\frac{nox^{n-1} + \frac{nn-n}{2} oox^{n-2} + \&c}{o}$ is equal to $nx^{n-1} + \frac{nn-n}{2} ox^{n-2} + \&c.$ On the contrary, when Newton cancels the terms with o as a factor in the second ratio, so as to obtain the result $1 : (nx^{n-1})$, the symbol "o" means "nothing", that is, it denotes zero. For Berkeley, this implies that Newton still subscribes to the idea of the infinitely small in the *De quadratura*, albeit referred to as "evanescent quantity", which besides cannot be distinguished from the Leibnizian notion of infinitesimal difference (Berkeley 1734, 26).

---

[5] In rebus mathematicis errores quam minimi non sunt contemnendi.

As could be expected, a vigorous reaction arose at that point against Berkeley's criticism. On the one hand, some of Newton's partisans engaged in open debate with Berkeley, like James Jurin (1684–1750) and Benjamin Robins (1707–1751) who ironically fought each other as rivals over the correct interpretation of Newton's ideas (Berkeley 1735; Cajori 1919b, 64–138; Jesseph 1993, 233–259; Jurin 1734; 1735; Robins 1735; 1736). On the other hand, and more importantly, Berkeley's criticism contributed to rekindle the interest in the theory of fluxions. Three new manuals by Thomas Bayes (1702–1761), James Hodgson (1672–1755), and John Muller (1699–1784) appeared in 1736 with the explicit objective of resolving Berkeley's objections (Bayes 1736; Hodgson 1736; Muller 1736).

Hodgson's manual is particularly interesting because it contains the first fluxional proof of the binomial theorem for rational exponents. The proof takes place within the context of the debate on Berkeley's objections discussed above, and depends on Hodgson's response. In an original tactic, Hodgson tried to answer Berkeley's first objection by assuming that it was not an objection at all. Indeed he proved the product rule by means of the same reasoning used by Berkeley (Hodgson 1736, XIII–XIV, 56–57). By supposing o to be "a Quantity infinitely small" and taking $xy = v$, he got:

$$v + \dot{v}o = (x + \dot{x}o)(y + \dot{y}o) = xy + y\dot{x}o + x\dot{y}o + \dot{y}\dot{x}oo.$$

Then, after division by o and after reduction, he obtained:

$$\dot{v} = y\dot{x} + x\dot{y} + \dot{y}\dot{x}o.$$

Finally he canceled the term $\dot{y}\dot{x}o$ because of the infinitely small factor o:

$$\dot{v} = y\dot{x} + x\dot{y}.$$

It must be recognized that Hodgson's proof presupposes both the existence of the infinitely small and Berkeley's claim according to which it is not possible to operate with a half-moment. As Hodgson deduces the power rule from the product rule, in the same way that Newton did it in the *Principia*, the problem posed by Berkeley's second objection disappears (Hodgson 1736, 60–61).

Hodgson's proof of the binomial theorem for rational exponents depends on the power rule (Hodgson 1736, 24–26). Hodgson presupposed that the binomial power $\overline{1+x}^n$, where n is a fractional number, can be expanded in a power series of the form:

$$\overline{1+x}^n = 1 + Ax + Bxx + Cx^3 + Dx^4 + Ex^5, \&c., \tag{2.2}$$

where $A, B, C, \ldots$ are undetermined coefficients. The proof consists then in finding such coefficients. By calculating fluxions on both sides of Eq. 2.2, he got:

$$n \times \overline{1 + x}^{n-1} \dot{x} = A\dot{x} + 2Bx\dot{x} + 3Cx^2\dot{x} + 4Dx^3\dot{x} + 5Ex^4\dot{x}, \&c. \tag{2.3}$$

Then, by dividing this equation by $\overline{1 + x}^n \times \dot{x}$, he got:

$$\frac{n}{1 + x} = \frac{A + 2Bx + 3Cx^2 + 4Dx^3 + 5Ex^4, \&c.}{1 + Ax + Bxx + Cx^3 + Dx^4 + Ex^5, \&c.}.$$

And it follows from this that:

$$n + nAx + nBxx + nCx^3 + nDx^4, \&c. =$$
$$A + (2B + A)x + (3C + 2B)x^2 + (4D + 3C)x^3, \&c.$$

Comparing the coefficients of equal powers of x, coefficients $A, B, C, \ldots$ can be determined:

$$A = n, \quad B = \frac{n - 1}{2}A, \quad C = \frac{n - 2}{3}B, \quad D = \frac{n - 3}{4}C, \ldots.$$

It is convenient to underline at this stage that the procedure consisting in comparing the coefficients of equal powers of x was known as the "method of undetermined coefficients", which was largely utilized to develop the theory of series in the eighteenth century (Ferraro and Panza 2003, 23–27). After substituting the coefficients obtained by the method of undetermined coefficients in Eq. 2.2, Hodgson got the expression of the binomial theorem for a rational exponent n:

$$\overline{1 + x}^n = 1 + nx + n \times \frac{n - 1}{2}x^2 + n \times \frac{n - 1}{2} \times \frac{n - 2}{3}x^3, \&c.$$

But he multiplied this expression by p^n, where p is a constant, in order to write the theorem in the form given by Newton in the *Epistola prior*, which completes Hodgson's proof.

For Hodgson, then, the binomial theorem's justification depends on the direct method of fluxions. A similar demonstration was proposed in 1742 by Colin Maclaurin (1698–1746), whose treatise on the theory of fluxions was more popular than Hodgson's manual (Maclaurin 1742, vol. 2: 607–608). Maclaurin makes it clear from the beginning that his treatise was motivated by Berkeley's criticism, and that he was familiar with the literature on the subject (Maclaurin 1742, vol. 1: iii). It is possible, then, that he had read Hodgson's manual. In any case, like Hodgson, he proved first the power rule and then obtained Eqs. 2.2 and 2.3. But he divided Eq. 2.3 by $n\dot{x}$, so he got:

$$\overline{1+x}^{\,n-1} = \frac{A}{n} + \frac{2Bx}{n} + \frac{3Cx^2}{n} + \frac{4Dx^3}{n} + \&c. \tag{2.4}$$

By taking $x = 0$, he determined the first coefficient $A = n$. By calculating fluxions on both sides of Eq. 2.4, one gets:

$$n - 1 \times \overline{1+x}^{\,n-2}\dot{x} = \frac{2B\dot{x}}{n} + \frac{6Cx\dot{x}}{n} + \frac{12Dx^2\dot{x}}{n} + \&c.$$

After division of this equation by $n - 1 \times \dot{x}$, the coefficient $B = n \times \frac{n-1}{2}$ is determined by taking $x = 0$. The values of the other coefficients can be obtained in an analogous manner.

Besides such manuals, Berkeley's criticism was also the direct cause of the posthumous edition of Newton's *De methodis*, translated and annotated by Colson, who contributed to the debate on the method of fluxions with the publication of "the only genuine and original Fountain of this kind of knowledge" (Newton 1736, x). But his own contribution consisted in his long annotations at the end of the volume, among which one can find a commentary on the binomial theorem (Newton 1736, 309–310). Colson thought that it would be a good exercise for students to find the power series expansion of the rational power $(a + x)^n$ by means of the direct and the inverse method of fluxions, and he actually showed how to combine these methods in order to obtain the binomial expansion. Colson's procedure consists in applying successively the power rule to the results obtained by appropriate algebraic manipulations and by integration. However, he was persuaded that a proper justification of the power rule should be grounded on the binomial theorem, for this is the way in which Newton justified the algorithm of the direct method of fluxions in the *De methodis*, as has been seen above. Therefore, for Colson, any proof of the binomial theorem relying on the direct method of fluxions should lead to a vicious circle, as he pointed out (Newton 1736, 310):

> Indeed it can hardly be said, that this, or any other that is derived from the Method of Fluxions, is a strict Investigation of this Theorem. Because that Method itself is originally derived from the Method of raising Powers, at least integral Powers, and previously supposes the Knowledge of the *Unciae*, or the numerical Coefficients. However it may answer the intention, of being a proper Example of this Method of Extraction, which is all that is necessary here.

Perhaps Colson did not have the opportunity to read and analyze Hodgson's proof of the binomial theorem, where no vicious circle is committed, or perhaps he considered that the power rule cannot be justified on the basis of the method of first and last ratios, as Hodgson did it in accordance with Newton's *Principia*.

Despite the existence of fluxional proofs of the binomial theorem where there is no vicious circle, Colson's observation that the principles of the method of fluxions depend on the binomial theorem will determine the shape and direction of the mathematical research related to the binomial theorem to be carried out. On the one hand, Colson's observation will be repeated by other authors within the context of the debate on the correctness of the method of fluxions. For instance, in the annotations to his English translation of the *De quadratura* and the *De analysi*, John Stewart (?–1766) claimed that particular caution is needed to ward off the risk of getting into a vicious circle when the method of fluxions is used to prove the binomial theorem (Newton 1745, 475). On the other hand, some mathematicians actually committed this logical fallacy, as Thomas Simpson (1710–1761) who transcribed, in 1750, Hodgson's proof of the binomial theorem, even though he had previously justified the power rule on the grounds of this theorem (Simpson 1750, 6, 110–112).

In sum, Newton used the binomial theorem to justify the algorithm of the direct method of fluxions in his *De analysi*, *De methodis*, and *De quadratura*, algorithm that was formulated in the form of what may be identified as the power rule. But he banished the method of infinite series from the *Principia*, and established the power rule using elementary algebraic operations. Both of these procedures were attacked by Berkeley's objections, and the efforts to reconcile the conflicting results of the method of fluxions led to two different conceptions of the binomial theorem. According to the first, which is based on the *Principia*, the method of fluxions is independent from the binomial theorem; moreover, this latter can be demonstrated by the method of fluxions. In contrast, the second conception rests on the idea that the direct method of fluxions is logically dependent on the method of infinite series, for the binomial theorem is needed to justify its fundamental algorithm.

2.1.3   Newton's Theorem in Germany

In the German tradition, Leibniz was the first to apprehend Newton's theorem and to use it for the solution of certain mathematical problems. In 1694 he applied it for example to the study of the equation of the elastic curve, emphasizing that the binomial theorem was a Newton's "artifice" (*artificium*), but without enunciating it explicitly (Leibniz 1694). Nevertheless, it was Christian Wolff (1679–1754) who introduced it into German academic circles. In his dissertation *Methodus serierum infinitarum* of 1705, Wolff summarizes the advances made in England toward developing the method of infinite series during the period between Mercator's *Logarithmo-technia* of 1668 and Hayes's *Treatise of fluxions* of 1704 (Wolff 1705). The binomial theorem is enunciated in the form Newton gave it in the *Epistola prior*, and is viewed as the keystone of the method of infinite series. For Wolff, this method not only characterizes the general structure of the *De quadratura*, but it also serves implicitly to shore up the mathematical construction of the *Principia* (Wolff 1705, 314): "In this regard, these series have been employed in the difficult work

Philosophiae naturalis principia mathematica [...]."[6] Wolff concludes his dissertation by observing that the concepts of the method of fluxions and those of differential calculus are equivalents, in the sense that what Newton was able to accomplish with his method of infinite series, upon which the method of fluxions would be built, could be achieved with the differential calculus of Leibniz.

Further, in 1710, Wolff incorporated the binomial theorem in his *Anfangsgründe aller mathematischen Wissenschaften*, and later in the Latin translation *Elementa matheseos universae* of 1713 (Wolff 1710, 48–57; Wolff 1713, vol. I: 264–269). This manual had a very large success among German scholars, ensuring the dissemination of the theorem. But no proof of it was offered in Germany until the decade of the 1740s.

This section reviews the demonstrations of the binomial theorem for rational exponents proposed in German countries that had some influence on the emergence and theoretical organization of the German combinatorial analysis. First, we will focus on the proofs relying on differential calculus. Next, we will turn our attention to some proofs based on functional equations. Finally, we will examine other proofs that depend on the multinomial theorem for positive integer exponents.

2.1.3.1 Differential Proofs of the Binomial Theorem for Rational Exponents

Leonhard Euler (1707–1783) states the binomial theorem for rational exponents in his *Introductio in analysin infinitorum*, but his first attempt to prove it is found in the *Institutiones calculi differentialis* (Euler 1748, vol. 1: 55; Euler 1755, 359–361). However, it is difficult to tell exactly when this took place because the composition of the *Institutiones* began approximately in 1741 and was completed about 1750 (Jushkevich 1983, 161). This was an unsuccessful attempt to prove the binomial theorem, since Euler engaged in the circular reasoning to which Colson referred in 1736. Further, Euler became aware of his error and wrote at least three other proofs that were no more based on differential calculus (Coolidge 1949, 155–156; Dhombres 1986, 151–153; Euler 1775; 1789; 1813; Pensivy 1988, 100–102, 112–114, 124–127). We will present here the details of his differential proof, and we will study in Sect. 2.1.3.2 his 1775 proof, which was published before the emergence of the German combinatorial analysis.

In Chapter I of his *Institutiones*, Euler defined the finite differences of the first order as follows (Euler 1755, 3–22):

$$\Delta y = y^{\mathrm{I}} - y, \quad \Delta y^{\mathrm{I}} = y^{\mathrm{II}} - y^{\mathrm{I}}, \quad \Delta y^{\mathrm{II}} = y^{\mathrm{III}} - y^{\mathrm{II}}, \quad \Delta y^{\mathrm{III}} = y^{\mathrm{IV}} - y^{\mathrm{III}}, \ldots, \tag{2.5}$$

where

$$y = (x), \quad y^{\mathrm{I}} = (x + \omega), \quad y^{\mathrm{II}} = (x + 2\omega), \quad y^{\mathrm{III}} = (x + 3\omega), \ldots$$

---

[6] Opportune quoque iisdem seriebus usus est in abstruso opere *Principiorum Philosophiae Naturalis Mathematicorum* [...].

and where ω is a finite increment for the value of x. But if ω is supposed to be an infinitesimal increment, then the finite differences of 2.5 become infinitesimal differences. Thus, by choosing $y = x^n$ and $\omega = dx$, and by applying the binomial theorem, Euler got (Euler 1755, 124):

$$y^{\mathrm{I}} = (x + dx)^n = x^n + nx^{n-1}dx + \frac{n(n-1)}{1.2}x^{n-2}dx^2 + \&c.$$

Then, by the first equation of 2.5, it follows that:

$$dy = \left(y^{\mathrm{I}} - y\right) = (x + dx)^n - x^n = nx^{n-1}dx + \frac{n(n-1)}{1.2}x^{n-2}dx^2 + \&c.$$

Euler established the power rule by observing that, in this latter equation, the terms in $dx^2, dx^3, \ldots$ will vanish before the term in dx does, so $dy = dx^n = nx^{n-1}dx$. Then, the binomial theorem was proved by means of Taylor's theorem, which was enunciated in the following form:

$$z = y + \frac{\omega dy}{dx} + \frac{\omega^2 ddy}{1.2dx^2} + \frac{\omega^3 d^3 y}{1.2.3dx^3} + \frac{\omega^4 d^4 y}{1.2.3.4dx^4} + \&c.$$

By taking $z = (x + \omega)^n$ and $y = x^n$, Euler obtained the binomial expansion:

$$z = (x + \omega)^n = x^n + \frac{n}{1}x^{n-1}\omega + \frac{n(n-1)}{1.2}x^{n-2}\omega^2 + \&c.$$

Euler did not specify the nature of the exponent n, but it is suggested in his text that the exponent is not restricted to positive integer values. In any case, Euler's circular reasoning is evident because of the successive differentiation of $y = x^n$, which depends on the power rule.

Euler's error was soon noticed and corrected by Abraham Gotthelf Kästner (1719–1800), ordinary professor of natural philosophy and mathematics at the University of Göttingen from 1756 to 1800, who remarked with deference in 1758 (Kästner 1758, 13):

> By means of higher-order fluxions, and of the fluents derived from these fluxions, John Colson explains how to investigate the binomial theorem, but only as an exercise of calculus and not as a way to show the truth. I think it was for the same purpose that the illustrious Euler included a similar investigation in the *Institutiones calculi differentialis*. It is not possible to call proofs those that presuppose the laws of differential calculus, but in which it is not shown that these are true independently of what is derived from them.[7]

---

[7] Quaesitis fluxionibus superiorum graduum, et fluentibus ex fluxionibus, theorema binomiale investigare docet Io. COLSON, sed calculi exercendi gratia, non veritatis ostendendae. Eodem fini insertam crediderim similem investigationem institutionibus calculi differentialis a Cel. EVLERO.

In 1758, Kästner offered a demonstration of the binomial theorem for rational exponents, which was based on differential methods and which was free of logical fallacies.

The demonstration was structured in three parts. First, Kästner proved by induction on the exponent m of $(a + b)^m$ that the binomial theorem is true for any positive integer exponent. For the base case, he verified the correctness of the theorem for $m = 1, 2, 3,$ and 4. Then, he gave a very explicit formulation of the principle of induction (Kästner 1758, 7):

> If the binomial theorem is true for any [positive] integer exponent m, then it is true for the same exponent m incremented by one unit.[8]

Thus, he assumed the claim for:

$$(a + b)^m = a^m + ma^{m-1}b + \frac{m\,(m-1)}{1.2}a^{m-2}b^2 + \dots$$

And by multiplying this expression by $(a + b)$, he got:

$$(a + b)^{m+1} = a^{m+1} + (m + 1)\,a^m b + \frac{(m+1)\,m}{1.2}a^{m-1}b^2 + \dots$$

Kästner also confirmed by induction the law of formation of the coefficients. By hypothesis, the general coefficient of the binomial expansion is:

$$P = \frac{m\,(m-1)\,(m-2)\dots(m-k+1)}{1.2.3\dots k}.$$

Therefore:

$$\left(\frac{m+1}{k+1}\right) \times P = \frac{(m+1)\,m\,(m-1)\,(m-2)\dots(m-k+1)}{1.2.3\dots k.\,(k+1)}.$$

Consequently, the general term of the binomial expansion for any positive integer exponent m is:

$$\frac{P\,(m-k)}{k+1}a^{m-k-1}b^{k+1}.$$

Non possunt hae demonstrationes appellari quae leges calculi differentialis sumunt, neque illas independenter ab eo quod ex iis deducitur veras esse ostendunt.

[8] Si theorema binomiale verum est, notante m exponentem aliquem integrum, verum etiam est, notante m, eudem exponentem, vnitate auctum.

This first part of the proof was originally published in 1745 and was later included, in 1760, in Kästner's *Anfangsgründe der Analysis endlicher Grössen* (Kästner 1745; 1760, 52–68).

The second part of the proof consists in justifying the power rule. In this case, Kästner's strategy also has two parts. First, he established the power rule for a positive integer power using a similar argument to that given by Euler in his *Institutiones*. However, unlike Euler's, Kästner's argument is not circular because the binomial theorem for positive integer exponents has been demonstrated independently of the power rule. Next, Kästner generalized the power rule to rational powers in the same way Newton did it in the *Principia mathematica*.

The third part of the proof concerns the case in which m is a rational number in $(a + b)^m = a^m \left(1 + \frac{b}{a}\right)^m$. Given the following equation:

$$(1 + y)^m = 1 + Ay + By^2 + \cdots + Py^k + Qy^{k+1} + \cdots = w, \tag{2.6}$$

where $A, B, C, \ldots$ are undetermined coefficients and where $y = \frac{b}{a}$, the proof consists in determining the coefficients of the series expansion. After differentiation of Eq. 2.6, Kästner got:

$$m(1 + y)^{m-1} dy = dw. \tag{2.7}$$

And by multiplying this by $(1 + y)$, he obtained $m(1 + y)^m dy = (1 + y) dw$, that is, $mwdy = (1 + y) dw$. Hence he got:

$$(1 + y)\frac{dw}{dy} - mw = 0.$$

By Eqs. 2.6 and 2.7, it follows from this that:

$$(1 + y)\left(A + 2B + \cdots + kP^{k-1} + (k+1)Q^k + \ldots\right) -$$

$$m\left(1 + Ay + By^2 + \cdots + Py^k + Qy^{k+1} + \ldots\right) =$$

$$(A - m) + (2B + A - mA)y + (3C + 2B - mB)y^2 + \ldots$$

$$\cdots + ((k+1)Q + kP - mP)y^k + \cdots = 0.$$

By the method of undetermined coefficients, Kästner derived from this the coefficients $A = m$, $B = \frac{(m-1)A}{2}$, $C = \frac{(m-2)B}{3}$, ..., and stated the general coefficient of the binomial expansion:

$$Q = \frac{(m-k)\,P}{k+1}.$$

This completes Kästner's demonstration of the binomial theorem for rational exponents.

In 1761 a German translation of this demonstration was printed in Kästner's *Anfangsgründe der Analysis des Unendlichen*, one of the most popular mathematical manuals of the period in German countries (Kästner 1761, 29–33). It is possible to identify the influence of Kästner on the work of other mathematicians. For instance, Wenceslaus Johann Gustav Karsten (1732–1787) exposed in 1760 the Leibnizian differential calculus in accordance with Euler's presentation, but the proof of the binomial theorem was structured in three parts, following Kästner's approach (Karsten 1760, 428–430, 547, 550–551, 567–568).

In general, the strategy for avoiding circular reasoning consisted in proving the power rule without using the binomial theorem for rational exponents, although different means were employed to achieve the same end. In some cases this rule was settled by using differential techniques. For example, in 1770 Georg Friedrich von Tempelhoff (1737–1807) obtained the power rule by applying logarithmic differentiation as follows (von Tempelhoff 1770, 198, 301–307):

$$\omega = x^m$$
$$l\omega = mlx$$
$$\frac{d\omega}{\omega} = \frac{mdx}{x}$$
$$d\omega = \frac{m\omega dx}{x} = \frac{mx^m dx}{x}$$
$$d\omega = mx^{m-1}dx,$$

where l stands for natural logarithm and where m is a positive integer. Perhaps this approach was inspired by the work *Elementa analyseos infinitorum* by Johann Andreas von Segner (1704–1777), who proposed this procedure 9 years earlier and with whom Tempelhoff studied mathematics at Halle (Anonym 1807, 537; Segner 1761, 115–116).

2.1.3.2 Functional Proofs of the Binomial Theorem for Rational Exponents

Originally the functional method was thought of as a means of avoiding circular reasoning in proving the binomial theorem for rational exponents. Unlike differential proofs, most functional proofs are built around the idea that differential calculus should be logically dependent on the binomial theorem, an insightful idea suggested, as seen above, by some

English mathematicians as a consequence of the debate on the correctness of the method of fluxions.

In the German tradition this idea was strongly expressed by Franz Ulrich Theodor Aepinus (1724–1802) (Aepinus 1763, 27–28):

> Even in the case of pure analysis, there are some demonstrations that have been profoundly weakened by sophistry; as when, for instance, it is said of many truths regarding quantities that they are general truths, although their demonstration is suitable only for whole numbers. *Newton*'s theorem is particularly confronted with such allegations [...]. But so remote is the possibility of questioning its truth in the remaining cases, where [the exponent] *m* is either a fractional number, or an irrational number, or a transcendental number, or even an imaginary number, that the general analysis of the infinite has been built upon this theorem, understood in its broadest sense. Those who then tried to demonstrate its truth in the general case committed an evident vicious circle in their reasoning, since they used the analysis of the infinite. Thus, the author undertook the significant task of proving, by using only ordinary algebra, the fundamental principle of the whole of the analysis [of the infinite].[9]

The coherence problems associated with the method of fluxions were explicitly interpreted by Aepinus as a foundational problem for differential calculus. And his approach suggests that he was not merely interested in giving a non-circular demonstration of the binomial theorem, but more ambitiously in providing a way of grounding differential calculus in ordinary algebra.

With the support of Euler, Aepinus left Berlin, where he was working from 1755 to 1757, and settled in Saint Petersburg, where he was elected a member of the Imperial Academy of Sciences and Arts. The paragraph cited above comes from the abstract of an article appeared in the Memoirs of the Academy of 1760–1761 (published in 1763). The aim of the article is certainly ambitious, but the demonstration Aepinus provided is based on some conceptual confusions about numbers and is expressed in a rather unsuitable notation. Notwithstanding its inadequacy, we will retain his notation with the intention of showing some of the typical problems that the German combinatorial school faced and tried to solve some years later.

---

[9] In Analysi etiam pura eiusmodi occurrunt demonstrationes, quas cauillandi studium non medio-criter labefactare est annisum, cum circa quantitates plurimae veritates tanquam generales admitti soleant, etiamsi demonstratio tantum ad numeros integros sit accomodata. Huiusmodi obtrectationes imprimis expertum est Theorema *Newtonianum* [...]. Tantum autem abest, vt pro reliquis casibus, quibus m est vel numerus fractus, vel irrationalis, vel transcendens, vel adeo imaginarius, de eius veritate dubitetur, vt potius huic Theoremati in latissimo sensu accepto vniuersa Analysis infinitorum sit superstructa. Hinc ii, qui eius veritatem in genere, Analysi infinitorum in subsidium vocata, demonstrare sunt conati, manifesto vitiosissimun circulum in ratiocinando commiterunt. Non inutiliter itaque collocauit laborem Auctor, cum demonstrationem fundamentalis huius totius Analyseos principii, idque per sola Algebra communis elementa, condere aggressus est.

The proof starts by assuming that:

$$(1 + x)^m = Ax^m + Bx^{m-1} + Cx^{m-2} + \ldots, \tag{2.8}$$

where $A, B, C, \ldots$ are undetermined coefficients and m is a real number (Aepinus 1763, 170; Dhombres and Pensivy 1988; a French translation of Aepinus's paper can be found in Pensivy 1988, 187–197). Then Aepinus supposes that the undetermined coefficients can be expressed as functions of m, which he symbolizes by superscripts. Thus the expressions $A^m, B^m, C^m, \ldots$ do not denote powers, but functions (i.e. a combination of symbols that refers to some quantity, which in this case can be expressed in terms of m). From Eq. 2.8, it follows that:

$$(2 (1 + x))^m = 2^m A^m x^m + 2^m B^m x^{m-1} + 2^m C^m x^{m-2} + \ldots,$$

and

$$((2x + 1) + 1)^m = A^m (2x + 1)^m + B^m (2x + 1)^{m-1} + C^m (2x + 1)^{m-2} + \ldots$$
$$= 2^m A^m A^m x^m + \left(2^{m-1} A^m B^m + 2^{m-1} B^m A^{m-1} \right) x^{m-1} +$$
$$\left(2^{m-2} A^m C^m + 2^{m-2} B^m B^{m-1} + 2^{m-2} C^m A^{m-2} \right) x^{m-2} + \ldots$$

By the method of undetermined coefficients, Aepinus deduces from this the following equalities:

$$2^m A^m = 2^m A^m A^m,$$
$$2^m B^m = 2^{m-1} \left(A^m B^m + B^m A^{m-1} \right),$$
$$2^m C^m = 2^{m-2} \left(A^m C^m + B^m B^{m-1} + C^m A^{m-2} \right),$$
$$2^m D^m = 2^{m-3} \left(A^m D^m + B^m C^{m-1} + C^m B^{m-2} + D^m A^{m-1} \right),$$
$$\vdots$$

On the base of the first equality, and by remarking that the value of the coefficient A is constant, Aepinus establishes that $A^m = A^{m-1} = A^{m-2} = \cdots = 1$. From this and from the equalities above, the remaining coefficients can be expressed in terms of the coefficient B:

$$C^m = \frac{B^m B^{m-1}}{1.2}, \quad D^m = \frac{B^m B^{m-1} B^{m-2}}{1.2.3}, \ldots$$

And the general coefficient is then:

$$T^m = \frac{B^m B^{m-1} B^{m-2} \ldots B^{m-k+1}}{1.2.3 \ldots k}.$$

As we can see, the Newtonian alphabetical notation chosen by Aepinus does not allow to show the mathematical dependency between T and k.

Now the whole demonstration depends on the possibility of determining the coefficient B. By rewriting Eq. 2.8 as $(1 + x)^{r+s} = x^{r+s} + B^{r+s} x^{r+s-1} + \ldots$ and $(1 + x)^r (1 + x)^s = x^{r+s} + (B^r + B^s) x^{r+s-1} + \ldots$ (with $m = r + s$), and by the method of undetermined coefficients, Aepinus obtains the functional equation:

$$B^{r+s} = B^r + B^s,$$

that is:

$$B^s = B^{r+s} - B^r.$$

Then Aepinus argues that the change of the quantity B^m (when m is increased or decreased) can be expressed as a ratio of B^m and s. He stated his idea in the following form (Aepinus 1763, 174):

Let s be infinitely small and let

$$\cdots - 3s, -2s, -s, 0, +s, +2s, +3s \ldots$$

be a progression continued to infinity on both sides, which takes all the real values of m while continuously progressing. As a result, it is possible to assert that if m is a real number, then B^m and m will have a given ratio to one another.[10]

In other words, Aepinus thought erroneously that the set of real numbers could be constructed by multiplying an infinitely small quantity s by each element of the set of integers, and in so doing the proportion between B^m and m would remain constant. After verifying that, for $m = 1$ in Eq. 2.8, one gets $B^1 = 1$, and on the basis of his mistaken idea about real numbers, Aepinus concluded that $B^m = m$ for any real number m.

Aepinus's argument based on a confusing conception of real numbers and on the unjustified generalization from $B^1 = 1$ to $B^m = m$ was criticized by Euler in his second attempt to prove the binomial theorem, presented to the Academy of Saint Petersburg in

---

[10] Sit s infinite paruum, et progressio $\cdots - 3s, -2s, -s, 0, +s, +2s, +3s \ldots$ vtrimque in infinitum continuata, transeundo per continuum, comprehendet omnes valores reales ipsius m, quapropter generatim affirmari potest, si m fuerit numerus realis quicunque, fore B^m ad m in ratione data.

1774 (published the next year). Euler remarked (Euler 1775, 106):

> Indeed I have proposed in the past a demonstration derived from the analysis of the infinite; but as this analysis itself rests upon our theorem, I realize now that it must be rejected because of *petitio principii*. But Aepinus, an illustrious member of our Academy, gave a demonstration which is not concerned by this vicious in Tome VIII of *Novi Commentarii*, where [...] he obtained the value of some coefficients A, B, C, D, etc. by means of the most ingenious method [...]. However this remarkable demonstration depends a lot on induction. Moreover, it is convenient to point out that the determination of the second coefficient B is not obtained by this method, but by other presuppositions that are very mysterious and abstruse.[11]

It is convenient to say here that Euler insists that the binomial theorem "is the foundation of the sublime analysis" (*fundamentum constituit universae analyseos sublimioris*) (Euler 1775, 103). In his second attempt to prove the binomial theorem for rational exponents, Euler will try to clarify the method used by Aepinus.

In his proof, Euler assumes the validity of the binomial theorem for positive integer exponents and writes the binomial expansion as follows (Dhombres 1986, 151–152; Euler 1775; Netto 1908, 203–204; Panza 1992, vol. 2: 671–672; Pensivy 1988, 124–127):

$$[n] = 1 + \frac{n}{1}x + \frac{n\,(n-1)}{1.2}x^2 + \&c.$$

Then, considering another positive integer m, Euler calculates the following product:

$$[n][m] = \left(1 + \frac{n}{1}x + \frac{n\,(n-1)}{1.2}x^2 + \&c.\right)\left(1 + \frac{m}{1}x + \frac{m\,(m-1)}{1.2}x^2 + \&c.\right)$$

$$= 1 + \frac{n+m}{1}x + \frac{(n+m)\,(n+m-1)}{1.2}x^2 + \&c.$$

From this, he deduced the following equation:

$$[n]\,[m] = [n+m].\tag{2.9}$$

---

[11] Equidem olim demonstrationem ex analysi infinitorum petitam tradideram; sed quia ipsa haec analysis nostro theoremate innititur, eam tanquam petitionem principii penitus reiiciendam nunc agnosco ; ab hoc vitio autem immunem demonstrationem dedit Illustr. Academiae nostrae Socius Aepinus in Tomo VIII Nouor. Commentar., vbi [...] methodo maxime ingeniosa elicuit valores aliquot coefficientium A, B, C, D, etc. [...]; interim tamen egregia ista demonstratio plurimum inductione innititur, praeterea vero etiam notari conuenit secundum coefficientem B ex hac methodo determinationem non accepisse, sed ex aliis conditionibus haud parum absconditis et abstrusis repetiisse.

It follows from this that, for a positive integer a:

$$[m]^a = [am].$$

If i is a positive integer such that $i = am$, then one gets:

$$\left[\frac{i}{a}\right]^a = [i].$$

By the binomial theorem for positive integer exponents, Euler obtains:

$$\left[\frac{i}{a}\right]^a = [i] = (1+x)^i.$$

The binomial theorem for positive rational exponents follows immediately from this equality, so:

$$\left[\frac{i}{a}\right] = (1+x)^{\frac{i}{a}}.$$

Concerning negative exponents, Euler took $n = -m$ in Eq. 2.9, so:

$$[m][-m] = [0] = (1+x)^0 = 1.$$

And from this, he gets:

$$[-m] = \frac{1}{(1+x)^m} = (1+x)^{-m}.$$

This completes Euler's proof.

In the same vein, Johann Andreas von Segner figured out a proof of the binomial theorem for rational exponents that seems to be influenced by Euler, though no reference whatsoever was made to Euler's work (Netto 1908, 204–205; Panza 1992, vol. 2: 673; Pensivy 1988, 128; Segner 1779). It was presented to the Academy of Science at Berlin in 1777 (published in 1779). Born in Pressburg (Bratislava), Segner spent most of his life in Germany, where he was appointed to the new chair of mathematics at the University of Göttingen in 1735, which will be occupied by Kästner in 1755 after Segner's departure to the University of Halle, where he lectured the rest of his life. For this reason Segner's work is usually associated with the German mathematical tradition.

As Euler, but without making any remark about the *petitio principii* or the foundation of the sublime analysis, Segner starts his demonstration by assuming the validity of the binomial theorem for a positive integer exponent n and he writes the binomial expansion as follows:

$$a^n S = a^n \left(1 + \frac{n}{1} \cdot \frac{b}{a} + \frac{n(n-1)}{1.2} \cdot \frac{b^2}{a^2} + \frac{n(n-1)(n-2)}{1.2.3} \cdot \frac{b^3}{a^3} + \&c. \right)$$

$$= a^n + \frac{n}{1} a^{n-1} b + \frac{n(n-1)}{1.2} a^{n-2} b^n + \frac{n(n-1)(n-2)}{1.2.3} a^{n-3} b^3 + \&c.$$

After calculating the product $(a^m S)(a^r S)$, where $n = m + r$ (with m and r positive integers), Segner gets:

$$a^n S = \left(a^m S \right) \left(a^r S \right) = a^{m+r} S. \tag{2.10}$$

By assuming the validity of the binomial theorem for positive integer exponents, and on the basis of the equality $aS = a \left(1 + \frac{b}{a} \right) = a + b$, Segner obtains:

$$a^m S = (a+b)^m . \tag{2.11}$$

By Eqs. 2.10 and 2.11, Segner established the binomial theorem for integer exponents (positive or negative) as:

$$a^m S = \frac{a^n S}{a^r S} = \frac{(a+b)^n}{(a+b)^r} = (a+b)^{n-r} = (a+b)^m .$$

From this equation, the binomial theorem for rational exponents was established as follows:

$$\sqrt[t]{a^m S} = a^{\frac{m}{t}} S = (a+b)^{\frac{m}{t}} ,$$

where t is a positive integer.

In order to justify the binomial theorem for real exponents, Segner proposed the following argument:

Inasmuch as irrational numbers, provided that they are real [numbers], can always be considered as a fraction with an infinitely large denominator, n can also denote any irrational number in this equation.[12]

It goes without saying that this argument cannot be accepted as correct.

---

[12] Numerus autem irrationalis, dummodo realis sit, cum semper spectari possit tanquam fractus denominatoris infinite magni: poterit n in eadem aequatione etiam numerum quemuis irrationalem denotare.

2.1.3.3 Proofs of the Binomial Theorem for Rational Exponents by the Method of Series

We are concerned here with proofs of the binomial theorem for rational exponents which depend upon the power series expansion of a given positive integer power of an "infinite" polynomial. As this proof technique consists in using the binomial theorem (restricted to the case where the exponent is a positive integer) to calculate this power series expansion, this method can be regarded as depending on the Newtonian method of infinite series.

In German countries, it was Segner who first proposed this kind of proof in 1758 in the second volume, devoted to "finite analysis" or algebra, of his *Cursus mathematici* (Segner 1758, 189–194). First he showed how to prove the binomial theorem for positive integer exponents by means of the arithmetical triangle (Segner 1758, 140, 145–147). In the case where the exponent is a negative integer, the problem was solved by division: $\frac{1}{(1+z)^m} = (1+z)^{-m}$.

Then he addressed the general case of rational exponents. By the binomial theorem for a positive integer exponent m, he obtained the following equation:

$$(1 + Z)^m = A + BZ + CZ^2 + DZ^3 + EZ^4 + \&c.,$$

where $Z = \alpha z + \beta z^2 + \gamma z^3 + \delta z^4 + \&c.$, and where $A, B, C, \ldots$ are the binomial coefficients:

$$A = 1, \quad B = \frac{m}{1}, \quad C = \frac{m(m-1)}{1.2}, \ldots$$

Then by calculating the positive integer powers of Z, he got:

$$BZ = B\alpha z + B\beta z^2 + B\gamma z^3 + \&c.$$
$$CZ^2 = C\alpha^2 z^2 + 2C\alpha\beta z^3 + \&c.$$
$$DZ^3 = D\alpha^3 z^3 + \&c.$$
$$\vdots$$

And hence:

$$(1 + Z)^m = A + B\alpha z + \left(B\beta + C\alpha^2\right) z^2 + \left(B\gamma + 2C\alpha\beta + D\alpha^3\right) z^3 + \&c. \tag{2.12}$$

From the equation:

$$\sqrt[m]{(1 + z)^n} = 1 + Z = 1 + \alpha z + \beta z^2 + \gamma z^3 + \delta z^4 + \&c., \tag{2.13}$$

where n is an integer, it follows:

$$(1 + z)^n = 1 + \frac{n}{1}z + \frac{n\,(n-1)}{1.2}z^2 + \&c. = (1 + Z)^m . \tag{2.14}$$

By the method of undetermined coefficients, Segner deduced from Eqs. 2.12 and 2.14 the coefficients of the power series expansion on the right-hand side of Eq. 2.13:

$$1 = 1,$$

$$\alpha = \frac{n}{m},$$

$$\beta = \frac{\frac{n}{m}\left(\frac{n}{m} - 1\right)}{1.2},$$

$$\gamma = \frac{\frac{n}{m}\left(\frac{n}{m} - 1\right)\left(\frac{n}{m} - 2\right)}{1.2.3},$$

$$\vdots$$

This completes Segner's proof.

Among the mathematicians to whom Segner turned when writing his proof were Charles Reyneau (1656–1728) and Alexis Claude Clairaut (1713–1765), who had already conceived a similar demonstration in 1708 and 1746 respectively (Clairaut 1746, 242–255; Reyneau 1708, vol. I, 408–422). In 1769, Segner's proof was transcribed by his former student Tempelhoff in the *Anfangsgründe der Analysis endlicher Größen*, and later reformulated by Georg Simon Klügel (1739–1812) in appendix to his *Analytische Trigonometrie* (Klügel 1770, 234–248; von Tempelhoff 1769, 352–361).

Klügel received his training in mathematics at the University of Göttingen under the direction of Kästner. In 1767 he was appointed professor of mathematics in Helmstedt, where he worked for about 20 years. In 1788 he was named to fill the chair of mathematics in the University of Halle, which had become vacant by the death of Karsten. Most of his mathematical research was done during these two periods. Regarding to the binomial theorem, his proof takes a dual perspective: first, he recognizes the influence of Kästner; second, he adopts an algebraic point of view in a foundational sense. Indeed, Klügel thought that the binomial theorem belonged to finite analysis or algebra and, as a result, its proof should be based as much as possible on algebraic methods.

For a correct justification of the binomial theorem for positive integer exponents, Klügel referred explicitly to Kästner's inductive proof and to Segner's combinatorial proof. The problem for negative integer exponents was solved by division: $\frac{(1+z)^n}{(1+z)^m} = (1 + z)^{n-m}$, where m and n are positive integers. For rational exponents, Klügel proceeded in two steps. The first step was to expand in series the particular case of:

$$\left(1 + az + bz^2 + cz^3 + \&c\right)^{\frac{1}{m}}$$

by applying successively the binomial theorem for positive integer exponents. To this end, he put:

$$1 + az + bz^2 + cz^3 + \&c = (1 + Z)^m = 1 + mZ + \frac{m\,(m-1)}{1.2}Z^2 + \&c.$$

And then, he got:

$$Z = \frac{az}{m} + \&c.$$

Taking now $\left(1 + \frac{a}{m}z\right) + Z'$, he obtained:[13]

$$1 + az + bz^2 + \&c = \left(\left(1 + \frac{a}{m}z\right) + Z'\right)^m$$

$$= \left(1 + \frac{a}{m}z\right)^m + m\left(1 + \frac{a}{m}z\right)^{m-1}Z' + \&c.$$

After some algebraic manipulations and by applying the binomial theorem for positive integer exponents, this leads to:

$$Z' = \frac{2mb - (m-1)\,a^2}{2mm}z^2 + \&c.$$

Continuing this procedure, he got the desired result:

$$\left(1 + az + bz^2 + \&c\right)^{\frac{1}{m}} = 1 + \frac{a}{m}z + \frac{2mb - (m-1)\,a^2}{2mm}z^2 + \&c. \qquad (2.15)$$

The second step was to generalize this result to arbitrary positive rational exponents. On the basis of the binomial theorem for positive integer exponents, Klügel put:

$$(1 + z)^{mn} = 1 + mnz + \frac{mn\,(mn-1)}{1.2}z^2 + \&c,$$

---

[13] Klügel uses only the notation Z, but we have introduced Z', following the reconstruction of (Pensivy 1988), in order to avoid possible confusions.

where m and n are positive integers. Then, by applying the result 2.15 to this last equation, he got:

$$\left(1 + mnz + \frac{mn\,(mn-1)}{1.2}z^2 + \&c\right)^{\frac{1}{m}} =$$

$$1 + \frac{mn}{m}z + \frac{2m\left(\frac{mn(mn-1)}{1.2}\right) - (m-1)\,(mn)^2}{2mm}z^2 + \&c =$$

$$1 + nz + \frac{n\,(n-1)}{1.2}z^2 + \&c = (1+z)^n.$$

Although this argument cannot be considered as a proof, Klügel was convinced that it provided a method for proving the binomial theorem for a positive rational exponent $\frac{n}{m}$. According to Klügel, this is a proof because the calculation of $\left((1+z)^{mn}\right)^{\frac{1}{m}} = \left((1+z)^m\right)^{\frac{n}{m}} = (1+mz+\&c)^{\frac{n}{m}}$ leads back to the expression of the binomial theorem for positive integer exponents. This is what he means when he says that the proof of the binomial theorem for rational exponents should consist in showing that its expansion in series has the same "form" as the expansion corresponding to the binomial theorem for positive integer exponents. Finally, the generalization of the theorem to negative rational exponents was done by division: $\frac{1}{(1+z)^{\frac{n}{m}}} = (1+z)^{-\frac{n}{m}}$.

*

In short, Wolff introduced the binomial theorem into German academic circles at the beginning of the eighteenth century. According to him, this theorem represents the principle upon which both the Newtonian method of series and the method of fluxions were founded. However, mathematical research on the binomial theorem was not conducted systematically in Germany until after the debate on the correctness of the method of fluxions had begun in England. Indeed, in Germany, all proofs of the binomial theorem for rational exponents appeared after the publication of Berkeley's *Analyst* and closely followed the development of the debate carried on in England, insofar as the aim of these proofs consisted in establishing the binomial theorem without incorrectly using the power rule. In this sense, the foundational problems of the theory of fluxions were transferred onto differential calculus in German mathematics, which gave rise to the question of whether the binomial theorem belonged to the realm of finite analysis or to the realm of infinite analysis. Two theoretical approaches have been identified. On the one hand some mathematicians thought that this theorem belonged to finite analysis, as Aepinus, Euler (after his 1775 proof), and Klügel. But other mathematicians did not take a clear position on the issue, though they were interested in solving the problem of circularity, as Kästner, Segner, or Tempelhoff. Among the first group, it is possible to identify a more radical position held by Aepinus and Euler, according to which both finite analysis and infinite analysis are to be founded on the binomial theorem.

2.2   The Multinomial Theorem

The history of the multinomial theorem runs in parallel with that of the binomial theorem. Indeed, before the dissemination of Newton's theorem, problems such as finding the square of a trinomial are very rarely discussed in the manuals of mathematics. For instance, William Oughtred (1574–1660) did not present the problem of calculating the square of $(a + b + c)$ until the third edition of his *Clavis mathematicae* (Oughtred 1652, 62). Pierre Hérigone (?–1643?) gave this same example in 1634 (Hérigone 1634, 19). However it would be an exaggeration to say that this example expresses the multinomial theorem. This theorem deals with the problem of determining the law of formation of the coefficients in the multinomial expansion, and it was certainly Newton's work on infinite series that drew the attention of mathematicians to this problem.

In a letter of May 16, 1695 to Johann Bernoulli (1667–1748), Leibniz claims to have discovered this law regarding first-degree polynomials (Leibniz 1695, 47):

> Once upon a time I invented an extraordinary rule for [calculating] not only the coefficients of powers of a given binomial $x + y$, but also of a given trinomial $x + y + z$ as well as those of any polynomial; a rule such that, given any power, for instance the tenth power, and considering all the terms contained in it, as $x^5 y^3 z^2$, I could immediately determine the coefficient [of this term] without previously having calculated any table [...].[14]

Bernoulli responded to Leibniz saying that, given a positive integer power m, the general coefficient of the series expansion was given by (Bernoulli 1695, 54–55):

$$\frac{m \, (m - 1) \, (m - 2) \, (m - 3) \, (m - 4) \dots (p + 1)}{1.2.3 \dots q \times 1.2.3 \dots r \times 1.2.3 \dots s, \, \&c},$$

which can be written as:

$$\frac{1.2.3 \dots m}{1.2.3 \dots p \times 1.2.3 \dots q \times 1.2.3 \dots r \times \dots}, \tag{2.16}$$

---

[14] Excogitavi autem olim mirabilem regulam, pro numeris coefficientibus potestatum, non tantum a binomio $x + y$, sed & a trinomio $x + y + z$, imo a polynomio quocunque, ut data potentia gradus cujuscunque, verbi gratia, decimi, & potentia in ejus valore comprehensa, ut $x^5 y^3 z^1$, possim statim assignare numerum coefficientem, quem habere debet, sine ulla Tabula jam calculata [...]. (In the translation, the misprint "$x^5 y^3 z^1$" has been corrected.)

where $p, q, r, \ldots$ are positive integers such that $p + q + r + \ldots = m$. Leibniz had arrived at the same result in the early 1680s (Knobloch 1976, 217). But formula 2.16 was introduced in Europe by Marin Mersenne (1588–1648) in his *Harmonicorum libri*, though this formula was used in the context of music theory for calculating the number of permutations with repetition (Coumet 1972, 7–9; Mersenne 1635, 118–119, 133). In other traditions, however, this formula for calculating the number of permutations with repetition was found earlier; for instance, Bhāskara II (1114–1185?) gave it in his *Līlāvatī* (Bhāskara 1150, 125; Srinivasiengar 1967, 79–94). Thus, Leibniz's discovery must be understood in the sense that it shows the relation between the formula for calculating the number of permutations with repetition and multinomial coefficients.

However, Leibniz's approach does not provide a general solution to the problem concerning the law of formation of the coefficients in the multinomial expansion, since it is restricted to the case of positive integer powers of first-degree polynomials. In eighteenth century mathematics, the multinomial theorem concerns the calculation of coefficients in the multinomial expansion of a rational power of a given power series. There are three points in this characterization which we should take note of in order to understand properly the nature of the problem in eighteenth century mathematics. First, the terms "multinomial" and "infinitinomial" are used to refer to polynomials which can have an infinite number of terms. Second, the problem consists in raising an infinitinomial to a rational power, or to a real power in some cases. Finally, in eighteenth century mathematics, mathematicians know that an approach like that of Leibniz does not provide a general solution to this problem and, as a result, the term "infinitinomial" does not refer in general to "infinite" first-degree polynomials, but to polynomials of "infinite" degree (or infinite series in our contemporary terminology).

In this section, we present the proofs of the multinomial theorem that were formulated before 1778, the year in which Hindenburg proposed his own proof. Thus, the aim of this section is to provide a historical background to Hindenburg's work.

2.2.1   Moivre's Multinomial Theorem

In 1697, Abraham de Moivre (1667–1754) published his paper *A method of raising an infinite multinomial to any given power, or extracting any given root of the same* in the *Philosophical Transactions* (de Moivre 1697). This paper was the starting point for practically all research on this issue that came after. In turn, the starting point for Moivre's multinomial theorem was the binomial theorem of Newton (de Moivre 1697, 691):

> 'T is about two Years since, that considering Mr. Newton's Theorem for Raising a Binomial to any given Power, or Extracting any Root of the same; I enquired, whether what he had done for a Binomial, could not be done for an infinite multinomial. I soon found the thing was possible [. . .].

Investigating Moivre's formulation of the multinomial theorem will help us to a better understanding of what Moivre means by saying that, concerning an infinite multinomial, it is possible to achieve a similar result to that obtained by Newton.

The aim of Moivre was to give a general rule for calculating the coefficient of the term $z^{m+\mu}$ in the series expansion of $\left(az + bz^2 + cz^3 + \&c.\right)^m$, where $\mu = 1, 2, 3, \ldots$ and m is a positive integer. To this end, he pointed out that the general coefficient of the term $z^{m+\mu}$ is constituted by two kinds of coefficients. On the one hand, it is constituted by the sum of literal coefficients of the form $a^p b^q c^r \ldots$, where $p, q, r, \ldots$ are positive integers such that $p + q + r + \ldots = m$. On the other hand, it is constituted by numerical coefficients which depends on each literal coefficient, and which were called "*unciae*" by Moivre. Thus, there are two specific rules for respectively calculating the sum of literal coefficients and the numerical coefficients.

In order to find the sum of literal coefficients corresponding to the term $z^{m+\mu}$, Moivre establishes the following classes of literal coefficients $a^p b^q c^r \ldots$, for $p \neq 0$:

1. $a^p b^q c^r d^s \ldots$ is said to be of the first class if $q \neq 0$.
2. $a^p b^q c^r d^s \ldots$ is said to be of the second class if $q = 0$ and $r \neq 0$.
3. $a^p b^q c^r d^s \ldots$ is said to be of the third class if $q = 0$, $r = 0$ and $s \neq 0$.
4. And so on.

The sum of literal coefficients corresponding to the term $z^{m+\mu}$ is then calculated recursively from the sum of literal coefficients corresponding to the terms $z^{m+\mu-\nu}$ ($\nu = 1, 2, 3, \ldots$) as follows. Let a^m be the literal coefficient of z^m in the series expansion of $\left(az + bz^2 + cz^3 + \&c.\right)^m$, then the terms of the sum of literal coefficients corresponding to the term $z^{m+\mu}$ in this series expansion are the following literal coefficients:

- All the literal coefficients of the term $z^{m+\mu-1}$, each one multiplied by $\frac{b}{a}$.
- All the literal coefficients of the term $z^{m+\mu-2}$ that are not of the first class, each one multiplied by $\frac{c}{a}$.
- All the literal coefficients of the term $z^{m+\mu-3}$ that are not of the first class or of the second class, each one multiplied by $\frac{d}{a}$.
- All the literal coefficients of the term $z^{m+\mu-4}$ that are not of the first class, or of the second class, or of the third class, each one multiplied by $\frac{e}{a}$.
- And so on.

This algorithm allows to find the sum of literal coefficients corresponding to the term $z^{m+\mu}$.

The numerical coefficient attached to each literal coefficient of the term $z^{m+\mu}$ is then calculated by applying formula 2.16. Following this procedure, Moivre was able to determine the first terms in the series expansion of:

$$\left(az + bz^2 + cz^3 + \&\text{c.}\right)^m = a^m z^m + \frac{m}{1} a^{m-1} b z^{m+1}$$

$$+ \left(\frac{m}{1} \times \frac{m-1}{2} a^{m-2} b^2 + \frac{m}{1} a^{m-1} c\right) z^{m+2}$$

$$+ \left(\begin{array}{c} \frac{m}{1} \times \frac{m-1}{2} \times \frac{m-2}{3} a^{m-3} b^3 \\ + \frac{m}{1} \times \frac{m-1}{2} a^{m-2} bc \\ + \frac{m}{1} a^{m-1} d \end{array}\right) z^{m+3} + \&\text{c.} \qquad (2.17)$$

It is easy to see that Moivre did not exactly achieve a similar result for the multinomial theorem to that obtained by Newton regarding the binomial theorem. Moivre's algorithm enables us to calculate the coefficient of the term $z^{m+\mu}$ only if we have previously calculated the coefficients of the terms $z^{m+\mu-\nu}$, whereas Newton's formula enables us to calculate the coefficient of any term in the binomial expansion directly. As will be seen later, a formula like that of Moivre was referred to as a "recursive formula" (*recurrirende Form*) by the combinatorial school, since the calculation of each term of the series expansion depends on our knowledge of the previous ones, while any formula similar to the binomial formula was a "non-recursive formula" or a "direct formula". In what follows, we will adopt this terminology.

Although this algorithm was only proposed for a positive integer m, Moivre claimed that it was valid for any rational number too, and promised to prove it in a subsequent paper. In 1730, he showed that the algorithm works correctly for the negative integer -1, but that was the only example not related to positive integers he ever gave (de Moivre 1730, 89).

On the other hand, Moivre considered that the multinomial theorem could be applied to solve the same problems treated by Newton in his *Epistola posterior* (de Moivre 1697). In fact, he wrote a paper in 1698 concerning the theorem on reversion of series, which was enunciated without proof in the *Epistola posterior* (de Moivre 1698). Moivre's aim was to prove the theorem on reversion of series by using the multinomial theorem. Given the power series:

$$az + bzz + cz^3 + dz^4 + \&\text{c.} = gy + hyy + iy^3 + ky^4 + \&\text{c.}, \qquad (2.18)$$

Moivre puts:

$$z = Ay + Byy + Cy^3 + Dy^4 + \&\text{c.}, \qquad (2.19)$$

where $A, B, C, \ldots$ are undetermined coefficients. From 2.18 and 2.19, he gets:

$$a(Ay + Byy + Cy^3 + \&c.) + b(Ay + Byy + Cy^3 + \&c.)^2 +$$
$$c(Ay + Byy + Cy^3 + \&c.)^3 + \&c. = gy + hyy + iy^3 + ky^4 + \&c.$$

By applying successively the multinomial theorem, it follows:

$$aAy + (aB + bAA)yy + (aC + 2bAB + cA^3)y^3 + \&c. = gy + hyy + iy^3 + ky^4 + \&c.$$

And by the method of undetermined coefficients, he obtains:

$$A = \frac{g}{a}, \qquad B = \frac{h - bAA}{a}, \qquad C = \frac{i - 2bAB - cA^3}{a}, \ldots$$

The result follows by replacing these coefficients in Eq. 2.19:

$$z = \frac{g}{a}y + \frac{h - bAA}{a}y^2 + \frac{i - 2bAB - cA^3}{a}y^3 + \&c., \qquad (2.20)$$

Clearly, this is a recursive formula.

Thus, Moivre's solutions were based on recursive formulae. Taken as a generalization of the binomial theorem, the multinomial theorem was supposed to be equally efficient as the binomial theorem in solving problems in Newton's theory of series. However, no more examples were given to test and refine this idea, and, more important, the validity of the multinomial theorem for fractional exponents remained an open question.

2.2.2   The Reception of the Multinomial Theorem

In the European tradition Moivre's theorem gradually came to be accepted as an important result in mathematics, and many mathematicians tried to prove its validity with respect to fractional exponents from the beginning of the eighteenth century. Methodologically, it is possible to identify three types of proofs. First, there are proofs that are grounded in the binomial theorem. Second, there is another group characterized by a fluxional or differential method of proof. Finally, one can distinguish a combinatorial approach.

Here we will deal with these different kinds of proofs, but we will restrict the scope of the present survey to researches carried out elsewhere than in Germany. Concerning German mathematicians, we will study their ideas in Sect. 2.2.3.

2.2.2.1 Proving the Multinomial Theorem by the Binomial Theorem
The first proof of this kind is probably due to Jones in 1706, who included it in his *Intro-duction to the mathematics* after enunciating without proof the binomial theorem (Jones

1706, 171–173). In accordance with the suggestion advanced by Newton in his *Epistola prior*, Jones simply regroups the terms of a given multinomial and then applies the binomial theorem. So he puts:

$$(a + bz + cz^2 + \&c.)^m = (a + (bz + cz^2 + \&c.))^m,$$

where m is a rational number, and then, by the binomial theorem:

$$(a + bz + cz^2 + \&c.)^m = (a + (bz + cz^2 + \&c.))^m$$
$$= a^m + \frac{m}{1}a^{m-1}(bz + cz^2 + \&c.) + \&c.$$

And then:

$$(a + bz + cz^2 + \&c.)^m = (a + (bz + cz^2 + \&c.))^m$$
$$= a^m + \frac{m}{1}a^{m-1}bz + \left(\frac{m}{1} \times \frac{m-1}{2}a^{m-2}b^2 + \frac{m}{1}a^{m-1}c\right)z^2 + \&c.$$

By multiplying this equation by z^m, he obtains Moivre's formula 2.17. Jones also included in his book the recursive formula 2.20, which concerns the theorem on reversion of series.

Much later, in 1736, the same argument was used by Hodgson to prove the multinomial theorem, and Moivre's formula 2.20 was applied to Eq. 2.1, which was an example given by Newton about the technique of reversion of series (Hodgson 1736, 32–38, 40–43).

A more complicated and less general proof was proposed by the Franch mathematician Reyneau in 1708 (Reyneau 1708, 408–426). The proof consists of three parts. First, the justification of the multinomial theorem for positive integer exponents as a corollary of the binomial theorem for positive integer exponents. Second, the use of this corollary to prove the validity of the binomial theorem for fractional exponents. Finally, Reyneau justifies the multinomial theorem for a fractional exponent of the form $\frac{1}{m}$ as follows. For every strictly positive integer m, he puts:

$$(a + bz + cz^2 + \&c.)^{\frac{1}{m}} = g + hz + iz^2 + \&c.,$$

and then:

$$(g + hz + iz^2 + \&c.)^m = a + bz + cz^2 + \&c.,$$

where $g, h, i, \ldots$ are undetermined coefficients. By the multinomial theorem for positive integer exponents, he gets:

$$g^m + \frac{m}{1}g^{m-1}hz + \left(\frac{m}{1} \times \frac{m-1}{2}g^{m-2}h^2 + \frac{m}{1}g^{m-1}i\right)z^2 + \&c. = a + bz + cz^2 + \&c.$$

Finally, by the method of undetermined coefficients and after some algebraic manipulations, he finds the coefficients:

$$g = a^{\frac{1}{m}}, \qquad h = \frac{1}{m}a^{\frac{1-m}{m}}b, \qquad i = \frac{1}{m} \times \frac{1-m}{2m}a^{\frac{1-2m}{m}}b^2 + \frac{1}{m}a^{\frac{1-m}{m}}c, \ldots$$

Reyneau repeats this procedure to calculate the expansion of $(az + bz^2 + cz^3 + \&c.)^{\frac{1}{m}}$.

In this kind of proofs, the multinomial theorem is not a generalization, as Moivre thought, of the binomial theorem, but a consequence. In particular, this character of being a mere corollary is more evident in Reyneau's proof, who explicitly insists on the dependence of the multinomial theorem on the binomial theorem. This idea will be echoed in the treatise *Calcul différentiel et intégral* by Deidier (1698–1746), who carefully reproduces Reyneau's demonstration (Deidier 1740, 31–47). Without referring to Reyneau, Johann Castillon (1709–1791), Stewart and Clairaut demonstrated the theorem in a similar manner, but they stopped at the second step of Reyneau's demonstrative scheme and, as a consequence, the multinomial theorem is mostly conceived as a mathematical demonstrative tool (Clairaut 1746, 242–255; Castillon 1742; Stewart 1745, 470–475).

2.2.2.2 Fluxional Proofs of the Multinomial Theorem

In 1736, Colson uses for the first time the method of fluxions to construct the multinomial expansion (Newton 1736, 311–312). As in the case of the binomial theorem, Colson presents the construction of the multinomial expansion rather as an exercise to help students master the method of fluxions than as a proof of the theorem. For Colson, the only true demonstration of this theorem is due to Moivre since his proof shows the law of formation of coefficients.

Colson proceeds as follows. He puts:

$$y = \overline{a + bz + cz^2 + \&c.}|^m$$

where the nature of m is not specified, but it could be a rational number. If one takes $v = a + bz + cz^2 + \&c.$, then the fluxion of y is:

$$\dot{y} = m\dot{v}v^{m-1},$$

where $\dot{v} = b + 2cz + \&c.$ Now, as long as one assumes that it is always possible to know the first term of a given multinomial expansion: $\overline{a + bz + cz^2 + \&c.}|^\alpha = a^\alpha + \&c.$, it follows that:

$$\dot{y} = m\dot{v}v^{m-1} = ma^{m-1}b + \&c.$$

Returning back to the fluents, one obtains:

$$y = ma^{m-1}bz + \&c.$$

From this expression, Colson deduces the term in z of the multinomial expansion:

$$y = \overline{a + bz + cz^2 + \&c.}|^m = a^m + ma^{m-1}bz + \&c.$$

By calculating the successive fluxions and then the fluents in this way, Colson suceeds in determining, in principle, all the terms of the multinomial expansion. Finally, he multiplies the expansion so obtained by z^m in order to get Moivre's formula 2.17.

Six years later Maclaurin deals with this same question (Maclaurin 1742, 608–609). As usual, the demonstration begins with the equation:

$$\overline{a + bz + cz^2 + \&c.}^m = A + Bz + Cz^m + \&c.,$$

where $A, B, C, \ldots$ are undetermined coefficients. If one takes $z = 0$ in the above equation, one gets the first coefficient $A = a^m$. By calculating the fluxions and dividing by $\dot{z}$, he gets:

$$\overline{a + bz + cz^2 + \&c.}^{m-1} \times \overline{mb + 2mcz + 3mdz^2 + \&c.} = B + 2Cz^m + 3Dz^2 + \&c.$$

The second coefficient $B = ma^{m-1}b$ follows by taking $z = 0$. Again, by calculating the fluxions, dividing by $2\dot{z}$ and taking $z = 0$, the coefficient $C = m \times \frac{m-1}{2}a^{m-2}b^2$ is determined. This completes Maclaurin's proof.

As we can see, and as Colson pointed out, it is difficult to give a precise definition of the law of formation of coefficients by using the method of fluxions.

2.2.2.3 A Combinatorial Proof of the Multinomial Theorem

Before the emergence of the German combinatorial school, there is only one single example of a combinatorial proof of the multinomial theorem, which was offered by Roger Joseph Boscovich (1711–1787) in 1747. In 1725, Boscovich left his native city of Dubrovnik and settled in Rome, where he was ordained to the priesthood nineteen years later. But during his ecclesiastical training, he studied mathematics and was such a brilliant student that he became professor of mathematics at Collegium Romanum from 1740 to 1757. His work on the multinomial theorem belongs to this period and consists of three papers (Boscovich 1747; 1748a; 1748b). The first presents his original combinatorial method of proof, which is discussed by Boscovich himself in his two subsequent papers.

In his first paper, Boscovich emphasizes the importance of the multinomial theorem for the infinitesimal calculus, and regrets that all known formulae for calculating the multinomial expansion are recursive formulae. His purpose is to provide a non-recursive or a direct formula, based on a combinatorial method.

Boscovich's combinatorial method consists in representing, through a combinatorial table, the partition of a given positive integer. In fact, he invented an algorithm for calculating the partition of a given positive integer. Although the algorithm was only partially formulated in natural language, the idea is pretty clear. Given the positive integer t, its partition table is obtained by the rules:

1. Write a string of t units, this is the first row.
2. If the row just written has a string of k units, with $k \geq 2$, write a new row of $k - 2$ units and then a 2, leaving the rest of the original row unchanged. Then apply rule 2 to the new row so obtained.
3. If a previous row has a string of k units, with $k \geq 3$, and its rightmost element is different from 2, write a new row of $k - 3$ units and then a 3, leaving the rest of the original row unchanged. Then, apply rule 2. (The treatment of the 'previous rows' begins with the first row, and if rule 3 has been applied to a certain row, it cannot be applied to the same row again.)
4. If a previous row has a string of k units, with $k \geq 4$, and its rightmost element is different from 2 and 3, write a new row of $k - 4$ units and then a 4, leaving the rest of the original row unchanged. Then, apply rule 2, and then rule 3. (The treatment of the 'previous rows' begins with the first row, and if rule 4 has been applied to a certain row, it cannot be applied to the same row again.)
5. And so on, until writing the last row composed by the only element t.

Using his method, Boscovich constructed the partition table of several integers. Figure 2.1 shows an example of these tables.

A partition table is used to determine the coefficient of the term $z^{mn+t\rho}$ in the expansion of:

$$\left(az^n + bz^{n+\rho} + cz^{n+2\rho} + dz^{n+3\rho} + \&c. \right)^m,$$

Fig. 2.1 Boscovich's partition table of 6

```
1 1 1 1 1 1
1 1 1 1 1 2
    1 1 2 2
      2 2 2
    1 1 1 3
      1 2 3
        3 3
      1 1 4
        2 4
        1 5
          6
```

where m, n, ρ, and t are positive integers. The utility of a partition table depends on a substitution rule:

$$\begin{pmatrix} b \; c \; d \; \ldots \\ 1 \; 2 \; 3 \; \ldots \end{pmatrix},$$

where b, c, d, $\ldots$ are the coefficients of the original multinomial and the numbers 1, 2, 3, $\ldots$ are the elements of a given partition table. By means of this substitution rule, each row of a partition table can be interpreted as a product of the form $b^q c^r d^s \ldots$, where q, r, s, $\ldots$ represent the number of terms of that kind in the row. Boscovich points out that the coefficient of $z^{mn+t\rho}$ is the sum of monomials of the form $a^{m-p} b^q c^r d^s \ldots$, where $m = (m - p) + q + r + s + \cdots$, i.e. $p = q + r + s + \cdots$. Then, all these monomials can be calculated by means of the partition table of t and the substitution rule. The last step to find the coefficient of $z^{mn+t\rho}$ consists in applying formula 2.16 to each monomial.

As an example, the coefficient of the term $z^{mn+6\rho}$ can be calculated as follows. By applying the substitution rule to the partition table of 6 in Fig. 2.1, Boscovich gets the products b^6, $b^4 c$, $b^2 c^2$, c^3, $b^3 d$, bcd, d^2, $b^2 e$, ce, bf, g. By completing these products in monomials of the form $a^{m-p} b^q c^r d^s \ldots$ and by applying formula 2.16 to each monomial, Boscovich gets the result:

$$\begin{pmatrix} \frac{m(m-1)(m-2)(m-3)(m-4)(m-5)}{1.2.3.4.5.6} a^{m-6} b^6 + \\ \frac{m(m-1)(m-2)(m-3)(m-4)}{1.2.3.4} a^{m-5} b^4 c + \\ \frac{m(m-1)(m-2)(m-3)}{1.2.1.2} a^{m-4} b^2 c^2 + \\ \frac{m(m-1)(m-2)(m-3)}{1.2.3} a^{m-4} b^3 d + \\ \frac{m(m-1)(m-2)}{1.2} a^{m-3} b^2 e + \\ m(m-1)(m-2) a^{m-3} bcd + \\ \frac{m(m-1)(m-2)}{1.2.3} a^{m-3} c^3 + \\ m(m-1) a^{m-2} bf + \\ m(m-1) a^{m-2} ce + \\ \frac{m(m-1)}{1.2} a^{m-2} d^2 + \\ m a^{m-1} g \end{pmatrix} z^{mn+6\rho} \qquad (2.21)$$

However, Boscovich's method only justifies the multinomial theorem for positive integer exponents. As Moivre, Boscovich thought that the binomial theorem was a corollary of the multinomial theorem and promised to generalize his method to a rational exponent (Boscovich 1748b, 88–99), but this never happened.

*

In short, the first kind of proof is the most common in the literature. In these proofs the multinomial theorem is considered as a corollary of the binomial theorem, and, in several cases, the multinomial theorem is only used as a tool to generalize the binomial theorem. Fluxional proofs are more concerned with showing the power of the method of fluxions for expanding a function into an infinite series rather than with establishing the law of formation of the coefficients in the expansion. These two kinds of proof have in common the recursive character of Moivre's formula. It is only in the third kind of proof that a direct formula (or method) is proposed and the law of formation of coefficients is precisely established. Moreover, in this kind of proof, the multinomial theorem is considered as a more general result than the binomial theorem.

2.2.3   The Reception of the Multinomial Theorem in Germany

With profound differences in some cases, it is nevertheless possible to identify these three kinds of proof in the German tradition. In Germany, as in the rest of Europe, the research on the multinomial theorem is guided by the interest in Moivre's work, except perhaps in the third kind of proof, which seems to be entirely inspired by Leibniz's ideas, as will be seen below.

2.2.3.1 Proving the Multinomial Theorem by the Binomial Theorem in Germany

As in the case of the binomial theorem, it was Wolff who introduced Moivre's multinomial theorem in Germany. Wolff mentions it for the first time in 1705 in his dissertation on the method of infinite series, and actually furnishes a proof in 1713 (Wolff 1705, 306; 1713, vol. I: 270–272). His proof is similar to that of Jones in all respects. However its importance lies not in its originality, but in its role in disseminating this kind of proof both in Germany and abroad. For instance, leaving aside Moivre's paper, Wolff's text is the only source actually cited by Boscovich (1748b, 93–94).

On the other hand, the history of this kind of proof in Germany follows a similar path to that of the rest of Europe. It is clear that in this context the multinomial theorem itself becomes much less important than the fact that it can be used as a tool for proving the binomial theorem for rational exponents, which is the case, for instance, in Segner's and Tempelhoff's work. In this respect, Klügel's position is more ambiguous, since he demonstrated both theorems for a rational exponent $\frac{1}{m}$ using alternatively each one of them to prove the other, as we have seen in Sect. 2.1.3.3. It is interesting to note here that Tempelhoff is the only author who mentions and demonstrates Moivre's theorem on reversion of series. His proof is identical to that of Moivre, except for the fact that he takes the following equation as his point of the departure (von Tempelhoff 1769, 605–609):

$$az^m + bz^{m+1} + cz^{m+2} + \&c. = gy^m + hy^{m+1} + iy^{m+2} + \&c.$$

2.2.3.2 Differential Proofs of the Multinomial Theorem

The merit of having used Leibnizian differential calculus for the first time to prove the multinomial theorem belongs to Jacob Bernoulli (1655–1705). His proof was written in 1701, but not published until 1744 (Bernoulli 1744; 1993, 175–180, 249–245). In fact, Bernoulli distinguishes between a multinomial of the form $a + b + c + \&c.$ and a multinomial of the form $az + bz^2 + cz^3 + \&c.$, and proposes a different method for raising each one of them to a certain power m (whose nature is not clearly defined but seems to be a positive integer): a combinatorial method in the first case, and a differential method in the second. The combinatorial method will be studied in Sect. 2.2.3.3.

In his differential proof, Bernoulli begins with the equation:

$$(az + bz^2 + cz^3 + \&c.)^m = pz^m + qz^{m+1} + rz^{m+2} + \&c.$$

where $p, q, r, \ldots$ are undetermined coefficients. By taking the logarithms of both sides, he gets:

$$ml(az + bz^2 + cz^3 + \&c.) = l(pz^m + qz^{m+1} + rz^{m+2} + \&c.),$$

where l represents the natural logarithm. By logarithmic differentiation, and after dividing by dz, it follows:

$$\frac{ma + 2mbz + 3mcz^2 + \&c.}{az + bz^2 + cz^3 + \&c.} = \frac{mpz^{m-1} + (m+1)qz^m + (m+2)rz^{m+1} + \&c.}{pz^m + qz^{m+1} + rz^{m+2} + \&c.}$$

And then:

$$mapz^m + (2mbp + maq)z^{m+1} + (3mcp + 2mbq + mar)z^{m+2} + \&c. =$$

$$mapz^m + ((m+1)aq + mbp)z^{m+1} + ((m+2)ar + (m+1)bq + mcp)z^{m+2} + \&c.$$

By the method of undetermined coefficients, one gets:

$$q = \frac{mbp}{a}, \qquad r = \frac{2mcp + (m-1)bq}{2a}, \ldots$$

Thus, supposing known the first coefficient $p = a^m$, which was customary since Moivre's proof publication, the coefficients q, r, etc. can be completely determined. This completes Bernoulli's proof. In 1755 Euler published a similar proof (Euler 1755, 519–520).

Another example of this method, but without using the logarithm, can be found in the work of Kästner (1759); (Panza 1992, vol. 2: 658). He puts $y = \alpha z + \beta z^2 + \gamma z^3 + \ldots$, and writes the equation:

$$(1 + y)^m = 1 + Az + Bz^2 + Cz^3 + \ldots = w,$$

where A, B, C, ... are undetermined coefficients and m is a rational or a real number. From this equation, he gets: $m(1 + y)^{m-1}dy = dw$. By multiplying this equality by $(1 + y)$ and by dividing by dz, he obtains:

$$mw\frac{dy}{dz} = (1 + y)\frac{dw}{dz}.$$

And then:

$$m\alpha + (2m\beta + mA)z + (3m\gamma + 2mA\beta + mB\alpha)z^2 + \ldots =$$

$$A + (A\alpha + 2B)z + (A\beta + 2B\alpha + 3C)z^2 + \ldots$$

By the method of undetermined coefficients, it follows:

$$A = m\alpha, \qquad B = \frac{(m - 1)A\alpha + 2m\beta}{2}, \ldots,$$

which completes Kästner's proof.

In his text Kästner made a modest but interesting technical improvement regarding mathematical notations. He proposed to replace the Newtonian alphabetical notation for coefficients by a more accurate indicial notation (even if he used the Newtonian notation in his proof!), so he wrote

$$y = (\lambda 1)z + (\lambda 2)z^2 + (\lambda 3)z^3 + \ldots + (\lambda n)z^n + \ldots$$

and

$$w = 1 + (L1)z + (L2)z^2 + (L3)z^3 + \ldots + (Ln)z^n + \ldots,$$

where the numbers juxtaposed to λ and L stand for indexes. This is one of the first examples of the use of indicial notation in this context.

Kästner transcribed this proof in 1761 in his *Anfangsgründe der Analysis des Unendlichen*, and Karsten included it in his *Mathesis theoretica elementaris*, where it is possible to find several passages copied from Kästner's work (Karsten 1760, 568–572; Kästner 1761, 37–45).

2.2.3.3   A Combinatorial Proof of the Multinomial Theorem in Germany

As pointed out in Sect. 2.2.3.2, Jacob Bernoulli used a combinatorial method to justify the multinomial theorem. His proof concerns polynomials of the first-degree. Bernoulli knew that one can find directly the coefficient of any term in the multinomial expansion of $(a + b + c + \&c.)^m$, with m a positive integer, by means of formula 2.16, as suggested by his brother Johann and by Leibniz. However, in order to find the terms themselves of the multinomial expansion, Bernoulli proceeded recursively, as can be seen in the following lexicographical arrangement:

$$1 \times a^m$$

$$b + c + d + e + f + g \quad \&c. \times a^{m-1}$$

$$\left. \begin{array}{l} bb + bc + bd + be + bf \quad \&c. \\ +cc + cd + ce \quad \&c. \\ +dd \quad \&c. \end{array} \right\} \times a^{m-2}$$

$$\left. \begin{array}{l} b^3 + b^2c + b^2d + b^2e \quad \&c. \\ +bc^2 + bcd \quad \&c. \\ \&c. \end{array} \right\} \times a^{m-3} \qquad (2.22)$$

$$\left. \begin{array}{l} b^4 + b^3c + b^3d \quad \&c. \\ +b^2c^2 \quad \&c. \\ \&c. \end{array} \right\} \times a^{m-4}$$

$$\&c. \times \&c.$$

By calculating all products of this arrangement and by applying formula 2.16 to each one of them, one gets the multinomial expansion.

The question of raising a first-degree polynomial to a positive integral power is more related, as can be seen here, to Leibniz's interests. However Leibniz never published his results on the multinomial theorem, but he was forced to make public, in 1700, a detailed account of his research on Moivre's theorem on reversion of series as a response to Fatio's accusations of plagiarism (Leibniz 1700, 206–208). Fatio claimed that Leibniz had plagiarized major portions of Moivre's papers published in the *Philosophical Transactions*. Leibniz denied the accusations made against him on the grounds that he knew nothing about Moivre's papers and, in particular, offered his own version of the theorem on reversion of series, which was "infinitely more general" than that of Moivre —according to Leibniz.

Leibniz's version of this theorem was indeed a little more general, but the real innovation was in his new mathematical notation. Leibniz puts the equation:

$$0 = (01Y + 02Y^2 + 03Y^3\&c) + (-10 + 11Y + 12Y^2 + 13Y^3\&c)Z +$$
$$(20 + 21Y + 22Y^2 + 23Y^3\&c)Z^2 + \&c,$$

where the two-digit numbers are determined coefficients named "small numbers" by Leibniz. If we take $11 = 12 = \cdots = 21 = 22 = \cdots = 0$, we get Moivre's equation 2.18. Now the question is to express Z in function of Y, i. e. we have to find the coefficients of the equation:

$$Z = 101Y + 102Y^2 + 103Y^3 + \&c,$$

where the three-digit numbers are undetermined coefficients named "major numbers" by Leibniz. This equation corresponds to Moivre's equation 2.19. By replacing Z in the first equation, one gets:

$$0 = (01 - 10.101)Y + (02 + 11.101 + 20.101^2 - 10.102)Y^2 +$$

$$(03 + 11.102 + 12.101 + (2)20.101.102 + 21.101^2 + 30.101^3 - 10.103)Y^3 + \&c.$$

And by the method of undetermined coefficients, it follows:

$$101 = \frac{01}{10},$$

$$102 = \frac{02 + 11.101 + 20.101^2}{10},$$

$$103 = \frac{03 + 11.102 + 12.101 + (2)20.101.102 + 21.101^2 + 30.101^3}{10},$$

$$104 = \frac{\begin{array}{c}04 + 11.103 + 12.102 + 13.101 + (2)20.101.103 + (2)21.101.102 + \\ 22.101^2 + (3)30.101^2.102 + 31.101^3 + 40.101^4\end{array}}{10},$$

$$\&c,$$

which completes Leibniz's proof. A digit in parentheses, like (2), is an "actual number", which indicates the number of possible permutations of the major numbers contained as a factor in the corresponding product. If we take $11 = 12 = \cdots = 21 = 22 = \cdots = 0$ in these coefficients, we get Moivre's formula 2.20.

Leibniz explains how the formation of coefficients obeys a combinatorial law (*lex combinationis*), which consists in four rules:

1. Denominator is always 10.
2. Numerator's first term is a small number equal to the two rightmost digits of the coefficient we are trying to determine. For instance, if we are trying to find the value of 103, numerator's first term is 03.

3. Each term of the numerator has at most one small number, whose left digit indicates the number of major numbers contained in that term. For instance, if a term begins with 21, there are two major numbers in that term.
4. In each term of the numerator, the sum of the small number's right digit and of the rightmost digits of the major numbers is equal to the rightmost digit of the coefficient we are looking for. For example, concerning coefficient 103, the sum is equal to 3 in the term 21.101^2.

These rules show some relations between the elements of the coefficients, but it is not possible to use them in order to find the coefficients. In fact, Leibniz's formula is, as Moivre's, a recursive formula, and the law of formation of coefficients is as clear as in Moivre's work.

On the other hand, from our contemporary point of view, it could be said that Leibniz's notation is elegant, concise, and expressive enough for mathematical purposes: it is in fact an indicial notation *avant la lettre*. However, in his time no one was convinced by the enthusiasm of Leibniz, who thought his notation could mean real progress for analysis (Leibniz 1700, 208), not even Kästner who tried later to improve Newton's notations, and Leibniz's innovation was severely criticized by different mathematicians in the eighteenth century, as will be seen in the following chapters.

*

In sum, the first kind of proof reproduces in Germany the same image of the multinomial theorem as a corollary of the binomial theorem and as a demonstrative tool. Differential proofs are more diversified than fluxional proofs, and first timid attempts were made to overcome alphabetical notations. The only combinatorial proof pertains to the Leibnizian tradition, and there is no other proof of the multinomial theorem in this tradition. In Germany, none of the methods provides a direct formula for calculating the general coefficient of the multinomial expansion. There is, in general, little interest in Moivre's theorem on reversion of series, even if Leibniz saw it as an example of the active role played by combinatorics in the field of analysis.

The Emergence of the German Combinatorial Analysis

3

3.1   Hindenburg on Mathematical Tables

Son of a merchant, Carl Friedrich Hindenburg was born on July 13, 1739, in Dresden
and died on March 17, 1808, in Leipzig.[1] He received his first education at home, and
continued the whole of his secondary school education in Freiberg. In 1757 he enrolled
at the University of Leipzig, studying medicine, philosophy, Greek, Latin, mathematics,
and physics. The field of mathematics did not attract his attention for many years, his
real interests being centered on literary and rhetorical studies. It was by accident that
Hindenburg came to appreciate mathematics. In 1763 his professor of rhetorics Christian
Fürchtegott Gellert (1715–1769) introduced him to a student named Schönberg, who had a
strong predilection for mathematical studies and who often traveled to Göttingen in order
to satisfy his academic curiosity. Hindenburg became the private teacher of Schönberg
and, as a consequence, was motivated to go deeply into mathematical questions. But his
early vocation for humanities still remained somewhat intact, as shown by the subject
matter of his two first books published in 1763 and 1769, books in the fields of philology
and philosophy (Hindenburg 1763, 1769). It seems that the turning point in his career
occurred when he met Kästner at the University of Göttingen, during the travels he made

---

[1] It is not unusual to find some sources that indicate that Hindenburg was born on July 13, 1741. This
is the date Moritz Cantor gave in his entry for the *Allgemeine deutsche Biographie* (von Liliencron &
von Wegele 1912, vol. 12: 456), which depends on Stimmel's entry for the *Allgemeine Encyklopädie
der Wissenschaften und Künste* (Ersch et al. 1831, 252). In turn, Stimmel consulted (Hamberger &
Meusel 1797, 335), which is the original source. But Baur claims that Hindenburg's birthday given
in (Hamberger & Meusel 1797, 335) is wrong and that the correct year is 1739 (Baur 1816, 621). In
the twentieth century, Netto used the correct year given by Baur in his historical reconstruction of
the German combinatorial school (Netto 1908, 202), but Cantor's entry seems to be the main source
of information about biographical issues, thus perpetuating the error.

© The Author(s), under exclusive license to Springer Nature Switzerland AG 2022   67
E. Noble, *The Rise and Fall of the German Combinatorial Analysis*, Frontiers
in the History of Science, https://doi.org/10.1007/978-3-030-93820-8_3

as a private tutor of Schönberg. In Leipzig, Hindenburg continued to study mathematics with Georg Heinrich Borz (1714–1799), ordinary professor of mathematics and a former student of Wolff. Then, from 1771 to 1781, he worked as Privatdozent at the University of Leipzig, until he became extraordinary professor of philosophy and physics in 1781. Finally, in 1786, he was appointed to the ordinary chair of physics, which he held until his death.

The German combinatorial analysis was created by Hindenburg. He defined the scope and objectives of this new branch of mathematics. In this chapter, and the chapters that follow, we are going to discuss in detail how this new theory emerged and how it soon came to be considered, at least in Germany, one of the most promising theories in eighteenth century mathematics. But before entering on this inquiry, it is quite relevant to ascertain what relation Hindenburg's work bears to mathematical tables.

Mathematical tables have been a decisive element in the scientific advance of the past two millennia. This is particularly true for "pure" and "applied" mathematics during the eighteenth century, when the construction of tables remains a fundamental aid in handling several mathematical objects, like logarithms, trigonometric functions, or numbers. Somme leading mathematicians worked hard to accomplish this routine task. For instance, Johann Heinrich Lambert (1728–1777) thought that the progress of mathematics depended substantially on tables and did everything in his power to ensure that tables would be computed. He convinced the Berlin Academy to support a corpus of astronomical tables, and even without the participation of the Academy, he was engaged on another ambitious project to construct, compile, and edit all sort of mathematical tables, although particular attention was given to factor and prime number tables (Bullynck 2010; Dikson 1923, vol. 2, 347–352; Glaisher 1878, 108–123). Lambert himself contributed to this large project by calculating, for instance, a table of factors of all integers from 1 to 10 200 not divisible by 2, 3, and 5, which was extended up to 102 000 the same year of 1770 (Lambert 1770a, 42–53, 1770b, 11–18, supplementa). However he was conscious of the fact that such a project could not be the work of one person only. The publication of these tables was then accompanied by an open invitation to the mathematical community to cooperate with this project extending the table of factors to 1 020 000, and eternal fame— said Lambert—would be the reward (Lambert 1770a, 49, 1770b, 9).

In search of immortality or persuaded by the relevance of the project, some potential collaborators came up with answers to such tantalizing appeal. One of these collaborators was Carl Friedrich Hindenburg, who entered into contact with Lambert in early August 1776 and was determined to extend the table to 5,000,000 (Hindenburg 1776b). However, in January of the same year the Austrian mathematician Anton Felkel (1740–1800?) had communicated to Lambert his intention to participate in the project, but later he changed his mind and chose to go his own way. This decision certainly disappointed Lambert and created some kind of a rivalry between Felkel and Hindenburg. In separate publications, one in Latin and the other in German, Felkel announced his plan to calculate the factors of all integers from 1 to 10,000,000 not divisible by 2, 3, and 5 (Felkel 1776a, 1776b). On the other hand, at the age of thirty-seven, Hindenburg published his first book on mathematics titled *Beschreibung einer ganz neuen Art, nach einem bekannten Gesetze fortgehende Zahlen, durch Abzählen oder Abmessen bequem und sicher zu finden*, where he exposed his method to calculate the table of factors (Hindenburg 1776a). The competition between Hindenburg and Felkel continued until the 1780s, but none of them was able to fulfill their promises.

Nonetheless, the scope of Hindenburg's book is not restricted to the computation of this factor table, but seeks to show how a general method, inspired by the sieve of Eratosthenes, could be implemented in order to calculate different kinds of tables by using a mechanical device. Hindenburg analyzes, for example, the cases of tables of primes, divisors, multiples, remainders, and powers. The aim of the book is then twofold: from a practical point of view, to provide a comprehensive guide to construct a calculator, a mechanical instrument; and from a theoretical point of view, to understand how various tables can be related to each other in order to establish a single method of calculation. It is in accordance with this theoretical goal that Hindenburg briefly discusses some properties of permutations and combinations, since combinatorics could be useful for standardizing numerical tables. In this sense, Hindenburg writes (Hindenburg 1776a, 97):

> So far, it has been assumed that the number of things or signs, which are shifted and put together by combinations and permutations, is ten. The more things there are, or the fewer things there are, the more signs, or the fewer signs, are required for these things. When these signs are combined and considered as numbers, they produce, however, a representation of numbers that is different from the decimal system; and thus these considerations lead naturally to the invention of different structures of numbers, even though no one has ever thought of it before.[2]

Thus Hindenburg's basic idea here is that the structure of the decimal system depends on the notions of "combination" and "permutation". In other words, Hindenburg emphasizes the combinatorial nature of numbers. And more generally, he thinks that all numbers can be defined by means of a reduced set of symbols since any positional number system would be grounded in some elementary combinatorial properties. This fundamental combinatorial character of number systems refers to the fact that combinatorics allows us to manipulate more general objects, like non-interpreted symbols. These symbols "become" (or denote) numbers only after an interpretation takes place in which each symbol means a digit. As a consequence, it is possible to mechanize the construction of number systems by taking advantage of their combinatorial properties and leaving aside

---

[2] Hier ist durchgängig vorausgesetzt worden, daß der gegebenen, durch Combination und Permutation zu verändernden und zusammen zu setzenden Dinge oder Zeichen, in allem zehn sind. Wären dieser Dinge mehr oder weniger, so würde man für sie auch mehr oder weniger Zeichen brauchen. Die Verbindung dieser Zeichen, wenn man sie als Zahlen betrachtete, würde jedoch die Zahlen alsdenn anders geben, als in dem Decimalsystem; und so würde diese Betrachtung, auf eine ungezwungene Art, zur Erfindung verschiedener Zahlengebäude führen, wenn man auch noch nie darauf gedacht hätte.

	1	10
0	00	01
	10	11
1	00	01
	10	11
10	00	01
	10	11
11	00	01
	10	11
100	00	01
	10	11
101	00	01
	10	11
110		

	1	2	10
0	00	01	02
	10	11	12
	20	21	22
1	00	01	02
	10	11	12
	20	21	22
2	00	01	02
	10	11	12
	20	21	22
10	00	01	02
	10	11	12
	20	21	22
11			

	1	2	3	10
0	00	01	02	03
	10	11	12	13
	20	21	22	23
	30	31	32	33
1	00	01	02	03
	10	11	12	13
	20	21	22	23
	30	31	32	33
2	00	01	02	03
	10	11	12	13
	20	21	22	23
	30	31	32	33
3				

Fig. 3.1 Hindenburg's representation of binary, ternary, and quaternary systems

any other question. In accordance with this idea, Hindenburg used some tables to represent the binary, ternary, and quaternary systems, as shown in Fig. 3.1.

Given a basic matrix, one can easily move from one number system to another by adding a new symbol and enlarging the matrix with a gnomon:

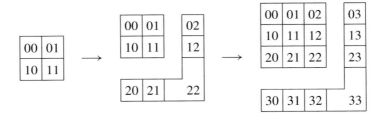

This procedure allows us to obtain a sequence of symmetric matrices in which the digits of the symmetric elements under the diagonal have been permuted. It is not hard to mechanize this procedure, and one of these matrices is all that is needed to determine any number in a given number system. For Hindenburg, the construction of tables is a good way of simplifying certain problems, especially when combinatorial methods are available.

Hindenburg shows, with an example, that his idea applies not only to the construction of whole numbers but also to that of fractional numbers (Hindenburg 1776a, 107–109). Consider the fraction $\frac{1}{7}$. Then the problem consists in finding a method based on "combinatorial" tables to express the decimal expansion of this fraction. Hindenburg's table is shown in Fig. 3.2. The rows and columns to the right of the column labeled "7" contain the decimal expansion of $\frac{1}{7}$. In fact, this table is a sort of Euclidean division table, where the labels of the rows are the dividends, 7 is the divisor, the labels of the columns are the remainders, and the corresponding quotients are the numbers placed at the intersection of a dividend's row and a remainder's column. For instance, the table shows that 4 is the

9	8	7	6	5	4	3	2	1	0	
						1				10
			2							20
3							4			30
				5						40
	6							7		50
					8					60
		9							10	70

Fig. 3.2 Hindenburg's decimal representation of $\frac{1}{7}$

quotient and 2 is the remainder of dividing 30 by 7. Although Euclidean division is the theoretical background of Hindenburg's method, the use of tables simplifies this process. To construct the table, one draws ten columns and seven rows, then starting from the upper left box one counts seven boxes to the right and writes "1", from there one counts seven boxes to the right and writes "2", and so on. To calculate the first decimal of $\frac{1}{7}$, one takes the fraction's numerator (which is "1"), and goes to the row labeled "10": the first decimal is then the number in this row which appears to be more to the right than any other number in this row, so it is 1. To calculate the second decimal of $\frac{1}{7}$, one takes the label of the column of the first decimal (which is "3"), and goes to the row labeled "30": the second decimal is the number in this row which appears to be more to the right than any other number in this row, so it is 4. To calculate the third decimal of $\frac{1}{7}$, one takes the label of the column of the second decimal (which is "2"), and goes to the row labeled "20": the third decimal is the number in this row which appears to be more to the right than any other number in this row, so it is 2. And so on. Notice that with Hindenburg's table method, no Euclidean division has been actually performed. Much of what has been done here consists in combining the digits contained in the table according to a simple rule: once you have picked up a digit from a column labeled "k", go to the row labeled "$k0$" in order to pick up the next digit. It seems that Hindenburg considered this somewhat vague statement as a kind of underlying combinatorial rule, which allows to automate all aspects of the process on a mechanical computing device. Indeed, Hindenburg imagined such a device, and described in his book how such a machine could be built.[3]

Tables like those shown in Figs. 3.1 and 3.2 provide representative examples of what Hindenburg wanted to achieve in his *Beschreibung*. Certainly his book was meant to be a contribution to Lambert's project, but it turned out to be a significant source of inspiration for Hindenburg's future work. The possibility of using mechanical devices for computing arithmetical operations depends on the theoretical possibility of combining symbols in

---

[3] About Hindenburg's mechanical devices, see (Bullynck 2006, 236–238, 259; 2009a, 147; 2010, 183–189).

accordance with some rule. However, what should have been a technical improvement ended up in a general conception of mathematics, according to which mathematical objects can be manipulated by various combinatorial rules in order to obtain theoretical results. In other words, Hindenburg discovered, or so he thought, a universal method for mathematics. Tables are not just a valuable database, but more importantly they codify the rules of combinatorics and can be applied to solve different mathematical problems, like that of constructing a number system or that of calculating the decimal expansion of a rational number.

3.2   Hindenburg and the Theory of Series

Perhaps Lambert's death in 1777 discouraged Hindenburg from pursuing his own project of calculating factor tables. Or perhaps, without Lambert's coordination, the lack of cohesion in the group of those who have decided to participate in this task pushed Hindenburg to modify his scientific interests. In any case, he will not address himself again to the problem of factor tables, at least not in a systematic way; on the contrary, he will never abandon the method of using tables to elucidate any mathematical question. As of 1778, his main research interests are in the fields of combinatorics and the theory of series. The two seminal works for the German combinatorial analysis emerged in this period under the titles *Infinitinomii dignitatum indeterminatarum leges ac formulae* and *Methodus nova et facilis serierum infinitarum exhibendi dignitates exponentis indeterminati* (Hindenburg 1778a; 1778b). The latter is in fact a dissertation defended by Hindenburg at the University of Leipzig to obtain the distinction of *magister legens*, which allowed him to become extraordinary professor of philosophy and physics in 1781.

  Both texts touch on the same subject, namely, that of showing that combinatorial methods can be successfully applied to the study of the multinomial theorem. Hindenburg systematized his discussion of this subject by summarizing it into two options in explicit accordance with Jacob Bernoulli's classification: a multinomial is either of the form $a + b + c + \&c.$ or of the form $az + bz^2 + cz^3 + \&c..$ The *Infinitinomii* deals with the multinomial theorem expressed under the first form, whereas the *Methodus nova* is concerned with multinomials of the second form. One year later, in 1779, Hindenburg put together both dissertations in a single volume and also incorporated some new paragraphs into the original text. This volume appeared under the similar title *Infinitinomii dignitatum exponentis indeterminati historia leges ac formulae* (Hindenburg 1779).

  In this section, we will first give an overview of the content and organization of both dissertations, and then we will discuss the modifications of the 1779 edition. Since developing a new mathematical symbolism was a major goal of the German combinatorial analysis, we will pay special attention to the use we make of mathematical symbols in our historical reconstruction: we will always use the original notation, but we will insert parentheses into some formulae to facilitate reading.

3.2.1   Hindenburg's First Approach to the Multinomial Theorem

3.2.1.1 The First *Infinitinomii*

At the moment when Hindenburg undertakes to write his *Infinitinomii dignitatum indeter-minatarum leges ac formulae*, the proofs of the binomial theorem for rational exponents have already a long history behind them, dating back to the early years of the eighteenth century, as we have seen in Chap. 2. Among all these proofs, and particularly in the German tradition related to Euler and Aepinus, as we have underlined before, there are some that promote the idea of the binomial theorem as the cornerstone of the organization of mathematical theories, especially in regard to mathematical analysis. In 1778, Hindenburg's understanding of how analysis is structured fits with this conception, as can be seen in this paragraph full of enthusiasm (Hindenburg 1778a, 1):

> Among the many excellent and remarkable discoveries by means of which the divine genius of Newton has nurtured the field of analysis and broadened its limits, and even if there are a lot of discoveries that deserve to be taken into consideration, none of them can compare with the theorem for the general expansion of the powers of a binomial. In addition to showing the universal generation of powers by expanding functions into finite or infinite series according to the number of terms, which largely benefits and widens the theory of series, this theorem is the basis upon which the whole sublime analysis takes its stand. This is why it should be considered, without doubt, as the most important theorem.[4]

While this paragraph establishes a link between Hindenburg's thought and the functional proofs of Euler and Aepinus, quoted by Hindenburg later in his text, it does not only highlight the idea that the binomial theorem is the most important theorem of "sublime analysis", but also suggests that its proof needs to come from the field of finite analysis, in a similar manner to that of Euler's and Aepinus's proofs. When reading the *Infinitinomii*, one realizes that Hindenburg has in fact a deep knowledge of the history of the binomial theorem, although he is rather selective in the sources he quotes. He mentions explicitly Kästner's differential proof, as well as that of Euler, and makes reference to Wolff's work. It is then clear that, on the one hand, all authors named here belong to the German tradition, and, on the other hand, there is a sort of opposition between differential proofs and proofs based on the methods of finite analysis. To forestall misunderstanding, it should be added that Hindenburg always speaks in a positive manner about Kästner's work, including his differential proof of the binomial theorem, but it should also be noted that

---

[4]   Inter praeclara multa et eximia inventa, quibus Analytices et egregie ornauit opes, et latius prorogauit fines, diuinum NEWTONI ingenium, nullum extare reperietur, quod possit, si commoda quae inde redundent plurima spectentur, Theoremati de generali potestatum a Binomio euolutione aequiparari. Praeterquam enim quod vere exhibeat dignitatum genesin vniueralem, functiones per series explicando, terminorum numero vel finitas vel infinitas: unde doctrina serierum mirifice aucta est et amplificata: ei etiam, quod maximum haud dubie censeri debet, Analysis sublimior, quam late patet, innititur veluti fundamento.

Hindenburg never employs differential methods comparable to those of his former teacher. Since Kästner himself raises the question of the relationship between the foundations of differential calculus and the binomial theorem by making reference to Colson's work, there is no doubt that Hindenburg is well aware of the objections to which differential methods are exposed regarding binomial theorem's proofs. It therefore seems likely that Hindenburg's predilection for proofs based exclusively on finite methods is rooted in the controversy about the foundations of the analysis of the infinite.

Hindenburg starts his examination of the multinomial theorem from the elementary way in which some authors have addressed this question before, namely, by associating the terms of the multinomial in order to get a binomial, and then by expanding this function into a power series by means of successive applications of the binomial formula. According to this approach, which became widely popular thanks to its dissemination through mathematical manuals, the multinomial theorem occupies the place of a simple corollary within the general hierarchy of mathematical statements. Although this kind of solution can be very attractive and very natural from a theoretical point of view, Hindenburg points out that the procedure is computationally very expensive, as it requires an enormous amount of labor and time to calculate every single coefficient of the multinomial expansion. Indeed, as he indicates at the beginning of his book, his goal consists in finding a remedy to this calculation problem, and, to do so, he should reshape the proof by using "the most remarkable methods" that mathematicians have created (Hindenburg 1778a, 3). Thus, Hindenburg's aim is in fact rather modest in the sense that he sees his own contribution as an improvement, certainly, but he knows that it is only a practical improvement in which not even the "remarkable methods" of the proof have been developed by him. If to this is added the fact that Hindenburg agrees with the idea that the binomial theorem is the most important theorem of analysis, it is not hard to see that this first dissertation of Hindenburg belongs to the lineage of proofs going back to the beginning of the century and that subordinates the multinomial theorem to the binomial theorem.

Those notable methods correspond, on the one hand, to those that were used by Wolff and Segner in their respective proofs, which have been analyzed in Sect. 2.2.3.1, and, on the other hand, to those of Jacob Bernoulli studied in Sect. 2.2.3.3. In his text, Hindenburg presents two demonstrations of the multinomial theorem which are based respectively on Wolff's and Bernoulli's proofs. Let's examine these demonstrations in detail.

Hindenburg conceives a new mathematical notation to be included in his interpretation of Wolff's proof, which consists of the following symbols:

1. Capital Greek or Latin letters are used to represent blocks of sums:

$$A = a,$$
$$B = (a + b),$$

$$\vdots$$

$$\Psi = (a + b + \cdots + \chi + \psi),$$

$$\Omega = (a + b + \cdots + \psi + \omega).$$

Greek capital letters play the role of some kind of metavariables in the sense that a Greek capital letter Ω can take the place of any Latin capital letter, and the same applies to Greek lowercase letters with respect to Latin lowercase letters. Thus, the sum $a + b + \cdots + \psi + \omega$ could eventually contain an infinite number of terms, in the same way that a multinomial could eventually be infinite concerning the number of its terms. Hindenburg expresses this idea by saying that there could be an infinite number of terms between a and ω.[5]

2. Lowercase letters, either Greek or Latin, with exponents to the left are used as follows:

$$^m b = (A + b)^m - A^m,$$

$$^m c = (B + c)^m - B^m,$$

$$^m d = (C + d)^m - C^m,$$

$$\vdots$$

$$^m \omega = (\Psi + b)^m - \Psi^m.$$

where m can be any real number. Lowercase letters with exponents to the left are in fact conventions to abbreviate the expression of the power series expansion of a given power of a possibly infinite sum. For instance, this is the expression of the mth power of a binomial:

$$(A + b)^m = B^m$$

$$= A^m + mA^{m-1}b + \frac{m(m-1)}{1.2}A^{m-2}b^2 + \&c.$$

$$= \left(A^m + {}^m b\right) = \left(a^m + {}^m b\right).$$

3. Gothic capital letters are used as abbreviations for binomial coefficients:

$$\mathfrak{A} = m, \quad \mathfrak{B} = \frac{m(m-1)}{1.2}, \quad \mathfrak{C} = \frac{m(m-1)(m-2)}{1.2.3}, \ldots,$$

---

[5] Hindenburg writes (Hindenburg 1778a, 4–5): "interpositis nimirum inter terminos extremos a et ω quotuis et quibuslibet intermediis, numero finitis siue etiam infinitis".

and the general expression of the coefficient is:

$$\mathfrak{N} = \frac{m\,(m-1)\cdots(m-n+1)}{1.2.3\ldots n}. \tag{3.1}$$

With these new symbols, Hindenburg can write the binomial expansion of Ω^m as follows:

$$\Omega^m = (\Psi + \omega)^m = \left(\Psi^m +^m \omega\right) = a^m +^m b + \ldots +^m \psi +^m \omega.$$

By substituting the lowercase letters with a left exponent for their respective values, and then by applying the binomial theorem to each of these values, one gets the desired result:

$$\begin{aligned}
\Omega^m =\;& A^m + \mathfrak{A}A^{m-1}b + \mathfrak{B}A^{m-2}b^2 + \cdots + \mathfrak{N}A^{m-n}b^n \\
& + \mathfrak{A}B^{m-1}c + \mathfrak{B}B^{m-2}c^2 + \cdots + \mathfrak{N}B^{m-n}b^n + \cdots \\
& + \mathfrak{A}\Psi^{m-1}\omega + \mathfrak{B}\Psi^{m-2}\omega^2 + \cdots + \mathfrak{N}\Psi^{m-n}\omega^n.
\end{aligned} \tag{3.2}$$

It should be noted that this is a recursive formula because of the definition of the terms $A^m = a^m$, $B^{m-k} = \left(A^{m-k} +^{m-k}b\right)$, $C^{m-k} = \left(B^{m-k} +^{m-k}c\right)$, ..., $\Psi^{m-k} = \left(X^{m-k} +^{m-k}\psi\right)$, where $k = 1, 2, 3, \ldots$.

Although Hindenburg presents his formula 3.2, as well as the method by means of which it has been determined, as an extension of the original research carried out by former mathematicians, it seems at least strange, if not disconcerting, that all his references to this kind of proofs are to Wolff's and especially to Segner's versions, and all the more so since Segner makes explicit mention of, for instance, Reyneau's proof. There is no doubt, therefore, that Hindenburg was well aware of the existence of foreign sources, but he decided not to quote them in his narrative. This stance could be explained away as a historiographical issue. As in the case of the binomial theorem, one can recognize a certain predisposition in Hindenburg to privilege the work of German savants concerning the attribution of scientific discoveries.

Let's now turn our attention to Hindenburg's interpretation of Jacob Bernoulli's proof. Even though, as in the previous case, the main modifications that were made by Hindenburg to Bernoulli's work may be summarized as the use of a new mathematical notation, Hindenburg seeks to justify the validity of the multinomial theorem for real exponents, which is indeed an actual generalization of Bernoulli's proof for positive whole exponents. Hindenburg finds another way to write Bernoulli's lexicographical arrangement 2.22, which has been presented in Sect. 2.2.3.3. Hindenburg's arrangement is shown in Fig. 3.3. For Hindenburg this is in fact a mathematical table consisting of many rows and one column. Given n objects $a,b, \ldots, \omega$, the table contains all combinations of n objects taken r at a time with repetition; that is to say, each row r contains all combinations of the n elements $a,b, \ldots, \omega$ taken r at a time with repetition. For example,

$$
\begin{array}{llll}
1) & a+b+c+d+\cdots & = a+b+c+d\cdots+\omega & = \Omega \\
\hline
2) & aa+ab+ac+ad+\cdots & = a[a+b+c+d\cdots+\omega] & = {}'A \\
 & bb+bc+bd+\cdots & = b[b+c+d\cdots+\omega] & = {}'B \\
 & cc+cd+\cdots & = c[c+d\cdots+\omega] & = {}'C \\
 & dd+\cdots & = d[d\cdots+\omega] & = {}'D \\
 & \&c & \&c & \&c \\
\hline
\end{array}
$$

$$
\begin{aligned}
3)\quad
&\left.\begin{array}{l}
aaa+aab+aac+aad+\cdots \\
abb+abc+abd+\cdots \\
acc+acd+\cdots \\
add+\cdots \\
\&c
\end{array}\right\} = a['A+'B+'C+'D\cdots+'\Omega] \;=\; ''A \\[2ex]
&\left.\begin{array}{l}
bbb+bbc+bbd+\cdots \\
bcc+bcd+\cdots \\
bdd+\cdots \\
\&c
\end{array}\right\} = b['B+'C+'D\cdots+'\Omega] \;=\; ''B \\[2ex]
&\left.\begin{array}{l}
ccc+ccd+\cdots \\
cdd+\cdots \\
\&c
\end{array}\right\} = c['C+'D\cdots+'\Omega] \;=\; ''C \\[2ex]
&\left.\begin{array}{l}
ddd+\cdots \\
\&c
\end{array}\right\} = d['D\cdots+'\Omega] \;=\; ''D \\[1ex]
&\qquad\qquad \&c \qquad\quad \&c \qquad\qquad\quad \&c
\end{aligned}
$$

Fig. 3.3 Hindenburg's interpretation of Bernoulli's lexicographical arrangement

the combinations of the letters a, b, and c taken 2 at a time with repetition are: aa, ab, ac, bb, bc, and cc, as can be seen in row 2 of the table. Evidently the sum of these combinations $aa+ab+ac+bb+bc+cc$ correspond to the terms one gets by calculating the square of the trinomial $a+b+c$ if one does not take into account the coefficient of each term. Thus, each row r can also be interpreted as a way of obtaining the terms that constitute the expansion in series of the rth power of a given multinomial, where the coefficient of each term is not taken into account.

To express the multinomial theorem, Hindenburg uses the following notations, which he created:

1. Capital letters with exponent to the left have the meaning that they have in Fig. 3.3, so:
   a. In particular,

$$
'A = a\,(a+b+\cdots\omega)
$$

$$
'B = b\,(b+c+\cdots\omega)
$$

$$
\vdots
$$

b. In general,

$$^n\mathrm{A} = a\left(^{n-1}\mathrm{A} + {}^{n-1}\mathrm{B} + {}^{n-1}\mathrm{C} + \cdots + {}^{n-1}\Omega\right),$$

$$^n\mathrm{B} = b\left(^{n-1}\mathrm{B} + {}^{n-1}\mathrm{C} + \cdots + {}^{n-1}\Omega\right),$$

$$\vdots$$

2. To designate the sum of terms of the combinations represented by capital letters with left exponent, Hindenburg introduced a symbol, inspired by the integral operator as he pointed out, which resembles the letter "s". In the absence of a better symbol, we have chosen "$\int$" which is pretty close to the original. For instance, given the objects a, b, c, the expression $\int\left(^2\mathrm{A}\right)$ is defined by $\int\left(^2\mathrm{A}\right) = a\left('\mathrm{A} +' \mathrm{B} +' \mathrm{C}\right) = (aaa + aab + aac + abb + abc + acc)$. More generally, Hindenburg defines the following expressions:

$$\int\left(\frac{'\mathrm{A}}{a}\right)^n = \int\frac{^n\mathrm{A}}{a} = {}^{n-1}\mathrm{A} + {}^{n-1}\mathrm{B} + \cdots + {}^{n-1}\Omega,$$

$$\int\left(\frac{'\mathrm{B}}{b}\right)^n = \int\frac{^n\mathrm{B}}{b} = {}^{n-1}\mathrm{B} + {}^{n-1}\mathrm{C} + \cdots + {}^{n-1}\Omega,$$

$$\vdots$$

3. Finally, the symbol $n\int$ denotes the number of terms that must be contained in the sum $\int$. In other words, an expression such as $n\int(^n\mathrm{A})$ represents the number of combinations with repetition that can be obtained from $^n\mathrm{A}$.

Thus, given a positive whole number n and a multinomial $\Omega = a + b + c + d \cdots + \omega$, Hindenburg gives the following formula:

$$\int\Omega^n = \int\left(\frac{'\mathrm{A}}{a}\right)^n = \int\frac{^n\mathrm{A}}{a} = {}^{n-1}\mathrm{A} + {}^{n-1}\mathrm{B} + \cdots + {}^{n-1}\Omega,$$

which allows to find all the terms (without coefficients) of the power series expansion of Ω^n.

Hindenburg calculates the number of terms of this expansion by adding the figurate numbers that are contained in the columns of his rearrangement of the arithmetical triangle, or Pascal's triangle, as can be seen in Fig. 3.4. The last row of this table shows how the numbers of terms of a given expansion can be calculated recursively on the basis of the

$n\!\int\!\Omega^{0}$	$n\!\int\!\Omega^{1}$	$n\!\int\!\Omega^{2}$	$n\!\int\!\Omega^{3}$	$n\!\int\!\Omega^{4}$	&c	$n\!\int\!\Omega^{n}$
$a=1$	1	1	1	1		1
$b=1$	2	3	4	5	&c	$n+1$
$c=1$	3	6	10	15		"
$d=1$	4	10	20	35	&c	
$e=1$	5	15	35	70		"
.	.	.	.	.	&c	
.	.	.	.	.		"
.	.	.	.	.	&c	
$\omega=1$	$\frac{r}{1}$	$\frac{r+1}{2}n\!\int\!\Omega^{1}$	$\frac{r+2}{3}n\!\int\!\Omega^{2}$	$\frac{r+3}{4}n\!\int\!\Omega^{3}$	&c	$\frac{r+n-1}{n}n\!\int\!\Omega^{n-1}$

Fig. 3.4 Hindenburg's arithmetical triangle

preceding column. On the other hand, Hindenburg recalls the following formula, which is not a recursive formula and serves to calculate the number of combinations with repetition:

$$n\!\int\!\Omega^{n} = \frac{r\,(r+1)\,(r+2)\cdots(r+n-1)}{1.2.3\ldots n} \tag{3.3}$$

He takes this formula from Bernoulli's *Ars conjectandi* (Bernoulli 1713, 115; Hindenburg 1778a, 24).

In Sect. 2.2.3.3, we have seen that Bernoulli determines the coefficient of a given term $a^{p}b^{q}c^{r}\ldots$, where $p+q+r+\cdots = n$, of the power series expansion of Ω^{n} by means of formula 2.16, and we pointed out that this is a combinatorial formula for calculating the number of permutations with repetition. However, the procedure for obtaining the multinomial coefficients based only on the notion of permutation appears to be too restrictive since, in that case, the power n of Ω^{n} must be a positive whole number. This is why Hindenburg abandons at this point Bernoulli's method, which he had scrupulously followed until now. In order to justify the validity of the multinomial theorem for any rational number, Hindenburg proposes a hybrid approach by combining the combinatorial method and the method of grouping terms based on the application of the binomial theorem. In fact, the binomial theorem is the keystone of Hindenburg's procedure (Hindenburg 1778a, 40):

This formula can also be extended in a *convenient* and *fruitful* way to the *remaining* exponents, whether *fractional* or *negative*. Bernoulli does not seem to have thought through the matter, and he could not even think about it, since his attention was *completely* turned toward *polynomial coefficients*, which are deduced from the *variation of permutations*, from where it follows the restriction of their domain to *positive whole* numbers *exclusively*. He did not take into account the *binomial coefficients*, which can be put together with polynomial coefficients, and for which the *renowned* Kästner has recently established a *law immune* to the *variation* of permutations that applies to any number, *positive* and *negative*, *natural* and *fractional*, etc.[6]

Here Hindenburg is hinting at the proof of Kästner that he mentioned at the beginning of his book, and that has been analyzed in Sect. 2.1.3.1. The end of the paragraph suggests that the proof is valid for any real exponent, but in fact it was originally conceived for fractional exponents.

Hindenburg proposes, then, to expand into series the function $(a + (b + c + \&c.))^m$, where m can be any fractional number, by applying the binomial theorem:

$$(a + (b+c+\&c.))^m = a^m + \mathfrak{A}a^{m-1}(b+c+d+\cdots) + \mathfrak{B}a^{m-2}(b+c+d+\cdots)^2$$
$$+ \mathfrak{C}a^{m-3}(b+c+d+\cdots)^3 + \cdots + \mathfrak{N}a^{m-n}(b+c+d+\cdots)^n.$$

The last term does not mean that the expansion is necessarily finite, on the contrary it represents the general term of the expansion, and so this could be infinite. From this equality, it is now possible to apply Bernoulli's combinatorial method in order to achieve the final result:

$$(a + (b+c+\&c.))^m = a^m + \mathfrak{A}a^{m-1}\mathfrak{a}(b+c+d+\cdots)$$
$$+ \mathfrak{B}a^{m-2}\mathfrak{b}('B +'C +'D + \cdots)$$
$$+ \mathfrak{C}a^{m-3}\mathfrak{c}(''B +''C +''D + \cdots)$$
$$+ \cdots + \mathfrak{N}a^{m-n}\mathfrak{n}(^{n-1}B +^{n-1}C +^{n-1}D + \cdots)$$
$$= a^m + ma^{m-1}(b+c+d+\cdots)$$
$$+ \frac{m(m-1)}{1.2}a^{m-2}(bb + 2bc + 2bd + \cdots + cc + \cdots) + \cdots,$$

where Gothic lowercase letters represent multinomial coefficients in general, that is to say they are coefficients calculated by means of formula 2.16.

---

[6] ea etiam ad *reliquos* exponentes, *fractos* et *negativos*, *recte* et cum *fructu* extendi potest; de quo non videtur cogitasse BERNOULLIVS, neque etiam cogitare poterat, qui ad *solos* respexerit *coefficientes polynomiales*, ex *permutationum varietate* deductos, eamque ob causam numeris *integris positiuis unice* adscritos, non item ad *Binomiales*, qui cum polynomialibus coniungi possunt, et quorum nostris temporibus *immunem* a permutationum *vicissitudinibus legem* demostrauit, *Vir* multis nominibus *Illustris*, KAESTERVS, numeros quoscunque, *positiuos* et *negatiuos*, *integros* et *fractos* &c, complectentem.

3.2.1.2 The New Method of Hindenburg

In his *Methodus nova*, Hindenburg deals with multinomials of the form $1 + az + bz^2 + cz^3 + \&c.$, and his objective is to find a non-recursive formula for determining the general coefficient of a term in the series expansion of such a multinomial raised to a fractional exponent. Indeed, he announced a proof, as many others did at that time, for real exponents, but no argument was offered in that regard. As in the case of the second proof of the *Infinitinomii*, the general strategy in the *Methodus nova* consists in using the binomial theorem, and then applying a combinatorial method to the result. However, in this case Hindenburg uses what he calls a new method instead of Bernoulli's combinatorial method. To have a better understanding of Hindenburg's dissertation, it seems obvious to start with a description of his new method.

The new method depends on the resolution of the following combinatorial problem (Hindenburg 1778b, 6–7): given a positive whole number n, to find all combinations of positive whole numbers that satisfy the condition that the sum of its elements is equal to n. In other words, the problem is to find all k-combinations with repetition taken k elements at a time from the first n positive whole numbers and, for each k-combination, the sum of its k elements must be equal to n. Perhaps it would be easier to simply say that Hindenburg is looking for the partitions of the integer n, but he wants to emphasize the combinatorial aspects of the problem. Hindenburg's principal source on this topic is (Euler 1748, vol. 1: 253–275), though he does not adopt Euler's terminology of "partition of numbers" (*partitio numerorum*) (Hindenburg 1778b, 15). We will not discuss Euler's work on partitions, but (Andrews 2007; Bell 2010; Panza 2007), for instance, can be consulted on this subject.

Instead of explaining his method in all its generality, Hindenburg gives the example of the partitions of the integer 10 to show how his method works. In (Panza 1992, vol. 2: 665–666), there is an interesting reconstruction, written in modern mathematical notations, of Hindenburg's method in all its generality. Here we have decided to follow Hindenburg's choice of giving an account of his method by an example. Hindenburg's table of the partitions of 10 can be seen in Fig. 3.5. The construction of this kind of tables to compute the coefficients of a power series expansion constitutes the most innovative part of his technique. In the table of Fig. 3.5, each row is called a "class" by Hindenburg, which is designated by a capital letter. However, later in his text, Hindenburg writes capital letters for classes with a superscript to the left: ^{10}A, ^{10}B, ^{10}C, ..., which recalls the integer that has been partitioned. More generally, he argues that each class is composed by all k-combinations with repetition taken k elements at a time from the first n positive integers and, for each k-combination, the sum of its elements is equal to n; that is to say, ^nA corresponds to the class of 1-combinations, ^nB corresponds to the class of 2-combinations, ^nC corresponds to the class of 3-combinations, and so on, where, for every element of every class, the sum of its terms is equal to n. It should be noted that Hindenburg claims that his method is a result directly derived from his early work on numerical tables (Hindenburg 1778b, 11–12). On the other hand, one needs to be careful not to confuse the meaning of capital letters with left superscripts in the

Fig. 3.5 Hindenburg's classes
of partitions of 10

A	10
	1, 9
	2, 8
B	3, 7
	4, 6
	5, 5
	1, 1, 8
	1, 2, 7
	1, 3, 6
C	1, 4, 5
	2, 2, 6
	2, 3, 5
	2, 4, 4
	3, 3, 4
	1, 1, 1, 7
	1, 1, 2, 6
	1, 1, 3, 5
	1, 1, 4, 4
D	1, 2, 2, 5
	1, 2, 3, 4
	1, 3, 3, 3
	2, 2, 2, 4
	2, 2, 3, 3
	1, 1, 1, 1, 6
	1, 1, 1, 2, 5
	1, 1, 1, 3, 4
E	1, 1, 2, 2, 4
	1, 1, 2, 3, 3
	1, 2, 2, 2, 3
	2, 2, 2, 2, 2
	1, 1, 1, 1, 1, 5
	1, 1, 1, 1, 2, 4
F	1, 1, 1, 1, 3, 3
	1, 1, 1, 2, 2, 3
	1, 1, 2, 2, 2, 2
	1, 1, 1, 1, 1, 1, 4
G	1, 1, 1, 1, 1, 2, 3
	1, 1, 1, 1, 2, 2, 2
H	1, 1, 1, 1, 1, 1, 1, 3
	1, 1, 1, 1, 1, 1, 2, 2
I	1, 1, 1, 1, 1, 1, 1, 1, 2
K	1, 1, 1, 1, 1, 1, 1, 1, 1, 1

Infinitinomii with the meaning of these letters in the *Methodus nova*, especially because the difference between them is not normally indicated in the specialized literature devoted to mathematical notations or to the history of the German combinatorial school (Cajori 1929, vol. 2: 65; Netto 1908, 207). Again they represent classes of partitions in the *Methodus nova*, while in the *Infinitinomii* they do not represent any class at all.

Let's see how this applies to the multinomial theorem. Hindenburg affirms that his proof follows quite closely that of Kästner, which has been analyzed in Sect. 2.2.3.2, even though he does not use any tool of differential calculus. In common with Kästner, Hindenburg starts from the following equation:

$$1 + az + bz^2 + cz^3 + \cdots + (\lambda n) z^n = 1 + y,$$

from which he gets:

$$\left(1 + az + bz^2 + cz^3 + \cdots + (\lambda n) z^n\right)^m = (1 + y)^m$$

$$= 1 + Az + Bz^2 + \cdots + (Ln) z^n \qquad (3.4)$$

$$= 1 + W,$$

where A, B, $\ldots$, (Ln) are undetermined coefficients. Thus, Hindenburg is using Kästner's expressions and, mainly, the symbols (λn) and (Ln). Nevertheless, no advantage is gained by using Kästner indicial notations (λn) and (Ln), since in formula 3.4 these symbols play the role of a general capital letter, so Hindenburg continues to give higher priority to Newton's alphabetical notation for coefficients, instead of taking advantage of Kästner's system of indexes n juxtaposed to the letters λ and L.

By the binomial theorem, it follows from Eq. 3.4 that:

$$(1 + y)^m = 1 + \mathfrak{A} y + \mathfrak{B} y^2 + \mathfrak{C} y^3 + \cdots + \mathfrak{N} y^n$$

$$= 1 + \mathfrak{A} \left(az + bz^2 + \cdots\right) + \mathfrak{B} \left(az + bz^2 + \cdots\right)^2 \qquad (3.5)$$

$$+ \mathfrak{C} \left(az + bz^2 + \cdots\right)^3 + \cdots + \mathfrak{N} \left(az + bz^2 + \cdots\right)^n$$

$$= 1 + Z,$$

where Gothic capital letters are the binomial coefficients given by formula 3.1. From Eqs. 3.4 and 3.5, Hindenburg deduces the equality $W = Z$ and points out that, in the expansions of the successive powers y, y^2, y^3, $\ldots$, the exponents of z grow respectively according to the following pattern:

$$1, 2, 3, 4, \ldots, n$$
$$2, 3, 4, \ldots, n$$
$$3, 4, \ldots, n$$
$$\vdots$$
$$n$$

For Hindenburg, this leads to the determination of the general coefficient (Ln) of the power series expansion, which is given by the formula:

$$(Ln)\, z^n = \mathfrak{A}\,(tn) + \mathfrak{B}\left({}^2t(n-1)\right) + \mathfrak{C}\left({}^3t(n-2)\right) + \cdots + \mathfrak{M}\left({}^{n-1}t2\right) + \mathfrak{N}\left({}^nt1\right),$$

where the expresion $({}^{\nu}t\mu)$, for $\mu, \nu = 1, 2, 3, \ldots, n$, represents the μth term of the expansion of y^{ν}.

The term $({}^{\nu}t\mu)$ is defined by means of the classes of partitions that have been discussed above. Given a power $y^{\nu} = \left(az + bz^2 + \cdots\right)^{\nu}$ of Eq. 3.5, the problem is to express the coefficient of any term in z of the expansion of y^{ν} by means of classes of partitions. To this end, Hindenburg establishes the following correlation between terms in z of the expansion of y^{ν} and classes:

1. For the expansion of y, z^{μ} is associated with ${}^{\mu}\mathrm{A}$.
2. For the expansion of y^2, $z^{\mu+1}$ is associated with ${}^{\mu+1}\mathrm{B}$.
3. For the expansion of y^3, $z^{\mu+2}$ is associated with ${}^{\mu+2}\mathrm{C}$.
4. In general, for the expansion of y^{ν}, $z^{\mu+\nu-1}$ is associated with ${}^{\mu+\nu-1}\mathcal{N}$, where $\mathcal{N}$ is a general symbol that represents any class of partitions.

Given $y^{\nu} = \left(az + bz^2 + \cdots\right)^{\nu}$, the coefficients of the terms in z are sums of monomials of the form $a^p b^q c^r \ldots$, with $p + q + r + \cdots = \nu$. Thus, in order to determine these sums of monomials by using classes, it is necessary to relate coefficients a, b, c, etc., to the elements of partitions. Hindenburg gives then the following substitution rule:

$$
\begin{array}{cccc}
a & b & c & \ldots \\
1 & 2 & 3 & \ldots
\end{array}
\tag{3.6}
$$

Finally, one gets the definition:

$$({}^{\nu}t\mu) = \mathfrak{n}({}^{\mu+\nu-1}\mathcal{N}), \tag{3.7}$$

where $\mathfrak{n}$ is the multinomial coefficient of a given monomial $a^p b^q c^r \ldots$, which is calculated by using formula 2.16.

An example of Hindenburg will allow us to gain a better understanding of the way in which all these formulae should be applied. Consider the power $y^6 = (az + bz^2 + \cdots)^6$, and suppose that one is trying to calculate the coefficient corresponding to the fifth term in the series expansion of y^6, i.e. one is looking for $(^6t5)$. By formula 3.7, one gets $(^6t5) = \mathfrak{f}(^{10}F)$ which is associated to $z^{\mu+\nu-1} = z^{10}$. The class ^{10}F has been determined in the table of Fig. 3.5, and this class can be rewritten as:

$$\mathfrak{f}(^{10}F) = \mathfrak{f}\left\{ \begin{array}{ccc} 1^5 & 5 & \\ 1^4 & 2 & 4 \\ 1^4 & 3^2 & \\ 1^3 & 2^2 & 3 \\ 1^2 & 2^4 & \end{array} \right\}$$

By using substitution rule 3.6 and by using formula 2.16 to calculate the multinomial coefficients of the monomials obtained from substitution, one gets the result:

$$\begin{pmatrix} 6a^5e \\ +30a^4bd \\ +15a^4c^2 \\ +60a^3b^2c \\ +a^2b^4 \end{pmatrix} z^{10}.$$

Thus, Hindenburg can write down the general formula for the multinomial expansion in terms of classes of partitions:

$$(1+y)^m = \left(1 + az + bz^2 + \cdots + (\lambda n)\, z^n \right)^m$$

$$= 1 + \mathfrak{Aa}(^1A)z + \left(\mathfrak{Aa}(^2A) + \mathfrak{Bb}(^2B)\right) z^2$$

$$+ \left(\mathfrak{Aa}(^3A) + \mathfrak{Bb}(^3B) + \mathfrak{Cc}(^3C)\right) z^3 + \cdots$$

$$+ \left(\mathfrak{Aa}(^nA) + \mathfrak{Bb}(^nB) + \cdots + \mathfrak{Nn}(^n\mathcal{N})\right) z^n.$$

Hindenburg emphasizes the fact that this formula enables one to directly determine any term of the multinomial expansion, that is to say, in order to find the coefficient of the term z^n in the multinomial expansion, there is no need to compute all previous coefficients from z to z^{n-1}, as was customary at the time. The general term $\left(\mathfrak{Aa}(^nA) + \mathfrak{Bb}(^nB) + \cdots + \mathfrak{Nn}(^n\mathcal{N})\right) z^n$, whose coefficient is independent of any other previous term, provides the expected solution. Hindenburg seemed very proud of his accomplishments and confident that he was the first to discover a non-recursive formula for the multinomial theorem (Hindenburg 1778b, 23). Moreover, Hindenburg was convinced that no one had ever implemented a method similar to his combinatorial tables of classes. However, as early as 1748, Boscovich advanced a non-recursive formula for calculating the coefficients of the multinomial expansion, which depends on a very similar method to that of Hindenburg, since it is based on partitions tables too, as has been noted in Sect. 2.2.2.3. In spite of this resemblance, Boscovich is not among the authors quoted by Hindenburg in the *Methodus nova*, though he quotes other foreign authors, like Moivre and Colson. It is difficult to tell whether Hindenburg knew of Boscovich's papers at the moment when he wrote the *Methodus nova*. The first mention of Boscovich's ideas dates from 1781, when Hindenburg published another text in which he simply declared that Boscovich had previously employed a combinatorial method to prove the multinomial theorem, but he did not reveal that both his method and that of Boscovich depend on partitions of numbers and that the use made of these partitions is strikingly similar in both cases (Hindenburg 1781b, IX). Noting that, during the decade of the 1790s, Hindenburg will acknowledge the merits of Boscovich's contributions to mathematics, it is possible that he would not have attentively studied Boscovich's papers in 1781. In fact, Hindenburg claims in 1795 that he found out about Boscovich's method only after the second *Infinitinomii* got published in 1779, so after the publication of the *Methodus nova* (Hindenburg 1795a, 420, footnote 40). In that case, Hindenburg and Boscovich arrived at equivalent results independently. However, since Boscovich's proof was conceived for positive whole numbers only, Hindenburg's proof is more general: it is valid for any fractional number inasmuch as Kästenr's proof of the binomial theorem, on which it depends, has been established for fractional numbers. In this sense, Hindenburg was the first mathematician to propose a non-recursive formula for directly calculating any coefficient in the power series expansion of a multinomial raised to a rational exponent.

*

In sum, it is possible to identify four common elements that are present in all three Hindenburg's demonstrations, namely: the use of the binomial theorem, the use of mathematical tables, the use of combinatorial techniques, and the development of new mathematical notations. The binomial theorem is considered the most important theorem of sublime analysis, on which the multinomial theorem depends. Mathematical tables are a central element of Hindenburg's procedures, which comes from his early work on Lambert's general project concerning numerical tables. Developing new mathematical notations is regarded as a major contribution to mathematics, so much so that this is the most relevant aspect of the first two demonstrations of Hindenburg. Hindenburg's new method is similar to that of Boscovich, although there seems no reason to doubt that they were independently worked out, and, indeed, the conclusions reached by Hindenburg were more general. Mathematicians who had a real influence on the way in which Hindenburg interpreted the question of the multinomial theorem are Moivre, Bernoulli, Kästner, Wolff, Euler, Segner, and Aepinus, with a clear predominance of German scholars.

3.2.2 The Second *Infinitinomii*, and the Sudden Apparition of Leibniz

In 1779 the *Infinitinomii* and the *Methodus nova* were reprinted and completed in one volume under the quasi identical title *Infinitinomii dignitatum exponentis indeterminati historia leges ac formulae* (Hindenburg 1779). The introduction of "*exponentis*" is rather a matter of style. On the contrary, the addition of "*historia*" reflects, as we shall see shortly, a growing awareness from Hindenburg that the method of partitions tables could be useful for more general purposes than proving a simple corollary of the binomial theorem. Arguably, the key to the organization of the book lies on this fact. The general plan of the book is as follows:

1. Preface (added to this new edition).
2. From paragraph I to paragraph XV, the first *Infinitinomii* without any modification.
3. From paragraph XVI to paragraph XIX, new text to make the transition from the first *Infinitinomii* to Hindenburg's new general method.
4. From paragraph XX to paragraph XXV, the *Methodus nova* with substantial modifications.
5. From paragraph XXVI to paragraph XXVIII, new mathematical problems solved by Hindenburg's method.
6. New addendum of combinatorial tables.

Paragraphs XVI to XIX offer a summary of the arguments tackled and deployed in Moivre's original proof of the multinomial theorem, Bernoulli's differential proof, Colson's educational exercise that shows how the multinomial expansion can be calculated by the method of fluxions, and Kästner differential proof, which have been reviewed in Sects. 2.1.3, 2.2.2.2, and 2.2.3.2. This account attempts neither to interpret any of these proofs in terms of Hindenburg's combinatorial method, nor to translate them into the new mathematical notation developed in his two previous dissertations, in contrast to the perspective assumed in the first *Infinitinomii* where Wolff's and Bernoulli's proofs were rethought along those lines. It is important to remember, however, that the first *Infinitinomii* was meant to be understood as a renewal of a mathematical question, and taken as a new answer to this question; that is to say, it was an original mathematical paper in which Hindenburg deals with a problem of the theory of series and proposes a solution, and, as modest as it was, this solution could be considered as a genuine contribution to the field since a mathematical method, that of Bernoulli, was successfully generalized. This is where the word "*historia*" of the title comes in. These transitional paragraphs narrate a history of the multinomial theorem from its invention by Moivre, to its inflection point in Bernoulli's procedures, and then to its culmination at the works of Colson and Kästner. Throughout these paragraphs, Hindenburg becomes a historian of his own scientific field of research, with the goal in mind of bringing historical depth to his new general combinatorial method. In fact, the plan of Hindenburg's book given above depicts its plan as seen through the eyes of someone interested in scrutinizing the details of the evolution of Hindenburg's thought, but from the view of a reader interested in the subject of the book, the organization is much simpler: the book comprises two parts, one historical part and one mathematical part. For instance, this is the point of view assumed in (Panza 1992, vol. 2: 651), where Panza points out that the book is divided into a historical part (paragraphs I to XX) and a mathematical part (from paragraph XXI to the end). Thus, Hindenburg intentionally relegated his first *Infinitinomii*, which was originally conceived as a mathematical paper, to a supporting role of historical background. This shift in perspective involves a more fundamental shift in his understanding of the multinomial theorem and the methods used to demonstrate it. In 1778, Hindenburg believed, as most scholars at the time, that there were two different approaches to the study of the multinomial theorem, depending on whether the multinomial is of the form $(a+b+c+\&c.)^m$ or of the form $(1+az+bz^2+cz^3+\&c.)^m$. Moreover, each of these approaches had its own method of finding the corresponding power series expansion. In the second *Infinitinomii*, this belief was abandoned in the face of a unified method and a unified approach: that of the classes of partitions. Therefore, the second *Infinitinomii* can be regarded as a generalization of the work done in the *Methodus nova*, and the extensive historical part suggests that Hindenburg placed his work as the denouement of almost a century of attempts to find a non-recursive formula for the multinomial theorem.

This generalization is a consequence of extending the techniques based on formula 3.7 to a wider range of problems concerning the theory of series. In this sense, the second *Infinitinomii* was intended to provide a working (but still limited) combinatorial method for solving some questions in the field of mathematical analysis. This objective is certainly more ambitious in scope than that pursued by Hindenburg in his previous dissertations, but it does not imply any kind of reorganization of mathematical analysis. Let's look more closely at his method, which was named by Hindenburg "the method of powers" ("*methodus potentiarum*") (Hindenburg 1779, 100). Given a possibly infinite multinomial $\alpha z + \beta z^2 + \gamma z^3 + \&c. = y$, and its successive powers $y, y^2, y^3, \ldots$, Hindenburg infers the following equation by applying formula 3.7 in a similar manner as described above:

$$ay + by^2 + cy^3 + \cdots + ny^n = a\mathfrak{a}(^1A)z + \left(a\mathfrak{a}(^2A) + b\mathfrak{b}(^2B) \right) z^2$$

$$+ \left(a\mathfrak{a}(^3A) + b\mathfrak{b}(^3B) + c\mathfrak{c}(^3C) \right) z^3 + \cdots \qquad (3.8)$$

$$+ \left(a\mathfrak{a}(^nA) + b\mathfrak{b}(^nB) + \cdots + n\mathfrak{n}(^n\mathcal{N}) \right) z^n.$$

where the coefficients a, b, c, ..., n are known. The general idea of Hindenburg's method consists in calculating the power series expansion of a function composed with the multinomial y by applying Eq. 3.8. It should be noted, however, that Hindenburg's method does not allow to calculate the series expansion of the given function itself, but only the expansion of the composition of the function with the multinomial y. The method of powers is, in fact, dependent upon the knowledge of the series expansion of the given function. On the other hand, Hindenburg emphasizes the fact that his method of powers always leads to a non-recursive formula for calculating the general coefficient in the expansion of such a composition of functions, which is a consequence of Eq. 3.8, where the coefficient of the general term $\left(a\mathfrak{a}(^n A) + b\mathfrak{b}(^n B) + \cdots + n\mathfrak{n}(^n \mathcal{N})\right) z^n$ is independent of any other previous coefficient of the series.

For instance, given the following quotient:

$$\frac{1}{1 - \alpha z - \beta z^2 - \gamma z^3 - \&c.}$$

Hindenburg put $\alpha z + \beta z^2 + \gamma z^3 + \&c. = y$ in order to use the expansion of $\frac{1}{1-y}$, which allows the application of Eq. 3.8:

$$\frac{1}{1 - \alpha z - \beta z^2 - \gamma z^3 - \&c} = \frac{1}{1 - y} = 1 + y + y^2 + y^3 + \&c.$$

$$= 1 + \mathfrak{a}(^1 A)z + \left(\mathfrak{a}(^2 A) + \mathfrak{b}(^2 B)\right) z^2 + \&c.$$

Other problems treated with the method of powers are:

- A quotient of two possibly infinite multinomials: $\frac{a+bz+cz^2+\&c.}{\alpha+\beta z+\gamma z^2+\&c.}$.
- Reversion of series. Indeed the question of reversion of series would be a consequence of solving the following problem. Given a multinomial

$$ay + by^2 + cy^3 + \&c.,$$

rewrite it under the form

$$\frac{\mathfrak{A}y}{(\alpha + \beta y)} + \frac{\mathfrak{B}y^2}{(\alpha + \beta y)^2} + \frac{\mathfrak{C}y^3}{(\alpha + \beta y)^3} + \&c.,$$

where $a, b, c, \ldots, \alpha$, and β are known, and where $\mathfrak{A}, \mathfrak{B}, \ldots$, are the coefficients to be determined. Using his method, Hindenburg got the following table:

a	b	c	$\cdots$	n
α^1	α^2	α^3	$\cdots$	α^n
β^{n-1}	β^{n-2}	β^{n-3}	$\cdots$	1
1	$\frac{n-1}{1}$	$\frac{(n-1)(n-2)}{1\cdot 2}$	$\cdots$	1

where the symbol "n" in the first row is ambiguous because it represents at the same time both a coefficient and the index of a coefficient. This table gives us the general coefficient

$$a\alpha^1\beta^{n-1} + \frac{n-1}{1}b\alpha^2\beta^{n-2} + \frac{(n-1)(n-2)}{1\cdot 2}c\alpha^3\beta^{n-3} + \cdots + n\alpha^n$$

of the general term $\frac{y^n}{(\alpha+\beta y)^n}$. Then Hindenburg pointed out that, from the binomial $\frac{y}{(\alpha+\beta y)}$, it is possible to solve the equation $\frac{y}{(\alpha+\beta y+\gamma y^2+\delta y^3+\&c)} = z$ by using the series reversion theorem. Moreover, he claimed that it was not necessary to apply this theorem since the coefficients of the reversed series could be easily calculated by means of his combinatorial method (Hindenburg 1779, 108), which implies that he believed his method did not merely justify the theorem on reversion of series, but also offered a non-recursive formula for the reversion. Nonetheless, he did not show how could be constructed such a non-recursive formula.

- Exponential expansions: $e^{\alpha z+\beta z^2+\gamma z^3+\&c}$.
- Logarithm expansions: $lh(1+\alpha z+\beta z^2+\gamma z^3+\&c)$ and $lh\left(\frac{1+\alpha z+\beta z^2+\gamma z^3+\&c}{1-\alpha z-\beta z^2-\gamma z^3-\&c}\right)$, where lh stands for natural logarithm.
- Trigonometric function expansions, as that of $\cos(\alpha z + \beta z^2 + \gamma z^3 + \&c)$ and $\sin(\alpha z + \beta z^2 + \gamma z^3 + \&c)$. He also applied his method to arcsine, but made a mistake in his calculations.

All these examples seem to indicate that the flexibility of Hindenburg's method increased his interest in the field of the theory of series, and stimulated the idea of consolidating combinatorial methods into a new, coherent approach to the field.

Besides this global mathematical reorientation, which arises from his early work, more substantive theoretical changes were set out in the Preface and the Addendum. The Preface of the second *Infinitinomii* have been organized in two parts. In the first part, Hindenburg provides a good description of the content of the book. In the second part, he discuses some issues pertaining to the theorem on reversion of series, which was justified by Moivre on the basis of the multinomial theorem. In his discussion, Hindenburg pays particular attention to the priority quarrel between Moivre and George Cheyne (1671–1743) about the theorem on reversion of series. Cheyne was a Scottish physician and mathematician, and acquired an unfortunate popularity in mathematical circles because of several priority controversies. In 1703, he published a treatise on the inverse method of fluxions, taking credit for several discoveries of other mathematicians, including Moivre's theorem on reversion of series (Cheyne 1703). Moivre complained in 1704 about this academic plagiarism, and Cheyne responded bitterly to Moivre's criticism. As the dispute became rather personal than academic, Moivre decided not to pursue it any further. For more details about this quarrel, see (Schneider 1968, 204–205).

In order to better understand the interest of Hindenburg in this subject, let me recall two remarks that Moivre made in his complaint about Cheyne's plagiarism of the theorem on reversion of series (de Moivre 1704, 74):

> Second, the illustrious Leibniz published this theorem long before this author [Cheyne] [...].
> Third, I was the first to bring this theorem to the court of public opinion, and the illustrious Leibniz is willing to recognize it [...].[7]

Hindenburg quotes this excerpt and makes the following comment (Hindenburg 1779, XIV):

> Leibniz has been identified here as the author of the theorem on reversion of series, which is however not identical to the theorem proposed by Moivre. [... A]nd I did not harbor any real interest in discovering it, for I was convinced that the theorem of Leibniz could not possibly be replaced by that of Moivre, which is the only one in use today, if that of Leibniz were better; unless our mind were persuaded, to which I paid attention, by Moivre's words according to which the acquisition of *knowledge out of the ordinary about the Doctrine of combinations is indispensable* for obtaining the successive coefficients of Leibniz.[8]

It is clear that Hindenburg offers a misleading interpretation of Moivre's remarks in declaring that Leibniz is the author of the theorem on reversion of series. In this passage, Moivre makes plain that his position on this subject comes down to two main points. The first is that even other scholars have published papers related to this theorem before Cheyne, such as Leibniz. The second is that he was the first to published the theorem. It is interesting to note that he did not even claim to be the author of the theorem, maybe because he knew that this achievement belonged to Newton alone. Indeed, Leibniz himself was drawn by Fatio into a controversy over priority of this theorem, since Leibniz plagiarized it too, even if he introduced a new mathematical notation as has been seen in Sect. 2.2.3.3. Thus, Hindenburg's extrmely unfortunate interpretation of Moivre's observations is really strange, and it is perhaps symptomatic of the untouchable stature that Leibniz's intellectual reputation had gained in German academic circles.

The conviction that Leibniz was the author of the theorem on reversion of series, besides being wrong, led Hindenburg in the direction of asking himself questions about

---

[7] Secundo, Celeberrimus *Leibnitius* longe jam tempore ante Authorem hoc ipsum Theorema exhibuit [...].

Tertio, Ego primus omnium Theorema in hunc finem publici juris feci, quod Cl. *Leibnitius* agnoscere non dedignatus est [...].

[8] Hic LEIBNITIVS nominatur auctor Theorematis de Serierum Regressu, ab eo tamen, quod MOIVRAEVS proposuit, diversi. [... A]tque ego etiam parum cupidus eius cognoscendae fuissem, persuasus, Theorema Leibnitiamun, si praestitisset, nullo modo a Moivraeano, quod vnice nunc in usu est, potuisse abrogari: nisi animum, ut attenderem, excitassent verba MOIVRAEI, quibus affirmat, ad continuandos coefficientes Leibnitianos *peritia haud vulgari in Combinationum doctrina opus esse.*

the mathematical merits of Leibniz's version of this theorem. For Hindenburg, if this theorem had its roots in Leibniz's thought, mathematical practice should be based on the original Leibnizian formulation of the theorem. Since Moivre's version imposed itself in mathematical practice as the standard tool for calculating the coefficients of reversions of series, Hindenburg concluded that the formulation of Leibniz could conceal unexpected difficulties, at least on the practical level. Hindenburg was indeed mainly concerned with the possibility that those difficulties could arise, as Moivre pointed out, from the "doctrine of combinations" (Hindenburg 1779, XIV):

> What I strongly longed for was to understand how Leibniz had used the *Doctrine of combinations* in order to determine the law of progression [of coefficients], so as to be able to compare it with mine.[9]

As we have seen in Chap. 2, the only relevant difference between Leibniz's and Moivre's formulation of this theorem is the numerical notation used by Leibniz for the coefficients. However, Leibniz claimed that the introduction of numerical coefficients allowed to establish a combinatorial law for calculating the series reversion coefficients, and this law had been captured, according to Leibniz, in the four rules that he proposed and that can be consulted in Sect. 2.2.3.3 on page 65. This is a very important point because Leibniz's four rules are supposed to be independent of the recursive algorithm given by Moivre (and copied by Leibniz) for calculating the coefficients of the power series expansion. As a consequence of their independence, Leibniz's four rules are supposed to provide an alternative and purely combinatorial method for reversing a series. Moivre's criticism of Leibniz's use of the doctrine of combinations refers to these rules, and then to Leibniz's alternative combinatorial method.

All four rules take advantage of the inherent arithmetical properties of numbers to find out the order in which digits or groups of digits that are used in the construction of a series reversion coefficient should properly follow each other. The different places that digits can occupy result in different orders or combinations, that is to say, in different sequences of digits or groups of digits. Leibniz's rules are supposedly helpful in identifying the exact combinations of digits that correspond to the series reversion coefficients. As Hindenburg's combinatorial method depends on the arithmetical properties of digits to calculate the coefficients of a series expansion via the rule of substitution 3.6, it is similar to Leibniz's idea in this very general sense. However, instead of remarking on this slight similarity, Hindenburg directed, despite all odds, strong criticism against Leibniz's numerical notation and combinatorial law (Hindenburg 1779, XVIII–XIX):

> However, I am afraid that this very example [Leibniz's theorem on reversion of series] runs counter to the progress of the analytic method, progress about which Leibniz was concerned.

---

[9] Quomodo igitur LEIBNITIVS, in formanda progressus lege, *Combinationum doctrina* usus fuerit, hoc erat, quod auide cupiebam discere, ut comparare cum meis possem.

Since: 1) the riddle of the Sphinx could have been expressed as: what is the form of the desired coefficients?; 2) the progression law [of coefficients] has been presented in a rather obscure form, and there is no indication of how it could be implemented, or of how the obtained terms have been determined; finally 3) even if there were some indication, those who would have wanted to use it in order to obtain the next terms would not easily find their way out of the labyrinth, where one keeps going in circles, without Ariadne's thread: to succeed, as Moivre pointed out, the acquisition of *knowledge out of the ordinary* about *the Doctrine of combinations* is certainly indispensable.[10]

After these sarcastic metaphors, instead of reinforcing his criticism, Hindenburg moderated the tone of his discourse, putting Leibniz's work in perspective. Regardless the technical and practical failures of Leibniz's combinatorial rules, the general idea of applying the doctrine of combinations to the field of analysis headed in the right direction, which becomes evident, according to Hindenburg, in the light of his method of powers. It should be noted that this is not a historical but a theoretical perspective based on a virtually nonexistent affinity between Leibniz's and Hindenburg's methods, which have only one thing in common: both depend on combinatorics. Thus, what Hindenburg learned from comparing his method with that of Leibniz was that he had fulfilled, without meaning to, one of the most cherished dreams of Leibniz, namely, that of applying successfully the doctrine of combinations to other areas of mathematics, and he had done it himself, without having been influenced by Leibniz at all. However, although Hindenburg's method works properly, he did not actually show how to apply the method of powers to the theorem on reversion of series. As has been seen above, he just claimed that it was possible to do so, but the fact is that the question of whether his tables of partitions lead to a non-recursive formula for calculating the series reversion coefficients will remain open for some time.

One might wonder whether Hindenburg was right in saying that his method of powers was superior to Leibniz's alternative combinatorial method and that Leibniz's method was in fact unintelligible and useless. Indeed, Hindenburg came to this conclusion but he did not show the actual comparison of both methods. In spite of this, there is no flaw in his argument because Leibniz's four rules do not constitute a mathematical method at all, and thus it is not possible to make a comparison. A mathematical method should provide an effective procedure to solve a given problem, but Leibniz's rules cannot be used to solve the problem of calculating any coefficient of a reversed series. For instance, adopting Leibniz's terminology given in Sect. 2.2.3.3, the numerator of the coefficient 104 contains the term 20.101.103 which is formed by the small number 20 and the major numbers 101

---

[10] Sed vereor, ne hoc ipsum exemplum damnum intulerit Artis Analyticae, quam LEIBNITIVS meditabatur, promotioni. Nam (1) Sphingis potius videtur αἴνιγμα referre, quam Coefficientium quaerendorum formam; (2) Progressus Lex verbis tantum non obscurissimis indicatur, ratio autem, quomodo applicari possit, non ostenditur, neque etiam ex terminis allatis apparet; denique (3) si vel maxime appareret, ea tamen, pro terminis paulo altioribus, qui uti vellent, mox in Labyrinthum abriperentur, vnde redire, absque filo Ariadneo haud facile possent: *peritia* enim, ut MOIVRAEVS ait, ad hoc efficiendum *haud vulgari* in *Combinationum doctrina* opus est.

and 103. As seen in Sect. 2.2.3.3, Leibniz determined the composition of the coefficient 104 by using actually Moivre's algorithm instead of his own method, but what happens when someone wants to apply Leibniz's method? Suppose that somehow we know that the small number 20 appears in the numerator of the coefficient 104 (for example, we can look at the result obtained by means of Moivre's algorithm). According to Leibniz's rule 3 (see page 65), the digit "2" in the small number 20 indicates that this small number should be multiplied by two major numbers, but rule 3 does not specify which ones. Rule 4 (see page 65) establishes that these two major numbers must satisfy the condition that the sum of their rightmost digits and the digit "0" of the small number 20 should be equal to 4, which is the rightmost digit of the coefficient 104. This condition is clearly satisfied by the major numbers 101 and 103, but it is satisfied by the major numbers 102 and 102 as well. Why does the term 20.101.103 appear in the numerator of the coefficient 104 instead of the term 20.102.102? Leibniz's rules do not answer this question, and thus they are useless. They are also unintelligible: for instance, on the basis of those rules, one does not understand how many terms should be contained in the numerator of the coefficient 104 (this difficulty is related to the problem of determining the small numbers that should appear in the numerator, problem that has been neglected by Leibniz, and this is why we were forced to assume above the hypothesis that the small number 20 appears in the numerator of 104). Hence, these useless and unintelligible rules cannot be compared with the effective method of powers invented by Hindenburg, who was then right in criticizing Leibniz for his lack of precision and clarity, and he was also right in underlining the mathematical superiority of his own invention.

To better understand the extent and nature of Hindenburg's fair and accurate criticism, it would be useful to consider what role Leibniz's four rules play in Leibniz's general conception of mathematics. Although Leibniz never elaborated a systematic exposition of his mathematical thought, he associated mathematics with several epistemological programs of his own, such as *mathesis universalis, lingua characteristica*, and *ars combinatoria*. None of these programs is restricted to the field of mathematics, but this science occupies a central place in each of them. These very ambitious programs never took a definitive form in Leibniz's writings, and much of what we know today about them comes from manuscripts that were unknown in the eighteenth century. In recent times, much ink has been spilled over those issues, and the interested reader can find a detailed treatment of those topics in the specialized literature (Couturat 1903; Mittelstrass & Schroeder-Heister 1986; Peckhaus 1997; Rabouin 2018). On the contrary, the content of Leibniz's programs was practically inaccessible to the scholars of the eighteenth century, even to those that were enthralled by Leibniz's general remarks on the subject. For instance, Bürmann, whose work will be analyzed in Chap. 5, was captivated by the Leibnizian idea of building a characteristic language and attempted to construct his own system inspired by Leibniz, but he complained about the absence of a concrete, tangible development of Leibniz's program, as can be seen in the following statement (Bürmann 1801b, 1):

I have named my system after the name Leibniz gave to a certain algebra of reasoning about which he talked so much and never published anything.[11]

Bürmann called his system "combinatorial characteristic". In the particular case of *ars combinatoria*, Leibniz published in 1666 his *De arte combinatoria*, but later he dismissed his own dissertation as an immature youth exercise that needed to be fully reworked, so it cannot be taken as an example of his program *ars combinatoria*, even if it contains some germinal ideas on the subject. In view of this situation, in which some scientists were inspired by Leibniz's programs even though no actual results of those programs were known by those scientists, historians have established a methodological distinction in order to explain this historical phenomenon. This distinction consists in the formulation of two ways of understanding Leibniz's general conception of mathematics: on the one hand, those general programs deprived of technical results should be conceived as a general reference framework within which mathematical research was carried out, and thus they functioned as an ideal that guided research; on the other hand, there are the technical aspects of Leibniz's mathematical works. This differentiation between ideal and actual realization helps to explain the influence of Leibniz on the work of some mathematicians without having to provide evidence of the use of a concrete mathematical result of Leibniz in the work of those mathematicians. For instance, this methodological distinction allows to say that Bürmann was influenced by Leibniz's programs even if he did not use any of Leibniz's mathematical results: Bürmann followed the Leibnizian ideal of a combinatorial characteristic language. Did Hindenburg follow Leibniz's ideals too? To answer this question, let's go back to our original concern. Despite Bürmann's opinion, Leibniz did publish at least one result related to his programs *lingua characteristica* and *ars combinatoria*, namely, his four rules for calculating the coefficients of a reversed series. In his 1700 article concerning the theorem on reversion of series, Leibniz does not elaborate on the theoretical aspects of his programs directly, but this he does through the example of the theorem on reversion of series. In his article, Leibniz points out that what he called small numbers and major numbers (see Sect. 2.2.3.3) are not just a simple mathematical notation, but they are "characteristics", that is to say, symbols belonging to his characteristic language and thus they express the very nature of things and the factual relations between the things they represent (*tanto utiliores esse notas, quanto magis exprimunt rerum relationes*) (Leibniz 1700, 208). And those relations have been captured by Leibniz's four combinatorial rules. Since Leibniz was convinced of the accuracy of his combinatorial rules and his characteristic symbols, he drew the following conclusion (Leibniz 1700, 208):

---

[11] Le titre que j'ai adopté est celui que Leibniz donnait à un algèbre de raisonnement dont il parla toujours et ne publia rien.

And thus finally, through the characteristic of the combinatorial art, algebra, which depends on it, has been improved.[12]

In the last sentence of his article, Leibniz reiterates the point that his dissertation *De arte combinatoria* was a juvenile work and that back then he failed to understand the deep implications of his *ars combinatoria* and *lingua characteristica*. In other words, that juvenile work contrasts with the intellectually mature vision that Leibniz offers of his science in 1700. And this is the important point to keep in mind: Leibniz is not interested in giving a "new" proof of the theorem on reversion of series or in formulating a new mathematical method, he wants to give us a taste of the power of his *lingua characteristica* and *ars combinatoria* as a science, not as a program or ideal, but as a reality. Leibniz's conclusion quoted above depends on that: algebra has been improved because of the power of his actual, existing science of *lingua characteristica* and *ars combinatoria*. In light of this, it is easy to see that Hindenburg's criticism makes a devastating impact because it suggests that, for Hindenburg, Leibniz's science of *lingua characteristica* and *ars combinatoria* is useless and unintelligible. Hindenburg even rejects Leibniz's conclusion explicitly when he points out in his sarcastic criticism, in which he equates Leibniz's combinatorial art with the riddle of the Sphinx and with an unsolvable labyrinth, that Leibniz's alleged improvements run in fact "counter the progress of the analytic method". Obviously, the methodological distinction explained above cannot be applied here to Hindenburg, since the target of his criticism is the Leibnizian science of *lingua characteristica* and *ars combinatoria* that, according to Leibniz himself, is a reality. It would be nonsense to say that Hindenburg criticized the science but believed in the ideal, since, according to Leibniz himself, Leibniz's four combinatorial rules and characteristic symbols (small and major numbers) embody that ideal. Hence, Hindenburg did not embrace Leibniz's ideals.

It is important to distinguish here between Leibniz's *ars combinatoria* and the theory of combinations. They are not the same thing. The theory of combinations is a scientific discipline, whereas Leibniz's *ars combinatoria* is a particular program, a personal interpretation of a scientific discipline. Thus, when Hindenburg claims that, in the work of Leibniz, the general idea of applying the theory of combinations to the field of analysis headed in the right direction, that does not mean that Leibniz's *ars combinatoria* was, for Hindenburg, the right way of understanding the theory of combinations. It means that, for Hindenburg, Leibniz could understand that the theory of combinations is an important scientific field, even though his interpretation of this field was useless and unintelligible. Instead of Leibniz's incomprehensible conception of the theory of combinations, Hindenburg proposes his own conception of this discipline, which evolved, as we have seen, independently of Leibniz's thought.

---

[12] Atque ita demum per Characteristicam ex Combinatoria arte, Algebra ei subordinata perficitur.

Hindenburg learned from his reading of Leibniz's paper on the theorem on reversion of series that the great philosopher of Leipzig had failed in his quest due to an obscure, abstruse conception of the theory of combinations, and he gained the confidence to embark on a more ambitious plan for his own combinatorial method and ideas. Hindenburg agreed with Leibniz in assuming that the theory of combinations, not Leibniz's *ars combinatoria*, is a fundamental branch of mathematics in the sense that its techniques can be expanded into other scientific areas if those techniques are correctly formulated, as he did in his *Infinitinomii*, but even this general assumption was the product of his own intellectual evolution, not of Leibniz's influence: as Hindenburg pointed out in his motivations for reading Leibniz's 1700 article, he wanted to compare his work on combinatorics and mathematical analysis with that of Leibniz, hence his general idea of justifying mathematical analysis by means of the theory of combinations was conceived before his discovery of Leibniz's combinatorial views and was already in operation when he read Leibniz's 1700 article. Going along with this general idea, Hindenburg closes the preface of his book with the statement that his combinatorial method is conducive to the progress of arithmetic, of common analysis, and of sublime analysis (Hindenburg 1779, XXI). Still and all, the scope of his treatise is restricted to the solution of some problems of the theory of series.

It is important to insist, however, on the fact that the development of Hindenburg's mathematical methods has no relation whatsoever to Leibniz's thought. As seen above, none of Hindenburg's mathematical works rely on any of Leibniz's mathematical results and, on the other hand, Hindenburg openly questions the viability of Leibniz's programs of *lingua characteristica* and *ars combinatoria*. He condemns those programs because of their conceptual obscurity and incomprehensible principles, assuming his own formulation and general conception of the theory of combinations as a better choice to use in the study of mathematical analysis. Nonetheless, Hindenburg extracts lines from Leibniz's paper on the theorem on reversion of series and from a letter of Leibniz (for the original letter, see (Leibniz 1705)), and places them into an epigraph to the mathematical tables Addendum of the second *Infinitinomii*. In those lines, Leibniz describes combinatorial tables as being a powerful analytical tool for the development of mathematics. This idea concurs with Hindenburg's view of mathematical tables, conveyed in his *Beschreibung einer ganz neuen Art*, and reinforced in his later books. Certainly, it would be nicer if Hindenburg had quoted Lambert instead of Leibniz, for it was Lambert's project that actually inspired Hindenburg's work on mathematical tables. This suggests that, in his choice of epigraph, Hindenburg was not so interested in being accurate with respect to his real sources and intellectual commitments as he was in underlining the pertinence of his methods by bringing up some passages where a great philosopher, such as Leibniz, had held a similar view concerning mathematical tables. One must not lose sight of the circumstantial nature of such an epigraph: it does not concern any of Leibniz's epistemological programs, since those programs have been dismissed by Hindenburg, it just draws attention to the adequacy of a given mathematical tool. Moreover, its circumstantial nature becomes still more evident when we take into account Hindenburg's treatment of Leibniz's thought before 1779. In the first *Infinitinomii* and the *Methodus nova*, there is hardly any mention of Leibniz's mathematical ideas. Perhaps the most important one occurs on paragraph XII of the first *Infinitinomii*, which remains unchanged in the 1779 edition (Hindenburg 1778a, 28–29; 1779, 28–29). There Hindenburg argues, somewhat loosely, that Leibniz was the original creator of formula 2.16 for calculating the multinomial coefficients. The aim pursued by Leibniz consisted in finding a formula such that, given any term of the multinomial expansion, "one could immediately determine the coefficient [of the term] without previously having calculated any table", phrase quoted by Hindenburg in 1778 and in 1779.[13] However, this phrase is not brought under the spotlight in the preface, or at the beginning of any section of the new edition of the *Infinitinomii* as an epigraph, since Hindenburg's method depends on computing combinatorial tables. Hindenburg does not even try to explain this apparent contradiction of Leibniz because he is not interested in Leibniz's ideas, but in his own. Just as an epigraph selected by a writer accomplishes the rhetorical effect of heightening the reader's empathy for the writer's subject, Hindenburg's epigraph intends to produce the rhetorical effect of heightening the reader's interest in mathematical tables. Here Hindenburg makes rhetorical use of Leibniz, as if to cover his own method of mathematical tables with the glory of an eminent philosopher, but really in order to gain sympathy for his views.

In sum, from a historical point of view, the 1779 edition of the *Infinitinomii* differs from the 1778 dissertations of Hindenburg in that it represents a radical restructuring in scope and style. Hindenburg's work evolved from research on a single theorem into research on the whole field of the theory of series. Moreover, Hindenburg is convinced that his method of powers has the potential to make a substantial contribution to the development of arithmetic, common analysis, and sublime analysis. Besides the actual applications of the method of powers, it is possible to identify a rhetorical use of Leibniz in the discourse of Hindenburg regarding the specific topic of mathematical tables. It is clear that Hindenburg discarded the general epistemological programs of Leibniz, such as *lingua characteristica* and *ars combinatoria*, and he decided to continue developing his own conception of the theory of combinations independently of Leibniz's ideas. On the other hand, none of Leibniz's mathematical results have played a relevant role, if any, in the evolution of Hindenburg's mathematical methods.

3.3   The German Combinatorial Analysis as a Research Project

On the occasion of his appointment as extraordinary professor of philosophy at the University of Leipzig, Hindenburg delivered a lecture, on 21 April 1781, called *Novi systematis permutationum combinationum ac variationum primas lineas et logisticae*

---

[13] (Hindenburg 1778a, 28, 1779, 28): "posset statim assignare numerum coefficientem, quem habere debet, sine vlla tabula iam calculata." For the original text of Leibniz, see footnote 14 in Chap. 2.

serierum formulis analytico-combinatoriis per tabulas exhibendae conspectum, which was also published that same year (Hindenburg 1781c). This booklet of thirty-two pages is composed of four paragraphs, which are preceded by a long epigraph by Leibniz and followed by an oath to Frederick Augustus III, Elector of Saxony. Another edition over three times longer, since it includes a preface, four additional paragraphs, and a final section of tables, appeared that same year under the imprint of other publisher (Hindenburg 1781b). There is no doubt that these are not two different drafts of the same piece, but a unique text from which Hindenburg took an excerpt that was presented before the faculty members and staff of the University of Leipzig. As the booklet does not differ in any respect from the integral text, all reference to this work of Hindenburg in this section will be to the long version.

One of the most striking features of the *Novi systematis* consists in the disparate assemblage of concepts and technical elements that Hindenburg borrowed from other mathematicians, but without actually attempting to bring all these concepts and elements into a coherent, unified whole. Another remarkable feature of the book, no less compelling than the previous one, is the lack of any new mathematical result, either in the field of combinatorics or in a field related to a subject matter previously studied by Hindenburg. Indeed, there has not been an attempt to address the question of developing a new combinatorial method, even though the title of the book promises to give us the general lines of a new system of combinations and permutations. The guidelines promised in the title are rather concerned with a research project for the foundation of mathematics on pure combinatorial grounds. It is not hard to identify four general spheres of the project: (1) a sphere concerning the conceptual delimitation of the new system, (2) a sphere concerning the elementary operations of the new system, (3) a sphere related to the development of mathematical symbolism, and (4) a work program concerning the potential applications of the system. The aim of the project is to establish a new mathematical theory, which was called by Hindenburg "combinatorial analysis", and later known as "German combinatorial analysis" in other countries. In this section, we present an account of the four areas of Hindenburg's project.

3.3.1   In Search of a Conceptual Background for the New System

Hindenburg claims that his new system has to be organized around two scientific disciplines: the art of combinations (*ars combinatoria*) and the logistic of series (*logistica serierum*), or the art of calculating series. The art of combinations is defined by Hindenburg as the discipline devoted to studying all the different ways in which different objects can be put together, transposed, or mixed, whereas the logistic of series is responsible for expressing in "analytic-combinatorial formulae" the results obtained by applying the art of combinations to the theory of series (Hindenburg 1781b, IV, XV). In addition, the system should be explicitly articulated using a simple combinatorial method, which is, certainly, Hindenburg's method of partition tables.

It should be noted that one of the main concerns of Hindenburg was to show that his project of basing mathematical analysis, or at least the theory of series, on combinatorics, was not unreasonable, since there are other cases involving comparable attempts to apply combinatorics to a large sector of science. It seems, in fact, that Hindenburg seeks to confer legitimacy on his project by recalling that renowned mathematicians had already explored in a preliminary way diverse areas of application of combinatorics. In this sense, he could say that his project continues research and development of an existing mathematical subject. However, his early sources on the doctrine of combinations are very scarce. In his dissertations of 1778, if Euler's work on partitions of numbers is not taken into account, Hindenburg only mentions the *Ars conjectandi* of Bernoulli and the *Elementa analyseos finitorum* of Segner. In 1781, there is no reference anymore to Segner's textbook. Instead, Hindenburg alludes to the *Ars conjectandi*, the *Dissertatio de arte combinatoria* of Leibniz, and the *Dissertatio inauguralis* of Nicolas Bernoulli (1687–1759). These three books share the same interest in using the doctrine of combinations as a tool to solve problems arising out of other branches of knowledge. In the fourth part of his treatise, Jacob Bernoulli proposes to apply his mathematical calculations concerning games of chance to decision problems that are prevalent in political, economic, and moral life (Bernoulli 1713, 210–239). His nephew Nicolas, who had received his education in combinatorics from his uncle Jacob, explores the possibility of applying the theory of probability to jurisprudence in his doctoral dissertation (Bernoulli 1709). Leibniz provides several examples of application of the art of combinations in different areas: in science, there are examples regarding metaphysics, physics, mathematics, and syllogistic; in art, there are examples regarding music and poetry; in human sciences, there are examples regarding jurisprudence and political philosophy (Leibniz 1666b). Although Hindenburg does not intend to treat such a wide range of applications, his project coincides with the idea of solving problems in different fields by means of combinatorial methods, even if the fields in which Hindenburg is interested are mathematical disciplines exclusively. Thus, Hindenburg presents those three books as a general theoretical framework for his own project. With certain reservations, Hindenburg also mentions the *Doctrine of Chances* of Moivre, but he does not place it in the same group because of its lack of applications outside the scope of games of chance (Hindenburg 1781b, XIV, footnote 1; de Moivre 1718).

In particular, Hindenburg considered Leibniz's *Dissertatio de arte combinatoria* and Bernoulli's *Ars conjectandi* to be the best examples available at his time of how the doctrine of combinations can be used to guide the development of human knowledge. The *Dissertatio de arte combinatoria* was an enlarged version of Leibniz's doctoral dissertation *Disputatio arithmetica de complexionibus* (Leibniz 1666a). In 1690, without Leibniz's consent, the *Dissertatio de arte combinatoria* was reprinted in Frankfurt and circulated widely. Leibniz complained about this unauthorized reprint in a letter addressed to André Morell, recognizing that several errors should be corrected before a new edition was brought out (Leibniz 1697). This never happened, but Leibniz reviewed some parts of his text in order to eliminate any error, and published his corrections in the *Acta eruditorum* (Leibniz 1691). More information about the history of the *Dissertatio de arte combinatoria* can be found, for example, in (Correia 2002; Knobloch & Berlin 1974; Serres 1982, 409 ff.). Apparently, Hindenburg did not know the letter of Leibniz, but he learned about Leibniz's negative opinion through Leibniz's paper on the theorem on reversion of series (Hindenburg 1781b, XIV, footnote n; Leibniz 1700, 208). On the other hand, it is well known that Bernoulli had completed the manuscript of the *Ars conjectandi* in 1688, but he never stopped trying to improve the fourth part of his book, convinced as he was that its content gave no completely satisfactory explanation of his applications of combinatorics and probability (Bernoulli 1975, 21 ff.; Sylla 2006). Hindenburg agreed with both Leibniz and Bernoulli that their texts were imperfect in various ways, emphasizing Leibniz's rejection of his own dissertation and Bernoulli's continuous dissatisfaction with his work. Indeed, Hindenburg attributed the failure of Leibniz and Bernoulli to lack of a sound comprehension of the basic principles involved in the doctrine of combinations, and claimed it was more much likely that his own project would be successful, for he had recently discovered the nature of those basic principles (Hindenburg 1781b, XVI):

> Thus, we have succeeded in reducing all questions to *the simplest laws of numeration*, which we use when writing down the *decimal, binary, quaternary, sexagesimal*, or any other *number system*. The explanation of this is so simple and obvious (since it is possible to write numbers *in order*, or *to skip a few according to a given progression* if necessary) that one might be surprised that neither Leibniz, nor Bernoulli, nor I myself at the moment when I wrote my book *Infinitinomii dignitatum exponentis indeterminati historia leges ac formulae* were able to understand it: *the foundation, as well as the proposed rules, of combinatorial operations consists solely in the way in which numbers are written* [...].[14]

Hindenburg is making reference here to his method for constructing number systems by means of combinatorial tables such as those shown in Fig. 3.1. As has been seen in Sect. 3.1, he rediscovers in 1776 the combinatorial nature of number systems by focusing on the positional properties of these systems. In his *Novi systematis*, Hindenburg understood that it would be helpful to look at the problem from a different perspective, namely: instead of supposing that the positional character of number systems and their combinatorial nature are two different aspects of numbers, one may suppose that both are identical. In that case, Hindenburg can hold the thesis that the foundation of the doctrine of combinations resides precisely in the most elementary arithmetic of natural numbers.

---

[14] Reduximus autem nunc totum negotium ad *simplicissimas numerandi Leges*, quibus utimur in *decadici, dyadici, tetradici, sexagesimalis*, et aliorum quorumque *Systematum numeris scribendis*; quae ratio tam est expedita, tamque facilis (nam numeros scribere *ex ordine*, atque etiam, ubi opus est, *per saltus progrediendo*, quilibet potest) ut mirum videri possit, neque LEIBNITIUM, neque BERNOULLIUM, neque me ipsum, cum scriberem de Infinitinomii Dignitatibus et Methodo Potentiarum libellum, vidisse: *Operationum combinatoriarum fundamentum et veluti normani propositani esse in numeros scribendi modo ac ratione unam fere atque unicam, h.e. praestantissimam* [...].

On the other hand, logistic of series is the result of applying the doctrine of combinations thus conceived to the theory of series. According to Hindenburg, once the theory of series has been rebuilt on this basis, it becomes natural to set common analysis in the same basis of combinatorics by using this theory to reinterpret the field of common analysis (Hindenburg 1781b, V). It is important to remember that, in 1779, the generalization of the combinatorial method of partitions to the method of powers persuaded Hindenburg into thinking that his method possesses the characteristics required to boost development in the areas of arithmetic, common analysis, and sublime analysis. At least on a conceptual level, Hindenburg postulates, in the *Novi systematis*, the principle that there exists an interdependence between the doctrine of combinations and arithmetic. So, in 1781, Hindenburg still believes in the positive contribution that his work offers to arithmetic and common analysis, but he says nothing about sublime analysis. One might ask then whether or not Hindenburg's new project is supposed to contribute toward improvements in sublime analysis. The answer to this question is not simple, and to begin to answer we must first consider an old debate about differential calculus.

In Chap. 2, we have seen that differential calculus (as well as the method of fluxions) was born surrounded by strong criticism concerning its concepts and methods. In France, one of the most severe opponents of differential calculus was Michel Rolle (1652–1719), who published in 1696 a bitter criticism of Leibniz's calculus in the *Journal de Savants*, though he did not use his real name (Rolle 1696; Blay 1986, 227–243; Costabel 1966; Mancosu 1989; Montucla 1758, 361–368). He argues that differential calculus does not contain any new element with regard to bringing mathematics into a fuller, greater, or better state; on the contrary, in all its artifice, it is in the end nothing but a cunning copy of elementary algebraic procedures. For Rolle, differential calculus coins "new terms" for describing old geometrical questions, such as that of determining the slope of a tangent at a given point, but behind these terms lie the assumptions, the concepts, the techniques of algebra, and nothing else. However, these words did not have the intended effect, and no one advanced arguments against Rolle's remarks. Instead of being discouraged, Rolle prepared to launch a new assault, which took place this time at the Royal Academy of Sciences in Paris.

By the end of the seventeenth century, the increasing dissemination of differential calculus in French mathematics had created an atmosphere of acute intellectual tension within the Royal Academy of Sciences. This tension manifested itself in struggles between two groups of the Academy, the first formed mainly by Pierre Varignon (1654–1722), the Marquis de l'Hospital (1661–1704), Nicolas Malebranche (1638–1715), and Louis Carré (1663–1711), and the second formed mainly by Jean Galloys (1632–1707), Philippe de La Hire (1640–1718), and Rolle. The first group is characterized by its support to Leibniz's calculus and its ontological commitment to the doctrine according to which there exists infinitesimal quantities, while the second group rejects the concepts and methods of differential calculus, especially the notion of infinitesimal quantity, and proclaims that the classical procedures of algebra created by Descartes and Fermat are adequate for dealing with any question about both geometric and mechanical curves. The conflict between the two groups finally broke out on 17 July 1700 when Rolle delivered a discourse against differential calculus to the members of the Academy, which marked the beginning of the debate between Rolle and Varignon. The Academy imposed a duty on the parties to keep confidential the discussion, but Varignon informed Leibniz and Johann Bernoulli about what was going on, sent them the reports of the debate, and asked them not to spread the information (Bernoulli 1988; Leibniz 1859, 97 ff.).

Although Rolle's speech was not recorded in the *Procès-Verbaux* of the Academy, Varignon gave a precise summary of its content in his reply, delivered during the meeting of the Academy held on 11 August 1700, and, some years later, Rolle published a recapitulation of his objections (Rolle 1703; Varignon 1700). On a conceptual level, Rolle criticized the incoherence and nonsense of the concepts and assumptions of differential calculus in a very similar manner as Berkeley will do thirty years later, pointing out that:

1. Differential calculus not only postulates the existence of infinitely small quantities and infinitely large quantities, but also postulates the existence of an infinite hierarchy of these quantities, where some are infinitely smaller or infinitely larger than others.
2. According to differential calculus, a given quantity remains unchanged after adding or subtracting an infinitesimal difference.
3. In differential calculus, the concept of infinitesimal difference is used incongruously, given that sometimes it means zero, and sometimes it means a nonzero quantity.

Point 1 addresses the notion of higher-order derivative, putting in sharp relief the problem of conceptualizing a quantity that is infinitely small (or large) with respect to a quantity that is already infinitely small (or large). Points 2 and 3 cast serious doubts on the validity of neglecting the instantaneous rate of change of a variable, procedure frequently used at that time, under the pretext of its infinitesimal nature, or under the pretext that one can arbitrarily set it equal to zero.

On a practical level, Rolle's attack consists in showing, by means of examples, that the strict application of differential calculus's rules leads to error. Let's look at one of these examples (Rolle 1703; Costabel 1966, 20–22; Mancosu 1989, 233–234; Montucla 1758, 362–363). One remarkable feature of Leibniz's new calculus is that it can be helpful for solving *maxima* and *minima* problems, but Rolle profoundly disagrees. He proposes to apply the rules of differential calculus to the following curve:

$$y = 2 + \sqrt{4x} + \sqrt{4 + 2x} \tag{3.9}$$

in order to find its eventual *maximum*. By using the rule given by l'Hopital in paragraph 47, section III, of the *Analyse des infiniment petits*, he gets the derivative:

$$dy = \frac{dx\sqrt{x} + dx\sqrt{4 + 2x}}{\sqrt{4x + 2x^2}}$$

Then, always following l'Hopital's treatise, one should put $dy = 0$ which gives us a *maximum* of the curve at $x = -4$. However, when $x = -4$, the value of y is an imaginary number, and therefore the curve has an imaginary *maximum*, which is absurd. According to Rolle, there is nothing in l'Hopital's book to prevent us from being drawn into this kind of fallacious conclusions. In contrast, Rolle claims that algebra provides a method to find the correct answer to this problem. He first writes Eq. 3.9 in the form:

$$y^4 - 8y^3 + 16y^2 - 12xy^2 + 48xy - 64x + 4x^2 = 0, \tag{3.10}$$

and then he uses Hudde's rule to find a *maximum* at the point $y = x = 2$. Moreover, he points out that if one applies differential calculus to Eq. 3.10, one gets the same "correct" result as one did by using the algebraic method of Hudde. Therefore, differential calculus leads not only to error, but also to inconsistent results. It was not hard for Varignon to show that Rolle was using Hudde's rule incorrectly, since $(2, 2)$ is not a *maximum* point but an intersection point of two branches of the solutions of Eq. 3.10.

At the end of 1701, the Academy tried to put an end to this controversy, and set up a scientific committee to review the arguments advanced respectively by Rolle and Varignon. Nevertheless, the committee could not reach any conclusion, and the intellectual confrontation between Rolle and Varignon ended abruptly in a climate of restlessness and dissatisfaction. As a consequence of this lack of conclusion, other mathematicians, whether members of the Academy or not, became involved in the controversy, which crossed the borders of French mathematics. On the one hand, the Abbé Thomas Gouye (1650–1725) continued to attack and discredit differential calculus as an unreliable theory in the *Journal de Trévoux*, and Leibniz himself fought back in the same scientific journal. On the other hand, Rolle resumed his line of argumentation in the *Journal de Savants*, with Joseph Saurin (1659–1737) as an occasional interlocutor. Let's examine how Rolle completed the argument given above against differential calculus.

In 1702, Rolle publishes a paper on the problem of finding the tangents of a curve at a multiple point (Rolle 1702). Even if he does not explicitly mention differential calculus, his text seems to be challenging the advocates of this new theory to figure out ways to solve this kind of problems by means of purely differential methods. On the contrary, he claims to have found a new algebraic method capable of solving these problems of tangents. He uses examples to explain his method, particularly the example of the curve represented by implicit Eq. 3.10. Figure 3.6 shows the way Rolle drew this curve, which is not an accurate representation since the branches of the curve should continue in fact beyond the point $G(2, 2)$. This lack of accuracy could explain his former mistake of thinking that G was a *maximum*. However, this time he is interested in finding algebraically the tangents to this curve at the double point G. To this end, he puts $y = OH$, $x = HF$, $zn = FE$,

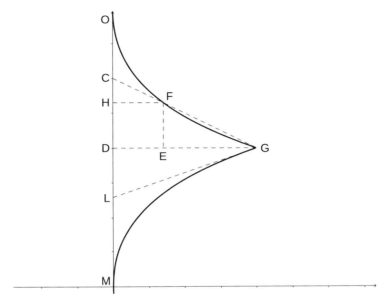

Fig. 3.6 Rolle's determination of a tangent at a double point

and $vn = EG$. Then he substitutes the value $vn + x$ for x and the value $zn + y$ for y in Eq. 3.10 to obtain a polynomial in n:

$$0 = z^4 n^4 + \left(4yz^3 - 8z^3 - 12vz^2\right) n^3 + \left(\begin{matrix} 6y^2z^2 - 24yz^2 - 24yvz \\ -12xz^2 + 16z^2 + 48vz + 4v^2 \end{matrix}\right) n^2$$

$$+ \left(\begin{matrix} 4y^3z - 24y^2z - 12y^2v - 24yxz \\ +32yz + 48yv + 48xz + 8xv - 64v \end{matrix}\right) n + A.$$

where $A = y^4 - 8y^3 + 16y^2 - 12xy^2 + 48xy - 64x + 4x^2$. Rolle assumes, without giving any proof or explanation, that the tangents at G can be determined by one of the coefficients of this polynomial. To find out which one, each coefficient has to be evaluated for the values $x = 2$ and $y = 2$, which are the coordinates of the point G. By this evaluation, several terms will cancel because several coefficients will be equal to zero, and the lowest-degree term that does not cancel gives us the coefficient that determines the tangents. In this case, after substitution, the coefficient of the quadratic term does not cancel and is equal to $\left(4v^2 - 32z^2\right)$. Then Rolle replaces v by $v = x = 2$ in this coefficient to get $16 - 32z^2$. By putting $16 - 32z^2 = 0$, he finds the subtangent $DL = z = \frac{1}{\sqrt{2}}$, which gives the slope $\frac{1}{\sqrt{8}}$ of the tangent line GL. Naturally, the slope $-\frac{1}{\sqrt{8}}$ of the second tangent can also be inferred from the subtangent.

In the same year of 1702, Saurin criticizes Rolle's paper (Saurin 1702). First, he emphasizes the inaccuracy of Rolle's graph shown in Fig. 3.6, recalling that this is only a representation of a small part of the curve. Although this is true, the partial representation of the curve does not really disrupt Rolle's argumentation on the problem of tangents. More importantly, Saurin accuses Rolle of cheating by presenting his solution as a new algebraic method, whereas the truth is that Rolle tried to hide the differential nature of his reasoning under the guise of algebraic expressions. Saurin argues that when Rolle replaces y by $zn + y$ and x by $vn + x$ in Eq. 3.10, Rolle's method consists in taking infinitesimal increments on both variables, so the expressions vn and zn are in fact algebraic disguises for the infinitesimal increments dx and dy. In consequence, this algebraic substitution tries to hide the fact that, by adding an increment to each variable, the point F of the curve in Fig. 3.6 gets closer to the point G, and thus the secant line GC will eventually become a tangent line at the point G, which is the central idea of differential calculus. Finally, Saurin explains how to calculate the slope of the tangents at G using the principles of differential calculus strictly, at least those given by l'Hopital. He takes the sum of the partial derivatives of Eq. 3.10, which he writes as follows:

$$4y^3dy - 24y^2dy - 12y^2dx - 24xydy + 32ydy + 48ydx + 48xdy + 8xdx - 64dx = 0$$

From this, he deduces the following relation:

$$\frac{dy}{dx} = \frac{3y^2 - 12y - 2x + 16}{y^3 - 6y^2 + 8y - 6xy + 12x},$$

which assumes an indeterminate form $\frac{dy}{dx} = \frac{0}{0}$ when $y = x = 2$. Thus, he applies l'Hopital's rule given in the *Analyse des infiniment petits* (de l'Hopital 1696, 145), taking the sum of the partial derivatives in the numerator and denominator:

$$\frac{dy}{dx} = \frac{6ydy - 12dy - 2dx}{3y^2dy - 12ydy + 8dy - 6xdy - 6xdx + 12dx}$$

which gives $\frac{dy}{dx} = \frac{dx}{8dy}$ for $y = x = 2$, i.e. $\left(\frac{dy}{dx}\right)^2 = \frac{1}{8}$. Saurin invokes then the principle of sectioin 2, article 9, of the *Analyse des infiniment petits* (de l'Hopital 1696, 25), according to which the subtangent of the curve is given by the quotient $\frac{xdy}{dx}$. It follows that the subtangent at G is $DL = \frac{1}{\sqrt{2}}$, as expected.

The constant accusations of disguising algebraic methods as differential methods, and vice versa, will cause the discussion to get bogged down in futile exchanges of contradictory positions. The debate between Rolle and Saurin will last about five years without reaching agreement. In 1706, the French Academy intervened again in the conflict and set up a new scientific committee, and again no scientific conclusion was reached, but the committee called for restraint and moderation on both sides. This institutional decision was called "*la Paix des infiniment petits*", the Peace of infinitesimals, by Bernard de Fontenelle (1657–1757) (de Fontenelle 1719, 122), who marginally participated in the war of infinitesimals by delivering, in 1704, an eulogy in honor of l'Hopital and his contributions to differential calculus (de Fontenelle 1704).

Sixty-three years after the Peace of infinitesimals, Borz presented a dissertation on the occasion of his appointment as ordinary professor at the University of Leipzig, entitled *De rationibus regularum, quas calculus differentialis in cunstituendis punctis curvarum multiplicibus, et subtangentibus in iis ad haec puncta ducendis offert*, and Hindenburg was entrusted with the duty of making a comment on his teacher's dissertation (Borz 1769). Notwithstanding the time elapsed and the progressive consolidation of differential calculus, Borz's dissertation is explicitly framed within the debate started by Rolle about the determination of multiple points and tangents at these points. In his dissertation, Borz makes direct reference to Saurin, to a 1704 report where Johann Bernoulli summarizes Saurin's response to Rolle's arguments exposed above, and to Cramer's book *Introduction à l'analyse de lignes courbes algébriques*, where Cramer analyzes Eq. 3.10 in a chapter on multiple points (Bernoulli 1742, 403–405; Cramer 1750, 417–418). The dissertation begins with the exposition of Saurin's calculation of the subtangent DL at the point G explained above, which Borz took from Bernoulli's report. Borz defines the objectives of his dissertation in accordance with this case, which are threefold and consist in answering the following questions (Borz 1769, V):

1. Why does the quotient $\frac{dy}{dx}$ calculated by Saurin in order to determine the subtangent DL assume an indeterminate form?
2. Why does the indeterminate form disappear after performing a subsequent differentiation according to l'Hopital's rule?
3. Why is l'Hopital's formula $\frac{xdy}{dx}$ for calculating the subtangent of a curve still valid after performing a second differentiation?

Indeed Borz generalizes these questions to other case studies, and so his general question consists in understanding the theoretical basis of differential calculus's methods.

Borz arrived at conclusions that were not really new for the time. The answers to his three questions can be summarized in the fact that the multiplicity of a root of a polynomial decreases one unit after differentiation. It is more interesting to consider the way in which he argues. Concerning Eq. 3.10, he uses Hudde's rule to elucidate the previous questions, emphasizing the algebraic nature of the method. First, he writes Eq. 3.10 as a polynomial in y and multiplies the result by a given arithmetic progression as follows:

$$
\begin{array}{cccccc}
y^4 & -8y^3 & -(12x-16)y^2 & +48xy & \underbrace{+4x^2-64x} & = & 0 \\[2mm]
\frac{4}{y} & \frac{3}{y} & \frac{2}{y} & \frac{1}{y} & \frac{0}{y} & & \\[2mm]
\hline
4y^3 & -24y^2 & -(12x-16)2y & +48x & & = & 0
\end{array}
$$

Then he writes Eq. 3.10 as a polynomial in x and multiplies the result by a given arithmetic progression as follows:

$$4x^2 \qquad +(-12y^2 + 48y - 64)x \qquad \underbrace{+y^4 - 8y^3 - 16y^2} \qquad = \qquad 0$$

$$\frac{2}{x} \qquad\qquad \frac{1}{x} \qquad\qquad \frac{0}{x}$$

$$8x \qquad +(-12y^2 + 48y - 64) \qquad\qquad\qquad = \qquad 0$$

If one sets $y = x = 2$, then the expressions $4y^3 - 24y^2 - (12x - 16)2y + 48x$ and $8x - 12y^2 + 48y - 64$ are equal to zero. Therefore, the quotient $\frac{dy}{dx}$ calculated by Saurin will assume an indeterminate form. By repeating this procedure for both $4y^3 - 24y^2 - (12x - 16)2y + 48x$ and $8x - 12y^2 + 48y - 64$, one obtains polynomials that do not admit $y = x = 2$ as a root, which means that $(2, 2)$ is a double point, and from there one derives obviously the same value of the subtangent DL calculated by Saurin.

Borz also explains what he considers to be the reasoning process behind differentiation of a given polynomial $x^m + ax^{m-1} + bx^{m-2}y^t - cy^{t-1} \ldots = 0$. To find the partial derivative with respect to x, he writes a polynomial in x and calculates the result as follows:

$$x^m \qquad\qquad +ax^{m-1} \qquad\qquad +by^t x^{m-2} \qquad \underbrace{-cy^{t-1}} \ldots \; = \; 0$$

$$\frac{mdx}{x} \qquad\qquad \frac{(m-1)dx}{x} \qquad\qquad \frac{(m-2)dx}{x} \qquad\qquad 0$$

$$mx^{m-1}dx \quad +(m-1)ax^{m-2}dx \quad +(m-2)by^t x^{m-3}dx \qquad\qquad = \; 0$$

The same procedure is used to find the partial derivative with respect to y:

$$bx^{m-2}y^t \qquad\qquad -cy^{t-1} \qquad\qquad \underbrace{+x^m + ax^{m-1}} \ldots \; = \; 0$$

$$\frac{tdy}{y} \qquad\qquad \frac{(t-1)dy}{y} \qquad\qquad 0$$

$$tbx^{m-2}y^{t-1}dy \quad -(t-1)cy^{t-2}dy \qquad\qquad = \; 0$$

It goes without saying that there is an astonishing parallelism between the algebraic method and the differential method from Borz's point of view.

Because of this parallelism, Borz concludes that algebra can accomplish the same tasks as would result from differential calculus (Borz 1769, XXII):

As the exponents of variables in a well-ordered equation are indeed in arithmetic progression, the differentials of the terms of the equation are then determined by multiplying these terms by the sequence of exponents, each of which is multiplied by the differential variable and divided by the variable: by means of this operation, one gets exactly the same result as that derived from Hudde's rule.[15]

At the beginning of the eighteenth century, the war of infinitesimals was brought to an end in France by the intervention of the Academy of sciences, but the scientific question of the legitimacy of differential calculus was left unsettled. The analysis carried out in his dissertation enabled Borz to form an opinion on this matter, which turned out to be more conciliatory than one could imagine. For him, differential calculus is a legitimate branch of mathematics, but algebra has appropriate tools for solving the same kind of problems with which differential calculus is concerned. It is worth mentioning that Borz was the first mathematician to deliver a speech on differential calculus in Leipzig (Krause 2003, 155), and that he was considered a specialist on this field of mathematics. Furthermore, his dissertation was very well received in German academic circles. For instance, in 1771, Kästner described it as an "erudite" inquiry into the question of multiple points, and even at a much later date, in 1831, it was still considered as a main reference source for this subject in the monumental work *Mathematisches Wörterbuch* (Kästner 1771, 81, footnote; Klügel et al. 1831, 868). Thus, in German academic circles, Borz's point of view about the relation between algebraic and differential methods was considered to be at least reasonable.

Although Hindenburg's reply to Borz's dissertation is lost, it is beyond doubt that the dissertation of his teacher not only provided him with a rich overview of the controversies surrounding differential calculus, but also gave him a certain understanding of algebraic alternative methods for solving problems that differential calculus is meant to solve. It seems plausible, then, to think that Hindenburg is willing to explore, in his 1781 project, algebraic procedures in order to rebuild sublime analysis. This could explain the absence of reference to sublime analysis in his *Novi systematis*. In effect, if differential techniques can be successfully replaced by algebraic procedures, as suggested by his former teacher, the field of sublime analysis can be seen as the theory of series free of all the conceptual difficulties involved in differential calculus. If one adds to this the fact that Hindenburg is convinced that his combinatorial method will contribute with new elements to develop algebra and the theory of series, one can understand that Hindenburg's project has been conceived as a promising reformulation of sublime analysis.

Unlike what Hindenburg intended to achieve with his method of powers in 1779, the conceptual delimitation of his new project implies a comprehensive restructuring of the entire organization of mathematics, and for this a solid theoretical foundation is needed. Hindenburg believes that this foundation can be supplied by his interpretation of the doctrine of combinations.

---

[15] Cum vero, vt iam monuimus, indices variabilium in bene ordinata aequatione progrediantur in progressione arithmetica; differentialia vero earundem prodeant multiplicando eas per seriem indicum, quorum singuli ducantur in variabilis differentiale diuisum per variabilem: idem sane efficitur hac operatione, quod praestat regula Huddenii.

3.3.2   The Combinatorial Operations of the New System

This area of Hindenburg's project deals with the notions of permutation, combination, and variation. Hindenburg's definition of these notions is much the same as today. Two types of problems are intended to be treated in later developments of the project. On the one hand, the aim is to develop methods, generally mathematical tables, with which permutations (with or without repetitions), combinations (with or without repetitions), and variations (with or without repetitions) can be actually enumerated. For instance, to enumerate all permutations of four distinct objects a, b, c, and d, Hindenburg suggests to write down these permutations in the following way (Hindenburg 1781b, XVII):

$abcd$	$bacd$	$cabd$	$dabc$
1234	2134	3124	4123
$\cdot\cdot 43$	$\cdot\cdot 43$	$\cdot\cdot 42$	$\cdot\cdot 32$
$\cdot 324$	$\cdot 314$	$\cdot 214$	$\cdot 213$
$\cdot\cdot 42$	$\cdot\cdot 41$	$\cdot\cdot 41$	$\cdot\cdot 31$
$\cdot 423$	$\cdot 413$	$\cdot 412$	$\cdot 312$
$\cdot\cdot 32$	$\cdot\cdot 31$	$\cdot\cdot 21$	$\cdot\cdot 21$
$adcb$	$bdca$	$cdba$	$dcba$

As shown in this table, the central idea of this enumeration is that the better way to represent a permutation of some given objects consists in associating natural numbers to those objects in order to express the permutation by means of numbers. This is the same idea behind Hindenburg's substitution rule 3.6, idea upon which his method of powers and his method of partitions were built. Combinations and variations are treated in a similar manner, that is to say, a substitution rule should be established first, and then one works with numbers, which are called "numeric elements" of a combination (variation) by Hindenburg.

The second type of problems stems from the question of counting the number of permutations, combinations, and variations, and the aim is to find formulae for a quick calculation of these quantities. In his 1779 *Infinitinomii*, Hindenburg was also interested in several calculating problems concerning partitions of numbers, but it is not clear whether these problems fall within the scope of his new project. In fact, all these calculating problems did not get very systematic treatment in Hindenburg's *Novi systematis*. He points out that fomula 2.16 gives the number of permutation with repetition, formula 3.3 gives the number of combinations with repetition, formula:

$$\frac{n(n-1)(n-2)\cdots(n-m+1)}{1\cdot 2\cdot 3\cdots m}$$

gives the number of combinations without repetition, and the number of variations with repetition of n objects taken m at a time is n^m. Hindenburg takes all these formulae from the *Ars conjectandi* of Bernoulli, and all of them, with the exception of the formula for calculating the number of variations, were already quoted in his 1779 *Infinitinomii*.

3.3.3 Creation of a New Mathematical Symbolism

Definitely, the most developed area of the project, as it was presented in the *Novi systematis*, is that related to the creation of a new mathematical symbolism. One can classify here Hindenburg's new mathematical symbols into three categories: (1) mathematical notations for series, (2) mathematical notations for coefficients (of series), and (3) mathematical notations for the doctrine of combinations.

1. Mathematical notations for series.
   a. Hindenburg uses the last letters of the alphabet (whether Greek or Latin: $p, q, r, \ldots$, and $\pi, \rho, \sigma, \ldots$) to denote mathematical series. For instance, the letter z can be used to denote the following series:

   $$\alpha y^m + \beta y^{m+d} + \gamma y^{m+2d} + etc = z.$$

   b. Expressions like tn and $\daleth n$ denote the nth term of a given series. For example, $zt3$ and $z\daleth 3$ are the third term γy^{m+2d} of series z given above. Although both expressions are equivalent, $\daleth n$ is used more commonly than tn by the combinatorial school and by Hindenburg himself. The symbol $\daleth$ is an approximate representation of the symbol used by Hindenburg, which is not always printed in exactly the same way, even on the same page, because of the typographical limitations of the time (Hindenburg 1781b, XXXIII).
   c. An expression like $zx(n + 1)$ denotes just the coefficient of the $(n + 1)$th term of series z. For instance, $zx(3)$ is the coefficient γ of the term γy^{m+2d}.
2. Mathematical notations for coefficients. In the *Novi systematis*, there is a significant proliferation of notations for coefficients, which can be organized in three groups: (a) general notations for coefficients, (b) notations for coefficients according to their place in a series, and (c) notations for specific coefficients.
   a. General notations for coefficients. Coefficients can be seen as determined, i.e. one knows the value of the coefficient, or undetermined, i.e. one does not know the value of the coefficient. Following the Newtonian tradition, determined and undetermined coefficients are represented by letters.
      i. Determined coefficients are represented by lowercase or capital letters of any alphabet, and one can use any kind of typography:

      $$A, B, C, \ldots; a, b, c, \ldots; \mathfrak{A}, \mathfrak{B}, \mathfrak{C}, \ldots; \mathfrak{a}, \mathfrak{b}, \mathfrak{c}, \ldots; \text{ etc.}$$

ii. Undetermined coefficients differ from determined coefficients in that they present a point over the letter:

$$\dot{A}, \dot{B}, \dot{C}, \ldots, \text{ etc.}$$

b. Notations for coefficients according to their place in a series.

   i. Hindenburg gives the following example:

$$11x + 12x^2 + 13x^3 + 14x^4 + etc = p$$

$$21x + 22x^2 + 23x^3 + 24x^4 + etc = q$$

$$31x + 32x^2 + 33x^3 + 44x^4 + etc = r$$

The digits to the left of the variable are indeed indexes. This notation is supposed to be used in systems of equations, where the first digit of an index corresponds to the row in the system, and the second digit corresponds to the column. Thus, the coefficients are expressed here by means of their indexes alone. Hindenburg takes this notation from (Leibniz 1768). Leibniz used this notation in the context of systems of linear equations (Knobloch 2001). Despite the inherent advantages of this Leibnizian notation, neither Hindenburg nor the combinatorial school will ever use it.

   ii. Expressions like $c^{(n)}$, $C^{(n)}$, $\gamma^{(n)}$, and $\Gamma^{(n)}$ represent the coefficient of the term in x^n of a given series. In fact, these are Kästner's notations $(\lambda n)\, x^n$ and $(Ln)\, x^n$ for coefficients, which were employed by Hindenburg in his *Methodus nova*, and respectively replaced later by $c^{(n)}$ and $C^{(n)}$ in his 1779 *Infinitinomi* (Hindenburg 1779, 69).

   iii. A coefficient can be designated by means of its place in a series with respect to a coefficient used as reference. If c is a coefficient chosen as reference, all the coefficients of a series can be written using c and a number over c that indicates the place in the series with respect to c. For example, given the polynomial $ax + bx^2 + cx^3 + dx^4$, one can write $\overset{0}{c} = c$, $\overset{-2}{c} = a$, $\overset{-1}{c} = b$, and $\overset{1}{c} = d$.

c. Notations for specific coefficients.

   i. The binomial coefficients are represented by Gothic capital letters with exponents to the left:

$$^m\mathfrak{A} = m, \qquad ^m\mathfrak{B} = \frac{m\,(m-1)}{1.2}, \qquad ^m\mathfrak{C} = \frac{m\,(m-1)\,(m-2)}{1.2.3}, \ \ldots$$

In the *Infinitinomi*, these coefficients do not have any exponent, but are represented by Gothic capital letters.

   ii. The multinomial coefficients are represented by Gothic lowercase letters $\mathfrak{a}$, $\mathfrak{b}$, $\mathfrak{c}$, $\ldots$, as in the *Infinitinomi*.

3. Mathematical notations for the doctrine of combinations.
   a. General notations for permutations, combinations, and variations.
      i. $P(1, 2, 3, \ldots)$, or $P(a, b, c, \ldots)$, means all permutations of the numeric elements 1, 2, 3, … (or of the objects $a, b, c, \ldots$).
      ii. $C(1, 2, 3, \ldots)$, or $C(a, b, c, \ldots)$, means all combinations of the numeric elements 1, 2, 3, … (or of the objects $a, b, c, \ldots$).

      $Con2nationes(1, 2, 3, \ldots)$ means the combinations of the numeric elements 1, 2, 3, … taken two at a time.

      $Con3nationes(1, 2, 3, \ldots)$ means the combinations of the numeric elements 1, 2, 3, … taken three at a time.

      In general, $Con(k)nationes(1, 2, 3, \ldots)$ means the combinations of the numeric elements 1, 2, 3, … taken k at a time.
      iii. $V(1, 2, 3, \ldots)$ means all variations of the numeric elements 1, 2, 3, ….
   b. Classes of combinations and variations.
      i. Capital letters in Roman font with an apostrophe to the left $'A$, $'B$, $'C$, … represent respectively the combinations of the numeric elements 1, 2, 3, … taken one at a time, two at a time, three at a time, and so on. Hindenburg calls them "classes of simple combinations" (*classes complexionum simpliciter*). For example, if one considers all combinations of the elements 1, 2, and 3, i.e. if one considers $C(1, 2, 3)$, then $'A = \{1, 2, 3\}$, $'B = \{12, 13, 23\}$ and $'C = \{123\}$.
      ii. Capital letters in Roman font with an exponent to the left ^nA, ^nB, ^nC, … represent the classes of partitions that we already know, and that are called "classes of combinations" (*classes complexionum*) by Hindenburg in his *Novi systematis*. Figure 3.5 shows an example of these classes.
      iii. Capital letters in italic font with an apostrophe to the left $'A$, $'B$, $'C$, … are the equivalent classes for variations with respect to the classes $'A$, $'B$, $'C$, … of combinations explained above. Hindenburg calls them "classes of simple variations" (*classes variationum simpliciter*).
      iv. Capital letters in italic font with an exponent to the left ^nA, ^nB, ^nC, … are the equivalent classes for variations with respect to the classes ^nA, ^nB, ^nC, … of combinations explained above. Hindenburg calls them "classes of variations" (*classes variationum*).

These are the mathematical notations that Hindenburg invented during the period from 1778 to 1781. Some of them will completely disappear, as the Leibnizian notation for the coefficients of a system of equations, others will be modified, and others will remain the same. Furthermore, due to typographic limitations, the look of the symbols is not always the same from one book to another. In any case, this area of Hindenburg's project will stir up mixed feelings in the mathematical community. It will become a source of inspiration for many, but it will draw a wave of criticism as well, not only from mathematicians of the time, but also from mathematicians of other periods. For instance, Cajori was a critic of the excessive proliferation of new mathematical notations in the work of Hindenburg, and perhaps this is the reason why he did not take into account Hindenburg's notations for combinations and permutations in his history of mathematical notations. The earliest source he cites for permutations and combinations is (Ohm 1829b, 33–38), where one can find the expression $\underset{(a,b,c,\ldots)}{\overset{k}{C}}$ for the combinations of a, b, c, ... taken k at a time, and the expression $\overset{n}{P}(a,b,c,\ldots)$ for the permutations of n objects a, b, c, ...(Cajori 1929, 79–80). It is evident that Ohm's notation takes its source from the work of Hindenburg, and this becomes more evident when taking into consideration the fact that Martin Ohm (1792–1872) was a student of Rothe, who was a student of Hindenburg and one of the principal members of the combinatorial school. For a study of Ohm's work in the context of the combinatorial school and the German algebraic analysis, see (Jahnke 1987; 1993).

3.3.4 Hindenburg's Scientific Work Program

Hindenburg enumerates ten types of problems pertaining to the theory of series which are especially well suited for the logistic of series. The aim consists in designing optimal combinatorial tables concerning these problems, which will serve as a basis for the creation of "analytic-combinatorial formulae", i.e. formulae written by means of the new combinatorial symbols of the system. Apart from the first two types of problems, for which examples of tables are offered at the end of the *Novi systematis*, Hindenburg does not give explicit solutions for those problems. His list must therefore be understood as an agenda of relevant issues and research questions in the logistic of series. Here is Hindenburg's agenda:

1. Products of series: the objective is to calculate the coefficients of the product using combinatorial tables.
2. Divisions of series: the objective is to calculate the coefficients of the quotient of two series using combinatorial tables.
3. Powers of series: the objective is to calculate the coefficients of any power of a given multinomial.
4. Series composition. Given two series such as:

$$p = ay^\alpha + by^\beta + cy^\gamma + \cdots$$
$$y = Az^\mu + Bz^{\mu+\Delta} + Cz^{\mu+2\Delta} + \cdots$$

   to find the coefficients of the series obtained by replacing the variable y of the series p by the series y. The nature of the exponents of series p and y is not specified.
5. Solutions of systems of linear equations. In fact, it seems that the linear equations of the system could eventually have an infinite number of terms, i.e. the system could involve an infinite set of variables, according to Hindenburg.

6. Elimination of "irrational functions" (i.e., roots or terms raised to a fractional power) that appear in a given series.
7. Series transformations. For instance, factorization.
8. Series interpolation.
9. The question of reversion of series.
10. Transcendental functions: trigonometric functions, the logarithm, etc.

While this is not an exhaustive list, it reveals the breadth of possible applications of the doctrine of combinations in the field of the theory of series. This suggests indeed that all questions concerning series fall into the scope of the program. In consequence, if the project were successful, the theory of series, which became the core of mathematical analysis with Euler's work, would be replaced by the logistic of series. It is worth noting that Hindenburg's work program includes all those problems of series that were faced by Newton in his *Epistola prior* and in his *Epistola posterior*, and that were solved using the binomial theorem (for the list of Newton's problems, see the end of Sect. 2.1.1). On the other hand, it seems that the inclusion of the question about the solution of a system of linear equations in Hindenburg's program was motivated by Leibniz's paper *Monitum de characteribus algebraicis* (Leibniz 1768). It is also worth noting that the multinomial theorem plays no role at all in Hindenburg's project, even if it is related to point (3). In fact, Hindenburg does not even mention it in his text. The multinomial theorem is just one proposition among others in the theory of series.

*

In short, the German combinatorial analysis emerges as a research project in 1781. From a historical perspective, this project arises from the proof of the multinomial theorem given by Hindenburg in his *Methodus nova*. In 1779, the method employed in this proof becomes a general method for solving a certain amount of problems in the theory of series. In 1781, Hindenburg is convinced that this general method is the key to succeed in achieving a new mathematical system. The different parts of the system reveal the foundational character of the project. The controversies surrounding differential calculus could explain Hindenburg's choice concerning the algebraic nature of his foundational project.

The Consolidation of the German Combinatorial Analysis

4

4.1 The Reversion of Series and the Rise of the German Combinatorial School

Hindenburg's combinatorial analysis was characterized in 1781 as a new branch of mathematics, whose explicit aim was to reformulate the theory of series and place it on what Hindenburg considered to be a firmer theoretical footing. However, in that year, this new branch of mathematics appears only in the form of a project. Many problems were left unsolved in the *Novi systematis*, and will remain unsolved for a long time. It will be necessary to wait until the 1790s before it could be said that the project outlined by Hindenburg has been launched. The decade between the drafting of the project and its implementation corresponds to a period during which more favorable conditions for developing Hindenburg's combinatorial analysis will be created. The project had, then, a long gestation period during which Hindenburg carefully disseminated his ideas among German academic circles through education and through his work as science editor. This meticulous endeavor to spread his ideas will finally start to bear fruit around 1794, when a particular community of mathematicians begins to form, gathered for the shared purpose of developing Hindenburg's combinatorial analysis. This group of mathematicians will be known under the name of "the German combinatorial school", or simply "the combinatorial school". The interest shown in the theorem on reversion of series will play a central role in the formation of the combinatorial school by motivating at the beginning the necessary cohesion in the group.

In this section, first a description of Hindenburg's editing work is presented, which should allow us to understand how Hindenburg managed to create a public space for discussing and promoting his ideas. Then, the theorem on reversion of series is analyzed in the context of a Lagrange's paper and the discussion of this question by the early members of the combinatorial school. Finally, the question of the relation between the theorem on reversion of series and the incipient formation of a working group is addressed.

4.1.1 Hindenburg, Science Editor

The Faculty of Philosophy of the University of Leipzig was composed of nine sections: Latin and Greek, History, Rhetoric, Theoretical Philosophy, Practical Philosophy, Political Philosophy, Physics, Mathematics, and History of Auxiliary Sciences (Gretschel 1830, 93–110). There were two categories of professors, that of ordinary professor and that of extraordinary professor, besides the nonofficial category of Privatdozent. Hindenburg spent all his professional life at this Faculty, mainly in the section of Physics. In 1780, his former teacher Borz was appointed rector of the University, and perhaps this helped Hindenburg to become, after teaching for ten years as Privatdozent, an extraordinary professor in 1781. From then on, Hindenburg's career progressed at a rapid pace, thanks to his innovative work in mathematics, certainly, but also to his capacity to develop strong professional relationships with other members of the Faculty. In particular, his relationship with Nathanael Gottfried Leske (1751–1786) will have significant positive consequences for him.

Leske was appointed extraordinary professor of natural history in 1775, and became professor of economic sciences three years later. In 1779, he tried his luck as editor and founded the scientific journal *Abhandlungen zur Naturgeschichte, Physik und Oekonomie aus der Philosophischen Transaktionen*, which provided in fact German translations of selected articles published originally in the first fourteen volumes of the *Philosophical Transactions* (Leske 1780). In the preface of the journal, Leske announces the publication of the second volume, but this will never happen. Instead, he threw himself into creating a new scientific journal in 1781, but this time he requested the collaboration of Christlieb Benedict Funk (1734–1786), ordinary professor of Physics at the University of Leipzig, and Hindenburg. Thus, under their editorial direction, the *Leipziger Magazin zur Naturkunde, Mathematik und Oekonomie* will appear quarterly from 1781 to 1785. The topics covered in the articles of the journal are so diverse that one can notice a certain lack of coherence. Hindenburg also addresses a wide range of topics relevant to applied mathematics, as his explanatory footnotes on life annuities and astronomy that one finds here and there in the pages of the journal, and relevant to pure mathematics, as an article on the fifth postulate of Euclid (Hindenburg 1781d). However, he is also interested in discussing subjects that are not necessarily related to mathematics but that are in vogue, as the case of the mechanical Turk constructed by Johann Wolfgang von Kempelen (1734–1804), an automaton that was supposed to play chess and that caused a sensation in Europe and America (Hindenburg 1785). The great diversity of topics discussed, as well as the different degree of rigor associated to each contribution to the journal, suggests that the journal did not target a specific audience, even if it was founded in accordance with the will of the editors to create a specialized publication. Apparently, the heterogeneity of the themes proved to be a difficult obstacle to overcome in the process of coordinating and organizing the different volumes of the journal, which led to the dissolution of the editorial team in 1785. In fact, since two years ago, the group had already fragmented, and Leske and Hindenburg were the only ones listed as editors on the front page of the journal. In the preface of the last volume, Leske and Hindenburg expressed the wish to enhance the thematic organization of the journal and announced that, for this purpose, the journal would be divided into two independent publications, namely, one publication devoted to economic sciences and natural history, and the other devoted to mathematics (Hindenburg and Leske 1785, 2). While Leske would continue to serve as editor in chief of the journal of economic studies and natural history, Hindenburg and Johann III Bernoulli (1744–1807) would coordinate the journal of mathematics.

The plan to publish simultaneously two scientific journals under the aegis of the University of Leipzig was suddenly compromised by Leske's death on November 25, 1786. However, Leske had still the strength to ensure that the publication process of the first volume took place as efficiently as possible, and afterward some of his colleagues took over the direction of the journal entitled *Leipziger Magazin zur Naturkunde und Oekonomie*, which appeared from 1786 to 1788.

On the other hand, Hindenburg begins his work as chief editor without the collaboration of Bernoulli, contrary to what had been announced, and the first volume of his journal entitled *Leipziger Magazin für die Mathematik* appears in 1786. The writings gathered in this volume are sometimes only tangentially related to mathematics. The title of the journal will be then changed accordingly from the second volume to *Leipziger Magazin für reine und angewandte Mathematik*, thereby enabling certain contributions, which bear little or no relation to the technical aspects of pure mathematics, to be included under the rubric of applied mathematics. Hindenburg thus becomes, in 1786, the science editor of the first specialized journal in mathematics that appeared in Germany. In fact, according to (Girlich 2009), where one can find further information about the process of specialization of scientific journals in Europe, Hindenburg's journal could be the first journal specialized in mathematics to be published in all of Europe, although Girlich points out that some scholars attribute the honor to the *Beyträge zur Aufnahme der theoretischen Mathematik*, edited by W. J. G. Karsten in Rostock from 1758 to 1761. However, for Girlich, the *Beyträge zur Aufnahme* is not a scientific journal but an anthology of mathematical papers. Be it as it may, this is a defining moment for Hindenburg's career since it offers him the chance to exert significant influence over the actual organization of German mathematics by choosing the topics that are open for public discussion. However, in 1787, Bernoulli decides to join the editorial committee, as planned from the beginning. At the same time, the Berlin Academy of Sciences entrusts Bernoulli with the task of editing the posthumous works of Lambert. Bernoulli proposes then that the journal serve as a platform for the publication of Lambert's unedited works, and Hindenburg gladly agrees. Thus, although the journal will certainly include some papers of Hindenburg on combinatorial analysis, its general orientation is not closely aligned with this new branch of mathematics. It is clear that Bernoulli's collaboration keeps in balance Hindenburg's tendency to privilege the combinatorial aspects of mathematics. Besides Lambert's writings, the journal embraces a rather broad conception of mathematics and welcomes, for instance, manuscripts of Bernoulli himself, Kästner, and Segner. Despite all its strengths, the last volume of the journal appears in 1789.

Hindenburg will suspend his editorial work for several years, until 1794 when he will launch a new scientific journal under the title *Archiv der reinen und angewandten Mathematik*, and eleven issues will appear in all between 1794 and 1800. This time, Hindenburg takes the reins of the journal alone and presents it, in the preface of the first volume, as if it were just the prolongation of the *Leipziger Magazin der Mathematik*, that is to say, both journals share the same objectives and interests. However, further in the preface, he adds the following clarification (Hindenburg 1795e, without numbering):

> At the request of several enlightened persons, I have incorporated in this volume some essays on combinatorial analysis in order to dispel the darkness that still looms, according to what I have been told verbally and in writing, over this matter. [...A]nd I am in no doubt that they will show, clearly and for every one to see, the richness of the combinatorial tools when applied to analytic problems and formulae, as well as they will show the deep connection between simplicity and generality, clarity of representation and facility of development. [...] I have gone into more detail than necessary about certain topics, at least as far as the system of science itself is concerned, system that I will present publicly in the near future. It will then be clear that it is possible to achieve the greatest geometric evidence and precision, and thus, even after completing this system, the analytic-combinatorial essays collected here will continue to be useful as long as I will constantly turn to them as a further proof.[1]

Let's put into perspective these words of Hindenburg. The first volume of his journal, in which "some essays on combinatorial analysis" can be found, is composed of four issues. Each issue contains three sections: one section of original articles, another section of book reviews, where one can find extracts from some books, and a final section of scientific correspondence. In all, there are twenty-eight articles in the first volume, eighteen of which are "analytic-combinatorial essays". It is plain that Hindenburg incorporated more than "some essays on combinatorial analysis" in the first volume. On the other hand,

---

[1] Ich habe, nach dem Wunsche mehrerer Kenner, einige Aufsätze über combinatorische Analysis in diesem Bande mit eingerückt, um die Dunkelheit zu zerstreuen, die über der Sache, wie man mich schriftlich und mündlich versichert hat, noch schwebte. [...U]nd ich zweifle nicht, sie werden den Reichthum der combinatorischen Hülfsmittel, auf analytische Probleme und Formeln angewendet, auch wie innig hier Simplicität und Allgemeinheit, Leichtigkeit in der Darstellung und Behendigkeit in der Entwickelung, mit einander verbunden sind, deutlich vor Augen legen. [...] Ich habe mich darinn über gewisse Gegenstände ausführlicher ausgebreitet, als es in einem Systeme der Wissenschaft selbst, das ich nächstens dem Publikum vorzulegen gedenke, nicht geschehen darf. Man wird finden, daß es der strengsten geometrischen Schärfe und Evidenz fähig sey; und so werden auch die hier beygebrachten combinatorisch-analytischen Abhandlungen, selbst nach Aufstellung eines solchen Systems, für das Publikum nicht verloren gehen, in so fern ich, als weitere Nachweisung, öfters daraufmich beziehen werde.

in the section of book reviews, the majority of texts deal with, or at least are related to, Hindenburg's combinatorial analysis. For instance, on can find, in the second issue of this volume, a review of the third edition of the *Anfangsgründe der Analysis endlicher Grössen* by Kästner published in 1794, but special emphasis is placed on those parts of Kästner's book which are related, according to Hindenburg, to his combinatorial analysis: on the last page of this review is explicitly mentioned the link between Kästner's work and Hindenburg's project exposed in the *Novi systematis* (Book review 1795, 235).

On the other hand, only one-third of the articles included in this volume does not address Hindenburg's combinatorial analysis, notable among these are those written by Kästner and those written by Johann Friedrich Hennert (1733–1813), professor of physics, mathematics, and philosophy at the University of Utrecht, both of whom regularly collaborate with the journal. Nevertheless, one should not forget that Hindenburg regarded the work of Kästner as a pillar of his combinatorial analysis in the sense that his combinatorial method was derived in part from Kästner's results. Indeed, Hindenburg had such admiration for his former teacher that he dedicated his 1779 *Infinitinomi*, where he presented his new combinatorial method, to him. On the other side, Hennert's articles are rather concerned with questions of physics, and, in fact, Hennert's collaboration with Hindenburg's journals dates back to the time of the *Leipziger Magazin zur Naturkunde, Mathematik und Oekonomie* and goes far beyond the dissolution of the *Archiv der reinen und angewandten Mathematik*. The first manuscript of Hennert to be published in one of Hindenburg's journals appeared in the *Leipziger Magazin* in 1782, and thereafter Hennert continued to send manuscripts to be published in the different journals Hindenburg ran. After the abrupt end of the *Archiv* in 1800, Hindenburg is still in possession of several manuscripts of Hennert and decides to put them together in a booklet appeared in 1805. In the preface, Hindenburg recalls that the writings included in this booklet were supposed to be published in the *Archiv*, and then he makes the following assessment of his journal (Hennert 1805, IV–V):

> And so, this was a singular journal especially conceived for the benefit of mathematics, was honored to have contributions from Hennert, Kästner, von Zach, Klügel, Pfleiderer, Kramp, Rothe, Pfaff, Bürmann, and Hauber, and made public several subtle investigations taken from the posthumous works of Lambert. This was a journal devoted to a much less known part and extension of knowledge, the combinatorial analysis; this was a journal that survived for several years and enjoyed the approval of enlightened public opinion, but that was, all of a sudden and in an instant, suppressed and stopped![2]

---

[2] Und so ward denn eine eigene, zum Gebrauch für die Mathematik besonders angelegte Zeitschrift, die Hennert und Kästner, v. Zach, Klügel und Pfleiderer, die Kramp und Rothe und Pfaff und Bürmann und Hauber mit ihren Beyträgen beehrten; die manche scharfsinnige Untersuchung aus Lamberts literarischen Nachlasse mittheilte; eine Zeitschrift, die zum Theil der Kenntniss und Erweiterung der viel zu wenig gekannten combinatorischen Analysis gewidmet war, und mehrere Jahre hindurch mit Beyfall der Kenner fortgedauert hatte, plötzlich und auf einmal in ihrem Laufe gehemmt und aufgehalten!

This assessment confirms that the goals and objectives Hindenburg was working toward support his vision of mathematics. On the one hand, the *Archiv* responds to the desire of creating a scientific journal specialized in mathematics, which is consistent with the aim of the *Leipziger Magazin für reine und angewandte Mathematik*. Among the authors cited here, Hennert, Kästner, von Zach (1754–1832), Pfleiderer (1736–1821), and Lambert never deal with questions related to the German combinatorial analysis. Theirs papers represent the most important part of the journal devoted to general mathematics. Furthermore, Johann III Bernoulli's idea of publishing the posthumous works of Lambert in the *Leipziger Magazin für reine und angewandte Mathematik* was subsequently taken up by Hindenburg, who allocated a section of the *Archiv* for this purpose. On the other hand, the other authors cited by Hindenburg conduct research into the German combinatorial analysis. In fact, despite Hindenburg's efforts made to ensure inclusion of other areas of mathematics, the vast majority of the pages of his journal are about combinatorial analysis. In this sense, the *Archiv* is not only a journal specialized in mathematics, but specialized mainly in the theory of Hindenburg, a theory that can aspire, according to Hindenburg's words, to the status of system of science. Despite his intentions, Hindenburg will never be able to accomplish the composition of a treatise on such a system of science, but the fundamental features of this system have been described, according to Hindenburg, in the papers on combinatorial analysis published in the *Archiv*. Thus, this journal will serve to frame the German combinatorial analysis in terms of a system of science.

Finally, Hindenburg launched another journal under the title *Sammlung combinato-risch-analytischer Abhandlungen*, which was issued on a very irregular basis and was exclusively devoted to the German combinatorial analysis. The first volume appeared in 1796, and the second one in 1800. A third volume was supposed to be published, but the journal was canceled because of the increases in printing costs as a consequence of war. However, as in the case of Hennert's manuscripts, the papers of the third volume were collected in a single book entitled *Über combinatorische Analysis und Derivations-Calcul*, which appeared in 1803. It seems that this journal was conceived as a supplement to the *Archiv* in order not to overload this latter with papers related to Hindenburg's combinatorial analysis. In any case, this is perhaps the first scientific journal specialized in a single (and new) branch of mathematics.

4.1.2   Lagrange's Inversion Formula and the Reversion of Series

Before going any further in this narrative about the consolidation of the German combinatorial school, it is necessary at this point to step aside and examine an old paper of Joseph-Louis Lagrange that will be decisive to bring together, in 1794, a group of former students of Hindenburg united around their former teacher's ideas. Lagrange's paper provides a differential formula for calculating, in form of a series, every root of what he calls "a literal equation", but it can be applied to transcendental equations as well (Lagrange 1770). Furthermore, Lagrange shows that Newton's theorem on reversion of series can be seen as a consequence of his formula. First let's see how Lagrange's differential formula was obtained.

4.1.2.1 Lagrange's Inversion Formula

Lagrange read his paper on the solution of "literal equations" before the Berlin Academy of Sciences on 18 January and on 5 April 1770. He begins his allocution by saying that he discovered a method capable to find the roots of any polynomial equation. Although the roots of a given polynomial equation will be expressed by means of a power series, he claims that this series obeys a known general law in such a way that the calculations can easily be carried out to the required accuracy. Additionally, his method can be used to calculate the series expansion of an equation's root raised to any power, and even to calculate the series expansion of any function composed with that root. Finally, he points out that his method can also be applied to transcendental equations (Lagrange 1770, 251; Lubet 1998, 79–84; Panza 1992, vol. 2, 550–569). However, he does not touch on this question in this paper but in another article in which he uses his formula to solve Kepler's equation (Lagrange 1771; Bottazzini 1989).

What Lagrange calls a literal equation is a polynomial equation of the form:

$$0 = a - bx + cx^2 - dx^3 + \&c.,$$

for polynomials of any degree. Then, he supposes that $p, q, r, \ldots$ are the solutions of this equation, and from the equality:

$$a - bx + cx^2 - dx^3 + \&c. = a \left(1 - \frac{x}{p} \right) \left(1 - \frac{x}{q} \right) \left(1 - \frac{x}{r} \right) \cdots$$

he gets, by dividing by $-bx$:

$$1 - \frac{a}{bx} - \frac{cx - dx^2 + \&c.}{b} = \frac{a}{bp} \left(1 - \frac{p}{x} \right) \left(1 - \frac{x}{q} \right) \left(1 - \frac{x}{r} \right) \cdots$$

Taking the natural logarithm of both sides, the equation becomes:

$$l \left(1 - \frac{a}{bx} - \frac{cx - dx^2 + \&c.}{b} \right) = l \left(\frac{a}{bp} \right) + l \left(1 - \frac{p}{x} \right)$$
$$+ l \left(1 - \frac{x}{q} \right) + l \left(1 - \frac{x}{r} \right) + \&c.,$$

where "l" is the original notation chosen by Lagrange and stands for natural logarithm. If one puts:

$$X = \frac{a}{x} + cx - dx^2 + ex^3 - \&c.,$$

so that one has:

$$l\left(1 - \frac{a}{bx} - \frac{cx - dx^2 + \&c.}{b}\right) = l\left(1 - \frac{X}{b}\right),$$

and if one expands the logarithms contained in the previous equation into infinite series, one gets:

$$\frac{X}{b} + \frac{X^2}{2b^2} + \frac{X^3}{3b^3} + \&c. = l\left(\frac{bp}{a}\right) + \left(\frac{p}{x} + \frac{p^2}{2x^2} + \frac{p^3}{3x^3} + \&c.\right) \tag{4.1}$$

$$+ x\left(\frac{1}{q} + \frac{1}{r} + \&c.\right) + \frac{x^2}{2}\left(\frac{1}{q^2} + \frac{1}{r^2} + \&c.\right) + \&c.$$

By computing all the expansions of X, X^2, X^3, ... in this equation, and then by applying the method of undetermined coefficients, one can express the value of the root p, of the logarithm of p, and of the positive powers of p in terms of the coefficients a, b, c, ... of the original polynomial $a - bx + cx^2 - dx^3 + \&c.$

By way of an example, the case of the trinomial equation $a - bx + cx^2 = 0$, Lagrange shows that the procedure described in the previous paragraph provides a method to determine, on a case by case basis, the values of the roots of a given equation. For instance, if one considers Lagrange's example of the trinomial equation, then $X = \frac{a}{x} + cx$ and, by using Eq. 4.1, one obtains the root $p = \frac{a}{b} + \frac{3a^2c}{3b^2} + \frac{5 \cdot 4a^3c^2}{2 \cdot 5b^5} + \cdots$. The greater the number of powers X, X^2, X^3, ... taken into account, the greater the number of terms in the infinite series of p, and so the greater the precision of the estimated value of p will become. However, Lagrange's aim is to have a general formula applicable directly to any literal equation, without having to go through the painful calculations imposed by Eq. 4.1 every time a different literal equation has to be solved. To improve the method, Lagrange will try to transform the expression on the left-hand side of Eq. 4.1 by putting:

$$\xi = \frac{cx - dx^2 + ex^3 - \&c.}{b},$$

so that:

$$\frac{X}{b} = \frac{a}{bx} + \xi,$$

and then, because the left-hand side of Eq. 4.1 is the logarithm expansion of $1 - \frac{X}{b}$ multiplied by -1, one gets:

$$\frac{X}{b} + \frac{X^2}{2b^2} + \frac{X^3}{3b^3} + \&c. = -l\left(1 - \frac{a}{bx}\right) - l\left(1 - \frac{\xi}{1 - \frac{a}{bx}}\right).$$

Or, what is the same thing, one gets the following identity because of Eq. 4.1:

$$-l\left(1 - \frac{a}{bx}\right) - l\left(1 - \frac{\xi}{1 - \frac{a}{bx}}\right) = l\left(\frac{bp}{a}\right) + \left(\frac{p}{x} + \frac{p^2}{2x^2} + \frac{p^3}{3x^3} + \&c.\right)$$

$$+ x\left(\frac{1}{q} + \frac{1}{r} + \&c.\right) + \frac{x^2}{2}\left(\frac{1}{q^2} + \frac{1}{r^2} + \&c.\right) + \&c.$$

After expanding the logarithms on the left-hand side into infinite series, it follows from this that:

$$\left(\frac{a}{bx} + \frac{a^2}{2b^2x^2} + \frac{a^3}{3b^3x^3} + \&c.\right)$$

$$+ \left(\frac{\xi}{1 - \frac{a}{bx}} + \frac{\xi^2}{2\left(1 - \frac{a}{bx}\right)^2} + \frac{\xi^3}{3\left(1 - \frac{a}{bx}\right)^3} + \&c.\right) = \tag{4.2}$$

$$l\left(\frac{bp}{a}\right) + \left(\frac{p}{x} + \frac{p^2}{2x^2} + \frac{p^3}{3x^3} + \&c.\right)$$

$$+ x\left(\frac{1}{q} + \frac{1}{r} + \&c.\right) + \frac{x^2}{2}\left(\frac{1}{q^2} + \frac{1}{r^2} + \&c.\right) + \&c.$$

Lagrange points out that the expressions $\xi, \xi^2, \xi^3, \ldots$ are in fact multinomial expansions that can be written in general as:

$$\xi = \pi + \pi'x + \pi''x^2 + \pi'''x^3 + \&c.$$

$$\xi^2 = \rho + \rho'x + \rho''x^2 + \rho'''x^3 + \&c. \tag{4.3}$$

$$\xi^3 = \sigma + \sigma'x + \sigma''x^2 + \sigma'''x^3 + \&c.$$

etc.,

where the apostrophes of the coefficients π, ρ, σ, ...are indexes. Similarly, the expressions $\left(1 - \frac{a}{bx}\right)^k$, for $k = 1, 2, 3, \ldots$ (this is not Lagrange's notation), are binomial expansions. Using these multinomial and binomial expansions, Lagrange calculates the power series expansions of the terms $\frac{\xi^k}{k\left(1 - \frac{a}{bx}\right)^k}$, for $k = 1, 2, 3, \ldots$ (this is not Lagrange's notation), in Eq. 4.2. After substituting these power expansions in Eq. 4.2, and after reducing terms, Lagrange deduces from Eq. 4.2, by means of the method of undetermined coefficients, the following formulae for the root p:

$$p = \frac{a}{b} + \left(\pi \frac{a}{b} + \pi' \frac{a^2}{b^2} + \pi'' \frac{a^3}{b^3} + \&c. \right)$$

$$+ \frac{1}{2} \left(2\rho \frac{a}{b} + 3\rho' \frac{a^2}{b^2} + 4\rho'' \frac{a^3}{b^3} + \&c. \right)$$

$$+ \frac{1}{2 \cdot 3} \left(2 \cdot 3\sigma \frac{a}{b} + 3 \cdot 4\sigma' \frac{a^2}{b^2} + 4 \cdot 5\sigma'' \frac{a^3}{b^3} + \&c. \right) + \&c.,$$

for any positive integer power of p:

$$p^2 = \frac{a^2}{b^2} + 2 \left(\pi \frac{a^2}{b^2} + \pi' \frac{a^3}{b^3} + \pi'' \frac{a^4}{b^4} + \&c. \right)$$

$$+ \frac{2}{2} \left(3\rho \frac{a^2}{b^2} + 4\rho' \frac{a^3}{b^3} + 5\rho'' \frac{a^4}{b^4} + \&c. \right)$$

$$+ \frac{2}{2 \cdot 3} \left(3 \cdot 4\sigma \frac{a^2}{b^2} + 4 \cdot 5\sigma' \frac{a^3}{b^3} + 5 \cdot 6\sigma'' \frac{a^4}{b^4} + \&c. \right) + \&c$$

etc.,

and for the logarithm of p:

$$l(p) = l\left(\frac{a}{b} \right) + \left(\pi + \pi' \frac{a}{b} + \pi'' \frac{a^2}{b^2} + \&c. \right)$$

$$+ \frac{1}{2} \left(\rho + 2\rho' \frac{a}{b} + 3\rho'' \frac{a^2}{b^2} + \&c. \right)$$

$$+ \frac{1}{2 \cdot 3} \left(1 \cdot 2\sigma + 2 \cdot 3\sigma' \frac{a}{b} + 3 \cdot 4\sigma'' \frac{a^2}{b^2} + \&c. \right) + \&c.$$

Because of Eqs. 4.3, it is easy to see that these formulae can be written in terms of derivatives at the point $x = \frac{a}{b}$ as follows:

$$l\,(p) = l\,(x) + \xi + \frac{d\xi^2 x}{2dx} + \frac{d^2 \xi^3 x^2}{2 \cdot 3 dx^2} + \&c.,$$

and for any strictly positive integer m:

$$p^m = x^m + m \left(\xi x^m + \frac{d\xi^2 x^{m+1}}{2dx} + \frac{d^2\xi^3 x^{m+2}}{2 \cdot 3dx^2} + \&c. \right) \qquad (4.4)$$

In particular, this latter formula provides a method to calculate the value of a root p of any literal equation.

Lagrange generalizes this result to cover equations of the form $\alpha - x + \phi(x) = 0$, where ϕ is any function of x, as can be seen in his general inversion theorem:

Consider the equation:

$$\alpha - x + \phi(x) = 0,$$

where $\phi(x)$ is any function of x. Let p be a root of this equation, that is to say, a value of x, and one wants to find the value of any function at p, like $\psi(p)$. For convenience, let's write the quantity $\frac{d\psi(x)}{dx}$ as $\psi'(x)$, and then I claim that, in general, one gets:

$$\psi(p) = \psi(x) + \phi(x)\psi'(x) + \frac{d(\phi(x))^2 \psi'(x)}{2dx}$$

$$+ \frac{d^2(\phi(x))^3 \psi'(x)}{2 \cdot 3dx^2} + \frac{d^3(\phi(x))^4 \psi'(x)}{2 \cdot 3 \cdot 4dx^3} + \&c.$$

where it is necessary to replace x by α after differentiation.[3]

This generalization of Lagrange depends on his (mistaken) belief according to which any function is expressible as a power series.

---

[3] Soit l'équation

$$\alpha - x + \phi x = 0$$

ϕx étant une fonction quelconque de x. Que p soit une des racines de cette équation, c'est à dire, une des valeurs de x, & qu'on demande la valeur d'une fonction quelconque de p comme ψp. Qu'on dénote, pour plus de simplicité, la quantité $\frac{d\psi x}{dx}$ par $\psi'x$, & je dis qu'on aura en général

$$\psi p = \psi x + \phi x \psi'x + \frac{d(\phi x)^2 \psi'x}{2dx}$$

$$+ \frac{d^2(\phi x)^3 \psi'x}{2 \cdot 3dx^2} + \frac{d^3(\phi x)^4 \psi'x}{2 \cdot 3 \cdot 4dx^3} + \&c.$$

où il faudra changer x en α après les différentiations.

4.1.2.2 Lagrange's Proof of Newton's Theorem on Reversion of Series

One of the strengths of Lagrange's previous results is their wide range of applications. Indeed, Lagrange contrived an elegant solution not only to the problem of approximating the roots of polynomial equations, but to a set of other problems. Among these, there is the problem of calculating the power series expansions of certain functions. At that time, one important example of this concerned the reversion of series, and Lagrange showed in his paper that Newton's theorem on reversion of series was a consequence of the application of formula 4.4.

Lagrange considers the following equation:

$$a - bx + cx^2 - dx^3 + \&c. = 0.$$

This is not necessarily a polynomial equation, since the expression on the left-hand side is an infinite series, even though this is not made explicit in Lagrange's paper. To solve this equation, according to Lagrange's method, one puts:

$$\xi = \frac{cx - dx^2 + ex^3 - \&c.}{b},$$

and then one calculates the following series expansions:

$$\xi^2 = \frac{c^2x^2 - 2cdx^3 + (d^2 + 2ce)x^4 - \&c.}{b^2}$$

$$\xi^3 = \frac{c^3x^3 - 3cdx^4 + \&c.}{b^3}$$

$$\xi^4 = \frac{c^4x^4 - \&c.}{b^4}$$

etc.

Thus, by taking $m = 1$ in formula 4.4, Lagrange obtains the following result:

$$p = \frac{a}{b} + \left(\frac{a^2c}{b^3} - \frac{a^3d}{b^4} + \frac{a^4e}{b^5} - \&c. \right)$$

$$+ \left(\frac{2a^3c^2}{b^5} - \frac{5a^4cd}{b^6} + \frac{3a^5(d^2 + 2ce)}{b^7} - \&c. \right)$$

$$+ \left(\frac{5a^4c^3}{b^7} - \frac{21a^5cd}{b^8} + \&c. \right)$$

$$+ \left(\frac{14a^5c^4}{b^9} - \&c. \right)$$

$$+ \&c.,$$

which is, says Lagrange, the well-known formula of Newton for calculating the reversion of a given series. For the sake of clarity, one can rearrange the terms on the right-hand side according to the increasing powers of a as follows:

$$p = \frac{a}{b} + \frac{c}{b^3}a^2 + \frac{2c^2 - db}{b^5}a^3 + \frac{5c^3 - 5bcd + b^2e}{b^7}a^4 + \&c.,$$

which is indeed the expression of p in terms of a, and corresponds to the formula given by Newton in his *Epistola posterior* (for Newton's formula, see Proposition 2.1 on page 21).

In his paper, Lagrange does not explain how the successive integer powers of the infinite multinomial ξ have been calculated. The standard procedure at that time consisted in applying the multinomial theorem of Moivre. As seen in Chap. 2, the multinomial theorem of Moivre, as well as the subsequent proofs of this theorem by other mathematicians until 1770, provides a recursive formula for calculating the coefficients of the multinomial expansion, with the sole exception of Boscovich's non-recursive method appeared in 1747. There is no evidence that Lagrange knew the method of Boscovich, but we know that Boscovich was not appreciated by many of the members of the Paris Academy of Sciences (Pappas 1996), including Lagrange whose correspondence with Jean le Rond d'Alembert (1717–1783) gives an impression of Lagrange's rather negative opinion on Boscovich (Serret 1882, 248, 276, 278, 280). Therefore, it is more likely that Lagrange used a recursive method for calculating those multinomial expansions.

4.1.3   The Rise of the German Combinatorial School: An Interpretation

It seems that Boscovich's non-recursive method for calculating the coefficients of the multinomial expansion was not widely known or disseminated in academic circles. As far as we know, none of the authors who intended to give a new proof of the multinomial theorem made any reference to Boscovich's method until 1781, when Hindenburg mentioned it briefly in passing in the preface of his *Novis systematis*. But even after 1781 this situation did not change, as we can see in the following passage where Sylvestre-François Lacroix (1765–1843) identifies only two categories of approach concerning the multinomial theorem before 1778, when Hindenburg offered his new non-recursive approach, and so ignoring Boscovich's work (Lacroix 1810, XXVIII):

> The theorem of Lagrange certainly offers a symmetric form with respect to the formulae about the reversion of series; but, concerning the powers of the indefinite polynomial that are used [in this theorem] and the calculation of their respective coefficients, we had no more resources at our disposal than the rules given by Moivre, or the successive relations obtained by Euler by means of differentiation. The first procedure rests on induction; the second one compels us to determine a given coefficient from the others, so that to reach the particular coefficient of a term, one has to pass through all the previous ones. This is why the *combinatorial Analysis* was created, a highly cultivated discipline in Germany, which was derived from certain ideas of Leibniz about the use of ordinal numbers, instead of letters, to designate the coefficients

of unknowns or variables, and which is used to construct, by regular operations, the series expansion coefficients of a polynomial raised to some power.[4]

Thus, Lacroix thinks that, prior to the formation of the German combinatorial school, there was no non-recursive formula for calculating the coefficients of the multinomial expansion, which is false since Boscovich's non-recursive formula dates from 1747, even if it concerns only positive integer exponents. Furthermore, he thinks that the German combinatorial analysis emerged from the attempts to ensure that Lagrange's inversion theorem does not depend on recursive formulae. As seen in Chap. 3, Hindenburg's combinatorial analysis emerged from a more complex historical process, in which Hindenburg never showed any interest in Lagrange's theorem, he never used any mathematical idea of Leibniz to establish his theory, and, even more, he severely criticized Leibniz's ordinal number notation for coefficients as being based on an abstruse and inadequate conception of the theory of combinations. This erroneous image of the emergence of the combinatorial analysis portrayed by Lacroix was in fact very popular, and it will remain the same over time, particularly with regard to the erroneous idea that Leibniz's work played a major role in the development of the German conbinatorial analysis. However, it is true that Lagrange's formula will be decisive for the future of Hindenburg's theory, but not in the direction Lacroix had in mind. Lagrange's formula will find itself in the midst of a priority conflict that does not concern Lagrange at all, but the combinatorial analysis of Hindenburg. In this conflict, what is at stake is a combinatorial formula equivalent to that of Lagrange. It is partially thanks to this conflict that the German combinatorial school will rise.

In the pages that follow, a reasonable interpretation of the events that marked the emergence of the German combinatorial school is provided. This historical interpretation is based on the thesis that the formation of the combinatorial school is the result, on the one hand, of the education oriented toward the combinatorial analysis carried out in the classroom by Hindenburg, and, on the other hand, of the conflict about the authorship of the combinatorial analysis, a conflict in which the former students of Hindenburg did not hesitate to come to the defense of their mentor, and in which the existence of a "neutral" space for disseminating the arguments of both sides was a key element in solving the conflict.

---

[4] Le théorème de Lagrange donne bien une forme symétrique aux formules du retour des suites ; mais les puissances du polynome indéfini s'y retrouvent, et pour en former les coefficiens, on n'avait que les règles données par Moivre, ou les relations successives obtenues par Euler, au moyen de la différentiation. Le premier de ces procédés n'est fondé que sur l'induction ; le second ne mène qu'à déterminer les coefficiens les uns par les autres, ensorte que pour parvenir à celui d'un terme, il faut passer par tous ceux qui le précèdent. Tels sont les motifs qui ont donné naissance à l'*Analyse combinatoire*, très-cultivée en Allemagne, déduite de quelques vues que Leibniz a proposées, sur l'emploi des nombres ordinaux à la place des lettres pour indiquer les coefficiens des inconnues ou des variables, et au moyen de laquelle on forme, par des opérations régulières, les coefficiens des termes du développement des puissances d'un polynome.

4.1.3.1 A Combinatorial and non-recursive Formula for the Reversion of Series

Trained in mathematics at the University of Leipzig, Hieronymus Christoph Wilhelm Eschenbach (1764–1797) was one of the first students of Hindenburg to rework mathematical problems by means of concepts proper to the combinatorial analysis. In 1791, he was engaged by a Dutch company as engineer and moved to the Dutch East Indies, but he was taken prisoner in 1796 during the conquest of Malacca by the British and died at the age of thirty-two on 7 March 1797, twenty-three days before his birthday. Because of his unfortunate death at a young age, he left behind very few writings. However, he had achieved a certain notoriety in Leipzig because of his doctoral dissertation, publicly defended on 30 May 1789, in which he proposed a new solution for the problem of series reversion (Eschenbach 1789). Eschenbach had already addressed this question in a previous paper (Eschenbach 1785), whose main subject matter was, however, a problem of tangents proposed by Kästner in his *Analysis des Unendlichen*. As the theorem on reversion of series is analyzed in a more comprehensive and coherent manner in his doctoral dissertation, the following discussion will be based on that particular text.

In the introduction to his dissertation, Eschenbach seems to be suggesting that he will attempt a historical reconstruction of the problem of series reversion. Nevertheless, instead of a historical reconstruction, one finds in his text only a description of Moivre's theorem and Tempelhoff's reinterpretation of Moivre's theorem. As seen in Sect. 2.2.3.1, Tempelhoff's proof follows the lines of Moivre's, except for the fact that the exponents of the series are written differently, i.e. Tempelhoff puts the equation:

$$az^m + bz^{m+1} + cz^{m+2} + \&c. = gy^m + hy^{m+1} + iy^{m+2} + \&c.$$

The use of a parameter m in this equation was perceived by Eschenbach as a generalization of Moivre's theorem, and this is why a whole section of his dissertation is devoted to a review of Tempelhoff's proof. Thus, Eschenbach is not really interested in history at all, but in the question of generalizing Moivre's theorem, and his dissertation is intended to be a contribution to this enterprise. To generalize this theorem, he chooses to follow a different path (Eschenbach 1789, 12):

> Although the successive terms [of the reversed series] can be derived from the previous ones by using [Moivre's] formula, it is also possible to obtain, by means of the analytic-combinatorial art, the truly general term of the desired series; i.e. it is possible to establish a formula for directly calculating any term by using the known coefficients and the exponents, without taking into account the order and without depending on the calculation of any other term.[5]

---

[5] [C]um enim per formulam [...] semper termini sequentes per antecedentes sint determinandi, dari etiam potest, artis analytico-combinatoriae ope, seriei quaesitae terminus vere generalis, i.e. per quem, quilibet terminus, extra ordinem et ab reliquis omnibus independenter, per ipsos datos coefficientes et exponentes, directe definitur.

Thus, the general aim of Eschenbach's dissertation is to establish a non-recursive formula for determining the general coefficient of a reversed series in terms of the known coefficients and exponents of the series taken as point of departure. Eschenbach's work falls within the mainstream tradition of Moivre's formula, but he made a brief and marginal comment on Leibniz's reformulation of the theorem, which consists in using numbers instead of letters to designate the coefficients of the reversed series, as seen in Sect. 2.2.3.3. This is what he says about Leibniz's reformulation (Eschenbach 1789, 5, footnote):

> Leibniz gave the first four terms of the desired series [...] and enunciated, though he did it using words, their law of progression in what may be the most obscure way, something which attracted criticism from Moivre, and later from Hindenburg. By means of Hindenburg's symbols, this law of Leibniz for a large number of things can be expressed in quite a simple way.[6]

Against the popular interpretation of the emergence of the combinatorial analysis as being the result of the reexamination of Leibniz's mathematical notation proposed in his paper on the reversion of series, this passage shows again that this is not the case and that Leibniz's notation was rejected not only by Hindenburg but also by other mathematicians close to him. Although Eschenbach points out that Leibniz's law can be formulated by using Hindenburg's theory, he is not interested in improving Leibniz's work since his objective is to generalize Moivre's theorem.

In his paper on reversion of series, Moivre summarizes his method for calculating the coefficients of the reversed series as follows (de Moivre 1698, 191): "Combine the Capital Letters as often as you can make the Sum of their Exponents Equal to the Index of the Power to which they belong." As seen in Sect. 2.2.1, capital letters represent the undetermined coefficients of the reversed series that have to be calculated recursively in terms of the known coefficients. Eschenbach quotes this sentence in his dissertation (Eschenbach 1789, 7), and it seems that he has set himself the goal of formalizing Moivre's somewhat vague rule by means of Hindenburg's theory. More specifically, he will apply the classes of partitions invented by Hindenburg to Moivre's recursive formula, and then he will generalize this idea in order to obtain a non-recursive formula. Thus, besides his general objective of creating a non-recursive formula, he wants to demonstrate that it is possible to translate Moivre's theorem itself into the language of Hindenburg's combinatorial analysis.

---

[6] Seriei quaesitae terminos quatuor priores proposuit LEIBNITIUS [...], legem etiam progressus pronunciavit, sed verbis, ut post MOIVRAEUM HINDENBURGIUS questus est, tantum non obscurissimis. Per signa, quae Clar. HINDENBURG introduxit, haec ipsa LEIBNITII lex formula, pro tanta rerum datarum diversitate satis simplici, exprimi potest.

To translate Moivre's theorem, Eschenbach puts, just as Moivre did, the following equation:

$$ax + bx^2 + cx^3 + \cdots = \alpha y + \beta y^2 + \gamma y^3 + \cdots,$$

and also:

$$y = \dot{\mathfrak{A}}x + \dot{\mathfrak{B}}x^2 + \dot{\mathfrak{C}}x^3 + \ldots,$$

where $\dot{\mathfrak{A}}$, $\dot{\mathfrak{B}}$, $\dot{\mathfrak{C}}$, ...are undetermined coefficients according to Hindenburg's notations introduced in the *Novis systematis* (see Sect. 3.3.3). Then, Eschenbach defines *a priori* the coefficients $a, b, c, \ldots$ as follows:

$$a = \alpha\mathfrak{a}(^1A)$$
$$b = \alpha\mathfrak{a}(^2A) + \beta\mathfrak{b}(^2B)$$
$$c = \alpha\mathfrak{a}(^3A) + \beta\mathfrak{b}(^3B) + \gamma\mathfrak{c}(^3C)$$

$$\text{etc.,}$$

and he gives the following general formula without any justification:

$$^{n-1}a = \alpha\mathfrak{a}(^nA) + \beta\mathfrak{b}(^nB) + \gamma\mathfrak{c}(^nC) + \ldots + {^{n-1}\alpha}\,\mathfrak{n}(^n\mathcal{N}). \tag{4.5}$$

As we know, Gothic lowercase letters represent multinomial coefficients in general, and capital letters with an exponent to the left represent Hindenburg's classes of partitions of a positive integer n. The symbol ^{n-1}a denotes the $(n - 1)$th coefficient of the series $ax + bx^2 + cx^3 + \cdots$, that is to say, one has $^1a = b$, $^2a = c$, $^3a = d$, etc., and the same applies to $^{n-1}\alpha$ with respect to $\alpha y + \beta y^2 + \gamma y^3 + \cdots$ (see Sect. 3.3.3).

For Eschenbach, formula 4.5 is an analytic-combinatorial reinterpretation of Moivre's law of formation of coefficients. In order to apply this formula to the undetermined coefficients $\dot{\mathfrak{A}}$, $\dot{\mathfrak{B}}$, $\dot{\mathfrak{C}}$, ..., Eschenbach establish the following substitution rule according to Hindenburg's method:

$$\begin{pmatrix} \dot{\mathfrak{A}} & \dot{\mathfrak{B}} & \dot{\mathfrak{C}} & \cdots \\ 1 & 2 & 3 & \cdots \end{pmatrix}$$

which allows to determine the classes of partitions:

$$^1A = (1) = \left(\dot{\mathfrak{A}}\right)$$
$$^2A = (2) = \left(\dot{\mathfrak{B}}\right)$$

$$^3\mathrm{A} = (3) = (\dot{\mathfrak{C}})$$

$$\vdots$$

$$^2\mathrm{B} = (1,\,1) = (\dot{\mathfrak{A}},\,\dot{\mathfrak{A}})$$

$$^3\mathrm{B} = (1,\,2) = (\dot{\mathfrak{A}},\,\dot{\mathfrak{B}})$$

$$\vdots$$

$$^3\mathrm{C} = (1,\,1,\,1) = (\dot{\mathfrak{A}},\,\dot{\mathfrak{A}},\,\dot{\mathfrak{A}})$$

$$\vdots$$

By replacing these values on formula 4.5, one finds:

$$a = \alpha\mathfrak{a}\dot{\mathfrak{A}}$$

$$b = \alpha\mathfrak{a}\dot{\mathfrak{B}} + \beta\mathfrak{b}\dot{\mathfrak{A}}^2$$

$$c = \alpha\mathfrak{a}\dot{\mathfrak{C}} + \beta\mathfrak{b}\dot{\mathfrak{A}}\dot{\mathfrak{B}} + \gamma\mathfrak{c}\dot{\mathfrak{A}}^3,$$

and therefore:

$$\dot{\mathfrak{A}} = \frac{a}{\alpha}$$

$$\dot{\mathfrak{B}} = \frac{b - \beta\dot{\mathfrak{A}}^2}{\alpha}$$

$$\dot{\mathfrak{C}} = \frac{c - 2\beta\dot{\mathfrak{B}}\dot{\mathfrak{A}} - \gamma\dot{\mathfrak{A}}^3}{\alpha}.$$

As can be seen here, except for the first, each coefficient has to be calculated from the previous ones, as originally proposed by Moivre. Thus, this is a recursive method that leads to the same results as those based on Moivre's formula. Therefore, Eschenbach has succeeded in translating Moivre's theorem into the theory of Hindenburg while respecting its original recursive nature.

However, Eschenbach's aim in writing his dissertation is not just to translate Moivre's theorem, it is to show that the classes of partitions, the multinomial coefficients, the analytic-combinatorial symbols, in short, the tools developed by Hindenburg allow better and faster techniques in mathematics. To remove the recursion from Moivre's formula, Eschenbach puts the equation:

$$z^\Lambda = \alpha y^\Pi + \beta y^{\Pi+\Delta} + \gamma y^{\Pi+2\Delta} + \delta y^{\Pi+3\Delta} + \cdots, \tag{4.6}$$

where Greek capital letters are numbers for which, according to Eschenbach, the binomial theorem is valid. This means that, based on the proofs of the binomial theorem proposed at that time, Greek capital exponents can be, at the very least, rational numbers. Then, with no justification and without giving any explanation of his reasoning, Eschenbach just stated his formula for the reversion of series:

$$
\begin{aligned}
y^\Sigma = {} & \left(\frac{z^\Lambda}{\alpha}\right)^m - m\,\frac{\mathfrak{a}(^1\mathrm{A})}{\alpha}\left(\frac{z^\Lambda}{\alpha}\right)^{^1 m} \\
& - m\left(\frac{\mathfrak{a}(^2\mathrm{A})}{\alpha} - \frac{^2m+1\,\mathfrak{Ab}(^2\mathrm{B})}{2\alpha^2}\right)\left(\frac{z^\Lambda}{\alpha}\right)^{^2 m} \\
& - m\left(\frac{\mathfrak{a}(^3\mathrm{A})}{\alpha} - \frac{^3m+1\,\mathfrak{Ab}(^3\mathrm{B})}{2\alpha^2} + \frac{^3m+2\,\mathfrak{Bc}(^3\mathrm{C})}{3\alpha^3}\right)\left(\frac{z^\Lambda}{\alpha}\right)^{^3 m} \\
& - m\left(\frac{\mathfrak{a}(^4\mathrm{A})}{\alpha} - \frac{^4m+1\,\mathfrak{Ab}(^4\mathrm{B})}{2\alpha^2} + \frac{^4m+2\,\mathfrak{Bc}(^4\mathrm{C})}{3\alpha^3} - \frac{^4m+3\,\mathfrak{Cd}(^4\mathrm{D})}{4\alpha^4}\right)\left(\frac{z^\Lambda}{\alpha}\right)^{^4 m} \\
& - etc.,
\end{aligned}
\tag{4.7}
$$

which depends on the following substitution rule to interpret the elements of the classes of partitions:

$$
\begin{pmatrix} \beta & \gamma & \delta & \cdots \\ 1 & 2 & 3 & \cdots \end{pmatrix},
\tag{4.8}
$$

and where $m = \frac{\Sigma}{\Pi}$, $^1m = \frac{\Sigma+\Delta}{\Pi}$, $^2m = \frac{\Sigma+2\Delta}{\Pi}$, $\ldots$, $^nm = \frac{\Sigma+n\Delta}{\Pi}$, and so on. The meanings of the other symbols are already known: Gothic capital letters with a left exponent represent binomial coefficients, Gothic lowercase letters represent multinomial coefficients, and capital letters in Roman font with a left exponent are Hindenburg's classes of partitions. The general term of the reversed series is, thus, given by:

$$
y^\Sigma \mathbin{\rceil} (n+1) =
\tag{4.9}
$$

$$
- m\left(\frac{\mathfrak{a}(^n\mathrm{A})}{\alpha} - \frac{^nm+1\,\mathfrak{Ab}(^n\mathrm{B})}{2\alpha^2} + \frac{^nm+2\,\mathfrak{Bc}(^n\mathrm{C})}{3\alpha^3} - \cdots \pm \frac{^nm+n-1\,\mathfrak{Nn}(^n\mathcal{N})}{n\alpha^n}\right)\left(\frac{z^\Lambda}{\alpha}\right)^{^n m},
$$

where the expression $y^\Sigma \mathbin{\rceil} (n+1)$ means, as we know, the $(n+1)$th term of the series y^Σ, and where the interpretation of the elements of the classes of partitions depends on the substitution rule 4.8. One can find an early version of formula 4.9 in (Eschenbach 1785, 11–13, footnote), though it is enunciated for positive integer exponents only.

In his dissertation, instead of proving his formula 4.9, Eschenbach illustrates it with concrete examples. Let us examine one of them and see how this formula works. Suppose we are given the series expansion of the arcsine function, which Eschenbach writes as follows:

$$z = y + \frac{1}{2 \cdot 3}y^3 + \frac{1 \cdot 3}{2 \cdot 4 \cdot 5}y^5 + \frac{1 \cdot 3 \cdot 5}{2 \cdot 4 \cdot 6 \cdot 7}y^7 + \frac{1 \cdot 3 \cdot 5 \cdot 7}{2 \cdot 4 \cdot 6 \cdot 8 \cdot 9}y^9 + \cdots .$$

We want to determine directly the fourth term in the series $y^{\frac{1}{6}}$, that is to say, we want to determine directly the fourth term in the series expansion of $(\sin(u))^{\frac{1}{6}}$ (this latter is not Eschenbach's notation). In accordance with the expansion of arcsin given above and with Eq. 4.6, we have:

$$\Lambda = \Pi = 1, \quad \Delta = 2, \quad \alpha = 1, \quad \beta = \frac{1}{2 \cdot 3}, \quad \gamma = \frac{1 \cdot 3}{2 \cdot 4 \cdot 5}, \quad \delta = \frac{1 \cdot 3 \cdot 5}{2 \cdot 4 \cdot 6 \cdot 7}.$$

Since we want to obtain the fourth term of $y^{\frac{1}{6}}$, we have:

$$n = 3, \quad {}^n m = {}^3 m = \frac{37}{6},$$

and according to the substitution rule 4.8:

$${}^3 A = (3) = (\delta), \quad {}^3 B = (1, 2) = (\beta, \gamma), \quad {}^3 C = (1, 1, 1) = (\beta^3).$$

The required multinomial and binomial coefficients are:

$$\mathfrak{a} = \frac{1}{1} = 1, \quad \mathfrak{b} = \frac{2 \cdot 1}{1 \cdot 1} = 2, \quad \mathfrak{c} = \frac{3 \cdot 2 \cdot 1}{3 \cdot 2 \cdot 1} = 1,$$

$${}^{3m+1}\mathfrak{A} = {}^{\frac{43}{6}}\mathfrak{A} = \frac{43}{6}, \quad {}^{3m+2}\mathfrak{B} = {}^{\frac{49}{6}}\mathfrak{B} = \frac{\frac{49}{6}\left(\frac{49}{6} - 1\right)}{2}.$$

Therefore, formula 4.9 gives the following result:

$$y^{\frac{1}{6}}14 = -\frac{1}{6}\left(\mathfrak{a}({}^3 A) - \frac{{}^{3m+1}\mathfrak{A}\mathfrak{b}({}^3 B)}{2} + \frac{{}^{3m+2}\mathfrak{B}\mathfrak{c}({}^3 C)}{3}\right)z^{3m}$$

$$= -\frac{359}{2^7 \cdot 3^7 \cdot 5 \cdot 7}z^{\frac{37}{6}},$$

which is effectively the fourth term in the series expansion of $(\sin(u))^{\frac{1}{6}}$.

Eschenbach's formula depends on Newton's binomial theorem and Hindenburg's method of partitions. Since none of these involves recursive techniques, this is a non-recursive formula. Furthermore, Eschenbach's dissertation contributes to develop Hindenburg's work program established in the *Novi systematis* inasmuch as it addresses points (9) and (10) of this program, concerning respectively series reversion and transcendental functions (see Sect. 3.3.4 for Hindenburg's program). Despite the fact that Eschenbach does not prove his formula, his elegant solution to the problem of the reversion of series will not go unnoticed among German mathematicians.

4.1.3.2 The Theory of Dimension Symbols

In 1792, a new mathematical theory was born in the city of Halle. It was named "the theory of dimension symbols" (*Theorie der Dimensionszeichen*) by its creator Ernst Gottfried Fischer (1754–1831), who studied mathematics and theology at the University of Halle. At the time that the theory of dimension symbols was published in that town, he was a professor of mathematics and physics at a *Gymnasium* in Berlin. Dimension symbols are mathematical notations developed by Fischer in order to express a "magnitude" (*Größe*) as "a simple product" (Fischer 1792, vol. 1, 7). More generally, the theory of dimension symbols aims to reorient the discussion of certain problems of finite analysis, and therefore also to reinterpret them, by proposing the use of a new mathematical notation. These problems concern mainly the theory of series, and, in particular, one of them has Fischer's full attention, namely, that of expressing the roots of an algebraic equation by means of a series, "which the most insightful analyst of the century, Mr. de la Grange, has solved in the most insightful and general way."[7] It is no accident that Fischer set aside a special place for this question in his theory, since he asserted that he had already discovered an innovative method for solving this problem when he found out about Lagrange's inversion theorem, as he puts it himself (Fischer 1792, vol. 1, IV–V):

> However, since I had arrived at the solution of this problem in a completely different manner than Mr. de la Grange did, and since science can only benefit from examining one and the same object from different points of view, I thought that it would be better to leave unchanged in my text those passages that are related precisely to this object [...].[8]

---

[7] (Fischer 1792, vol. 1, IV): daßder scharfsinnigste Analyst unseres Jahrhunderts, Herr de la Grange [...], auf eine höchst scharfsinnige und allgemeine Art aufgelöset habe.

[8] Da ich indessen auf einem ganz anderen Wege, als Herr de la Grange zu der Auflösung des Problems gelangt war, und es für die Wissenschaft nie anders als vortheilhaft seyn kann, wenn ein und derselbe Gegenstand aus verschiedenen Gesichtspuncten untersuchet wird, so hielt ich es für besser, diejenigen Abschnitte meines Werkes, welche eben den Gegenstand betreffen [...] ungeändert zu lassen [...].

Thus, according to Fischer, he was trying to work out how to get the solutions of algebraic equations using series, which is the main motivation of Lagrange's paper discussed above, and he arrived at a similar solution but in an independent manner.

The similarities between Fischer's work and the work of other mathematicians do not end there. While Fischer and Lagrange shared the same motivation for developing a new method, the method found by Fischer was similar to that of Hindenburg, and again he pretended to have discovered it in an independent manner (Fischer 1792, vol. 1, V):

> In the aftermath, when I thought I was pursuing this very path alone, I found another excellent and respectful geometer, professor Hindenburg, who has promised, eleven years ago now, in his *Novi systematis permutationum combinationum ac variationum primae lineae et logisticae serierum formulis analytico-combinatoriis per tabulas exhibendae conspectus et specimena, combinaisons et variations* (Leipzig 1781), to release to the public an analytic treatise, whose completion will certainly produce a notable analytic treatise and, perhaps, will make my modest work unnecessary. However, so far and as far as I know, this promise has not been fulfilled [...].[9]

This passage suggests that at least the general lines of Fischer's theory of dimension symbols were drawn up before Fischer was made aware of the existence of Hindenburg's combinatorial analysis. At this stage, there are the two following alternatives: either Fischer began working on his book before the publication of the *Novi systematis*, where Hindenburg presented his general project concerning the combinatorial analysis, or he began after 1781 without knowing anything about the existence of Hindenburg's work. However, given the great deal of surprise expressed by Fischer by the time elapsed since Hindenburg committed himself to complete a treatise on combinatorial analysis, the first alternative is unlikely. In consequence, the second scenario dictates that Fisher seized upon the idea of creating dimension symbols after 1781. In this case, since Fischer himself acknowledges the striking resemblance between his theory and that of Hindenburg, to such an extent that the publication of a treatise on combinatorial analysis would render unnecessary the theory of dimension symbols, Fischer should have been more circumspect in presenting his theory, instead of presenting it as if it were a new and original theory. In any case, his lack of prudence brought about a reaction of some former students of Hindenburg bent on preserving the integrity of their teacher's ideas.

Before recounting the story of this reaction, it would be appropriate to review some general aspects of the theory of dimension symbols, since this will give us a better sense

---

[9] In der Folge fand ich auf eben dem Wege, wo ich anfänglich ganz einsam zu geben wähnte, noch einen andern vortreflichen und achtungswürdigen Geometer, Herrn Prof. Hindenburg, der schon vor 11 Jahren in seinen *primis lineis novi systematis permutationum, combinationum et variationum (Lipsiae* 1781), dem Publicum zu einem analytischen Werke Hofnung machte, dessen Vollendung gewiß manches schätzbare analytische Werk, und vielleicht auch diese meine geringe Arbeit entbehrlich machen würde. Bis jetzt ist aber, so viel mir bekannt ist, diese Hofnung nicht erfüllt worden [...].

of what this theory looks like against Hindenburg's combinatorial analysis background. Fischer formulates an infinity of orders of dimension symbols as follows (Fischer 1792, vol. 1, 7–26):

1. Dimension symbols of the first order: $\overset{1}{\text{I}}, \overset{2}{\text{I}}, \overset{3}{\text{I}}, \ldots$.
2. Dimension symbols of the second order: $\overset{1}{\text{II}}, \overset{2}{\text{II}}, \overset{3}{\text{II}}, \ldots$.
3. Dimension symbols of the third order: $\overset{1}{\text{III}}, \overset{2}{\text{III}}, \overset{3}{\text{III}}, \ldots$.
4. And so on.

These symbols represent in fact the partitions of the positive whole number written above the Roman numerals, and the Roman numerals are used to indicate the "order" of the partitions, that is to say, they indicate the number of parts that each partition should contain. For instance, the dimension symbol $\overset{6}{\text{III}}$ designates the partitions of 6 that are formed by three parts, so $\overset{6}{\text{III}}$ contains the following partitions: $(1, 1, 4)$, $(1, 2, 3)$, $(2, 2, 2)$. This is exactly the same idea as that expressed by Hindenburg's classes of partitions, and therefore Fischer not only developed similar notations but an identical combinatorial method to that of Hindenburg too.

In the same way as Hindenburg's classes of partitions, Fischer's orders of dimension symbols depend on a substitution rule to interpret the elements of each order. Thus, as Hindenburg, Fischer associates letters to numbers. For example, for the elements of $\overset{6}{\text{III}}$, one gets:

1. from $(1, 1, 4)$, and after replacing 1 by a and 4 by d: (a, a, d),
2. from $(1, 2, 3)$, and after replacing 1 by a, 2 by b, and 3 by c: (a, b, c),
3. from $(2, 2, 2)$, and after replacing 2 by b: (b, b, b).

This serves to calculate the multinomial coefficient of each monomial aad, abc, and bbb, and so to reinterpret the symbol $\overset{6}{\text{III}}$ as a "quantity": $\overset{6}{\text{III}} = 3a^2d + 6abc + b^3$. Moreover, it is possible to express a given order of dimension symbols as a sum of products of dimension symbols of lower order. For instance, $\overset{6}{\text{III}} = \overset{1}{\text{I}} \cdot \overset{1}{\text{I}} \cdot \overset{4}{\text{I}} + \overset{1}{\text{I}} \cdot \overset{2}{\text{I}} \cdot \overset{3}{\text{I}} + \overset{2}{\text{I}} \cdot \overset{2}{\text{I}} \cdot \overset{2}{\text{I}} = 3a^2d + 6abc + b^3$.

When dimension symbols are used as coefficients of a given series, which is possible since they can be reinterpreted as quantities as explained in the previous paragraph, Fischer prefers to use Gothic capital letters with an exponent above it. This is, then, an alternative notation for the orders of dimension symbols:

1. Dimension symbols of the first order: $\overset{n}{\text{I}} = \overset{n+1}{\mathfrak{A}}$.
2. Dimension symbols of the second order: $\overset{n}{\text{II}} = \overset{n+2}{\mathfrak{B}}$.

3. Dimension symbols of the third order: $\overset{n}{\text{III}} = \overset{n+3}{\mathfrak{C}}$.

4. And so on.

Thus, the orders of dimension symbols correspond to what Hindenburg designated by means of a class of partitions preceded by a general multinomial coefficient, i.e. Fischer's orders represent respectively the same quantity that the analytic-combinatorial formulae $\mathfrak{a}(^nA)$, $\mathfrak{b}(^nB)$, $\mathfrak{c}(^nC)$, ..., of Hindenburg do, in such a way that one can establish identities between them: $\mathfrak{a}(^nA) = \overset{n}{\text{I}} = \overset{n+1}{\mathfrak{A}}$, $\mathfrak{b}(^nB) = \overset{n}{\text{II}} = \overset{n+2}{\mathfrak{B}}$, $\mathfrak{c}(^nC) = \overset{n}{\text{III}} = \overset{n+3}{\mathfrak{C}}$, and so on.

It must be admitted that the resemblance between Fischer's notations and Hindenburg's can raise serious suspicions about the origin of Fischer's theory. But an even more striking coincidence is the underlying idea upon which both theories are based, namely, the idea of conceiving the elements of partitions as general combinatorial symbols devoid of any particular meaning, which are able, however, to be associated to certain algebraic objects, and so able to be interpreted algebraically as the coefficients of a given series. As seen in Chap. 3, this idea was the most important consideration that led Hindenburg to the conception of the combinatorial analysis as a system of science.

Using his dimension symbols and this general idea on which his theory is based, Fischer addresses the problem that Lagrange solved "in the most insightful and general way", i.e. the problem of finding the roots of an algebraic equation using series (Fischer 1792, vol. 1, 67 ff.). He puts the equation:

$$y = x^m + \overset{2}{\mathfrak{A}}x^{m+r} + \overset{3}{\mathfrak{A}}x^{m+2r} + \overset{4}{\mathfrak{A}}x^{m+3r} + etc.,$$

and the problem consists in finding a series expansion in y to express the power x^t, where the nature of t is not specified. To this end, Fischer gives the following formula:

$$x^t = y^{\frac{t}{m}} - \frac{t}{m}\overset{2}{\mathfrak{A}}y^{\frac{t+r}{m}} \qquad\qquad (4.10)$$

$$- \left[\left(\frac{t}{m}\overset{3}{\mathfrak{A}}\right) - \left(\frac{t}{m}\cdot\frac{t+m+2r}{2m}\overset{4}{\mathfrak{B}}\right)\right]y^{\frac{t+2r}{m}}$$

$$- \left[\left(\frac{t}{m}\overset{4}{\mathfrak{A}}\right) - \left(\frac{t}{m}\cdot\frac{t+m+3r}{2m}\overset{5}{\mathfrak{B}}\right) + \left(\frac{t}{m}\cdot\frac{t+m+3r}{2m}\cdot\frac{t+2m+3r}{3m}\overset{6}{\mathfrak{C}}\right)\right]y^{\frac{t+3r}{m}}$$

$$- etc.$$

If one replaces $\overset{2}{\mathfrak{A}}$ by $\mathfrak{a}(^1A)$, $\overset{3}{\mathfrak{A}}$ by $\mathfrak{a}(^2A)$, ..., $\overset{4}{\mathfrak{B}}$ by $\mathfrak{a}(^2B)$, $\overset{5}{\mathfrak{B}}$ by $\mathfrak{b}(^3B)$, and so on, one gets Eschenbach's formula 4.7 (for $\alpha = 1$ and $\Lambda = 1$). However, Fischer makes no reference to Eschenbach's result, which was published seven years earlier if one takes into account the first version of Eschenbach's formula printed in his 1785 paper. Fischer gives also Eschenbach's formula 4.9 concerning the general coefficient of a reversed series, without mentioning its author and using his dimension symbols (Fischer 1792, vol. 1, 68). Therefore, the new solution discovered by Fischer for Lagrange's problem of finding the roots of an algebraic equation by means of series was the solution proposed by Eschenbach for generalizing Moivre's theorem on reversion of series.

Unlike Eschenbach, Fischer offers a proof of his formula 4.10, which consists in expanding into series each power of y that appears in it, as follows:

$$y^{\frac{t}{m}} = x^t + \frac{t}{m}\overset{2}{\mathfrak{A}}x^{t+r} + \left(\frac{t}{m}\overset{3}{\mathfrak{A}} + \frac{t}{m}\cdot\frac{t-m}{2m}\overset{4}{\mathfrak{B}}\right)x^{t+2r}$$

$$+ \left(\frac{t}{m}\overset{4}{\mathfrak{A}} + \frac{t}{m}\cdot\frac{t-m}{2m}\overset{5}{\mathfrak{B}} + \frac{t}{m}\cdot\frac{t-m}{2m}\cdot\frac{t-m}{3m}\overset{6}{\mathfrak{C}}\right)x^{t+3r} + etc.$$

$$\alpha y^{\frac{t+r}{m}} = \alpha x^{t+r} + \alpha\frac{t+r}{m}\overset{2}{\mathfrak{A}}x^{t+2r} + \alpha\left(\frac{t+r}{m}\overset{3}{\mathfrak{A}} + \frac{t+r}{m}\cdot\frac{t+r-m}{2m}\overset{4}{\mathfrak{B}}\right)x^{t+3r} + etc.$$

$$\beta y^{\frac{t+2r}{m}} = \beta x^{t+2r} + \beta\frac{t+2r}{m}\overset{2}{\mathfrak{A}}x^{t+3r} + etc.$$

$$\gamma y^{\frac{t+3r}{m}} = \gamma x^{t+3r} + etc. \tag{4.11}$$

$$etc.,$$

where $\alpha, \beta, \gamma, \ldots$ are abbreviations for the coefficients of formula 4.10, i.e. one has:

$$\alpha = -\frac{t}{m}\overset{2}{\mathfrak{A}}$$

$$\beta = -\left[\left(\frac{t}{m}\overset{3}{\mathfrak{A}}\right) - \left(\frac{t}{m}\cdot\frac{t+m+2r}{2m}\overset{4}{\mathfrak{B}}\right)\right]$$

$$\gamma = -\left[\left(\frac{t}{m}\overset{4}{\mathfrak{A}}\right) - \left(\frac{t}{m}\cdot\frac{t+m+3r}{2m}\overset{5}{\mathfrak{B}}\right) + \left(\frac{t}{m}\cdot\frac{t+m+3r}{2m}\cdot\frac{t+2m+3r}{3m}\overset{6}{\mathfrak{C}}\right)\right]$$

$$etc.$$

After replacing these values of $\alpha, \beta, \gamma, \ldots$ in Eqs. 4.11, and after adding these equations, one obtains the equality:

$$x^t = y^{\frac{t}{m}} + \alpha y^{\frac{t+r}{m}} + \beta y^{\frac{t+2r}{m}} + \gamma y^{\frac{t+3r}{m}} + etc.$$

which corresponds to formula 4.10, given the values of α, β, γ, etc. Fischer emphasizes that his method of proof follows the directions provided by Moivre in his original paper on the reversion of series, which means that his proof is based on the multinomial theorem. On the other hand, Fischer includes at the end of the second volume of his book Lagrange's general inversion theorem, although he fails to explain whether there is a correlation between his formula and Lagrange's theorem (Fischer 1792, vol. 2, 171–176).

4.1.3.3 Unmasking the Usurper

As usual in similar cases, the publication of Fischer's book was the subject of several brief notes describing its content, which appeared in what can be considered as rather large circulation journals in Germany. One can find this kind of notes, for example, in the *Neue Leipziger gelehrte Anzeigen*, issue 97 (1792), in an addendum to the *Gothaische gelehrte Zeitung*, issue 5 (1793), or in the *Allgemeine Literatur-Zeitung*, issue 102 (1793). Ignoring their small differences, the common denominator of these reviews is the remark about the similarities between Fischer's theory and that of Hindenburg. Nevertheless, the remark was presented in a neutral manner without giving any personal opinion and without implying that any plagiarism had been committed. On 25 November 1792, the review appeared in the first of those journals reached the ears of Heinrich August Töpfer (1758–1833), a former student of Hindenburg (Töpfer 1793, 2). Intrigued by the remark about the similarities between both theories, Töpfer sought to deepen his understanding of Fischer's work, and realized that the theory of dimension symbols pursues the same aims, covers the same mathematical phenomena, and uses the same methods as Hindenburg's combinatorial analysis. Shocked and outraged, Töpfer immediately begins drawing up a manuscript where he denounces Fischer's imposture. Five months later, on 1 May 1793, some two hundred pages that meticulously bring to light the ruses and stratagems of the usurper Fischer are ready for the printing (Töpfer 1793, XVI).

Töpfer does not hide his indignation and directly accuses Fischer of plagiarism. His aim is to prove it. To this end, he conducts an argumentative battle on three different fronts: Fischer's honesty, the originality of the dimension symbols, and the range of subjects covered by the theory of dimension symbols. First of all, Töpfer calls into question the veracity of Fischer's claims that he was completely unaware of the very existence of the combinatorial analysis at the moment when he came up with his own theory. As can be seen from the quotation above, Fischer remains to some extent ambiguous in this respect, but it really seems that he denies having read Hindenburg and having been influenced by his ideas, no matter how similar both theories otherwise are. For Töpfer, there are too many coincidences regarding the theory of dimension symbols and the combinatorial analysis, both in form and in content, to be a simple product of chance, which reflects Fischer's bad faith. Töpfer argues that Fischer acted in bad faith in asserting that his mathematical notations, his formula 4.10 concerning the reversion of series, and his general combinatorial method based on integer partitions were entirely new discoveries of his own, carried out independently of those of Hindenburg, and even of those of Lagrange. According to Töpfer, Fischer's alleged lack of knowledge of the combinatorial analysis is absolutely not credible, since Hindenburg's works had been the object of many reviews, published in several German journals. In particular, Töpfer points out that Hindenburg's works had been reviewed in 1783 by Johann III Bernoulli in one of the most prestigious German journals at the time, the *Nouveaux Mémoires de l'Académie Royale des Sciences et Belles-Lettres de Berlin*. Effectively, one can find in this journal a letter by Bernoulli where he gives a brief but comprehensive description of Hindenburg's work (Bernoulli 1783). Bernoulli talks about the *Novi systematis* and says that this book contains a new theory which has its roots in Hindenburg's work on mathematical tables; he adds that this new theory has a large range of important applications in analysis, particularly in the theory of series, where it can be used to solve the most difficult problems; finally, he underlines the combinatorial method developed by Hindenburg in the *Methodus nova*, the 1778 *Infinitinomii*, and the 1779 *Infinitinomii*. By the way, as far as we know, Bernoulli is the only author (outside the combinatorial school) who correctly links the origins of the combinatorial analysis to Hindenburg's work on mathematical tables and not to Leibniz. Furthermore, as Töpfer points out, even the *De serierum reversione* of Eschenbach had been reviewed, for instance, in the *Allgemeine Literatur-Zeitung* (Book review 1790). Thus, for Töpfer, even the lack of reference to Eschenbach's dissertation is suspicious. Given the large dissemination of all these reviews in German academic circles, Töpfer draws the conclusion that it is unimaginable to think that Fischer was unaware of all this, and then Fischer lied about his knowledge of Hindenburg's combinatorial analysis.

The second group of objections concerns the invention of dimension symbols. Töpfer looked at all these symbols, comparing them carefully with the combinatorial notations created by Hindenburg, and concluded that none of them could be trusted as a reliable, original symbol: all of them were gross imitations of Hindenburg's symbolism (Töpfer 1793, 52–68). At the end of his book, one can find a table that summarizes these comparisons. His arguments for these conclusions relied largely on the fact that a dimension symbol has the same abstract structure as Hindenburg's notations for classes of partitions, in the sense that, as the symbolic representation of Hindenburg's classes of partitions, a dimension symbol consists of a main symbol, whether a Gothic capital letter or a Roman numeral, which indicates the number of parts the partitions should have, and of a numerical superscript, which indicates the positive integer that has been partitioned. In addition, as pointed out in Sect. 4.1.3.2, dimension symbols have the same meaning concerning their application to series as Hindenburg's combinatorial formulae.

The third group of objections is directed toward the subject matters covered by Fischer in his book. Töpfer maintains that the body of Fischer's book is organized following closely the work plan sketched by Hindenburg in the *Novi systematis*. The main questions addressed in Fischer's book include the problem of dividing one series by another, the problem of composing one series with another, the study of transcendental functions such as the logarithm or the sine function, the problem of expanding into series a possibly infinite multinomial raised to an arbitrary power, and, of course, the question of reversing a given series. All of these form part of Hindenburg's work program, as pointed out in Sect. 3.3.4, and Töpfer seeks to convince his readership that Fischer even went so far as to imitate Hindenburg's style of writing, to such an extent that there is an astonishing parallelism between the way of posing the problems in the *Novi systematis* and the way of posing them in Fischer's book. Among these questions, particular attention has been paid to the theorem on reversion of series, whose discussion occupies virtually one third of Töpfer's text. Töpfer briefly revises the history of this theorem (Töpfer 1793, 124–136). His selective historical account begins with Newton's original formulation, then passes to the discussion of Moivre's reformulation, and talks extensively about the reception of

Moivre's theorem in the work of Hindenburg, who had paid, to be honest, little attention to the solution of this question. Naturally, Eschenbach's formula for the reversion of series takes a prominent place in this historical reconstruction. In this regard, Töpfer aims to prove that there is a historical continuity to the succession of theorems, or of the different versions of the same theorem, about the reversion of series that support Eschenbach's result beginning in the work of Newton. Töpfer points out that the history of this theorem starts with Newton's Proposition 1 stated in the *Epistola posterior*, which is the first known statement of this theorem, and then he argues that Moivre explicitly intended his formula 2.20 for the reversion of series to be based on Newton's ideas, which is a totally accurate description of events, as seen in Chap. 2. Töpfer recalls also that Tempelhoff's theorem on reversion of series was meant to be a reformulation of Moivre's. Therefore, since Eschenbach's dissertation concerns a generalization of Moivre's and Tempelhoff's versions of this theorem, the analytic-combinatorial solution given by Eschenbach for reversing a series must be classified as falling within the Newtonian tradition. On the contrary, Fischer's formula for the reversion of series constitutes a violation of historical continuity as long as the author of the theory of dimension symbols claims to have made the discovery of his formula on his own and independently of any mathematical tradition. Once again, Töpfer draws the conclusion that Fischer plagiarized the results obtained within the framework of Hindenburg's research program. In his exasperation, Töpfer goes so far as mentioning that other former student of Hindenburg, Heinrich August Rothe, had also addressed the question of the reversion of series, but this takes the argument too far since Rothe's work on this subject, which will be analyzed in Sect. 4.1.3.4, dates from 31 August 1793, that is, it dates from later than Fischer's book.

Töpfer's accusation of plagiarism leveled against Fischer produces the desired effect. The news about Fischer's unfaithful appropriation of Hindenburg's theory spreads in the German academic world. This is how the review of Töpfer's book appeared in the *Allgemeine Literatur-Zeitung* starts (Book review 1794):

> Professor Fischer from Berlin has published, as we know, a theory of dimension symbols, because of which he has been openly accused of plagiarizing the combinatorial analysis of Hindenburg.[10]

The attack launched against Fischer in order to prevent the theory of dimension symbols from taking the place of the combinatorial analysis, which has remained however in the project phase for more than eleven years, has hit its target. But the labor of dismantling the theory of dimension symbols rests not only on Töpfer's shoulders, as will be seen in what follows.

---

[10] Hr. Prof. Fischer in Berlin hat bekanntlich eine Theorie der Dimensionszeichen herausgegeben, wöruber er offentlich eines Plagiums der Hindenburgischen combinatorischen Analytik beschuldigt worden ist.

4.1.3.4 A Proof of Eschenbach's Formula

In 1787, Eschenbach stated his formula 4.9 without proof, five years later Fischer stole this formula but proved it. This is a point in favor of Fischer's allegations about the originality of his work since, in the world of mathematics, the one who proves a result can rightfully claim his portion of the ownership of the discovery. The former students of Hindenburg, however, were not willing to share such a remarkable formula. Instead of making a frontal attack, such as identifying a misstep in Fischer's proof, they opted to present a demonstration made with the tools developed by Hindenburg, as if a new proof could erase the existence of that earlier proof. The confection of the new demonstration was left in the hands of one of the most brilliant students of Hindenburg, Heinrich August Rothe, who finished writing a dissertation on this subject on 31 August 1793 (Rothe 1793). Rothe is convinced that his proof will be enough to discredit Fischer's, provided two conditions are met. First, his demonstration should be as general as possible. He recalls Fischer's remark that Lagrange proved in the most general way his formula 4.4, but this formula applies only to a positive integer exponent m, as Rothe points out (Rothe 1793, IV). In contrast, Eschenbach stated his formula for any exponent, or at least for any number for which the binomial theorem is valid, i.e. for at least rational numbers, and Rothe intends to prove this formula for the same kind of exponents. Second, his proof should bring out all the compelling aspects of the combinatorial analysis, especially those aspects connected with the advantages of the analytic-combinatorial notations over the dimension symbols (Rothe 1793, V).

The second condition of Rothe's proof depends substantially on the notion of "local symbol" (*signa localia* in Latin, or *Lokalzeichen* or *Localausdrücke* in German), which is introduced in his dissertation and which will be well received by the future members of the German combinatorial school. In general, a local symbol is a function used to express each term of a given series individually. The term "function" must be understood here in the sense in which it was understood in the eighteenth century, that is (Ferraro and Panza 2003, 19): "a function was not an association between the elements of two given sets: it was a symbolic notation (which was termed 'analytical expression', 'formula' or 'form') expressing a quantity in terms of another quantity." In particular, Rothe's local symbols are expressions of the form:

$$p^m \mathrm{x}(1), \quad p^m \mathrm{x}(2), \quad p^m \mathrm{x}(3), \quad \ldots, \quad p^m \mathrm{x}\,(n+1),$$

which denote respectively the first, the second, the third, etc. coefficient of the series p^m. The symbols employed in these notations were introduced by Hindenburg in the *Novi systematis*, as can be seen in Sect. 3.3.3. Indeed, Hindenburg had proposed to represent succinctly a given series using one of the final letters of the alphabet, in this case p, and had introduced the symbol x $(n + 1)$ as an alternative notation for coefficients. Thus, the meaning of such expressions as $p^m \mathrm{x}\,(n + 1)$ is simply "the coefficient of the $(n + 1)$th term of the series p^m". In Rothe's dissertation, these expressions have, however, a more specific meaning. Given a series

$$p = ax^s + bx^{s+d} + cx^{s+2d} + \cdots,$$

the series expansion of p^m can be written as:

$$p^m = p^m x(1) x^{sm} + p^m x(2) x^{sm+d} + p^m x(3) x^{sm+2d} + \cdots + p^m x(n+1) x^{sm+nd} + \cdots,$$
$$(4.12)$$

where m, s, and d can be, at least, rational numbers. To specify that local symbols are functions that express the quantity of a coefficient in terms of other given quantities, in these case in terms of the coefficients a, b, c, etc. of the series p, Rothe recommends the use of the notation $p[a, b, c, \ldots]$, called scale (*scala*), to indicate explicitly the quantities on which the local symbols depend, and so a formula containing local symbols should be always accompanied by the corresponding scale. For instance, the coefficients of the series expansion of p^m are, according to Hindenburg's multinomial theorem:

$$p^m x(1) = a^m,$$

$$p^m x(2) = {}^m \mathfrak{A} a^{m-1} \mathfrak{a} \left({}^1 A \right),$$

$$p^m x(3) = {}^m \mathfrak{A} a^{m-1} \mathfrak{a} \left({}^2 A \right) + {}^m \mathfrak{B} a^{m-2} \mathfrak{b} \left({}^2 B \right),$$

and in general:

$$p^m x(n+1) = {}^m \mathfrak{A} a^{m-1} \mathfrak{a} \left({}^n A \right) + {}^m \mathfrak{B} a^{m-2} \mathfrak{b} \left({}^n B \right) + \cdots + {}^m \mathfrak{N} a^{m-n} \mathfrak{n} \left({}^n \mathcal{N} \right) \quad (4.13)$$

$$p[a, b, c, \ldots], \qquad \begin{pmatrix} b & c & d & \cdots \\ 1 & 2 & 3 & \cdots \end{pmatrix},$$

where one can find at the end both the scale and the substitution rule to interpret the elements of the classes of partitions.

For Rothe, the use of local symbols is indissociable from his proof of the theorem on reversion of series, which depends also on the following identity:

$$\frac{s(m+1) + nd}{m+1} p^{m+1} x(n+1) = sap^m x(n+1) + (s+d) bp^m x(n) \quad (4.14)$$

$$+ (s+2d) cp^m x(n-1) + \cdots + (s+nd) \overset{n}{a} p^m x(1),$$

$$p[a, b, c, \ldots]$$

where $\overset{n}{a}$ represent the coefficient of the nth term of the series p. This polynomial identity expresses the conditions to be satisfied by the coefficients of the Eq. 4.12. To justify this relation, Rothe writes the series expansion of p^{m+1} using his local symbols:

$$\left(ax^s + bx^{s+d} + cx^{s+2d} + \cdots + \overset{n}{a}x^{s+nd} + \cdots \right)^{m+1} =$$

$$p^{m+1}x(1)x^{s(m+1)} + p^{m+1}x(2)x^{s(m+1)+d} + \cdots + p^{m+1}x(n+1)x^{s(m+1)+nd} + \cdots .$$

After differentiation and after dividing by the infinitesimal quantity dx, Rothe gets:

$$(m+1)\left(ax^s + bx^{s+d} + cx^{s+2d} + \cdots \right)^m \times \left(sax^{s-1} + (s+d)bx^{s+d-1} + \cdots \right) =$$

$$s(m+1)p^{m+1}x(1)x^{s(m+1)-1} + (s(m+1)+d)p^{m+1}x(2)x^{s(m+1)+d-1} + \cdots .$$

Because of Eq. 4.12, and after dividing by $(m+1)$, one gets:

$$\left(sax^{s-1} + (s+d)bx^{s+d-1} + \cdots \right) \times$$

$$\left(p^m x(1)x^{sm} + p^m x(2)x^{sm+d} + \cdots + p^m x(n+1)x^{sm+nd} + \cdots \right) =$$

$$sap^m x(1)x^{s(m+1)-1} + \left(sap^m x(2) + (s+d)bp^m x(1) \right) x^{s(m+1)+d-1} + \cdots$$

$$\cdots + \left(sap^m x(n+1) + (s+d)bp^m x(c) + \cdots + (s+nd)\overset{n}{a}p^m x(1) \right) x^{s(m+1)+nd-1} =$$

$$\frac{s(m+1)}{m+1}p^{m+1}x(1)x^{s(m+1)-1} + \frac{s(m+1)+d}{m+1}p^{m+1}x(2)x^{s(m+1)+d-1} + \cdots$$

$$\cdots + \frac{s(m+1)+nd}{m+1}p^{m+1}x(n+1)x^{s(m+1)+nd-1}.$$

By the method of undetermined coefficients, Rothe compares terms and deduces the relation of coefficients expressed by Eq. 4.14.

Once formula 4.14 has been established, Rothe proves the theorem on reversion of series. To this end, he puts the equation:

$$y^l = \alpha x^s + \beta x^{s+d} + \gamma x^{s+2d} + \cdots , \qquad \text{for } \alpha \neq 0,$$

and supposes that:

$$x^s = \dot{A}y^{\frac{ls}{r}} + \dot{B}y^{\frac{l(s+d)}{r}} + \dot{C}y^{\frac{l(s+2d)}{r}} + \cdots + \overset{n}{\dot{A}}y^{\frac{l(s+nd)}{r}} + \cdots ,$$

where capital letters with a point above them are undetermined coefficients, and $\overset{\scriptstyle n}{A}$ is a notation for the nth undetermined coefficient. Rothe's proof consists in determining the value of these coefficients.

For this purpose, Rothe writes the series expansion of $y^{\frac{ls}{r}}$ by means of local symbols as follows:

$$y^{\frac{ls}{r}} = q^{\frac{s}{r}}x(1)x^s + q^{\frac{s}{r}}x(2)x^{s+d} + q^{\frac{s}{r}}x(3)x^{s+2d} + \cdots + q^{\frac{s}{r}}x(n+1)x^{s+nd} + \cdots$$

$$\tag{4.15}$$

$$q\,[\alpha, \beta, \gamma, \ldots].$$

And he also writes the powers of x using local symbols:

$$x^s = \overset{\cdot}{A}y^{\frac{ls}{r}} + \overset{\cdot}{B}y^{\frac{l(s+d)}{r}} + \overset{\cdot}{C}y^{\frac{l(s+2d)}{r}} + \cdots + \overset{\scriptstyle n}{A}y^{\frac{l(s+nd)}{r}} + \cdots,$$

$$x^{s+d} = p^{\frac{s+d}{s}}x(1)y^{\frac{l(s+d)}{r}} + p^{\frac{s+d}{s}}x(2)y^{\frac{l(s+2d)}{r}} + p^{\frac{s+d}{s}}x(3)y^{\frac{l(s+3d)}{r}} + \cdots$$

$$+ p^{\frac{s+d}{s}}x(n)y^{\frac{l(s+nd)}{r}} + \cdots,$$

$$x^{s+2d} = p^{\frac{s+2d}{s}}x(1)y^{\frac{l(s+2d)}{r}} + p^{\frac{s+2d}{s}}x(2)y^{\frac{l(s+3d)}{r}} + \cdots + p^{\frac{s+2d}{s}}x(n-1)y^{\frac{l(s+nd)}{r}} + \cdots,$$

$$x^{s+3d} = p^{\frac{s+3d}{s}}x(1)y^{\frac{l(s+3d)}{r}} + \cdots + p^{\frac{s+3d}{s}}x(n-2)y^{\frac{l(s+nd)}{r}} + \cdots,$$

and in general:

$$x^{s+nd} = p^{\frac{s+nd}{s}}x(1)y^{\frac{l(s+nd)}{r}} + \cdots,$$

$$p\,[\overset{\cdot}{A}, \overset{\cdot}{B}, \overset{\cdot}{C}, \ldots].$$

From these series expansions and from Eq. 4.15, he gets:

$$0 = \left(\left(\overset{\cdot}{A}q^{\frac{s}{r}}x(1)\right) - 1\right)y^{\frac{ls}{r}} + \left(\left(\overset{\cdot}{B}q^{\frac{s}{r}}x(1)\right) + \left(p^{\frac{s+d}{s}}x(1) \cdot q^{\frac{s}{r}}x(2)\right)\right)y^{\frac{l(s+d)}{r}}$$

$$+ \left(\left(\overset{\cdot}{C}q^{\frac{s}{r}}x(1)\right) + \left(p^{\frac{s+d}{s}}x(2) \cdot q^{\frac{s}{r}}x(2)\right) + \left(p^{\frac{s+2d}{s}}x(1) \cdot q^{\frac{s}{r}}x(3)\right)\right)y^{\frac{l(s+2d)}{r}} + \cdots$$

$$+ \begin{pmatrix} \left(\overset{\scriptstyle n}{A}q^{\frac{s}{r}}x(1)\right) + \left(p^{\frac{s+d}{s}}x(n) \cdot q^{\frac{s}{r}}x(2)\right) \\ + \left(p^{\frac{s+2d}{s}}x(n-1) \cdot q^{\frac{s}{r}}x(3)\right) + \cdots \\ \cdots + \left(p^{\frac{s+nd}{s}}x(1) \cdot q^{\frac{s}{r}}x(n+1)\right) \end{pmatrix} y^{\frac{l(s+nd)}{r}} + \cdots.$$

The values of the desired coefficients follow from this equation by the method of undetermined coefficients:

$$\dot{A} = \alpha^{-\frac{s}{r}} = \frac{s}{s}q^{-\frac{s}{r}}x(1), \qquad \text{since} \ \left(\dot{A}q^{\frac{s}{r}}x(1)\right) - 1 = \dot{A}\alpha^{\frac{s}{r}} - 1 = 0,$$

$$\dot{B} = -\frac{s}{r}\alpha^{-\frac{s+d}{s}-1}\beta = \frac{s}{s+d}q^{-\frac{s+d}{r}}x(2),$$

since

$$\left(\dot{B}q^{\frac{s}{r}}x(1)\right) + \left(p^{\frac{s+d}{s}}x(1) \cdot q^{\frac{s}{r}}x(2)\right) = \dot{B}\alpha^{\frac{s}{r}} + \left(\dot{A}^{\frac{s+d}{s}}\right)\left(\frac{s}{r}\right)\alpha^{\frac{s}{r}-1}\beta$$

$$= \dot{B}\alpha^{\frac{s}{r}} + \left(\frac{s}{r}\right)\alpha^{-\frac{s+d}{s}}\alpha^{\frac{s}{r}-1}\beta = 0.$$

And:

$$\dot{C} = \frac{s}{s+2d}q^{-\frac{s+2d}{r}}x(3),$$

since:

$$\frac{s+2d}{s+2d}q^{-\frac{s+2d}{r}}x(1) = p^{\frac{s+2d}{s}}x(1),$$

and then:

$$0 = \left(\dot{C}q^{\frac{s}{r}}x(1)\right) + \left(\frac{s+d}{s+2d}q^{-\frac{s+2d}{r}}x(2)\right)\left(q^{\frac{s}{r}}x(2)\right) + \left(\frac{s+2d}{s+2d}q^{-\frac{s+2d}{r}}x(1)\right)\left(q^{\frac{s}{r}}x(3)\right)$$

$$= \left(\dot{C}q^{\frac{s}{r}}x(1)\right) + \left(p^{\frac{s+d}{s}}x(2) \cdot q^{\frac{s}{r}}x(2)\right) + \left(p^{\frac{s+2d}{s}}x(1) \cdot q^{\frac{s}{r}}x(3)\right),$$

and by formula 4.14, it follows that:

$$\left(\frac{s+d}{s+2d}q^{-\frac{s+2d}{r}}x(2)\right)\left(q^{\frac{s}{r}}x(2)\right) + \left(\frac{s+2d}{s+2d}q^{-\frac{s+2d}{r}}x(1)\right)\left(q^{\frac{s}{r}}x(3)\right) =$$

$$- \left(\frac{s}{s+2d}q^{-\frac{s+2d}{r}}x(3)\right)\left(q^{\frac{s}{r}}x(1)\right),$$

which implies:

$$\left(\dot{C}q^{\frac{s}{r}}x(1)\right) - \left(\frac{s}{s+2d}q^{-\frac{s+2d}{r}}x(3)\right)\left(q^{\frac{s}{r}}x(1)\right) = 0.$$

In general, the following equation holds for any undetermined coefficient $\overset{\scriptscriptstyle n}{A}$:

$$0 = \left(\overset{\scriptscriptstyle n}{A} q^{\frac{s}{r}} x(1)\right) + \left(\frac{s+d}{s+nd} q^{-\frac{s+nd}{r}} x(n)\right) \left(q^{\frac{s}{r}} x(2)\right)$$

$$+ \left(\frac{s+2d}{s+nd} q^{-\frac{s+nd}{r}} x(n-1)\right) \left(q^{\frac{s}{r}} x(3)\right) + \cdots$$

$$+ \left(\frac{s+(n-1)d}{s+nd} q^{-\frac{s+nd}{r}} x(2)\right) \left(q^{\frac{s}{r}} x(n)\right)$$

$$+ \left(\frac{s+nd}{s+nd} q^{-\frac{s+nd}{r}} x(1)\right) \left(q^{\frac{s}{r}} x(n+1)\right),$$

and then, by using formula 4.14, one gets:

$$- \left(\frac{s}{s+nd} q^{-\frac{s+nd}{r}} x(n+1)\right) \left(q^{\frac{s}{r}} x(1)\right) =$$

$$\left(\frac{s+d}{s+nd} q^{-\frac{s+nd}{r}} x(n)\right) \left(q^{\frac{s}{r}} x(2)\right) + \left(\frac{s+2d}{s+nd} q^{-\frac{s+nd}{r}} x(n-1)\right) \left(q^{\frac{s}{r}} x(3)\right) + \cdots$$

$$+ \left(\frac{s+nd}{s+nd} q^{-\frac{s+nd}{r}} x(1)\right) \left(q^{\frac{s}{r}} x(n+1)\right),$$

and from there it follows that:

$$\left(\overset{\scriptscriptstyle n}{A} q^{\frac{s}{r}} x(1)\right) - \left(\frac{s}{s+nd} q^{-\frac{s+nd}{r}} x(n+1)\right) \left(q^{\frac{s}{r}} x(1)\right) = 0,$$

which finally gives the expression of the general coefficient:

$$\overset{\scriptscriptstyle n}{A} = \frac{s}{s+nd} q^{-\frac{s+nd}{r}} x(n+1).$$

Using Hindenburg's notation for expressing the general term of a series, Rothe establishes the combinatorial formula:

$$x^s \, \daleth \, (n+1) = \left(\frac{s}{s+nd} q^{-\frac{s+nd}{r}} x(n+1)\right) y^{\frac{l(s+nd)}{r}}, \tag{4.16}$$

which serves to directly calculate the $(n+1)$th term of a reversed series.

Rothe points out that Eschenbach's formula 4.9 can be derived from formula 4.16 by setting $\overset{0}{m} = \frac{s}{r}$, $\overset{1}{m} = \frac{s+d}{r}$, $\overset{2}{m} = \frac{s+2d}{r}$, ..., $\overset{n}{m} = \frac{s+nd}{r}$, which gives:

$$x^s \, \daleth \, (n+1) = \left(\frac{\overset{0}{m}}{\overset{n}{m}} q^{-\overset{n}{m}} x(n+1)\right) y^{\overset{n}{m}l}$$

According to Rothe, this is exactly the same formula given by Eschenbach, but it has been expressed more succinctly. Indeed, in accordance with Eq. 4.13, the quantity $q^{-{}^{n}m}x(n+1)$ can be expressed as follows:

$$q^{-{}^{n}m}x(n+1) =$$

$$-{}^{n}m\mathfrak{A}\alpha^{-{}^{n}m-1}\mathfrak{a}\left({}^{n}A\right) + {}^{-{}^{n}m}\mathfrak{B}\alpha^{-{}^{n}m-2}\mathfrak{b}\left({}^{n}B\right) + \cdots + {}^{-{}^{n}m}\mathfrak{N}\alpha^{-{}^{n}m-n}\mathfrak{n}\left({}^{n}\mathcal{N}\right)$$

$$= \frac{-{}^{n}m}{\alpha^{n}m}\left(\frac{\mathfrak{a}({}^{n}A)}{\alpha} - \frac{{}^{n}m+1\mathfrak{A}\mathfrak{b}({}^{n}B)}{2\alpha^{2}} + \frac{{}^{n}m+2\mathfrak{B}\mathfrak{c}({}^{n}C)}{3\alpha^{3}} - \cdots \pm \frac{{}^{n}m+n-1\mathfrak{N}\mathfrak{n}({}^{n}\mathcal{N})}{n\alpha^{n}}\right)$$

$$q\,[\alpha,\beta,\gamma,\ldots],\qquad \begin{pmatrix}\beta & \gamma & \delta & \cdots \\ 1 & 2 & 3 & \cdots\end{pmatrix}.$$

Therefore, this is effectively the formula of Eschenbach.

Local symbols are the notational instrument giving Rothe the possibility to present calculations as a sequence of combinatorial formulae in his proof of Eschenbach's result, and this is so because those symbols are in fact combinatorial functions, taking the word "function" in the sense it had in the eighteenth century. From a notational point of view, Rothe's dissertation represents an improvement compared to Hindenburg's *Novi systematis* insofar as it develops a coherent notation for conveying the central ideas of the combinatorial analysis, and, consequently, it moves Hindenburg's project of creating a mathematical language forward from the point of being just a loose collection of notations. Rothe's local symbols provides Hindenburg's combinatorial analysis with its own combinatorial functions. From a theoretical point of view, Rothe's proof deeply depends on the multinomial theorem or, more precisely, on Hindenburg's version of the multinomial theorem, which has been used to establish Eq. 4.13. However, this proof does not depend only on the theory of the combinatorial analysis but also on differential calculus, since differential calculus has been used to justify formula 4.14, and so the proof does not rest only on the ground of Hindenburg's theory.

Rothe's dissertation completes Eschenbach's work not only in the sense that it provides a proof of Eschenbach's formula, but Rothe pays attention also to Eschenbach's claim according to which the mathematical notations of the combinatorial analysis can give us a good understanding of the complex law that Leibniz formulated for calculating the coefficients of a reversed series, and enables us to express it clearly and concisely. Here again Eschenbach does not show how this is to be done, but Rothe tries to do it at the end of his dissertation. As Leibniz, Rothe puts the equation:

$$0 = (01Y + 02Y^{2} + 03Y^{3}\&c) + (-10 + 11Y + 12Y^{2} + 13Y^{3}\&c)Z +$$

$$(20 + 21Y + 22Y^{2} + 23Y^{3}\&c)Z^{2} + \&c,$$

and then supposes that:

$$Z = \dot{A}Y + \dot{B}Y^2 + \dot{C}Y^3 + \cdots,$$

where $\dot{A}$, $\dot{B}$, $\dot{C}$, ... are undetermined coefficients. To find the value of these coefficients, Rothe calculates the positive integer powers of the possibly infinite multinomial Z as follows:

$$Z = \dot{A}Y + \dot{B}Y^2 + \dot{C}Y^3 + \cdots,$$
$$Z^2 = \mathfrak{b}(^2\text{B})Y^2 + \mathfrak{b}(^3\text{B})Y^3 + \mathfrak{b}(^4\text{B})Y^4 + \cdots,$$
$$Z^3 = \mathfrak{c}(^3\text{C})Y^3 + \mathfrak{c}(^4\text{C})Y^4 + \mathfrak{c}(^5\text{C})Y^5 + \cdots,$$

etc.

$$\begin{pmatrix} \dot{A} & \dot{B} & \dot{C} & \cdots \\ 1 & 2 & 3 & \cdots \end{pmatrix}.$$

From these powers and from the original equation, it follows that:

$$0 = \left(01Y + 02Y^2 + 03Y^3 + \cdots \right)$$
$$+ \left(-10 + 11Y + 12Y^2 + 13Y^3 + \cdots \right)\left(\dot{A}Y + \dot{B}Y^2 + \dot{C}Y^3 + \cdots \right)$$
$$+ \left(20 + 21Y + 22Y^2 + 23Y^3 + \cdots \right)\left(\mathfrak{b}(^2\text{B})Y^2 + \mathfrak{b}(^3\text{B})Y^3 + \mathfrak{b}(^4\text{B})Y^4 + \cdots \right)$$
$$+ \left(30 + 31Y + 32Y^2 + 23Y^3 + \cdots \right)\left(\mathfrak{c}(^3\text{C})Y^3 + \mathfrak{c}(^4\text{C})Y^4 + \mathfrak{c}(^5\text{C})Y^5 + \cdots \right)$$
$$= (01 - 10\dot{A})\,Y + \left(02 - 10\dot{B} + 11\dot{A} + 20\mathfrak{b}(^2\text{B}) \right)Y^2$$
$$+ \left(03 - 10\dot{C} + 11\dot{B} + 12\dot{A} + 20\mathfrak{b}(^3\text{B}) + 21\mathfrak{b}(^2\text{B}) + 30\mathfrak{c}(^3\text{C}) \right)Y^3 + \cdots.$$

And by the method of undetermined coefficients, one gets:

$$\dot{A} = \frac{01}{10},$$
$$\dot{B} = \frac{02 + 11\dot{A} + 20\mathfrak{b}(^2\text{B})}{10},$$
$$\dot{C} = \frac{03 + 11\dot{B} + 12\dot{A} + 20\mathfrak{b}(^3\text{B}) + 21\mathfrak{b}(^2\text{B}) + 30\mathfrak{c}(^3\text{C})}{10},$$

etc.,

which are the coefficients calculated by Leibniz, as can be seen in Sect. 2.2.3.3.

For Rothe, his reformulation renders comprehensible Leibniz's law of formation of coefficients (Rothe 1793, 26):

> The very difficult combinatorial law that regulates [the formation] of coefficients, which was enunciated in common language by Leibniz in the most obscure way, has been completely clarified by the use of analytic-combinatorial symbols. From this example, it becomes apparent how many advantages and benefits follow for Analysis from using the analytic-combinatorial symbols.[11]

From a our contemporary point of view, it could be said that Rothe's reformulation consists in replacing Leibniz's symbols by Hindenburg's notations, and nothing else. But it should be understood that this statement expresses Rothe's support to Hindenburg's criticism of Leibniz's conception of the theory of combinations discussed in Sect. 3.2.2. Rothe explicitly contrasts "the most obscure way" in which Leibniz's combinatorial art was formulated with the efficient and comprehensible combinatorial theory of Hindenburg, which allows to formulate complex mathematical expressions in simple analytic-combinatorial symbols. On the other hand, Rothe's reformulation makes clear the conviction that natural language is not suitable for mathematics, and reinforces the necessity for an ongoing research project on mathematical notations.

4.1.3.5 Eschenbach-Rothe's Formula and Lagrange's Inversion Theorem

In his treatise on dimension symbols, Fischer was the first to raise the question of the equivalence between his formula 4.10 and Lagrange's formula 4.4, but seemed unable to prove it. As seen in Sect. 4.1.3.2, Fischer's formula 4.10 was a translation of Eschenbach's formula 4.9 into the language of dimension symbols, thus if Fischer's claim about the equivalence of both formulae is right, Eschenbach's formula must be equivalent to Lagrange's as well. Töpfer, Rothe, and even Hindenburg accepted the weaker hypothesis that Eschenbach's formula could be derived from Lagrange's inversion theorem. Indeed, after the publication of Fischer's book, they make systematic reference to Lagrange's work when discussing the theorem on reversion of series, particularly the solution of Eschenbach. It is worth noting that, until 1794, Hindenburg never mentions the name of Fischer in his publications, not even in his booklet (Hindenburg 1793), which was devoted to the analysis of the problem of reversing series and where one can find, on page twenty-four, the statement that Eschenbach's formula can be derived from Lagrange's, but this is an issue to which we will return later. The correctness of the weaker hypothesis will be demonstrated in 1794 by Johann Friedrich Pfaff (1765–1825), a former student of Kästner

---

[11] Lex combinationis complicatissima, quam hic coefficientes servant a *Leibnitio* verbis perquam obscuris pronunciata, signis hic combinatorio-analyticis clarissima reddita est. Iudicari itaque potest, etiam ex hoc exemplo, qua vtilitate, quoque commodo, signa analytico-combinatoria in Analysi possint adhiberi.

and professor of mathematics in Helmstadt, in two articles which constitute his first contribution to the *Archiv der reinen und angewandten Mathematik* of Hindenburg (Pfaff 1794a, 1794b; Lubet 1998, 94–98). In the first article, Pfaff justifies Lagrange's inversion theorem using Taylor's theorem, and then, in the second article, he deduces Rothe's formula 4.16, which might more fairly be called the formula of Eschenbach-Rothe. In 1795, Rothe proposed a method to "deduce" Lagrange's theorem from the combinatorial analysis by using local symbols, and so Lagrange's formula 4.4 could be seen as a consequence of Eschenbach-Rothe's formula 4.16 (Rothe 1795). In what follows, Pfaff's and Rothe's articles will be examined in detail.

In his first article (Pfaff 1794b), following Lagrange's inversion theorem, Pfaff considers an equation of the form:

$$y = x - z\phi(x),$$

where ϕ can be any function of x. Then he considers other function ψ of x and assumes that it can be expressed as follows:

$$\psi(x) = \psi(y) + Y^I z + Y^{II} z^2 + \cdots + Y^N z^n + etc., \tag{4.17}$$

where Y^I, Y^{II}, ..., Y^N are undetermined coefficients, and the superscripts of these coefficients are indeed indexes. Pfaff's proof of Lagrange's theorem consists in finding a general formula to directly calculate the general coefficient Y^N, and Pfaff points out that this formula will be a function of y.

Since $x = y + z\phi(x)$, Pfaff uses Taylor's theorem in order to express the function ψ as a power series in the variable z:

$$\psi(x) = \psi(y + z\phi(x))$$

$$= \psi(y) + z\phi(x)\frac{d\psi(y)}{dy} + z^2(\phi(x))^2\frac{d^2\psi(y)}{1.2\,dy^2} + \cdots \tag{4.18}$$

$$+ z^n(\phi(x))^n\frac{d^n\psi(y)}{(1.2\ldots n)\,dy^n} + etc.$$

Since ϕ and ψ can be "any" functions of x, one can have in particular $\phi = \psi$, and so Eq. 4.17 becomes:

$$\phi(x) = \phi(y) + Y^I z + Y^{II} z^2 + \cdots + Y^N z^n + etc.$$

Replacing this value in 4.18, and by comparing coefficients with Eq. 4.17, it follows that:

$$Y^I = \phi(y)\frac{d\psi(y)}{dy}.$$

Pfaff also calculates the second coefficient Y^{II}. Given the value of Y^I, one deduces from Eq. 4.17 (taking $\phi = \psi$) that:

$$(\phi (x))^2 = (\phi (y))^2 + z\phi (y) \frac{d (\phi (y))^2}{dy} + \cdots$$

Replacing the expansions of $\phi(x)$ and $(\phi (x))^2$ in Eq. 4.18, and by comparing coefficients with Eq. 4.17, it follows that:

$$Y^{II} = \frac{d \left((\phi (y))^2 \frac{d\psi (y)}{dy} \right)}{1 \cdot 2 \, dy}.$$

Then, Pfaff explicitly advances an argument by mathematical induction by supposing that the coefficients $Y^I, Y^{II}, \ldots, Y^{N-1}$ are of the form:

$$Y^R = \frac{d^{r-1} \left((\phi (y))^r \frac{d\psi (y)}{dy} \right)}{(1.2 \ldots r) \, dy^{r-1}},$$

for all $R = r = 1, 2, 3, \ldots, N - 1$. Under this assumption, he can express the functions $(\phi (x))^3, \ldots, (\phi (x))^r$ as power series in the same way as the previous functions $\phi (x)$ and $(\phi (x))^2$. Then, replacing these expansions in Eq. 4.18, and by comparing coefficients with Eq. 4.17, he can prove the inductive step, and thus concludes that the general coefficient of the series 4.17 is given by:

$$Y^N = \frac{1}{(1.2.3 \ldots n) \, dy^{n-1}} \cdot d^{n-1} \left((\phi (y))^n \frac{d\psi (y)}{dy} \right), \tag{4.19}$$

where $N = n$ is any positive whole number. This completes Pfaff's proof of Lagrange's inversion theorem. A Latin version of this proof will be included in a brief treatise published in 1797, and where Pfaff will come back to the issue of the reversion of series and its applications to the problem of finding the roots of some equations (Pfaff 1797b, 227–231).

In his second article (Pfaff 1794a), Pfaff deduces Eschenbach-Rothe's formula 4.16 from Lagrange's inversion theorem. He puts the equation:

$$z = a^I x + a^{II} x^2 + a^{III} x^3 + etc.,$$

where the superscripts of the coefficients are indexes, and he writes $Fx = a^I x + a^{II} x^2 + a^{III} x^3 + etc.$, therefore:

$$0 = x - z \left(\frac{x}{Fx} \right).$$

He points out that, if one sets $y = 0$, $\phi(x) = \frac{x}{Fx}$, and $\psi(x) = (\phi(x))^s$, it becomes clear that one can apply Lagrange's theorem to this latter equation and to $\psi = \phi^s$, which gives the following series expansion, according to the general coefficient expressed by formula 4.19:

$$\frac{x^s}{(Fx)^s} = (\varphi(x))^s$$

$$= y^s (Fy)^{-s} + \frac{s}{s+1} \frac{d\left(y^{s+1}(Fy)^{-s-1}\right)}{dy} z + \frac{s}{s+2} \frac{d^2\left(y^{s+2}(Fy)^{-s-2}\right)}{1.2\,dy^2} z^2 + etc.$$

Pfaff does not specify the nature of s, but it is, at least, a rational number since he wants to prove Eschenbach-Rothe's formula. By changing x to y on the right-hand side of this equation, and thus the derivatives should be evaluated at $x = 0$, this equation can be rewritten as:

$$x^s = \frac{s}{s}\left(x^s z^{-s}\right) z^s + \frac{s}{s+1} \frac{d\left(x^{s+1} z^{-s-1}\right)}{dx} z^{s+1} + \frac{s}{s+2} \frac{d^2\left(x^{s+2} z^{-s-2}\right)}{1.2\,dx^2} z^{s+2} + \cdots$$

$$+ \frac{s}{s+n} \frac{d^n\left(x^{s+n} z^{-s-n}\right)}{(1.2.3\ldots n)dx^n} z^{s+n} + etc. \tag{4.20}$$

Then, he considers the series:

$$z^{-s-n} = a^I x^{-s-n} + a^{II} x^{-s-n+1} + \cdots + a^{III} x^{-s} + etc., \tag{4.21}$$

from which he gets:

$$x^{s+n} z^{-s-n} = a^I + a^{II} x + a^{III} x^2 + \cdots + a^{N+1} x^n + etc. \tag{4.22}$$

Pfaff says that, given this latter equation, one knows that the general coefficient of the series is given by:

$$a^{N+1} = \frac{d^n\left(x^{s+n} z^{-s-n}\right)}{(1.2\ldots n)\,dx^n},$$

where the derivatives are evaluated at $x = 0$. It is possible that he calculated this coefficient by applying Taylor's theorem to the left-hand side of Eq. 4.22 (taken z as a fixed parameter). Therefore, the coefficient of the term z^{s+n} in the series given by Eq. 4.20 is equal to:

$$\frac{s}{s+n} a^{N+1} = \frac{s}{s+n} z^{-s-n} \times (n+1),$$

where the expression on the right-hand side denotes, as we know, the $(n + 1)$th coefficient of the series z^{-s-n}.

If one sets $x = X^{\frac{1}{d}}$ and $z = Z^{\frac{d}{p}}$ in Eq. 4.20 (here again Pfaff does not specify the nature of p and d), the general term of its right-hand side is given by:

$$\frac{s}{s+nd}\; \frac{d^n\left(\left(X^{\frac{s}{d}+n}\right)\left(Z^{-\frac{s+nd}{p}}\right)\right)}{(1.2.3\ldots n)dx^n}\; Z^{\frac{s+nd}{p}}.$$

If one also sets $x = X^{\frac{1}{d}}$ and $z = Z^{\frac{d}{p}}$ in Eqs. 4.21 and 4.22, the general coefficient of the expression on the right-hand side of these equations becomes:

$$\mathfrak{a}^{N+1} = \frac{d^n\left(\left(X^{\frac{s}{d}+n}\right)\left(Z^{-\frac{s+nd}{p}}\right)\right)}{(1.2.3\ldots n)dx^n}.$$

Therefore, the general term of the expansion of $x^s = X^{\frac{s}{d}}$ in Eq. 4.20 can be written as follows:

$$x^s\;\daleth(n+1) = \frac{s}{s+nd}Z^{-\frac{s+nd}{p}} \times (n+1)\, Z^{\frac{s+nd}{p}},$$

which is effectively the formula 4.16 of Eschenbach and Rothe.

In a final comment, Pfaff notes that, although he already knew Lagrange's inversion theorem, it was the reading of Fischer's book on dimension symbols, particularly the remark about the equivalence between Fischer's formula for the reversion of series and Lagrange's, that led him to reconsider the question of the relation between Lagrange's inversion theorem and Eschenbach-Rothe's formula for the reversion of series (Pfaff 1794a, 87–88). After reflecting a while, he came up with the ingenious scheme of his proofs.

As pointed out above, in 1795 Rothe deduces Lagrange's inversion theorem from his formula 4.16 (Rothe 1795, 442 ff.). In fact, his proof consists in showing a sequence of symbolic transformations in which he uses his local symbols in order to obtain the consequence stated in Lagrange's inversion theorem. However, as will be seen below, the use of local symbols seems to be restrained to provide some notational abbreviations for the coefficients obtained by applications of Taylor's theorem. It is interesting to note that, in his article, Rothe follows Pfaff's rephrasing of Lagrange's inversion theorem, which suggests that Rothe considers that his proof and Pfaff's articles are complementary works, that is to say, his proof and those of Pfaff can be regarded as a collaborative work. Thus, as Pfaff, he considers an equation of the form $x = y + z\phi(x)$, and puts $x = y + v$, so he gets:

$$z^{-1} = \frac{\phi(y+v)}{v}$$

By using Taylor's theorem to expand $\phi(y + v)$ as a power series, he obtains:

$$z^{-1} = \phi(y) v^{-1} + \frac{d\phi(y)}{dy} v^0 + \frac{d^2\phi(y)}{1.2 \, dy^2} v^1 + \frac{d^3\phi(y)}{1.2.3 \, dy^3} v^2 + etc.$$

This series can be reversed by means of Eschenbach-Rothe's formula 4.16 (setting $d = 1$, $r = l = -1$ in this formula), which yields:

$$v^s \, \daleth(n+1) = \frac{s}{s+n} q^{s+n} x(n+1) z^{s+n}$$

$$q \left[\phi(y), \frac{d\phi(y)}{dy}, \frac{d^2\phi(y)}{1.2 \, dy^2}, \frac{d^3\phi(y)}{1.2.3 \, dy^3}, \cdots \right],$$

where the expression in the second row is the scale of the series q (for the meaning of a scale, see Sect. 4.1.3.4).

In another paper (Rothe 1794, 228–229), Rothe argues that, given a function ξ of $x = z + c$, one can calculate the following series using Taylor's theorem:

$$(\xi(x))^\beta = (\xi(z))^\beta + \frac{d\left((\xi(z))^\beta\right)}{dz} c^1 + \cdots + \frac{d^n\left((\xi(z))^\beta\right)}{(1.2\ldots n) dz^n} c^n + \cdots,$$

for some number β, whose nature is not specified by Rothe. On the other hand, considering the scale:

$$r \left[\xi(z), \frac{d\xi(z)}{dz}, \frac{d^2\xi(z)}{1.2 \, dz^2}, \frac{d^3\xi(z)}{1.2.3 \, dz^3}, \cdots \right],$$

ξ^β can be expressed in terms of local symbols as follows:

$$(\xi(x))^\beta = r^\beta x(1) + r^\beta x(2) c^1 + \cdots + r^\beta x(n+1) c^n + \cdots.$$

Therefore, by comparing the coefficients of these two expansions, it follows that:

$$r^\beta x(n+1) = \frac{d^n\left((\xi(z))^\beta\right)}{(1.2\ldots n) dz^n}.$$

Rothe's argument is based on the belief that any function can be expanded in power series, as it was common at the time. Thus, taking into account this result, he can write the general coefficient of q^{s+n} as follows:

$$q^{s+n} x(n+1) = \frac{d^n(\phi(y))^{s+n}}{(1.2.3\ldots n) dy^n}.$$

Hence, the general term of v^s is given by:

$$v^s \, \daleth \, (n+1) = \frac{s}{s+n} \frac{d^n \, (\phi \, (y))^{s+n}}{(1.2.3 \dots n) \, dy^n} z^{s+n}.$$

From this latter formula, Rothe deduces the expansions of the successive powers of v:

$$v = \frac{1}{1} \phi \, (y) \, z + \frac{1}{2} \frac{d \, (\phi \, (y))^2}{dy} z^2 + \frac{1}{3} \frac{d^2 \, (\phi \, (y))^3}{1.2 \, dy^2} z^3 + \cdots ,$$

$$v^2 = \frac{2}{2} \, (\phi \, (y))^2 \, z^2 + \frac{2}{3} \frac{d \, (\phi \, (y))^3}{dy} z^3 + \cdots ,$$

$$v^3 = \frac{3}{3} \, (\phi \, (y))^3 \, z^3 + \cdots ,$$

etc.,

which can be written succinctly as:

$$v \, \daleth n = \frac{1}{n} \frac{d^{n-1} \, (\phi \, (y))^n}{(1.2.3 \dots (n-1)) \, dy^{n-1}} z^n ,$$

$$v^2 \, \daleth \, (n-1) = \frac{2}{n} \frac{d^{n-2} \, (\phi \, (y))^n}{(1.2.3 \dots (n-2)) \, dy^{n-2}} z^n ,$$

$$v^3 \, \daleth \, (n-2) = \frac{3}{n} \frac{d^{n-3} \, (\phi \, (y))^n}{(1.2.3 \dots (n-3)) \, dy^{n-3}} z^n ,$$

$$\vdots$$

$$v^n \, \daleth 1 = \frac{n}{n} \, (\phi \, (y))^n \, z^n.$$

Then, given another function ψ of $x = y + v$, the theorem of Taylor leads to the power series:

$$\psi \, (x) = \psi \, (y) + \frac{d\psi \, (y)}{dy} v + \frac{d^2\psi \, (y)}{1.2 \, dy^2} v^2 + \frac{d^3\psi \, (y)}{1.2.3 \, dy^3} v^3 + \cdots .$$

Replacing in this equation the successive powers of v by the corresponding expansions calculated in the previous paragraph, and after regrouping terms, Rothe obtains:

$$\psi \, (x) = \psi \, (y) + \frac{1}{1} \frac{\phi \, (y) \, d\psi \, (y)}{dy} z$$

$$+ \left(\frac{1}{2} \frac{d \, (\phi \, (y))^2 \, d\psi \, (y)}{dy^2} + \frac{2}{2} \frac{(\phi \, (y))^2 \, d^2\psi \, (y)}{1.2 \, dy^2} \right) z^2 + \cdots ,$$

and, in general, the coefficient of the term z^n of this expansion is given by:

$$\psi(x) \times (n+1) = \begin{pmatrix} \frac{1}{n} \frac{d^{n-1}(\phi(y))^n d\psi(y)}{(1.2.3...(n-1))dy^n} \\ +\frac{2}{n} \frac{d^{n-2}(\phi(y))^n d^2\psi(y)}{1.2.3...(n-2)\times 1.2\, dy^n} \\ \vdots \\ +\frac{n}{n} \frac{(\phi(y))^n d^n \psi(y)}{(1.2.3...n)dy^n} \end{pmatrix}$$

Therefore, it follows that:

$$(1 \cdot 2 \cdot 3 \cdots n)\, \psi(x) \times (n+1) = \frac{1}{dy^n} \begin{pmatrix} \frac{1}{n} (^n\mathfrak{A})\, d^{n-1}\, (\phi(y))^n\, d\psi(y) \\ +\frac{2}{n} (^n\mathfrak{B})\, d^{n-2}\, (\phi(y))^n\, d^2\psi(y) \\ \vdots \\ +\frac{n}{n} (^n\mathfrak{N})\, (\phi(y))^n\, d^n\psi(y) \end{pmatrix}$$

$$= \frac{1}{dy^n} \begin{pmatrix} d^{n-1}\, (\phi(y))^n\, d\psi(y) \\ +\binom{n-1}{\mathfrak{A}} d^{n-2}\, (\phi(y))^n\, d^2\psi(y) \\ \vdots \\ +\binom{n-1\,\,\,\,-1}{\mathfrak{N}} (\phi(y))^n\, d^n\psi(y) \end{pmatrix}$$

$$= \frac{d^{n-1}\, (\phi(y))^n\, d\psi(y)}{dy^n},$$

where the Gothic capital letters with a left superscript are binomial coefficients (see Sect. 3.3.3 for their meaning). Thus, the general coefficient of the series expansion of ψ is:

$$\psi(x) \times (n+1) = \frac{d^{n-1}\, (\phi(y))^n\, d\psi(y)}{(1 \cdot 2 \cdot 3 \cdots n)\, dy^n},$$

which corresponds to the formula 4.19 given by Pfaff in his proof of Lagrange's inversion theorem. This completes Rothe's proof.

In short, taken together, the articles of Pfaff and Rothe suggest a way of overcoming the question posed by Fischer about the relationship between a combinatorial formula for the reversion of series and Lagrange's theorem, and, by the standards of the time, they successfully establish the equivalence between Eschenbach-Rothe's formula and Lagrange's inversion theorem. This will help put an end to the dispute concerning the authorship of the combinatorial formula for the reversion of series. In the general context of this dispute, Pfaff's and Rothe's articles show that the combinatorial analysis, with its notations and concepts, can reach remarkable results which the theory of dimension symbols was unable to prove. This will definitely jeopardize the future of Fischer's theory.

4.1.3.6 Hindenburg's Silence

In this story of suspicions and controversy, an impartial audience cannot help but notice the stunning absence of the main character along the narrative thread of the plot. As the creator of the combinatorial analysis, Hindenburg must have felt quite outraged by the publication of the theory of dimension symbols. Why does he not dare to speak out against the alleged originality of Fischer's theory? He shares, however, the same sentiment toward this situation as his former students, as is confirmed by a letter addressed to Georg Christoph Lichtenberg (1742–1799) on 24 May 1793 (Hindenburg 1992, 96):

> The present writing [Töpfer's *Combinatorische Analytik*], which I have the honor to send you, is certainly not mine, but concerns me in a very particular way, as you will see. The plagiarism committed against me by Fischer is much more serious than that committed by Achard against Gehler.[12]

Franz Carl Achard (1753–1821) was a German scientist who published in 1791 the book *Vorlesungen über die Experimentalphysik*, which contains long paragraphs copied from the reference dictionary *Physicalisches Wörterbuch*, written and published by Johann Samuel Traugott Gehler (1751–1795) between 1787 and 1795. This case of plagiarism caused quite a stir among German scientists, and Hindenburg insists, in the rest of his letter, that even if Achard's actions were highly dishonorable, stealing a bunch of paragraphs is a much less serious crime than stealing an entire theory. Thus, in the private sphere, Hindenburg complains about the shameless plagiarist Fischer and passes critical judgment upon him.

Nevertheless, in the public sphere, Hindenburg remains silent. In 1793, he delivered a speech at the University of Leipzig on the problem of the reversion of series, which in fact revolved around Eschenbach's formula 4.9 and Rothe's formula 4.16 (Hindenburg 1793). In his speech, Hindenburg endorses the notion of "local symbol" proposed by Rothe and uses it to translate the different versions of the theorem on reversion of series given by

---

[12] Gegenwärtige Schrift, die ich die Ehre habe, Ihnen zu übersenden, ist zwar nicht von mir, geht mich aber, wie Sie finden werden, sehr nahe an. Das Plagium, das Fischer an mir begangen hat, ist viel bedenklicher, als das Achard an Gehlern begieng.

Newton, Moivre, and Tempelhoff into the language of his combinatorial analysis. On the contrary, neither Lagrange's inversion theorem nor Lagrange's version of the theorem on reversion of series was reinterpreted in terms of the combinatorial analysis, but within Hindenburg's discourse, Lagrange's inversion theorem occupies the privileged place of being a more general proposition from which Eschenbach's and Rothe's formulae can be derived. Despite the fact that this remark was originally made by Fischer in his book on dimension symbols, Hindenburg does not breathe a word of this. Instead, Hindenburg continually refers to Töpfer's *Combinatorische Analytik* in order to support his ideas, but taking special care not to reveal anything about the theory of dimension symbols. This is even more surprising considering that the main objective of Töpfer in writing his book was to unmask the imposture of Fischer. However, Hindenburg decides to completely ignore this controversial issue and review Töpfer's, Rothe's, and Eschenbach's works as if the question of plagiarism did not exist.

In an article appeared one year later in the *Archiv*, Hindenburg takes up again the topic of the reversion of series and Lagrange's inversion theorem (Hindenburg 1794d). As in his speech of 1793, Hindenburg summarizes here Eschenbach's and Rothe's dissertations on the subject, but this time the focus is on Pfaff's deduction of Eschenbach-Rothe's formula 4.16 from Lagrange's inversion theorem, analyzed above in Sect. 4.1.3.5. Other than reviewing the work of his colleagues, the aim of Hindenburg's article is to highlight the mathematical virtues and qualities of his combinatorial analysis. It is in this context that Hindenburg finally makes reference to Fischer's book in two different passages. First, he floats the idea that the justification of the theorem on reversion of series carried out by means of the combinatorial analysis is more rigorous than the one proposed by Lagrange, even though some authors, like Fischer, affirm that Lagrange's argumentation meets the highest standards of mathematical rigor (Hindenburg 1794d, 90). The second allusion to Fischer's work bears less a character of anecdote and more that of a relevant scientific fact. In contrast with Lagrange's differential procedures, Hindenburg celebrates the efficiency of the analytic-combinatorial formulae, asserting that they can better express the law of formation concerning the general coefficient of a reversed series. To provide a sounder basis for this idea, Hindenburg quotes, defying all expectations, a passage from Fischer's book which supports, according to Hindenburg, his point of view on the advantages and strengths of combinatorial methods over differential methods (Hindenburg 1794d, 92). While, in the private sphere, Hindenburg complains about the gravity of the plagiarism committed by Fischer and spreads the news, in his academic writings either he does not even mention Fischer, or he makes reference to Fischer's work in a very positive way, to the point of using it to support his ideas, as if Fischer were one of his collaborators.

Hindenburg's behavior becomes even more complex. In the section devoted to book reviews of the first issue of the *Archiv*, one can find a note by Hindenburg subtitled "A contribution to the future history of this new science [the combinatorial analysis]", where Hindenburg explores prospective scientific developments of his theory, which was currently under construction (Hindenburg 1794b). Fischer's theory forms, however, the central axis around which Hindenburg's note revolves. Hindenburg has finally decided to face the problem head-on, or almost head-on. By contrast with his former students, he never talks of plagiarism but of "coincidence" (*Uebereinstimmung*) between the combinatorial analysis and the theory of dimension symbols. Faced with such a coincidence, he wonders about the origins of Fischer's theory, without venturing into any hypothesis in this regard, and remains fairly impartial. Indeed, he suggests that the reader should consult, on the one hand, Töpfer's and Rothe's works where the hypothesis of a suspicious origin has been discussed, and, on the other hand, a book by Fischer published in 1794 where Fischer tries to explain the origins of his theory in order to defend himself against the accusations of plagiarism (in fact, Hindenburg carefully enumerates in his note the reasons given by Fischer in favor of the originality of the theory of dimension symbols) (Hindenburg 1794b, 112–114; Fischer 1794). Hindenburg is careful not to draw any conclusion and merely outlines the arguments for and against an independent origin. In particular, his note draws special attention to Fischer's allegation according to which the conception of the dimension symbols occurred during a period previous to the date on which he found out about Hindenburg's notations. In this case too Hindenburg does not take a position, but this does not preclude him from reporting the opinions of others, especially those of Töpfer. This is part, apparently, of a well-conceived strategy designed to dismantle the theory of dimension symbols, a strategy in which Hindenburg must keep publicly the proper balance between impartiality and denunciation, while his collaborators openly accuse Fischer of plagiarism and supply supporting evidence.

In the end, the theory of dimension symbols could not resist the onslaught of this still small but well-organized group formed around Hindenburg, and died despite the fact that it was backed up by an actual treatise, whereas the awaited treatise on the combinatorial analysis was promised more than thirteen years ago. And as for Fischer, Hindenburg permitted him to publish a paper in his *Archiv* in 1797, in which Fischer himself replaces his theory of dimension symbols with Hindenburg's combinatorial analysis (Fischer 1797, 426–440).

*

It is difficult to measure the impact that this well-controlled silence of Hindenburg had on the extinction of the theory of dimension symbols. In contrast, Hindenburg was not wrong about predicting "the future history" of the combinatorial analysis. His strategy of leaving his defense in the hands of his former students led to their unification. They worked in coordination with each other to plan for their roles and operated as an integrated group to prevent Fischer's theory from taking hold in German mathematical communities. Regarding the theorem on reversion of series, this coordination is the only possible explanation of the cross-references between Töpfer's *Combinatorische Analytik* and Rothe's dissertation, since their publication dates fall very close to each other and, even more, Töpfer knows the content of Rothe's dissertation before it gets published. The oriented education carried out by Hindenburg in the classroom has paid off in a public debate won by his former students. They finally carry through the victory of this confrontation by proving the equivalence between Lagrange's inversion theorem and Eschenbach-Rothe's formula, event that takes place on the pages of the *Archiv*, the recently inaugurated mathematical journal of Hindenburg. Although this proof has been written by two different persons, it is presented in Hindenburg's reviews as the result of a group of mathematicians that shares a particular conception of mathematics, a specific mathematical language, a general mathematical method, and a common aim. The German combinatorial school emerges from this historical process and the *Archiv* shall be made available to its members in order to promote the progress of the combinatorial analysis, which will be the theory of the school from this moment on.

4.2   The Most Important Theorem in the Whole of Analysis

The formation of the German combinatorial school is the result of coordinated and orderly cooperation between Töpfer and Rothe, which was reinforced by the promotion of their work and the strategy of impartial silence conducted by Hindenburg. These three mathematicians can be considered a scientific team in the sense that they organize their tasks with the view to reaching a common goal, and their organization allows them to obtain new mathematical results, as the combinatorial formula 4.9 proved by Rothe in his dissertation or, even better, a legitimization for the combinatorial analysis as a mathematical theory in spite of the lack of an explicit and systematic presentation of it, which was the main objection advanced by Fischer to defend his theory. The high visibility the plagiarism conflict acquired among German scholars, and the presence of Lagrange's inversion theorem in the midst of it, attracted the attention of Pfaff. As can be seen in Sect. 4.1.3.5, it is clear that Pfaff was interested in solving a mathematical problem, that of deducing Eschenbach-Rothe's formula from Lagrange's theorem, but did not appear to be interested in the priority controversy. Indeed, Pfaff does not use the tools or the methods of the combinatorial analysis in his article. After writing his proofs, he will become acquainted with Hindenburg's project and will become certainly one of the most important members of the combinatorial school. Like him, other mathematicians will take interest in Hindenburg's mathematical project and will contribute to its development. All these contributions take place in the period after the priority dispute, which suggests that this conflict plays a major role in shaping the historical course of Hindenburg's project and in consolidating the combinatorial school. These contributions belong to a period characterized by the interest in solving the problems of Hindenburg's scientific work program established in the *Novi systematis*.

In this section, this period will be analyzed with the purpose of understanding how the new mathematical system of Hindenburg begins finally to take shape. It will be seen here that, in this historical process, the multinomial theorem becomes the cornerstone of the theoretical organization of the combinatorial analysis, to such an extent that Hindenburg calls it "the most important theorem in the whole of analysis". According to Netto, Hindenburg derived this new conviction from the fact that, in the proof of

Eschenbach's formula 4.9 given by Rothe in 1794, the application of the multinomial theorem is indispensable for the determination of the power series expansions used in the argumentation (Netto 1908, 216). Although Netto's explanation is possible and other scholars have subscribed to this view (Jahnke 1996, 146–147; Séguin 2005, 70), we disagree with such an interpretation because it seems to be an oversimplification of a more complex historical dynamics. In this section, we intend to show that this new conviction of Hindenburg depends mainly on three factors: first, a definitive return to the thought of Moivre and its reinterpretation as being the true source of the combinatorial analysis; second, the justification of the binomial theorem using the multinomial theorem; and third, the resolution of some of the programmatic problems indicated in the *Novi systematis* on the basis of the multinomial theorem, including the reversion of series.

4.2.1   Back to Moivre

In 1795, Hindenburg tackles again the study of Moivre's original article on the multinomial theorem. It is worth insisting that the *Methodus nova* was published seventeen years earlier and that, back then, Hindenburg was convinced that his method offered the best solution for justifying the multinomial theorem. Perhaps the recent events led him to reconsider and question whether the work of Moivre could reveal more secrets hidden in its pages; after all, Eschenbach had discovered his formula for the reversion of series by studying Moivre's work. In any case, after seventeen years, there is a revival of interest in the old articles of Moivre. This revival of interest, as well as its consequences for the combinatorial analysis, will be the subject matter of what follows.

4.2.1.1 The Method of Combinatorial Involutions

In 1794, Hindenburg publishes a new combinatorial method for his theory, which is based on his method of combinatorial tables and can be seen as an improvement of his partition tables (Hindenburg 1794c). The improvement consists in maximizing the use of the digits appearing in a table in order to simplify and reduce the very long partition tables. To this end, Hindenburg introduces the notion of "combinatorial involution", on which this technical improvement depends and which he draws from his own work on mathematical tables. As seen in Sect. 3.1, his *Beschreibung* contains a combinatorial interpretation of positional number systems, and some examples of this idea are illustrated in Fig. 3.1 by the tables of the binary, ternary, and quaternary systems. In 1794, Hindenburg explicitly constructs the table of the decimal system, which is shown in Fig. 4.1, even if this task can be easily done following the instructions of the *Beschreibung* (or the common sense). However, Hindenburg wants to show this trivial table to insist on the idea that, given a numeric base, the table corresponding to the number system in this base will be contained in all tables corresponding to number systems in bases higher than that (Hindenburg 1794c, 19 ff.). Thus, he insists that the first four boxes of the first table in Fig. 3.1 are included in the table of the decimal system shown in Fig. 4.1. Similarly, the first nine boxes of the

Fig. 4.1 Hindenburg's
decimal system

0	1	2	3	4	5	6	7	8	9	etc.
00	01	02	03	04	05	06	07	08	09	
10	11	12	13	14	15	16	17	18	19	
20	21	22	23	24	25	26	27	28	29	
30	31	32	33	34	35	36	37	38	39	etc.
40	41	42	43	44	45	46	47	48	49	
50	51	52	53	54	55	56	57	58	59	
60	61	62	63	64	65	66	67	68	69	
70	71	72	73	74	75	76	77	78	79	etc.
80	81	82	83	84	85	86	87	88	89	
90	91	92	93	94	95	96	97	98	99	

etc.   etc.   etc.   etc.

second table in Fig. 3.1 reappear in the table of the decimal system. Hence, the table of
the decimal system contains the elements of the tables corresponding to number systems
expressed in a base lower that ten.

Certainly, the meaning of the symbols contained in these tables varies from one table
to another since the meaning of a given sequence of digits varies from one number system
to another. When Hindenburg says that the table of a number system contains the tables
of the previous number systems, he means that it contains the same boxes with the same
sequences of symbols, putting aside the numerical meaning of the digits. In this sense,
all these tables are first and foremost tables of symbols devoid of any specific content and
meaning. To systematize the idea of constructing tables of symbols devoid of meaning,
Hindenburg reintroduces the notion of "gnomon" (Hindenburg 1794c, 20–21), which he
actually used in his *Beschreibung* to construct this kind of tables, as seen in Sect. 3.1, but
this time he returns to the Greek notion of "gnomon", from which he derives his method
of combinatorial involutions.

The Greek notion of "gnomon" has been extensively studied by many researchers in
different times (see, for instance, Heath 1921a,b; Knorr 1975; Michel 1958; Zhmud 1989).
Let's just remember that the term "gnomon" originally referred to the part of a sundial that
casts a shadow onto the dial, which generally has the shape of a right angle. Thus, a
gnomon was a kind of pattern used to measure time. In the Pythagorean tradition, the
notion evolved into the arithmetical concept that, if figurate numbers are constructed by
means of gnomons, one can calculate the sum of some basic arithmetic progressions. For
instance, in the following figure:

right angles represent gnomons, and taking the subsequent gnomons of the figure together, one obtains the sums 1, $1+3 = 4$, and $1+3+5 = 9$. It goes without saying that this figure leads to the interpretation of any square number as sum of odd numbers: $1+3+5+\ldots+(2n-1) = n^2$. There comes into view the Pythagorean arithmetical interpretation of this notion: the points are arithmetical unities and the parallelograms surrounded by gnomons are the sums of the respective unities. Furthermore, the use of gnomons provides the figure with a synthetic and constructive character: on the one hand, the same figure can express several sums at once, and, on the other hand, the figure is well-structured according to a clear pattern.

In a Pythagorean sense, Hindenburg suggests that the elements of the tables in Figs. 3.1 and 4.1 are first and foremost marks in the same sense that the points of figurate numbers are simply marks. The constructive character of gnomons allows him to arrange the chosen marks in a certain order within a given parallelogram. Besides, their synthetic character opens up the possibility of focusing individually on a specific parallelogram bounded by a particular gnomon. Just as the Pythagoreans affirm that the parallelogram

represents, once the interpretation of points as unities is made, the sum $1 + 3 = 4$, Hindenburg affirms that the marks "00", "01", "10", and "11" can be interpreted as digits in the parallelogram

00	01
10	11

but their numerical meaning depends on whether the focus is on this individual parallelogram, in which case these digits are numbers of the binary system, or on this parallelogram considered as part of a bigger parallelogram, in which case the digits are numbers of the respective number system. This interpretative flexibility is made possible by the notion of gnomon, and the possibility of bringing together different number systems under a single table is due to the particular structure of the figure. This structure encompasses the idea of "involution" (Hindenburg 1794c, 21):

> This figurative representation of number systems, with respect to others and with respect to themselves, shows a true involution.[13]

---

[13] Diese figürliche Darstellung der Zahlensysteme in und um einander, zeigt also eine wahre Involution.

Thus, the improvement of Hindenburg's combinatorial method originates from his early work on mathematical tables, as his original method of partition tables.

More precisely, Hindenburg offers the following definition of "combinatorial involution" (Hindenburg 1794c, 13):

I call combinatorial involutions (*Involutiones combinatoriae*) the procedures through which, for every desired term, coefficient, or value, and independently of its order, the arrangement of the given quantities will be determined by the digits or any other symbol of complexions in such a way that its expression will contain nothing superfluous, nothing that does not belong to the desired term, but at the same time the previous terms, which stand side by side, determine the following ones [...].[14]

Thus, Hindenburg calls "combinatorial involution" certain method used to calculate a particular value. This method consists of some simple rules and of a mathematical table constructed in accordance with those rules. In general, the table itself is called a "combinatorial involution", or simply an "involution", since the constructive rules are supposed to be implicitly reflected in the table.

Since the construction of a combinatorial involution depends on the rules governing it, it is possible to have as many different combinatorial involutions as there are different constructive rules. In his article, Hindenburg presents several examples of combinatorial involutions. Let's take a closer look at two of them to better understand this notion. The first example consists of combinatorial involutions of permutations. Hindenburg enumerates all the possible permutations of four objects a, b, c, and d as follows:

1234, *abcd*	2134, *bacd*	3124, *cabd*	4123, *dabc*
1243, *abdc*	2143, *badc*	3142, *cadb*	4132, *dacb*
1324, *acbd*	2314, *bcad*	3214, *cbad*	4213, *dbac*
1342, *acdb*	2341, *bcda*	3241, *cbda*	4231, *dbca*
1423, *adbc*	2413, *bdac*	3412, *cdab*	4312, *dcab*
1432, *adcb*	2431, *bdca*	3421, *cdba*	4321, *dcba*

According to Hindenburg, this table has been constructed following the rule that the whole numbers from 1234 to 4321 whose digits are exclusively 1, 2, 3, or 4, and there are no

---

[14] Combinatorische Involutionen (*Involutiones combinatoriae*) nenne ich diejenigen Verfahren, nach welchen, für jede ausser der Ordnung geforderte Glieder, Coefficienten oder Werthe, die Anordnung aus gegebenen Grössen, durch Zifern oder andere Zeichen zu Complexionen, so getroffen wird, daß in dem Ausdruke dafür nichts Ueberflüssiges enthalten ist, was zu dem geforderten Gliede nicht gehört, zugleich aber alle vorhergehende Glieder, [...] in und neben einander liegend, die folgenden bestimmen [...].

repeated digits, must be arranged in increasing order. This same example of enumeration can be found in the *Novi systematis* (Hindenburg 1781b, 17), and this same rule for the enumeration of permutations is explicitly stated in (Hindenburg 1784, XLVI). Thus, Hindenburg reworks here his old solutions to the problem of enumerating permutations, which belong to his project of creating a new mathematical system. But now he thinks that this particular solution to the problem of enumerating permutations can be improved by adding gnomons. For instance, the first and last columns of the previous table can be rewritten as:

1	2	3	4		4	1	2	3
1	2	4	3		4	1	3	2
1	3	2	4		4	2	1	3
1	3	4	2		4	2	3	1
1	4	2	3		4	3	1	2
1	4	3	2		4	3	2	1

A   *D*

These two tables are combinatorial involutions, and represent an improvement over the previous table because they enclose more information. The smallest gnomon of the involution *A* reassembles all possible permutations of one object represented by the digit 4, the following gnomon reassembles all possible permutations of two objects represented by the digits 3 and 4, and the last gnomon reassembles all possible permutations of three objects represented by the digits 2, 3, and 4. Furthermore, the elements of greater gnomons are based on the elements contained by smaller gnomons, which provides the table with the character of involution. Something similar happens in the case of the involution *D*. In this example, the constructive rule of the involutions is the same rule enunciated above for enumerating all permutations of four objects.

Another kind of combinatorial involutions can be seen in Fig. 4.2, which Hindenburg called "involution by cyclic periods". In this case, the constructive rule differs from that of involutions *A* and *D*. A cycle is a string of symbols, in this case a string of digits. For instance, the first involution in Fig. 4.2 is formed by three cycles, namely: the first column is composed of the cycle 1, the second column is composed of the cycle $1 - 2 - 3$, and the third column is composed of the cycle $1 - 2 - 3 - 4$. Every cycle contains a last element, which is the last element of the string. A period is a block of the table in which the last line of the block is the only line of the block whose elements are exclusively last elements of cycles. In Fig. 4.2, the different periods of these involutions are surrounded by gnomons. For instance, regarding the first involution, there are three periods highlighted by gnomons, although the last gnomon, which should surround the entire table, has not been drawn.

Fig. 4.2 Hindenburg's involutions by cyclic periods

```
 1 | 1 | 1       1 | 1 | 1 | 1      1 | 1 | 1
 1   2   2       1   2   2   2      2 | 2   2
 1   3   3       1   3   3   3      1   3   3
 1   1   4       1   1   4   4      2   1   4
 1   2   1       1   2   5   5      1   2   5
 1   3   2       1   3   6   6      2   3   1
 1   1   3       1   1   1   7      1   1   2
 1   2   4       1   2   2   8      2   2   3
 1   3   1       1   3   3   9      1   3   4
 1   1   2       1   1   4  10      2   1   5
 1   2   3       1   2   5  11      1   2   1
 1   3   4       1   3   6  12      u.  s.  w.
```

Such arrangements had been studied earlier by Hindenburg in his paper (Hindenburg 1786), though he did not used the notion of gnomon. In this paper, Hindenburg considers an arrangement of the form:

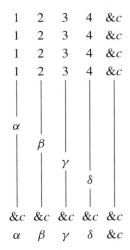

where Greek letters represent the last element of a cycle. Since the last line of the table is its only line formed exclusively by last elements of cycles, the four columns shown in the table are a cyclic period. This kind of arrangements, in the context of this earlier paper of Hindenburg, has been analyzed in (Bullynck 2009b, 60–63). Thus, in this case too, Hindenburg's example is based on his early work, which was intended to contribute to the development of the research project presented in the *Novi systematis*.

In principle, the method of combinatorial involutions is independent of any kind of symbols used in the table. Hindenburg constantly insists that, for instance, one can replace digits by letters in these tables. In general, one could replace digits with any other kind of symbols, provided that a constructive rule is given for those symbols. This goes along with the idea that the symbols in these combinatorial tables should be understood as marks that will be associated with some mathematical objects (like coefficients) in order to determine some quantity. Nevertheless, the use of digits is highly recommended because of the arithmetical properties of numbers, which facilitate the formulation of constructive rules for involutions. Hindenburg thinks that the versatility of the method of involutions comes from the multiplicity of constructive rules that can be formulated using the arithmetical-combinatorial properties of positive whole numbers. From a historical point of view, this inherent versatility represents, for Hindenburg, progress on the part of the combinatorial analysis regarding its contribution to the development of mathematical analysis in general (Hindenburg 1794c, 22):

> Ultimately, what is important above all else is to observe that the largest application of the theory of combinations to analysis was thus a natural and necessary consequence of such a transformation [i.e. the transformation of numerical tables into combinatorial tables], and that led to the creation of more convenient and harmonious symbols of combinatorial nature, or of any other nature, which are equally suitable for representing, in a brief and clear manner, combinatorial concepts, whether simple or complex, and mostly new, and which can be used in local or analytic-combinatorial formulae.[15]

Therefore, Hindenburg perceived his method of combinatorial involutions as the conclusion of a historical process in which this method becomes the most convenient tool to compute the values expressed by the specific functions of the combinatorial analysis, i.e. the local formulae created by Rothe.

However, Hindenburg does not believe that this historical process started with his own work and, in this sense, that his method could be considered as an isolated event in the history of mathematics. On the contrary, he suggests that the method's genesis can be traced back to the works of Moivre, Euler, Daniel Bernoulli (1700–1782), Lagrange, and Lambert (Hindenburg 1794c, 45). In fact, in another paper Hindenburg advances some arguments to support his view, though in this paper he focuses exclusively on the possible relationship between his work and that of Euler, Lambert, and Bernoulli (Hindenburg 1795c). It is worth noting that, in both cases, Leibniz does not appear among the authors that inspired the combinatorial methods of Hindenburg, according to Hindenburg himself.

---

[15] Was endlich über alles wichtig ist, die ausgedehnteste Anwendung der Combinationslehre auf die Analysis war nun eine natürliche und nothwendige Folge einer solchen Umänderung, und zog die Erfindung bequemer combinatorischer und anderer harmonirenden Zeichen herbey, die gleich geschickt sind, die, größtentheils neuen, combinatorischen, einfachen und zusammengesetzten, Begriffe kurz und deutlich darzustellen, und zu Local- und combinatorisch-analytischen Formeln sich anordnen zu lassen.

If Leibniz's *ars combinatoria* were a reference framework or an ideal for Hindenburg's research, then Leibniz should be placed at the top of both lists. His absence clearly indicates that the idea according to which Leibniz's *ars combinatoria* program guided Hindenburg's research is a fictional and misleading historical reconstruction. Moreover, Hindenburg actually makes a comment on Leibniz when discussing the reasons for choosing digits as the most appropriate symbols concerning the construction of involution tables. The comment concerns the numerical notation proposed by Leibniz to represent the coefficients of a reversed series, which had been severely criticized by Hindenburg, Eschenbach, and Rothe. In his comment, Hindenburg reiterates his criticism of Leibniz's notation (Hindenburg 1794c, 17–18):

> Digits and numbers instead of letters have already been used by Leibniz, who applied them to solve a very difficult problem; but he did not take as the basis of his method the first, simplest, and at the same time most widely applicable, combinatorial question, which contains the laws of counting for all number systems, and his new representation was far from finding, or from deserving, approval in its chaotic form, to the point that Moivre, who was pretty well versed in combinations subjects, described it as complicated, and even his friend Johann Bernoulli rejected it as abstract and useless![16]

Although this criticism dates from 1779, in this passage Hindenburg radicalizes his position. In his 1779 *Infinitinomii*, his criticism consisted in emphasizing the lack of clarity of Leibniz's interpretation of the theory of combinations, which led to a useless and unintelligible formulation of the law of formation of the coefficients of a reversed series, as seen in Sect. 3.2.2. On the contrary, Hindenburg points out now that Leibniz was not even able to grasp the simplest and most fundamental question of the doctrine of combinations. Since the method of combinatorial involutions depends on this fundamental aspect of the doctrine of combinations, which Leibniz did not understand, Hindenburg does not consider Leibniz's work as one of the sources of inspiration of his own work.

In the lines that follow the passage quoted above, Hindenburg tones down his negative remarks of Leibniz's obscure conception of combinatorics as not to offend the sensibilities of the reader, who, given the popularity of Leibniz in German-speaking countries, could be an enthusiastic partisan of Leibniz's thought, but he does not enter into the details of Leibniz's epistemological programs and the aspects of those programs that should be

---

[16] Zifern und Zahlen statt der Buchstaben, hat zwar Leibnitz schon gebraucht, und ihre Verbindung auf ein sehr verwickeltes Problem angewendet; aber die erste und einfachste, und in Absicht auf Anwendung zugleich ausgedehnteste aller combinatorischen Aufgaben, welche die Gesetze des Zählens nach allen Zahlensystemen enthält, hat er dennoch bey seinem Verfahren nicht zum Grunde gelegt, und es fehlte so viel daß seine neue Darstellung Beyfall fand, oder in ihrer chaotischen Gestalt finden konnte, daß selbst de Moivre, der sich doch sehr wohl auf Combinationen verstand, sie für schwierig erklärte, und sein Freund, Johann Bernoulli! selbige sogar als zu abstract und unbrauchbar verwarf.

analyzed or discarded.[17] However, there is another text by Hindenburg that goes further with this explanation; in fact, the text is a sort of introduction made by Hindenburg to an article of Christian Kramp (1760–1826), where Hindenburg returns again to the question of Leibniz's notation (Kramp 1796, 91–100). According to Hindenburg, Leibniz realized the importance of the combinatorial art while studying some mathematical problems, particularly the problem of reversing series. Somehow, his study of the theorem on reversion of series made him understand that the theory of combinations could become the absolute method (*absolute Art*) of human knowledge. After emphasizing the strong enthusiasm shown by Leibniz when talking about his absolute method, Hindenburg elaborates a little bit more on what Leibniz's project is:

> I am not talking here about Leibniz's remarks on *Analysis axiomatum* or on *Alphabetum cogitationum humanarum* according to which he hoped to establish *Analysis suprema* on the ground of the theory of combinations, remarks that border on fervor in some places. I simply understand the natural outburst of admiration provoked by the delightful views of a new land which he had already visited in his mind and which was for the rest of us *terra incognita*.[18]

Even if this passage gives no details about any precise element of Leibniz's epistemological program or programs, it clarifies that Hindenburg was not interested in the Leibnizian idea of constructing a universal alphabet of human thoughts in order to establish a universal characteristic from which one could derive a new, ultimate, mathematical analysis. It should be noted that Hindenburg elaborates these remarks in the context of Leibniz's work on the theorem on reversion of series. Thus, what he knows about Leibniz's epistemological programs comes mostly from Leibniz's 1700 paper on the reversion of series, which has been analyzed in Sect. 2.2.3.3. As seen in Sect. 3.2.2, Hindenburg severely criticized the conception of the theory of combinations presented by Leibniz in that paper, describing it as useless and unintelligible. If it is taken into account that, in that paper, Leibniz puts his program of universal characteristic at the center of his program

---

[17] Here are the textual lines of Hindenburg: "Es giengen nehmlich, außer der eben angezeigten Reduction aller übrigen auf ein einziges sehr einfaches combinatorisches Grundproblem, Leibnizen, wie Herr M. Toepfer sehr wohl erinnert, verschiedene nützliche Einrichtungen, Sätze und bequeme Zeichen (die ganze Zeichensprache) in der Combinationslehre noch ab, vornehmlich: genaue Abtheilung der zu verbindenden Dinge nach Classen zu bestimmten Summen, bequeme Bezeichnung solcher Classen, und promte Darstellung ihrer Werthe; wo man noch hinzusetzen kann: figürliche Anordnung der Zahlen- oder Buchstabencomplexionen neben und untereinander." Hindenburg's reference to Töpfer directs the reader to (Töpfer 1793, 9), where one can see, by the way, that Töpfer repeats Hindenburg's criticism of Leibniz's conception of combinatorics analyzed in Sect. 3.2.2.

[18] Ich rede hier nicht von Leibnitzens Aeußerungen über die *Analisin axiomatum* und das *Alphabetum cogitationum humanarum*, die er von einer, auf die Combinationslehre zu gündenden, *Analysi suprema* erwartete; Aeußerungen –die in einigen Stellen nahe an Schwärmerey gränzen. Ich verstehe blos die natürlichen Ausbrüche von Bewunderung, welche die entzückenden Aussichten in ein neues Land ihm veranlaßten, das er im Geiste schon ganz durchreist hatte, und das nur noch für Andere –eine *terra incognita* war.

of *ars combinatoria*, it is clear that Hindenburg distances himself from the Leibnizian program of *ars combinatoria* in the previous quotation. Hindenburg welcomes, in the passage quoted above, the general idea of basing analysis on the doctrine of combinations (not on Leibniz's *ars combinatoria* program), but that is all. It seems that this general idea is the only aspect of Leibniz's ambitious projects that merits the attention of Hindenburg, but his attention is conditioned by the fact that he had the same idea independently of Leibniz.

4.2.1.2 A Combinatorial Law of Nature

The retreat of Leibniz's image from Hindenburg's discourse was soon accompanied by a renewed interest in Moivre's work, even if its intellectual importance and its impact on the development of Hindenburg's theory had been rather limited so far. Indeed, one can find several laudatory passages in Hindenburg's writings where he underlines the mathematical merits of some authors, like Kästner, Lambert, and even Leibniz. In general, these words are motivated by the desire to link the works of those authors (as regards Kästner and Lambert) to the origins of the combinatorial analysis. Although Hindenburg acknowledges his debt to Moivre, whose version of the multinomial theorem is the starting point of the *Infinitinomii*, he does not compliment Moivre's work before the decade of the 1790s. Taking this into account, it is easy to see that the publication in 1795 of a booklet devoted to analyze Moivre's original article on the multinomial theorem strongly suggests that Hindenburg's thought had reached a turning point (Hindenburg 1795d).

Hindenburg's reinterpretation of Moivre is based on the method of combinatorial involutions, since a particular kind of involution will be designed to deal with the calculations of the coefficients belonging to multinomial expansions. Thus, Hindenburg defines the involution of what he calls "combinations of a sum n", which are indeed the partitions of the integer n, by means of two rules (Hindenburg 1795d, IV). These constructive rules have been enunciated in a very synthetic way, so for the sake of clarity it is convenient to break them dawn into the following more specific list of instructions to construct the partitions involution of a given integer n:

1. The first gnomon contains 1 only.
2. To construct a new gnomon from a given gnomon:
   a. Write the digit 1 to the left side of each line contained in the given gnomon.
   b. From the top down:
      i. If the first line of digits obtained by (a) contains a second and a third elements (from left to right) and they are different from each other, then write at the bottom the line whose first element (from left to right) is the sum of the first and second elements of the first line (from left to right) and whose other elements correspond to the sequence of elements of the first line from the third element onward.
      ii. If the second line of digits obtained by (a) contains a second and a third elements (from left to right) and they are different from each other, then write at the bottom the line whose first element (from left to right) is the sum of the first and

second elements of the second line (from left to right) and whose other elements correspond to the sequence of elements of the second line from the third element onward.

   iii. Go on like this until the penultimate line obtained by (a).

   iv. Write at the bottom the sum of the two elements of the last line obtained by (a).

3. Continue this process of constructing gnomons until the last line (from the top down) of a gnomon constructed by rules 1 and 2 is formed by the digit n alone.

This algorithm allows one to express the partitions of a given integer in the form of involution (for another historical reconstruction of this kind of involution, see Panza 1992, vol. 2, 676–680).

Hindenburg exemplifies his algorithm by constructing the partitions involution of the integer 7, which is shown in Fig. 4.3. In this case, according to rule 1 of the algorithm, the first gnomon is:

$$\lfloor 1$$

Then, by applying rules 3 and 2 of the algorithm to this gnomon, one obtains the following gnomon:

$$\begin{array}{c} 1 \;\rfloor\, 1 \\ 2 \end{array}$$

Fig. 4.3 Hindenburg's partitions involution of 7

```
&c   1   1   1   1   1   1 ⌐1
     1   1   1   1   1 ⌐2
     1   1   1   1 ⌐3
     1   1   1   2   2
     1   1   1 ⌐4
     1   1   2   3
     1   1 ⌐5
&c   1   2   2   2
     1   2   4
     1   3   3
     1 ⌐6
     2   2   3
     2   5
     3   4
     7
   &c    &c          &c
```

By applying rules 3 and 2 of the algorithm to this new gnomon, one obtains this other gnomon:

$$
\begin{array}{l}
1 \mid 1 \mid 1 \\
1 \mid 2 \\
3
\end{array}
$$

The iterative process of the algorithm leads to the construction of the desired involution as shown in Fig. 4.3. The symbol "&c" inserted in this figure means that it is possible to continue producing bigger gnomons by applying the algorithm to the involution of 7, in which case those bigger gnomons will correspond to the partitions of integers grater than 7. Thus, the involutive character of these tables consists in the fact that the table of a given integer n not only contains the partitions of that integer, but also the partitions of all integers less than n. Hindenburg calls these tables "alphabetical involutions" or "lexicographical involutions" since, if one replaces the digits "1", "2", "3", etc. by the letters "a", "b", "c", etc. respectively, each line of the table is arranged in alphabetical order.

In the hope of elucidating the relationship between his lexicographical involutions and the multinomial theorem, Hindenburg analyzes the rules proposed by Moivre to determine the coefficients of the multinomial expansion. As seen in Sect. 2.2.1, Moivre formulated a set of rules to determine the coefficient of a given term of the series expansion of $\left(az + bz^2 + cz^3 + \&c.\right)^m$, and this coefficient can be expressed as a sum of monomials in which each monomial is a product of the known coefficients a, b, c, etc. and of a numerical value calculated by formula 2.16 and called "*unciae*" by Moivre. In an unexpected twist, Hindenburg claims that this set of rules is an actual law of Nature (*LEX quam obseruat NATVRA*) with respect to the formation of this type of coefficients, though he attributes this idea to Moivre himself (Hindenburg 1795d, VII). Hindenburg does not elaborate on what he means by this strong assertion, but one can find some further explanations in (de Moivre 1730, 87) and more explicitly in (de Moivre 1704, 97–98). In these texts, Moivre explains that his method to raise a multinomial to a given power is neither mechanical, nor experimental, but it obeys the natural law of progression of a series (de Moivre 1704, 97–98). As an example of mechanical and experimental methods, Moivre refers to the method employed by Cheyne to find the multinomial expansion, which can be classified within the category of proofs of the multinomial theorem based on the binomial theorem, such as those studied in Sects. 2.2.2.1 and 2.2.3.1. For Moivre, the technique of grouping the terms of a multinomial in order to form a binomial does not help us to understand the law of formation of the desired coefficients, though this technique is naturally a valid method of proof. This technique consists in applying mechanically and repeatedly the binomial theorem, but the procedure does not serve to make the conceptual construction of the multinomial expansion explicit. On the contrary, Moivre asserts that his constructive rules enable us to express the inherent relationships among the different elements that are involved in the power series expansion justified by the multinomial theorem, since the application of these rules shows how a given term of this expansion has been constructed on the basis of the previous ones. Furthermore, Moivre suggests that it is not because of this constructive aspect that those relationships exist, but because of the mathematical objects on which they rest, and so his constructive rules describe the behavior of these objects. From here, Hindenburg concludes that these rules express a law of Nature.

Hindenburg is thus convinced that Moivre discovered not only an important theorem of analysis, but also a law that transcends the borders of theory and takes root on Nature. However, as it happens normally in pioneering discoveries, this law did not reach one of its best expressions in the work of its discoverer (Hindenburg 1795d, IX):

1. Since, even if the rules proposed [by Moivre] in natural language are pretty clear, they have not been presented in the form of a convenient analytic-combinatorial formula.
2. Since, in the case of a *positive integer exponent m* [i.e., the power m to which $az + bz^2 + cz^3 +$ &c. must be raised], coefficients are determined by lengthy computations. [...]
3. Since no coefficient can be found without previously calculating *every preceding coefficient*, from the first to a^m, and no formula has been given for determining the *general coefficient*.[19]

It is clear that Hindenburg is comparing his method with that of Moivre in these three points. It comes out from this comparison that the combinatorial analysis holds the technical capabilities to capture this combinatorial law of Nature in its finest expression. It holds the analytic-combinatorial formulae based on the mathematical notations developed by Hindenburg and Rothe, the lexicographical involutions just created by Hindenburg to avoid lengthy computations, and the non-recursive formula invented by Hindenburg long time ago to directly determine the general coefficient of multinomial expansions.

Certainly, these remarks are not meant to be a criticism of Moivre. On the contrary, they summarize the way in which Moivre's multinomial theorem is related to the German combinatorial analysis, and place the original formulation of the theorem in the history of Hindenburg's theory as its most important precursor. The translation of Moivre's rules for the multinomial theorem into the combinatorial analysis is condensed in the lexicographical involution shown on the right-hand side in Fig. 4.4. Below this involution, one can find the substitution rule established by Hindenburg to interpret the digits of the involution. The result of this interpretation corresponds to the table on the left-hand side in Fig. 4.4. In this interpreted involution, each line should be read as a product of the form $b^q c^r d^s \cdots$, and each product should be multiplied by $a^{m-\star}$, where $q, r, s, \ldots$ are natural

---

[19] 1° quod leges propositae, verbis quidem satis claris per se, non autem formula idonea analytico-combinatoria exponantur. 2° quod Coefficientes, vbi *Exponens m numerus* est *integer positiuus*, prolixius determinentur. [...] 3° quod Coefficiens exhiberi nullus possit, nisi simul et *praecedentes singuli expositi* sint, ad primum vsque a^m, inclusive, et formula adeo *Coefficientem* sistens *generalem* nulla appareat.

```
                      a^m 1
 a^{m-*}  ...  | b | b | b | b | b | b |            &c    | 1 | 1 | 1 | 1 | 1 | 1 |
 a^{m-*}  ...  | b | b | b | b | c |                      | 1 | 1 | 1 | 1 | 2 |
 a^{m-*}  ...  | b | b | b | d |                          | 1 | 1 | 1 | 3 |
 a^{m-*}  ...  | b | b | c | c |                          | 1 | 1 | 2 | 2 |
 a^{m-*}  ...  | b | b | e |                              | 1 | 1 | 4 |
 a^{m-*}  ...  | b | c | d |                              | 1 | 2 | 3 |
 a^{m-*}  ...  | b | f |                          &c      | 1 | 5 |
 a^{m-*}  ...  | c | c | c |                              | 2 | 2 | 2 |
 a^{m-*}  ...  | c | e |                                  | 2 | 4 |
 a^{m-*}  ...  | d | d |                                  | 3 | 3 |
 a^{m-*}  ...  | g |                                      | 6 |
     &c                        &c        &c                                  &c
```

$$\begin{pmatrix} 1 & 2 & 3 & 4 & 5 & 6 & \ldots \\ b & c & d & e & f & g & \ldots \end{pmatrix}.$$

Fig. 4.4 Hindenburg's lexicographical involution

numbers such that $q + r + s + \ldots = \star$. In what follows, the symbol $\star$ will be replaced by k. By taking advantage of the lexicographical order of this kind of involutions, Hindenburg defines the following classes of partitions:

1. All lines whose first element is 1 (from left to right) belong to the class $^n\mathbb{A}$.
2. All lines whose first element is 2 (from left to right) belong to the class $^n\mathbb{B}$.
3. All lines whose first element is 3 (from left to right) belong to the class $^n\mathbb{C}$.
4. And so on.

$^n\mathbb{A}$, $^n\mathbb{B}$, $^n\mathbb{C}$, ... are classes of partitions of the positive integer n. Although this is not Hindenburg's terminology, these classes will be called "lexicographical classes" in order to avoid any confusion between these classes and the classes of partitions defined by Hindenburg before 1795. For instance, from the involution shown in Fig. 4.4, one can obtain the following lexicographical classes:

$$^6\mathbb{A} = \left\{ \begin{array}{c} (1, 1, 1, 1, 1, 1), (1, 1, 1, 1, 2), (1, 1, 1, 3), \\ (1, 1, 2, 2), (1, 1, 4), (1, 2, 3), (1, 5) \end{array} \right\}$$

$$^6\mathbb{B} = \{(2, 2, 2), (2, 4)\} \tag{4.23}$$

$$^6\mathbb{C} = \{(3, 3)\}$$

$$^6\mathbb{F} = \{(6)\}$$

On the other hand, by examining the interpreted involution on the left-hand side in Fig. 4.4, Hindenburg points out that lexicographical classes can also be defined as follows:

1. If $q \neq 0$, then the product $b^q c^r d^s \cdots$ belongs to the first class $^n\mathbb{A}$.
2. If $q = 0$ and $r \neq 0$, then the product $b^q c^r d^s \cdots$ belongs to the second class $^n\mathbb{B}$.
3. If $q = 0$, $r = 0$, and $s \neq 0$, then the product $b^q c^r d^s \cdots$ belongs to the third class $^n\mathbb{C}$.
4. And so on.

These defining rules are exactly the same rules established by Moivre to determine the coefficients of multinomial expansions, as can be seen in Sect. 2.2.1. This is the reason why Hindenburg claims that Moivre was one of the first mathematicians to use lexicographical combinatorial methods to solve problems of mathematical analysis (Hindenburg 1795d, XIII).

Since Moivre's rules can be translated into the language of the combinatorial analysis by means of rules (1)–(4) given above, Moivre's formulation of the multinomial expansion can be written too in terms of analytic-combinatorial formulae as follows:

$$(az + bz^2 + cz^3 + \&c.)^m = a^m z^m + \binom{k-1}{m\mathfrak{Aa}} a^{m-k} \left(^1\mathbb{A}\right) z^{m+1} \tag{4.24}$$

$$+ \binom{k-1}{m\mathfrak{Aa}} a^{m-k} \left(^2\mathbb{A} + {}^2\mathbb{B}\right) z^{m+2} + \binom{k-1}{m\mathfrak{Aa}} a^{m-k} \left(^3\mathbb{A} + {}^3\mathbb{C}\right) z^{m+3} + \cdots$$

$$+ \binom{k-1}{m\mathfrak{Aa}} a^{m-k} \left(^{2n}\mathbb{A} + {}^{2n}\mathbb{B} + \cdots + {}^{2n}\mathcal{N} + {}^{2n}\overset{n}{\mathcal{N}}\right) z^{m+2n}$$

$$+ \binom{k-1}{m\mathfrak{Aa}} a^{m-k} \left(^{2n+1}\mathbb{A} + {}^{2n+1}\mathbb{B} + \cdots + {}^{2n+1}\mathcal{N} + {}^{2n+1}\overset{n+1}{\mathcal{N}}\right) z^{m+2n+1}$$

$$\begin{pmatrix} 1 & 2 & 3 & \cdots \\ b & c & d & \cdots \end{pmatrix}$$

Other than the notations for lexicographical classes, the symbols used in this formula are already known: Gothic capital letters with a left superscript represent multinomial coefficients, Gothic lowercase letters represent multinomial coefficients, and the use of indexes over coefficients has been explained in Sect. 3.3.3. Contrary to Moivre's formula, which was established for a positive integer m only, this reinterpreted formula holds for any rational (or even real) number m, since m appears exclusively in the binomial coefficients. According to Hindenburg, this analytic-combinatorial interpretation allows to complete Moivre's proof of the multinomial theorem, given that Moivre asserted that his method held for any fractional number but he never proved it.

Furthermore, this reinterpreted formula not only completes Moivre's proof, but also improves Moivre's formula 2.17, since the introduction of lexicographical classes transforms the recursive formula of Moivre into a non-recursive formula. As an example of the advantages of this non-recursive formula, Hindenburg shows how it can be applied to directly determine the coefficient of the term z^{m+6} of the power series expansion of $p^m = (az + bz^2 + cz^3 + \&c.)^m$. According to his method, the first step consists in constructing the lexicographical involution of the integer 6, from which the required lexicographical classes can be derived. This involution is shown in Fig. 4.4 and the corresponding classes are listed in equalities 4.23. Therefore, by applying his non-recursive formula, Hindenburg obtains the following result:

$$p^m \daleth 7 = \binom{k-1}{m\mathfrak{Aa}} a^{m-k} \left({}^6\mathbb{A} + {}^6\mathbb{B} + {}^6\mathbb{C} + {}^6\mathbb{F} \right) z^{m+6}$$

$$= \begin{pmatrix} ({}^m\mathfrak{Ff})\, a^{m-6}b^6 + ({}^m\mathfrak{Ee})\, a^{m-5}b^4c + ({}^m\mathfrak{Dd})\, a^{m-4}b^3d \\ + ({}^m\mathfrak{Dd})\, a^{m-4}b^2c^2 + ({}^m\mathfrak{Ec})\, a^{m-3}b^2e + ({}^m\mathfrak{Ec})\, a^{m-3}bcd \\ + ({}^m\mathfrak{Bb})\, a^{m-2}bf + ({}^m\mathfrak{Ec})\, a^{m-3}c^3 + ({}^m\mathfrak{Bb})\, a^{m-2}ce \\ + ({}^m\mathfrak{Bb})\, a^{m-2}d^2 + ({}^m\mathfrak{Aa})\, a^{m-1}g \end{pmatrix} z^{m+6}, \qquad (4.25)$$

which is indeed the desired coefficient.

4.2.1.3 Corroborating Evidence for the Combinatorial Law of Nature

In the same year of 1795, Hindenburg publishes in his *Archiv* a German version of the booklet analyzed in the previous section (Hindenburg 1795a). In this paper, some small corrections have been made to the original Latin version, but they do not alter its content or structure, and a new paragraph has been included, which is devoted to Boscovich's proof of the multinomial theorem (for this proof, see Sect. 2.2.2.3). This paragraph begins with the reminder that Boscovich's method was conceived independently of the former method of Moivre (Hindenburg 1795a, 402):

> Fifty years later than Moivre, and initially without knowing anything about his multinomial theorem, Boscovich also proposed a similar method to raise a multinomial to a given power.[20]

Indeed, Boscovich himself claims not to have read Moivre's paper on the multinomial theorem until after he discovered his own method (Boscovich 1748b, 94). For Hindenburg, it is important to emphasize this point because the independence of their results implies

---

[20] Boscovich hat 50 Jahre später als de Moivre, und ohne von dessen Polynomialtheorem anfänglich etwas zu wissen, auch ein diesem ähnliches Verfahren angegeben, ein Infinitinomium zu einer verlangten Potenz zu erheben.

that the method of lexicographical involutions is not only a calculation procedure invented by someone, but the mathematical expression of a combinatorial law that has been discovered twice by two scientists with no relationship to each other. According to Hindenburg, although Moivre was the first to discover this law, it was more clearly enunciated in Boscovich's lexicographical procedures (Hindenburg 1795a, 403).

In his article, Hindenburg describes Boscovich's method of partitions which is based on the construction of tables of partitions, where the parts of each partition are arranged in increasing order. To express Boscovich's tables in terms of lexicographical classes, Hindenburg proposes to change the arrangement of the partitions from an increasing order to a decreasing order. This can be easily done by following the same set of rules given by Boscovich (see Sect. 2.2.2.3), but instead of writing the elements of the lines of the table from left to right, they must be written from right to left. For instance, Fig. 2.1 shows Boscovich's table of the partitions of the integer 6 in increasing order, and Fig. 4.5 shows the same table in decreasing order. From the table in decreasing order, Hindenburg obtains the following lexicographical classes:

$$^6\mathbb{A} = \{(1, 1, 1, 1, 1, 1)\}$$

$$^6\mathbb{B} = \{(2, 1, 1, 1, 1), (2, 2, 1, 1) (2, 2, 2)\}$$

$$^6\mathbb{C} = \{(3, 1, 1, 1), (3, 2, 1), (3, 3)\} \tag{4.26}$$

$$^6\mathbb{D} = \{(4, 1, 1), (4, 2)\}$$

$$^6\mathbb{E} = \{(5, 1)\}$$

$$^6\mathbb{F} = \{(6)\}$$

In general, by inverting the order of the lines of Boscovich's tables, it is possible to find all the lexicographical classes required to calculate the coefficients of multinomial expansions. Therefore, by using his notion of "lexicographical class" and the set of rules

Fig. 4.5 Lexicographical arrangement in decreasing order

1	1	1	1	1	1
2	1	1	1	1	
2	2	1	1		
2	2	2			
3	1	1	1		
3	2	1			
3	3				
4	1	1			
4	2				
5	1				
6					

that defines Boscovich's method, Hindenburg can rewrite the multinomial theorem as follows:

$$p^m = \left(az + bz^2 + cz^3 + \&c.\right)^m \tag{4.27}$$

$$= a^m z^m + \binom{k-1}{m \mathfrak{Aa}} a^{m-k} \left(^1 \mathbb{A}\right) z^{m+1}$$

$$+ \binom{k-1}{m \mathfrak{Aa}} a^{m-k} \left(^2 \mathbb{A} + ^2 \mathbb{B}\right) z^{m+2} + \binom{k-1}{m \mathfrak{Aa}} a^{m-k} \left(^3 \mathbb{A} + ^3 \mathbb{B} + ^3 \mathbb{C}\right) z^{m+3} + \cdots$$

$$+ \binom{k-1}{m \mathfrak{Aa}} a^{m-k} \left(^n \mathbb{A} + ^n \mathbb{B} + ^n \mathbb{C} + \cdots + ^n \mathcal{N}\right) z^{m+n}$$

$$\begin{pmatrix} 1 & 2 & 3 & \cdots \\ b & c & d & \cdots \end{pmatrix}$$

In this formula, the definition of lexicographical classes depends on Boscovich's tables, hence these classes are different from Hindenburg's lexicographical classes defined in the previous Sect. 4.2.1.2, which depends on Hindenburg's tables of lexicographical involutions. However, it is evident that this formula and formula 4.24 based on lexicographical involutions are equivalent.

For instance, by using formula 4.27 to calculate the coefficient of the term z^{m+6} in the series expansion of p^m, one gets:

$$p^m \mathsf{1} 7 = \binom{k-1}{m \mathfrak{Aa}} a^{m-k} \left(^6 \mathbb{A} + ^6 \mathbb{B} + ^6 \mathbb{C} + ^6 \mathbb{D} + ^6 \mathbb{E} + ^6 \mathbb{F}\right) z^{m+6}.$$

where $^6 \mathbb{A}$, $^6 \mathbb{B}$, $^6 \mathbb{C}$, $^6 \mathbb{D}$, $^6 \mathbb{E}$, and $^6 \mathbb{F}$ are the classes listed in equalities 4.26. After interpreting these classes by means of the substitution rule indicated below formula 4.27, one can see that this term is equal to the term expressed by Eq. 4.25, which correspond to formula 2.21 given by Boscovich as an example of application of his method (setting $n = \rho = 1$ in Boscovich's example).

After 1795, Boscovich's articles on the multinomial theorem will be regarded as one of the major references of the combinatorial analysis by the German combinatorial school. In 1796, Klügel explains Hindenburg's method of involutions, but his explanation consists in comparing Hindenburg's method and that of Boscovich (Klügel 1796, 57–62). In 1797, Pfaff does the same thing, but he places the accent on the relationship between Moivre's and Hindenburg's works, and goes as far as referring to the method of lexicographical involutions as "the method of Moivre-Hindenburg" (Pfaff 1797b, 260–321). This method becomes the foundation of the combinatorial analysis, as Hindenburg himself puts it at the end of *Der polynomische Lehrsatz das wichtigste Theorem der ganzen Analysis* (Hindenburg 1796a, 304):

> The combinatorial analysis has finally removed its veil, and, henceforth, it is no longer left to blind approximation whether and how one can reach *Legem Naturae*.[21]

Thus the multinomial theorem states the regularity imposed by Nature on mathematical objects, and the method of lexicographical involutions depicts such a regularity through its tables.

*

In sum, the method of involutions consists of mathematical tables constructed on the basis of appropriate algorithms. Lexicographical involutions were conceived as a way of exploring and focusing on Moivre's formulation of the multinomial theorem. The application of lexicographical involutions to the reconstruction of Moivre's rules for determining the coefficients of multinomial expansions leads to the discovery of what is considered to be the foundation of the combinatorial analysis. Since this foundation is conveyed by the multinomial theorem, this theorem must become the principle upon which the combinatorial analysis rests.

Assuredly, so far the discovery by Hindenburg of the combinatorial law of Nature formulated by Moivre is the most important event in the historical evolution of the German combinatorial analysis. This law provides Hindenburg's theory with a conceptual framework within which the theory will be constructed, while incorporating an ideal to be pursued: the goal of shaping mathematical analysis according to this law. So far Hindenburg had been struggling to understand the theoretical and programmatic stability of his views by attempting to emphasize the differences between his position and that of other mathematicians. Notably, this struggle was tinged with dramatism in the case of Leibniz. In German-speaking countries, Leibniz was indissolubly associated with the theory of combinations, to the point of considering him the real creator of that theory. On the contrary, up until the decade of the 1790s Hindenburg had had a rather modest academic career, marked, to be honest, by a scarce mathematical research. When Hindenburg realized, in 1779, that a modest Privatdozent like him had accomplished on his own what Leibniz could not, he had to contend with the problem of how to publicly present his innovative views on mathematical analysis without crashing into the fame of Leibniz. There are documents that attest to the existence of this concern; for instance, as seen in Sect. 3.2.2, in his 1779 *Infinitinomii* Hindenburg was torn between his fair criticism of Leibniz's conception of the theory of combinations and the untouchable reputation of

---

[21] Die combinatorische Analysis hat endlich den Schleyer aufgedeckt, und es bleibt hinfort nicht mehr dem blinden Ungefähr überlassen, ob und wenn es die *Legem Naturae* herbeyführen will.

Leibniz. His solution back then consisted in making rhetorical use of some passages of Leibniz by putting them as epigraph to a section of his book, whereas he severely criticized Leibniz's *ars combinatoria* in his theoretical exposition of the theory of combinations. It is true that playing such a double game could be confusing. So, in order to settle the question of Leibniz, let's recapitulate what we know about this issue. First, the historical analysis of Hindenburg's early work shows that Hindenburg developed the first stage of his combinatorial method, called "method of powers" in 1779, independently of Leibniz's *ars combinatoria* program. This analysis is corroborated by Hindenburg's claim according to which he knew nothing about Leibniz's combinatorial ideas before 1779, when he got curious about Leibniz's combinatorial method for the theorem on reversion of series; therefore, Hindenburg had already created his combinatorial method when he decided to study Leibniz's paper on the reversion of series, that is to say, in order to formulate his combinatorial method for the study of mathematical analysis, Hindenburg did not follow Leibniz's ideas, he followed his own ideas. Second, in 1779, Hindenburg criticized Leibniz's epistemological programs of *lingua characteristica* and *ars combinatoria*, describing them as useless and unintelligible, and he rejected them as an adequate approach to the theory of combinations in that they are conceptually obscure and rest on incomprehensible principles. In particular, this implies that Hindenburg did not consider Leibniz's epistemological programs as a framework or ideal for his own mathematical research. Later Hindenburg not only reaffirmed his criticism but also claimed that, in fact, Leibniz even failed to grasp the most fundamental principle of the theory of combinations. Additionally, Hindenburg explicitly distanced himself from Leibniz's universal characteristic program (therefore from Leibniz's *ars combinatoria* as well) by pointing out that he was not interested in such a program, which also implies that Hindenburg did not consider Leibniz's epistemological programs as a framework or ideal for his own mathematical research. Furthermore, when Hindenburg lists the mathematicians that have been of influence on the development of his own conception of combinatorics, Leibniz is not included in the list. Moreover, the second and last stage of development of Hindenburg's combinatorial method (i.e. the method of combinatorial involutions) was explicitly based on Moivre's mathematical work, not on Leibniz's works and not on Leibniz's programmatic ideals. On the other hand, Eschenbach, Rothe, and Töpfer explicitly supported Hindenburg's criticism of Leibniz's epistemological programs. And, as far as we know, no other member of the combinatorial school mentions Leibniz's epistemological programs in the context of his work on the German combinatorial analysis. Finally, no concrete mathematical results of Leibniz will ever be used by the combinatorial school. We think that all these points provide enough evidence to conclude that Leibniz's thought cannot be considered as a historical antecedent of the German combinatorial analysis. Perhaps someone would disagree with this conclusion, in which case it would be imperative to explain how all these points can be reconciled with the idea that Leibniz's thought played an influential role in the historical development of the German combinatorial analysis. But all these points cannot just be ignored or overlooked by historians as has been the case thus far. In addition, Moivre's combinatorial law of Nature will take the vacant place of conceptual framework and ideal to pursue for the scientific research of the combinatorial school. The discovery of this law seems to have convinced Hindenburg of the theoretical soundness and programmatic stability of his views. For Hindenburg, this law represents the missing conceptual foundation of his theory, which is then linked to the mathematical tradition of Moivre.

4.2.2   Independence, Primacy, and the Multinomial Theorem

Until the end of the 1780s, Hindenburg was convinced that the most important theorem in the field of analysis was the binomial theorem of Newton, and he shared this belief with other mathematicians like Euler and Aepinus. As seen above, this is no longer the case in 1795, at least with respect to the combinatorial analysis in which the multinomial theorem took the place of the most fundamental theorem inasmuch as it conveys the most essential law of combinations. However, the multinomial theorem cannot be regarded as the most important theorem in the whole of analysis since, even within the field of the combinatorial analysis itself, it depends on the binomial theorem, according to Hindenburg's proof given in the *Methodus nova*. In 1796, Klügel works out a new proof of the multinomial theorem that is meant to be both combinatorial and independent of the binomial theorem. In this context of conceptual structuring of the theory, Klügel's proof can be seen as an attempt to assign the multinomial theorem the place it merits, namely, the place of a theorem that depends exclusively on a combinatorial law of Nature. At the same time, Rothe demonstrates the binomial theorem by means of the tools and concepts of the combinatorial analysis. In this same context, Rothe's proof can be seen, in its turn, as an attempt to show that the binomial theorem is subordinated to the combinatorial principles conveyed by the multinomial theorem. In consequence, taken together, the aim of both proofs is to establish the primacy of the multinomial theorem over every proposition of analysis, particularly over the binomial theorem.

4.2.2.1 The Independence of the Multinomial Theorem

In 1796, Klügel sets the explicit goal of proving the multinomial theorem without using the binomial theorem and by applying the combinatorial tools developed by the combinatorial school (Klügel 1796, 48). His proof is organized in four steps: in the first step, the theorem is justified for a positive integer exponent; then, it is justified for a negative integer exponent; later, it is justified for a fractional positive exponent; and, finally, it is justified for a negative fractional exponent. It is worth saying that Klügel's argumentation does not stand up to a close scrutiny, as will be seen below, but despite this it played a significant role in the configuration of the combinatorial analysis. We will come back to this issue in Sect. 5.1.1, where some objections to this proof will be reviewed.

As usual at that time, the proof starts with the supposition that the multinomial series expansion exists:

$$\left(a + bz + cz^2 + dz^3 + \ldots\right)^m = A + Bz + Cz^2 + Dz^3 + \cdots,$$

where A, B, C, etc. are unknown coefficients. In the first step of the proof, the exponent m is supposed to be a positive whole number. Following Moivre, Klügel seeks to express these unknown coefficients as a sum of monomials of the form $a^p b^q c^r \cdots$, where p, q, r, etc. are positive integers such that $p + q + r + \cdots = m$ and each monomial have a numerical coefficient, called *uncia* by Moivre. To determine the monomials corresponding to the coefficient of a given term z^n of the multinomial expansion, Klügel constructs the lexicographical involution of the integer n, assuming that if n is zero, one gets $A = a^m$. If n is greater than zero, the lexicographical involution is given by the following table:

$$
\begin{array}{llll}
1 & \cdots & 1 & 1 \;| 1 \\
1 & \cdots & 1 \;| 2 \\
1 & \cdots & 3 \\
\vdots & \vdots & \vdots \\
n
\end{array}
$$

$$
\begin{pmatrix}
1 & 2 & 3 & \cdots \\
b & c & d & \cdots
\end{pmatrix}
$$

In fact, since this involution contains the lexicographical involutions of the integers from 1 to $n - 1$, this table allows us to calculate at once, and independently of one another, all the monomials belonging to the unknown coefficients of the terms $z, \ldots, z^{n-1}$. Thus, after replacing digits with letters according to the substitution rule indicated below the table, the first gnomon gives the sum b of monomials corresponding to the coefficient B, the second gnomon gives the sum $(b^2 + c)$ of monomials corresponding to the coefficient C, the third gnomon gives the sum $(b^3 + bc + d)$ of monomials corresponding to the coefficient D, and so on. Each term of these sums is of the form $b^q c^r \cdots$ for certain integers $q, r, \ldots$, and each must be multiplied by a^{m-k}, where $q + r + \ldots = k$, which gives the following sums of monomials:

- For B: $a^{m-1}b$.
- For C: $a^{m-2}b^2 + a^{m-1}c$.
- For D: $a^{m-3}b^3 + a^{m-2}bc + a^{m-1}d$.
- And so on.

At this point, to fully determine the unknown coefficients, it suffices to calculate the corresponding *unciae*.

To this end, Klügel proposes to follow the usual procedure of applying the well-known formula for multinomial coefficients, which is a legitimate procedure since m is supposed to be a positive integer. Thus, given a monomial $a^{m-k}b^q c^r \cdots$, its numerical coefficient or *uncia* can be calculated by means of the following formula:

$$\frac{m\,(m-1)\,(m-2)\ldots(m-k+1)}{1.2.3\ldots q \times 1.2.3\ldots r \times 1.2.3\ldots s \times etc}.$$

However, instead of using this formula, Klügel proposes to determine the required numerical coefficients by the following equivalent formula:

$$\left(\frac{m\,(m-1)\ldots(m-k+1)}{1.2.3\ldots k}\right)\left(\frac{1.2.3\ldots k}{1.2.3\ldots q \times 1.2.3\ldots r \times 1.2.3\ldots s \times etc}\right),$$

$$(4.28)$$

in such a way that the required numerical coefficients will be composed of a binomial coefficient as a first factor and of a multinomial coefficient as a second factor. The insistence on writing these coefficients in this particular way is guided by the desire to find an exact match with respect to Hindenburg's formulae. Indeed Hindenburg calculates these numerical coefficients or *unciae* by means of a product of a binomial coefficient and a multinomial coefficient, respectively represented by a Gothic capital letter and a Gothic lowercase letter:

$$^{m}\mathfrak{N} = \frac{m\,(m-1)\,(m-2)\ldots(m-k+1)}{1.2.3\ldots k}$$

and

$$\mathfrak{n} = \frac{1.2.3\ldots k}{1.2.3\ldots q \times 1.2.3\ldots r \times 1.2.3\ldots s \times etc}.$$

However, in the case of Hindenburg's multinomial theorem, the validity of the binomial coefficients rests on the binomial theorem of Newton, as seen in Sect. 3.2.1.2. This is why Hindenburg's formula for the multinomial series expansion depends on Newton's theorem when the exponent m is a rational (or even a real) number. On the contrary, Klügel thinks that formula 4.28, which has been deduced from the usual formula for multinomial coefficients, shows that the binomial coefficients can be obtained from the multinomial coefficients. As long as m is supposed to be a positive integer all goes well, but problems arise when m is no longer supposed to be a positive integer, and then Klügel will introduce some fuzzy arguments in his proof.

Adopting Hindenburg's notation for the *unciae*, the application of formula 4.28 to the monomials obtained by the method of involutions leads to the determination of the desired coefficients:

$$B = ma^{m-1}b = {}^{m}\mathfrak{A}\mathfrak{a}a^{m-1}b,$$

$$C = \frac{m(m-1)}{1.2}a^{m-2}b^2 + ma^{m-1}c = {}^{m}\mathfrak{B}\mathfrak{b}a^{m-2}b^2 + {}^{m}\mathfrak{A}\mathfrak{a}a^{m-1}c,$$

$$D = \frac{m(m-1)(m-2)}{1.2.3}a^{m-3}b^3 + \frac{m(m-1)}{1.2}\cdot\frac{2}{1}a^{m-2}bc + ma^{m-1}d$$

$$= {}^{m}\mathfrak{C}\mathfrak{c}a^{m-3}b^3 + {}^{m}\mathfrak{B}\mathfrak{b}a^{m-2}bc + {}^{m}\mathfrak{A}\mathfrak{a}a^{m-1}d,$$

etc.

This concludes the proof for a positive integer exponent.

In the case of a negative integer exponent, the problem is solved by division:

$$\frac{\left(a + bz + cz^2 + dz^3 + \cdots\right)^m}{\left(a + bz + cz^2 + dz^3 + \cdots\right)^n} = \left(a + bz + cz^2 + dz^3 + \cdots\right)^{m-n},$$

where m and n are positive integers.

The third step of the proof corresponds to the case of a positive fractional number. Since Klügel wants to avoid the use of the binomial theorem to justify the binomial coefficients that appear in the *unciae*, he invokes an argument that can be characterized as an argument from analogy. He argues that if one is given two powers:

$$\left(a + bz + cz^2 + dz^3 + \ldots\right)^m$$

and

$$\left(a + bz + cz^2 + dz^3 + \ldots\right)^n,$$

where m and n are positive whole numbers, one can assert that there is an equality relation between their series expansions, though this equality relation does not concern the quantity but the "form" of the coefficients belonging to the respective series expansions. According to Klügel, this equality of form is ensured by the first step of the proof in which it has been demonstrated that the coefficients of the multinomial expansion are sums of monomials of the form ${}^{m}\mathfrak{N}\mathfrak{n}a^{m-k}b^qc^r\cdots$, and so there is a similarity concerning the coefficients of the series expansions of $\left(a + bz + cz^2 + dz^3 + \ldots\right)^m$ and $\left(a + bz + cz^2 + dz^3 + \ldots\right)^n$. In a sort of analogical reasoning, Klugel claims that (Klügel 1796, 73):

Thus, $(a + bz + cz^2 + dz^3 + \ldots)^{\frac{m}{n}}$ and $(a + bz + cz^2 + dz^3 + \ldots)^{\frac{n}{m}}$ are of the same form, and it follows too that $(a + bz + cz^2 + dz^3 + \ldots)^m$ and $(a + bz + cz^2 + dz^3 + \ldots)^{\frac{1}{m}}$ are of the same form, that is to say, these quantities differ from each other only by the replacement of m with $\frac{1}{m}$.[22]

Therefore, Klügel's argument consists in inferring a similarity concerning the form of the series expansions of $(a+bz+cz^2+dz^3+\ldots)^m$ and $(a+bz+cz^2+dz^3+\ldots)^{\frac{1}{m}}$ from the similarity concerning the form of the series expansions of $(a+bz+cz^2+dz^3+\ldots)^{\frac{1}{n}}$ and $(a+bz+cz^2+dz^3+\ldots)^{\frac{n}{m}}$, and so the general coefficient of a series expansion for a fractional exponent $\frac{1}{m}$ and the general coefficient of a series expansion for a positive integer exponent m are of the same form, i.e. are expressed by the same formula. This argument was roundly criticized at the time, as will be seen later.

From here, Klügel generalize the multinomial theorem to the case of any fractional exponent $\frac{i}{m}$. He claims that, given that the series expansions of $(a+bz+cz^2+dz^3+\ldots)^m$ and $(a+bz+cz^2+dz^3+\ldots)^{\frac{1}{m}}$ are of the same form, one can raise these expressions to the integer power i without modifying the form of the general coefficient of the respective series expansions, and so both general coefficients are of the same form. In the case of a negative fractional exponent, the problem is solved by division: $\dfrac{1}{(a+bz+cz^2+dz^3+\ldots)^{\frac{i}{m}}} =$ $(a+bz+cz^2+dz^3+\ldots)^{-\frac{i}{m}}$.

Convinced of the soundness of his reasoning, Klügel deduces a very important corollary (Klügel 1796, 73):

> The binomial theorem is a corollary of the multinomial theorem according to the method used here.[23]

In a footnote to this corollary, Hindenburg, who used to add explanatory remarks to the texts printed in his anthologies and journals, underlines this point by saying that, thanks to Klügel's proof, the multinomial theorem has become a universal theorem in the sense that it is no longer dependent on the binomial theorem (Klügel 1796, 73).

4.2.2.2 A Foundational Combinatorial Proof of the Binomial Theorem
In the same year of 1796, Rothe publishes a proof of the binomial theorem for any real exponent (Rothe 1796). The general plan of the proof is to build first the binomial coefficients by using the doctrine of combinations, and, once this has been achieved, to deduce the binomial expansion from this combinatorial construction. It is worth insisting upon the originality of this plan compared to the current proofs at the time, in which the construction of the binomial coefficients is a consequence of the construction of the

---

[22] Also hat $(a+bz+cz^2+dz^3+\ldots)^{\frac{m}{n}}$ einerley Form mit $(a+bz+cz^2+dz^3+\ldots)^{\frac{n}{m}}$, folglich auch $(a+bz+cz^2+dz^3+\ldots)^m$ einerley Form mit $(a+bz+cz^2+dz^3+\ldots)^{\frac{1}{m}}$, das heißt, beyde Größen sind nur darin verschieden, daß m und $\frac{1}{m}$ in ihnen vertauscht sind.

[23] Der binomische Lehrsatz ist nach der hier gebrauchten Methode ein Corollarium des polynomischen.

binomial expansion itself. This difference is important in the historical context of the proof because it means that the theoretical foundation of the binomial theorem belongs to the doctrine of combinations, or at least this seems to be Rothe's purpose in writing a proof of the binomial theorem in 1796 despite the fact that he thinks that the validity of this theorem had been successfully established long time ago, particularly by Kästner who is constantly mentioned in Rothe's booklet. This new combinatorial proof thus aims not so much to demonstrate the binomial theorem as to persuade us that its theoretical justification rests on the principles of combinatorics. To this end, Rothe deploys what can be qualified as an "algebra of combinations", to borrow the expression coined by Panza to describe Rothe's proof in (Panza 1992, vol. 2, 673–675), where one can find an interesting reconstruction of this proof placed in the context of the development of analysis and algebra in the eighteenth century.

Although Rothe does not derive the binomial theorem from the multinomial theorem, as Klügel did, he begins his booklet by pointing out that a binomial coefficient can be viewed as a particular instance of multinomial coefficients, since, from a combinatorial perspective, a binomial coefficient $\frac{n(n-1)(n-2)\cdots(n-p+1)}{1\cdot 2\cdot 3\cdots p}$ express the number of permutations with repetition of two objects a and b, where the object a appears $n - p$ times and the object b appears p times, which is a particular instance of permutations with repetition of k objects where each object appears a certain number of times and the number of these permutations is calculated by a multinomial coefficient. There is some similarity between Rothe and Klügel in this regard in the sense that both are interested in showing that the general combinatorial law for this kind of permutations concerns multinomial coefficients, and binomial coefficients are nothing more than a particular example of this law. When n is a positive whole number, Rothe points out that the binomial theorem is a direct consequence of this law, since the expansion of $(a + b)^n$ can be obtained by calculating, for each $k = 0, 1, 2, \ldots, n$, the number of permutations with repetition of $a^{n-k}b^k$, where the object a appears $n - k$ times and the object b appears k times, which gives the corresponding coefficients of the terms $a^{n-k}b^k$ belonging to this finite expansion. However, this combinatorial law is stated for a positive whole number n in the handbooks of the time, and Rothe intends to prove that this law holds in fact for any real number n.

To express his results, Rothe creates some analytic-combinatorial formulae by modifying Hindenburg's notations. Thus, Rothe's proof not only contributes to the consolidation of the combinatorial analysis by showing that an important theorem of the theory of series rests on the doctrine of combinations, but also by pursuing the development of more convenient mathematical notations. Let's look at these analytic-combinatorial formulae in detail. To be able to operate inside his algebra of combinations, Rothe considers the following sequence of numbers:

$$\frac{f}{1}, \quad \frac{f(f-1)}{1\cdot 2}, \ldots, \frac{f(f-1)\cdots(f-m+1)}{1\cdot 2\cdots m}, \ldots, \frac{f(f-1)\cdots(f-m-r+1)}{1\cdot 2\cdots(m+r)}, \ldots$$

where f is a real number, and m and r are positive integers. Then, he denotes by $^f\mathfrak{M}$ its general term, so:

$$^f\mathfrak{M} = \frac{f\,(f-1)\cdots(f-m+1)}{1\cdot 2\cdots m},$$

and the $(m+r)$th term of the sequence is denoted by:

$$^f\overset{r}{\mathfrak{M}} = \frac{f(f-1)\cdots(f-m-r+1)}{1\cdot 2\cdots(m+r)}.$$

Rothe's notation is clearly inspired by Hindenburg's notations for binomial coefficients. As seen in Sect. 3.3.3, for Hindenburg, a Gothic capital letter with a superscript to the left represents a binomial coefficient, and if one adds a digit over a given coefficient, this digit indicates the place of the coefficient in the polynomial expression to which the coefficient belongs. On the contrary, $^f\overset{r}{\mathfrak{M}}$ is not a coefficient in principle but a quantity expressed in terms of f, m, and r, that is to say, $^f\overset{r}{\mathfrak{M}}$ is a general function in the sense accorded to this term in the eighteenth century, and this function is not necessarily attached to any polynomial, whether finite or infinite, as it is the case for coefficients. One should be careful not to confuse Rothe's functions with Hindenburg's binomial coefficients, especially because the digit over Rothe's functions or formulae enters into the calculation of the quantity they represent, which is not the case for Hindenburg's binomial coefficients. In fact, even when there is no digit over a Rothe's formula, as in $^f\mathfrak{M}$, one should consider that there is a digit equal to zero over this formula. Hence, the Gothic letter and the digit over this letter are a sort of index number that indicates the number of the term in the sequence given above to which the formula refers. For instance, the Gothic letter "$\mathfrak{M}$" in $^f\mathfrak{M}$ means the mth term of the sequence, for an arbitrary positive integer m; the Gothic letter "$\mathfrak{N}$" in $^f\mathfrak{N}$ means the nth term of the sequence, for an arbitrary positive integer n; the Gothic letter "$\mathfrak{P}$" in $^f\mathfrak{P}$ means the pth term of the sequence, for an arbitrary positive integer p. However, when one uses the first letters of the alphabet, "$\mathfrak{A}$" means the first term of the sequence, "$\mathfrak{B}$" means the second term of the sequence, and so on. This can certainly be regarded as an inconsistency in Rothe's notation, which is due to the persistence of the alphabetical notation for coefficients inherited from the Newtonian tradition.

Rothe establishes the following elementary relations, which can be easily confirmed by the simple definition of his formulae:

$$^f\overset{-1}{\mathfrak{M}} = \frac{m}{f-m+1}(^f\mathfrak{M}),$$

$$^f\overset{-2}{\mathfrak{M}} = \frac{m(m-1)}{(f-m+2)(f-m+1)}(^f\mathfrak{M}),$$

and in general:

$$^f\mathfrak{M}^{-r} = \frac{m(m-1)\cdots(m-r+1)}{(f-m+r)\cdots(f-m+2)(f-m+1)}(^f\mathfrak{M}),$$

for a positive integer r. From here, it follows in particular that $^f\mathfrak{A}^{-1} = 1$, and $^f\mathfrak{A}^{-r} = 0$ for every integer $r \geq 2$. It follows also that:

$$m\left(^f\mathfrak{M}\right) = (f-m+1)\left(^f\mathfrak{M}^{-1}\right). \tag{4.29}$$

Then he proves the relation:

$$n\left(^f\mathfrak{M}\right)\left(^g\mathfrak{N}^{-m}\right) = (f-m+1)\left(^f\mathfrak{M}^{-1}\right)\left(^g\mathfrak{N}^{-m}\right) + (g-n+m+1)\left(^f\mathfrak{M}\right)\left(^g\mathfrak{N}^{-m-1}\right),$$

$$\tag{4.30}$$

where g is a real number, as follows. By multiplying Eq. 4.29 by $^g\mathfrak{N}^{-m}$, one gets:

$$\left(m^f\mathfrak{M}\right)\left(^g\mathfrak{N}^{-m}\right) = (f-m+1)\left(^f\mathfrak{M}^{-1}\right)\left(^g\mathfrak{N}^{-m}\right).$$

But by replacing m by $n-m$, f by g, and g by f in this equation, one obtains:

$$(n-m)\left(^g\mathfrak{M}\right) = (n-m)\frac{g(g-1)\cdots(g-(n-m)+1)}{1\cdot 2\cdots(n-m)} = (n-m)\left(^g\mathfrak{N}^{-m}\right),$$

$$^f\mathfrak{N}^{m-n} = \frac{f(f-1)\cdots(f-m+1)}{1\cdot 2\cdots m} = {}^f\mathfrak{M},$$

$$(g-n+m+1)\left(^g\mathfrak{M}^{-1}\right) = (g-n+m+1)\frac{g(g-1)\cdots(g-(n-m)+2)}{1\cdot 2\cdots(n-m-1)}$$

$$= (g-n+m+1)\ {}^g\mathfrak{N}^{-m-1},$$

and so:

$$(n-m)\left(^g\mathfrak{N}^{-m}\right)\left(^f\mathfrak{M}\right) = (g-n+m+1)\left(^g\mathfrak{N}^{-m-1}\right)\left(^f\mathfrak{M}\right).$$

from which it follows immediately the desired Eq. 4.30 by using equality 4.29.

Rothe uses formulae 4.29 and 4.30 to justify the following combinatorial equation:

$$
{}^g\mathfrak{P} + \left({}^f\mathfrak{A}\right)\binom{-1}{{}^g\mathfrak{P}} + \left({}^f\mathfrak{B}\right)\binom{-2}{{}^g\mathfrak{P}} + \cdots + \binom{-2}{{}^f\mathfrak{P}}\left({}^g\mathfrak{B}\right) + \binom{-1}{{}^f\mathfrak{P}}\left({}^g\mathfrak{A}\right) + {}^f\mathfrak{P} = {}^{f+g}\mathfrak{P},
$$

$$(4.31)$$

which is known nowadays as the "Vandermonde's identity" and can be written as $\sum_{k=0}^{P}\binom{f}{k}\binom{g}{p-k} = \binom{f+g}{p}$ using our contemporary notation $\binom{f+g}{p}$ for binomial coefficients, since the symbol ${}^f\overset{r}{\mathfrak{P}}$ invented by Rothe means the same thing as $\binom{f}{p+r}$. Rothe proves this equation by induction, verifying first that it holds for $p = 1$ and $p = 2$. Then, from formulae 4.29 and 4.30, it follows that:

$$
n\left({}^g\mathfrak{N}\right) = (g - n + 1)\binom{-1}{{}^g\mathfrak{N}},
$$

$$
n\left({}^f\mathfrak{A}\right)\binom{-1}{{}^g\mathfrak{N}} = f\binom{-1}{{}^g\mathfrak{N}} + (g - n + 2)\left({}^f\mathfrak{A}\right)\binom{-2}{{}^g\mathfrak{N}},
$$

$$
n\left({}^f\mathfrak{B}\right)\binom{-2}{{}^g\mathfrak{N}} = (f - 1)\left({}^f\mathfrak{A}\right)\binom{-2}{{}^g\mathfrak{N}} + (g - n + 3)\left({}^f\mathfrak{B}\right)\binom{-3}{{}^g\mathfrak{N}},
$$

$$
\vdots
$$

$$
n\binom{-1}{{}^f\mathfrak{N}}\left({}^g\mathfrak{A}\right) = (f - n + 2)\binom{-2}{{}^f\mathfrak{N}}\left({}^g\mathfrak{A}\right) + g\binom{-1}{{}^f\mathfrak{N}},
$$

$$
n\left({}^f\mathfrak{N}\right) = (f - n + 1)\binom{-1}{{}^f\mathfrak{N}}.
$$

By adding these identities and by assuming that Eq. 4.31 holds for $p = n - 1$, he gets:

$$
n\left({}^{f+g}\mathfrak{N}\right) = (f + g - n + 1)\binom{-1}{{}^{f+g}\mathfrak{N}}
$$

$$
= (f + g - n + 1)\left[\binom{-1}{{}^g\mathfrak{N}} + \left({}^f\mathfrak{A}\right)\binom{-2}{{}^g\mathfrak{N}} + \cdots + \binom{-2}{{}^f\mathfrak{N}}\left({}^g\mathfrak{A}\right) + \binom{-1}{{}^f\mathfrak{N}}\right]
$$

$$
= n\left[{}^g\mathfrak{N} + \left({}^f\mathfrak{A}\right)\binom{-1}{{}^g\mathfrak{N}} + \cdots + \binom{-1}{{}^f\mathfrak{N}}\left({}^g\mathfrak{A}\right) + {}^f\mathfrak{N}\right]
$$

which completes the proof of the Vandermonde's identity.

To deduce the binomial theorem from the Vandermonde's identity 4.31, Rothe assumes that one is given the following two series:

$$^f S = y^f + {}^f\mathfrak{A} y^{f-1} z + {}^f\mathfrak{B} y^{f-2} z^2 + \cdots ,$$

and:

$$^g S = y^g + {}^g\mathfrak{A} y^{g-1} z + {}^g\mathfrak{B} y^{g-2} z^2 + \cdots .$$

By multiplying these series and by applying the Vandermonde's identity, he shows that the general coefficient of the product $\left(^f S\right)\left(^g S\right)$, i.e. the coefficient corresponding to the term $y^{f+g-p} z^p$ of this product, is given by $^{f+g}\mathfrak{B}$, so:

$$\left(^f S\right)\left(^g S\right) = {}^{f+g} S.$$

From here, it follows that:

$$\left(^f S\right)^\alpha = {}^{\alpha f} S, \tag{4.32}$$

for any positive integer α. By setting $f = 1$ in this equation, it follows that:

$$^\alpha S = \left(^1 S\right)^\alpha = (y + z)^\alpha ,$$

which gives us the expansion of $(y + z)^\alpha$ for a positive integer α. The validity of the binomial theorem for a negative exponent is justified by division: $\frac{1}{\alpha S} = {}^{-\alpha} S$. Setting $f = \frac{\beta}{\alpha}$ in Eq. 4.32, for positive integers α and β, one gets $\left(^{\frac{\beta}{\alpha}} S\right)^\alpha = {}^\beta S = {}^1 S^\beta$, and therefore $\sqrt[\alpha]{^1 S^\beta} = {}^1 S^{\frac{\beta}{\alpha}} = {}^{\frac{\beta}{\alpha}} S$, which proves the binomial theorem for a positive rational exponent. The case of a negative rational exponent is solved by division: $\frac{1}{^{\frac{\beta}{\alpha}} S} = {}^{-\frac{\beta}{\alpha}} S$. Concerning the case of a real exponent, Rothe proposes the following explanation (Rothe 1796, 11):

> Irrational numbers and transcendental real numbers can be expressed by means of rational numbers as accurately as desired, so that the error is less than any given quantity.[24]

Although this cannot be considered as a proof, it is worth noting that, at the time, all mathematicians who intended to demonstrate the binomial theorem for a real exponent

---

[24] Numeri irrationales atque transcendentes reales, per numeros rationales tam accurate exprimi possunt, vt error quacunque quantitate minor euadat.

advanced this kind of vague arguments, which is understandable because the concept of "real number" has not yet been clarified.

In a final footnote, Rothe says he hesitated to publish his proof because, after finishing a draft of his booklet, he realized that Segner had already used a similar method for demonstrating the binomial theorem in 1777 (Rothe 1796, 16). As seen in Sect. 2.1.3.2, it seems that Segner's method was inspired by the functional proof of the binomial theorem given by Euler in 1775. Rothe does not mention Euler's proof, and it is hard to tell whether he had read it or not before writing his own proof. It is beyond any doubt that he was at least aware of the existence of Euler's proof, since Hindenburg mentions it in his *Infinitinomii*, but Rothe had no reason to hide this information, in the event that he had read it. In fact, Rothe methodically quotes his sources and the works he discovered after finishing his booklet, like Segner's proof but also a proof given by Simon Antoine Jean L'Huilier (1750–1840) in 1795 which is very similar to that of Segner (L'Huilier 1795, V–XI). Thus, it is likely that this is an original proof of Rothe, conceived independently of those other demonstrations.

It is also intriguing to question whether formula 4.31 was an original result of Rothe or not. The only source for the doctrine of combinations to which he makes reference in his booklet is the exposition given by Kästner in the *Anfansgründe der Analysis endlicher Grössen*, from which he derives formulae 4.29 and 4.30. However, Kästner's exposition of the doctrine of combinations does not contain any hint from which it may be inferred the Vandermonde's identity 4.31, which was named after the French mathematician Alexandre-Théophile Vandermonde (1735–1796) who first published it in Europe in 1772, or at least his article (Vandermonde 1772) is the earliest European known source where this expansion of a binomial coefficient appears. However, it is unlikely that Rothe took this formula directly from Vandermonde's article, since no reference is made to Vandermonde in his writings or in the writings of the combinatorial school published until 1796. The fist mention of Vandermonde's article by a member of the combinatorial school appears in a note added by Kramp to his *Analyse des réfractions astronomiques* in which he points out that he knew nothing about Vandermonde's article before his book went to press (Kramp 1799, xiii). This identity appears implicitly, as a consequence of the binomial theorem, in the 1775 paper of Euler devoted to the binomial theorem (Euler 1775), but apparently Rothe had not read this paper. The Vandermonde's identity is explicitly stated as a theorem by Euler in (Euler 1784, 93–94), although it is written as:

$$
\begin{bmatrix} n \\ 0 \end{bmatrix}\begin{bmatrix} p \\ q \end{bmatrix} + \begin{bmatrix} n \\ 1 \end{bmatrix}\begin{bmatrix} p \\ q+1 \end{bmatrix} + \begin{bmatrix} n \\ 2 \end{bmatrix}\begin{bmatrix} p \\ q+2 \end{bmatrix} + \begin{bmatrix} n \\ 3 \end{bmatrix}\begin{bmatrix} p \\ q+3 \end{bmatrix} + \cdots = \begin{bmatrix} p+n \\ q+n \end{bmatrix},
$$

for positive integers p, q, and n, and where $\begin{bmatrix} p \\ q \end{bmatrix}$ is Euler's notation for a binomial coefficient $\binom{p}{q}$, expressed in our contemporary notation. However, there is also no mention of this article of Euler in Rothe's booklet. Perhaps Rothe rediscovered this formula himself.

In any case, from a theoretical point of view, Rothe's proof is different from those of Euler and Segner. In these latter proofs, the relation expressed by formula 4.31 follows from identifying the coefficients of the expansions of powers of the form $(a+b)^m (a+b)^n$ and $(a+b)^{m+n}$. On the contrary, Rothe proves formula 4.31 independently of any polynomial expansion and using exclusively combinatorial identities. In a second step, the binomial theorem is deduced from this combinatorial formula. Thus, this thoerem conveys above all a combinatorial law, which is in fact the same combinatorial law conveyed by the multinomial theorem, since Rothe characterized the binomial coefficients as a particular kind of multinomial coefficients. In this sense, Rothe's proof aims to show that one of the most important theorems of analysis, according to the characterization of the binomial theorem given by talented mathematicians like Euler, has its roots in the doctrine of combinations.

4.2.3   Hindenburg's Project Implementation

The central problems to be addressed by the combinatorial analysis have been listed in the programmatic text *Novi systematis*. Up until 1795 development of Hindenburg's program proceeded slowly; furthermore, no systematic attempt has as yet been undertaken to implement this program. Hindenburg himself has not yet started the composition of the long-promised system of science which should be based on the combinatorial analysis. On the other hand, the process of dissemination of his ideas through teaching and publishing in his specialized journals was vital to increase the number of scholars interested in developing his theory. Thanks to the formation of the combinatorial school, some of those problems will be solved by means of the techniques and concepts of the combinatorial analysis. Perhaps the discovery of the ultimate principle of the combinatorial analysis reinforced the idea of building a system of science and persuaded the members of the combinatorial school of the feasibility of such a project. In any case, the first methodical studies of these problems were conducted after Hindenburg announced in 1795 that the multinomial theorem had been identified as the ultimate principle of the combinatorial analysis. In this section, some of these studies will be reviewed.

4.2.3.1 Toward an Algebra of Involutions
In Sect. 4.2.1.2, it was argued that the multinomial theorem was catapulted from near anonymity to being the centerpiece of Hindenburg's scientific project because a combinatorial law of Nature was supposed to be contained in Moivre's rules for determining the multinomial expansion. The discovery of this law in Moivre's writings motivated Hindenburg to introduce his method and concept of lexicographical involutions in order to capture this law. However, in this context, lexicographical involutions are used as a tool to prove the multinomial theorem. In 1796, Hindenburg strives to put together what can be described as an algebra of lexicographical involutions in an effort to show that these are not merely mathematical tools but objects belonging to the doctrine of combinations, and,

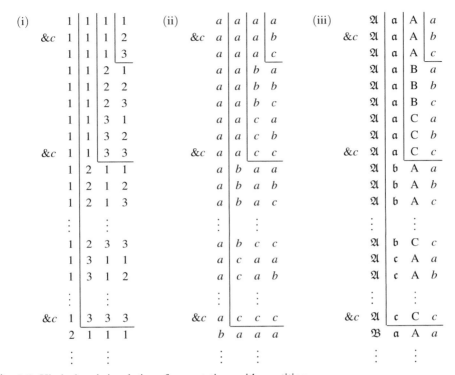

Fig. 4.6 Hindenburg's involution of permutations with repetition

as a consequence, they can be studied independently of the theory of series (Hindenburg 1796a). For Hindenburg, this is a very important step for the success of the long-promised system of science, which is supposed to be based on the doctrine of combinations: from a logical point of view, as the basis of the entire edifice of analysis, the doctrine of combinations must be an autonomous mathematical theory whose only foundation is a combinatorial law of Nature.

In his article (Hindenburg 1796a, 168 ff.), Hindenburg shows first that permutations and combinations with repetition in general can be enumerated by means of involutions. In Fig. 4.6, one can see the general pattern of involutions for enumerating k-permutations of three objects a, b, and c, where repetition of each object is allowed and where k is a positive integer and stands for the number of elements of the permutation. The involution on the left-hand side uses the uninterpreted symbols "1", "2", and "3", whereas in the involution (ii) these symbols have been interpreted according to the following substitution rule:

$$\begin{pmatrix} 1 & 2 & 3 \\ a & b & c \end{pmatrix}$$

In this figure only four columns of involution (i) are displayed but one can add as many columns as desired following the obvious pattern. In fact, this is a particular example of involution by cyclic periods, which was described in Sect. 4.2.1.1. Each column is constructed in accordance with an obvious cycle and the periods are clearly indicated by gnomons. For k corresponding to the number of a given column (from right to left), all k-permutations with repetition of a, b, and c are listed in the corresponding periods of involution (ii). Thus, there are three 1-permutations (listed in the first gnomon or period) which belong to the first class of permutations with repetition:

$$'A = \{a,\ b,\ c\},$$

where $'A$ is the notation used by Hindenburg since 1779 to denote this class of permutations (he does not use braces to enclose the members of the class). Similarly, there are nine 2-permutations with repetition of a, b, and c (listed in the second gnomon or period) which belong to the second class of permutations with repetition:

$$'B = \{aa,\ ab,\ ac,\ ba,\ bb,\ bc,\ ca,\ cb,\ cc\},$$

where $'B$ is the notation used by Hindenburg since 1779 to denote this class of permutations (he does not use braces to enclose the members of the class). And so on. This can easily be generalized to cover k-permutations with repetition of the objects a, b, c, d, etc. by adding the corresponding objects to each cycle of involution (ii) and by modifying in consequence the substitution rule.

Upon this basis Hindenburg defines a sum $'J$ for classes of permutations as follows:

$$'J = {}'A +{}' B +{}' C +{}' D + \cdots +{}'\mathcal{N}$$

$$(a \quad b \quad c \quad d \quad \ldots)$$

where $'\mathcal{N}$ stands for the nth class of permutations and the expression below this identity specifies the permuted objects. This sum operation means that $'J$ contains all things that are members of $'A$, or $'B$, $\ldots$, or $'\mathcal{N}$. Thus, this operation has a meaning very close to that of our set-theoretical operation of union. On the other hand, the poor choice of notation $'J$ for this sum can cause confusion, since it is identical to the notation of the classes to be added. This is why Hindenburg reserves the letter J to denote a sum of classes, and this letter should never be used for any other purpose.

In Fig. 4.6, involution (ii) is the result of interpreting "1", "2", "3" as the objects a, b, c respectively for the whole involution, but it is possible to assign different meanings to the same digits depending on other parameters. For instance, Hindenburg points out that if one is given several sequences of objects, one can decide to interpret digits by columns in such a way that each column of an uninterpreted involution would be linked to a particular sequence of objects and the meaning of the digits appearing in each column

would be defined by the objects of its corresponding sequence. He summarizes this idea
in the following table:

$$
\left\{
\begin{array}{llllll}
1, & 2, & 3, & 4, & \ldots & \\
a, & b, & c, & d, & \ldots & = \; p \\
A, & B, & C, & D, & \ldots & = \; q \\
\mathfrak{a}, & \mathfrak{b}, & \mathfrak{c}, & \mathfrak{d}, & \ldots & = \; r \\
\mathfrak{A}, & \mathfrak{B}, & \mathfrak{C}, & \mathfrak{D}, & \ldots & = \; s \\
\text{etc.} & & & & &
\end{array}
\right\}
\tag{4.33}
$$

Here one can see several sequences p, q, r, s, etc. of objects that are respectively linked
to the first, second, third, fourth, etc. columns of a given involution (from right to left),
and the first row of digits indicates the meaning of each digit according to the column
of the involution. For instance, if sequences p, q, r, and s have only three elements,
this compound substitution table yields the involution (iii) shown in Fig. 4.6. To signify
this particular interpretation of digits, Hindenburg invented the following notation for
the classes of permutations corresponding to the respective periods or gnomons of the
involution (he does not use braces to enclose the elements of the classes):

$$
{}^{p}_{\prime}A = \{a, \; b, \; c\},
$$

$$
{}^{qp}_{\prime}B = \{Aa, \; Ab, \; Ac, \; Ba, \; Bb, \; Bc, \; Ca, \; Cb, \; Cc\},
$$

$$
{}^{rqp}_{\prime}C = \{\mathfrak{a}Aa, \; \mathfrak{a}Ab, \; \mathfrak{a}Ac, \; \ldots, \; \mathfrak{b}Aa, \; \mathfrak{b}Ab, \; \mathfrak{b}Ac, \; \ldots,
$$
$$
\mathfrak{c}Aa, \; \mathfrak{c}Ab, \; \mathfrak{c}Ac, \; \ldots, \; \mathfrak{c}Ca, \; \mathfrak{c}Cb, \; \mathfrak{c}Cc\},
$$

etc.

where ${}^{p}_{\prime}A$ stands for the first class corresponding to the first gnomon, ${}^{qp}_{\prime}B$ stands for
the second class corresponding to the second gnomon, and so on. These classes are
called "classes of permutations" by Hindenburg, but they will be called here "classes of
compound permutations" to emphasize the fact that several sequences are involved and
to avoid any confusion with the classes ${}^{\prime}A, {}^{\prime}B$, etc., which are also called "classes of
permutations" by Hindenburg.

This allows Hindenburg to define a product of sequences in terms of the classes derived from his combinatorial involutions as follows:

$$qp = \overset{qp}{'B},$$

$$rqp = \overset{rqp}{'C},$$

$$srqp = \overset{srqp}{'D},$$

and in general:

$$\ldots tsrqp = \overset{\ldots tsrqp}{'\mathcal{M}},$$

where $\overset{\ldots tsrqp}{'\mathcal{M}}$ can be any class of compound permutations. It should be noted that sequences p, q, r, etc. can have any number of elements, and in principle they can be infinite. However, the definition of this product depends on the method of involutions, which can only be applied to finite sequences. Hindenburg does not seem to be concerned with this issue, and the context of his work suggests that those sequences are not actually infinite but can be extended as long as desired.

Hindenburg is particularly interested in presenting his classes of partitions and his lexicographical involutions from the point of view of this incipient algebra of involutions. In the past he had been dealing with classes of partitions in terms of combinations exclusively, but now this notion is extended to permutations, adopting a more comprehensive approach to this topic. Given a positive integer n, one considers all partitions of n. Each partition is composed of k parts whose sum is equal to n. A permutation class of partitions is defined by Hindenburg according to the number k of parts of the partitions. Thus, the first permutation class of partitions ^nA contains all partitions of n composed of one part exactly and contains also all permutations of these partitions. The second permutation class of partitions ^nB contains all partitions of n composed of two parts exactly and contains also all permutations of these partitions. And so on. As can be seen, the letters A, B, etc. serve to indicate the number of the class (first, second, etc.) and must be written in italic type. The terminology "permutation class of partitions" adopted here does not correspond to that chosen by Hindenburg, who uses instead the term "class of complexions", which is ambiguous since he uses the same term to designate combination classes of partitions (these combination classes will be described below and correspond to the classes of partitions studied in Chap. 3). Hindenburg gives the following algorithm to construct these permutation classes of partitions concerning a given positive integer n:

1. The first class ^nA has only one element: n.

2. For k greater or equal to 2, the kth class is given by:
   a. For each partition contained in the $(k-1)$th class, subtract 1 from the last part (from left to right) of the partition, and:
      i. If the result of this subtraction is equal to zero, this sequence of digits does not belong to the kth class.
      ii. If the result of this subtraction is strictly greater than zero, write 1 to the left of the first part (from left to right) of the partition. This sequence of digits belongs to the kth class.
   b. For each partition of the kth class obtained in (2a), subtract 1 from the last part (from left to right) of the partition, and:
      i. If the result of this subtraction is equal to zero, this sequence of digits does not belong to the kth class.
      ii. If the result of this subtraction is strictly greater than zero, add 1 to the first part (from left to right) of the partition. This sequence of digits belongs to the kth class.
   c. For each partition of the kth class obtained in (2b), subtract 1 from the last part (from left to right) of the partition, and:
      i. If the result of this subtraction is equal to zero, this sequence of digits does not belong to the kth class.
      ii. If the result of this subtraction is strictly greater than zero, add 1 to the first part (from left to right) of the partition. This sequence of digits belongs to the kth class.
   d. And so on.

The algorithm stops after a finite number of steps since the successive subtractions lead necessarily to zero. Hindenburg enunciates the rules of this algorithm using letters a, b, c, etc. instead of digits 1, 2, 3, etc. (Hindenburg 1796a, 177–178), but it is easier and clearer to formulate these rules using numbers. In Fig. 4.7, table (I) shows the construction of all permutation classes of partitions concerning the integer 5, and table (II) shows the result of interpreting table (I) in accordance with the substitution rule given below the tables.

In this case, Hindenburg also defines a sum ^{n}J for permutation classes of partitions with respect to a positive integer n:

$$^{n}J = {}^{n}A + {}^{n}B + {}^{n}C + {}^{n}D + \cdots + {}^{n}N$$

$$\begin{pmatrix} 1 & 2 & 3 & \ldots \\ a & b & c & \ldots \end{pmatrix}$$

where ^{n}N stands for the nth permutation class of partitions. As in the case of the sum $'J$, the sum ^{n}J means that ^{n}J contains all things that are members of ^{n}A, or ^{n}B, …, or ^{n}N. For instance, the sum ^{5}J consists of all partitions listed in table (I) of Fig. 4.7. From this,

(I)

$$
\begin{array}{l}
5 \quad {}^{5}A \\ \hline
1\ 4 \\
2\ 3 \quad {}^{5}B \\
3\ 2 \\
4\ 1 \\ \hline
1\ 1\ 3 \\
1\ 2\ 2 \\
1\ 3\ 1 \quad {}^{5}C \\
2\ 1\ 2 \\
2\ 2\ 1 \\
3\ 1\ 1 \\ \hline
1\ 1\ 1\ 2 \\
1\ 1\ 2\ 1 \quad {}^{5}D \\
1\ 2\ 1\ 1 \\
2\ 1\ 1\ 1 \\ \hline
1\ 1\ 1\ 1\ 1 \quad {}^{5}E
\end{array}
$$

(II)

$$
\begin{array}{l}
e \quad {}^{5}A \\ \hline
a\ d \\
b\ c \quad {}^{5}B \\
c\ b \\
d\ a \\ \hline
a\ a\ c \\
a\ b\ b \\
a\ c\ a \quad {}^{5}C \\
b\ a\ b \\
b\ b\ a \\
c\ a\ a \\ \hline
a\ a\ a\ b \\
a\ a\ b\ a \quad {}^{5}D \\
a\ b\ a\ a \\
b\ a\ a\ a \\ \hline
a\ a\ a\ a\ a \quad {}^{5}E
\end{array}
$$

(III)

$$
{}^{5}A
\begin{bmatrix}
a & a & a & a & a \\
a & a & a & a & b \\
a & a & a & b & a \\
a & a & c & & \\
a & b & a & a & \\
a & b & b & & \\
a & c & a & & \\
a & d & & &
\end{bmatrix}
$$

$$
{}^{5}B
\begin{bmatrix}
b & a & a & a \\
b & a & b & \\
b & b & a & \\
b & c & &
\end{bmatrix}
$$

$$
{}^{5}C
\begin{bmatrix}
c & a & a \\
c & b &
\end{bmatrix}
$$

$$
{}^{5}D \quad d\ a
$$

$$
{}^{5}E \quad e
$$

$$
\begin{pmatrix}
1 & 2 & 3 & \cdots \\
a & b & c & \cdots
\end{pmatrix}
$$

Fig. 4.7 Hindenburg's permutation class of partitions

Hindenburg deduces the following algebraic relations:

$$'A = {}^{1}A + {}^{2}A + {}^{3}A + \cdots ,$$

$$'B = {}^{2}B + {}^{3}B + {}^{4}B + \cdots ,$$

$$'C = {}^{3}C + {}^{4}C + {}^{5}C + \cdots ,$$

etc.,

and hence:

$$'J = {}'A + {}'B + {}'C + \cdots = {}^{1}A + \left({}^{2}A + {}^{2}B\right) + \left({}^{3}A + {}^{3}B + {}^{3}C\right) + \cdots = {}^{1}J + {}^{2}J + {}^{3}J + \cdots .$$

It is worth insisting that these results form the basis for developing an algebra of classes.

Once the digits of partitions have been interpreted in terms of letters, Hindenburg says that a sequence of letters thus obtained is said to be of order a if its first element is a, or of order b if its first element is b, and so on. For instance, the sequence "$d\ a$" in table (II) of Fig. 4.7 is the only sequence in this table of order d, and the sequences "$c\ b$"

and "$c\ a\ a$" are all sequences of order c in this table. One can regroup these sequences by orders instead of permutations classes, which yields a lexicographical reorganization of sequences. In Fig. 4.7, table (III) is a lexicographical involution organized by orders of sequences. Here this kind of involution will be called "lexicographical permutation involution" to avoid confusion with the lexicographical involutions studied in Sect. 4.2.1.2, which do not contain permutations of their elements. In a lexicographical permutation involution, it is possible to identify different blocks of sequences according to their order, so that $^n\mathbb{A}$ is called "permutation order a" and contains the sequences of order a, $^n\mathbb{B}$ is called "permutation order b" and contains the sequences of order b, and so on. Table (III) of Fig. 4.7 shows these permutation orders for $n = 5$. Hindenburg gives the following algorithm to construct the lexicographical permutation involution of a positive integer n:

1. The first gnomon only contains the sequence of digits "1". If $n = 1$, this is the desired involution; if not, go to step (2).
2. For k greater or equal to 2, the kth gnomon is given by:
   a. Write "1" to the left of each row contained in the $(k - 1)$th gnomon.
   b. Below the rows obtained by (2a), write the following rows:
      i. First write the sequence whose first element (from left to right) is the sum of the first and second elements (from left to right) of the first row (from the top down) obtained by (2a) and whose other elements are the elements of the first row obtained by (2a) from the third element onward.
      ii. Then write the sequence whose first element (from left to right) is the sum of the first and second elements (from left to right) of the second row (from the top down) obtained by (2a) and whose other elements are the elements of the second row obtained by (2a) from the third element onward.
      iii. And so on until the last row obtained by (2a).
   c. If the first (and only) element of the last row obtained by (2b) is equal to n, this is the desired involution; if not, keep going and construct more gnomons by applying rule (2).

Hindenburg enunciates these rules using letters instead of digits (Hindenburg 1796a, 178), but it is better to use digits since letters are supposed to be a particular interpretation of digits. In Fig. 4.7, involution (III) has been constructed by means of this algorithm.

As in the case of permutation classes, there exists a sum operation for permutation orders:

$$^n\mathbb{J} = {}^n\mathbb{A} + {}^n\mathbb{B} + {}^n\mathbb{C} + {}^n\mathbb{D} + \cdots + {}^n\mathbb{N}$$

$$\begin{pmatrix} 1 & 2 & 3 & \cdots \\ a & b & c & \cdots \end{pmatrix}$$

where $^n.\mathbb{J}$ is the sum of the permutation orders indicated on the right-hand side of this identity, and this operation means that $^n.\mathbb{J}$ contains all things that are members of $^n A$, or $^n B$, ..., or $^n N$. For instance, $^5.\mathbb{J}$ contains all sequences corresponding to the rows of involution (III) in Fig. 4.7.

Given a table of permutation classes of partitions for a positive integer n, like table (I) shown in Fig. 4.7, its digits can be interpreted by columns according to the compound substitution rule 4.33, so that digits belonging to the first column (from right to left) are associated to the sequence p of this compound substitution rule, digits belonging to the second column (from right to left) are associated to the sequence q, and so on. This interpretation allows to define the following compound permutation classes of partitions:

$$ {}_n^p A = \left\{ {}_n^0 \right\}, $$

$$ {}_n^p B = \left\{ A{}_n^{-1}, \ B{}_n^{-2}, \ \ldots, \ N{}_a^{-1} \right\}, $$

$$ {}_n^p C = \left\{ aA{}_n^{-2}, \ aB{}_n^{-3}, \ \ldots, \ {}_n^{-2} Aa \right\}, $$

etc.,

where ${}_n^0$ stands for the nth element of the sequence p, ${}_n^{-1}$ for the $(n-1)$th element of the sequence p, ${}_n^{-2}$ for the $(n-2)$th element of the sequence p, ${}_n^{-3}$ for the $(n-3)$th element of the sequence p, $N{}^{-1}$ for the $(n-1)$th element of the sequence q, and ${}_n^{-2}$ for the $(n-2)$th element of the sequence r. In fact, Hindenburg does not define these classes in this general way, but gives particular examples of them. However, the definitions given above express Hindenburg's idea and have been formulated using Hindenburg's notations, except for the braces. In Fig. 4.8, table (I) shows the result of such an interpretation of digits for $n = 4$, and table (II) shows the corresponding lexicographical compound permutation involution. For these classes and for the compound substitution rule 4.33, Hindenburg defines the

Fig. 4.8 Hindenburg's compound permutation class of partitions

(I)

			d	^4A
		A	c	
		B	b	^4B
		C	a	
	a	A	b	
	a	B	a	^4C
	b	A	a	
a	a	A	a	^4D

(II)

$\mathfrak{A}$	a	A	a
	a	A	b
	a	B	a
	A	c	
	b	A	a
	B	b	
	C	a	
	d		

following sum:

$$\,^{...srqp}_{n}J = \,^{p}_{n}A + \,^{qp}_{n}B + \,^{rqp}_{n}C + \,^{srqp}_{n}D + \cdots,$$

which has a similar meaning to that of the sums of other classes defined above.

As a consequence of the above definitions of sum, Hindenburg obtains the following relations between two different kinds of classes of compound permutations:

$$\,^{p}_{\prime}A = \,^{p}_{1}A + \,^{p}_{2}A + \,^{p}_{3}A + \,^{p}_{4}A + \cdots,$$

$$\,^{qp}_{\prime}B = \,^{qp}_{2}B + \,^{qp}_{3}B + \,^{qp}_{4}B + \cdots,$$

$$\,^{rqp}_{\prime}C = \,^{rqp}_{3}C + \,^{rqp}_{4}C + \cdots,$$

$$\text{etc.,}$$

Therefore:

$$\,^{p}_{\prime}A + \,^{qp}_{\prime}B + \,^{rqp}_{\prime}C + \cdots = \,^{p}_{1}A + \left(\,^{p}_{2}A + \,^{qp}_{2}B\right) + \left(\,^{p}_{3}A + \,^{qp}_{3}B + \,^{rqp}_{3}C\right) + \cdots = \,^{p}_{1}J + \,^{qp}_{2}J + \,^{rqp}_{3}J + \cdots$$

These equations complete the incipient algebra of classes of permutations developed by Hindenburg in his paper.

A similar algebra of classes can be developed for combinations. Figure 4.9 shows an example of the application of the method of involutions to systematically enumerate k-combinations of objects a, b, c, etc., where repetition of each object is allowed and where

(i)

'A	1	2	3
	11	12	13
'B		22	23
			33
	111	112	113
		122	123
'C			133
		222	223
			233
			333
	1111	1112	1113
'D	:	:	:

(ii)

'A	a	b	c
	aa	ab	ac
'B		bb	bc
			cc
	aaa	aab	aac
		abb	abc
'C			acc
		bbb	bbc
			bcc
			ccc
	aaaa	aaab	aaac
'D	:	:	:

(iii)

	a	a	a	a
&c	a	a	a	b
	a	a	a	c
	a	a	b	b
	a	a	b	c
	a	a	c	c
&c	a	b	b	b
	a	b	b	c
	a	b	c	c
	a	c	c	c
	b	b	b	b
&c	b	b	b	c
		:	:	

Fig. 4.9 Hindenburg's involution of combinations with repetition

k is a positive integer that stands for the number of elements of the combination. This example displays the enumeration of k-combinations with repetition of three objects a, b, and c as kth classes of combinations, so that table (i) regroups the first class $'A$ of 1-combinations with repetition, the second class $'B$ of 2-combinations with repetition, and so on, while table (ii) shows these same classes but digits "1", "2", and "3" have been respectively interpreted as the objects a, b, and c. Given n objects "1", "2", ..., "n" to be combined, Hindenburg formulates the following algorithm to recursively construct these classes:

1. The first class $'A$ contains the sequence of digits "1", and the sequence "2", ..., and the sequence "n" as a combination.
2. For k greater or equal to 2, the kth class is given by:
   a. Write "1" to the left of each sequence of digits contained in the $(k-1)$th class. The sequences of digits thus obtained belong to the kth class.
   b. For each sequence obtained by (2a), if its first two elements (from left to right) are different from each other, add 1 to the first element. The sequences of digits thus obtained belong to the kth class.
   c. For each sequence obtained by (2b), if its first two elements (from left to right) are different from each other, add 1 to the first element. The sequences of digits thus obtained belong to the kth class.
   d. And so on.

Here again Hindenburg enunciates these rules using letters instead of digits (Hindenburg 1796a, 175). Table (iii) in Fig. 4.9 is the involution corresponding to these classes: each line of the first gnomon corresponds to a given element of the first class, each line of the second gnomon corresponds to a given element of the second class, and so on. Hindenburg does not give the algorithm to construct this involution, but the pattern is pretty clear.

The sum $'J$ of these classes is defined by:

$$'J = {}'A + {}'B + {}'C + {}'D + \cdots + {}'N$$

$$(a \quad b \quad c \quad d \quad \ldots)$$

where the expression below the equation indicates the objects to be combined and is an abbreviation of the substitution rule that associates "1" to "a", "2" to "b", and so on. This operation means that $'J$ contains all things that are members of $'A$, or $'B$, ..., or $'N$. From this definition, Hindenburg deduces the following relation:

$$'N = a^n + a^{n-1}('A) + a^{n-2}('B) + a^{n-3}('C) + \cdots + a^0('N)$$

$$\begin{pmatrix} 1 & 2 & 3 & \ldots \\ a & b & c & \ldots \end{pmatrix} \qquad \begin{pmatrix} 1 & 2 & 3 & \ldots \\ b & c & d & \ldots \end{pmatrix}$$

In this case, the class on the left-hand side of the equation must be interpreted according to the substitution rule on the left-hand side, and the classes on the right-hand side must be interpreted according to the substitution rule on the right-hand side. Expressions of the form a^k are characterized by Hindenburg as abbreviations for sequences of k repeated elements a, and expressions of the form $a^k({}'\mathrm{P})$, where ${}'\mathrm{P}$ is any class interpreted according to the substitution rule on the right-hand side, define a new class by means of a product operation. This product operation consists in multiplying the elements of the class ${}'\mathrm{P}$ by a^k, which means that the sequence a^k must be written to the left of each sequence contained in ${}'\mathrm{P}$. Thus, a class of the form $a^k({}'\mathrm{P})$ contains the sequences that result from performing this product operation. It is worth noting that, for a product thus defined, the empty finite sequence a^0, containing no terms, is a neutral element. Therefore, Hindenburg's equation expresses the class ${}'\mathcal{N}$ linked to the substitution rule on the left-hand side as a sum of classes of the form $a^k({}'\mathrm{P})$ linked to the substitution rule on the right-hand side, since the first term a^n of this sum can be understood as the class that results from multiplying the sequence a^n by the elements of the class that contains exclusively the neutral element.

Hindenburg also reworked his method of classes of partitions, studied in Chap. 3, and his lexicographical involutions, studied in Sect. 4.2.1.2, and tried to incorporate them into his new algebra of classes. Figure 4.10 shows Hindenburg's example for illustrating this algebra. As we know already, table (i) is a table of combination classes of partitions for

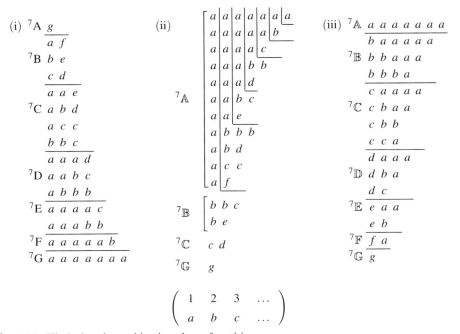

Fig. 4.10 Hindenburg's combination class of partitions

the integer 7, table (ii) is a lexicographical combination involution in increasing order, associated to Moivre's work, and table (iii) is a lexicographical combination involution in decreasing order, associated to Boscovich's work. In table (i), ^{7}A, ^{7}B, …, and ^{7}G are combination classes of partitions. In tables (ii) and (iii), classes are defined by the first term of their elements in such a way that ^{7}A is the lexicographical combination class of elements whose first term is a, or class of order a, ^{7}B is the lexicographical combination class of elements whose first term is b, or class of order b, and so on. For these kinds of classes, Hindenburg defines the following sums:

$$^n\mathrm{J} = {}^n\mathrm{A} + {}^n\mathrm{B} + {}^n\mathrm{C} + {}^n\mathrm{D} + \cdots + {}^n\mathcal{N}$$

$$\begin{pmatrix} 1 & 2 & 3 & \cdots \\ a & b & c & \cdots \end{pmatrix}$$

and:

$$^n\mathbb{J} = {}^n\mathbb{A} + {}^n\mathbb{B} + {}^n\mathbb{C} + {}^n\mathbb{D} + \cdots + {}^n\mathcal{N}$$

$$\begin{pmatrix} 1 & 2 & 3 & \cdots \\ a & b & c & \cdots \end{pmatrix}$$

The sum $^n\mathrm{J}$ means that $^n\mathrm{J}$ contains all things that are members of $^n\mathrm{A}$, or $^n\mathrm{B}$, …, or $^n\mathcal{N}$, where $^n\mathcal{N}$ stands for the nth class. The sum $^n\mathbb{J}$ has a similar meaning. It is worth noting that the sum $^n\mathbb{J}$ is defined independently of the arrangement chosen for the elements of the classes to be added. For instance, from table (ii) in Fig. 4.10, one can see that for lexicographical combination classes arranged in increasing order, this sum is given by:

$$^7\mathrm{J} = {}^7\mathrm{A} + {}^7\mathrm{B} + {}^7\mathrm{C} + {}^7\mathrm{G},$$

and from table (iii) in Fig. 4.10, one gets for lexicographical combination classes arranged in decreasing order:

$$^7\mathbb{J} = {}^7\mathbb{A} + {}^7\mathbb{B} + {}^7\mathbb{C} + {}^7\mathbb{D} + {}^7\mathbb{E} + {}^7\mathbb{F} + {}^7\mathbb{G}.$$

This definition and this example show that Hindenburg's algebra of classes presupposes the existence of empty classes: in the sum $^7\mathrm{J} = {}^7\mathrm{A} + {}^7\mathrm{B} + {}^7\mathrm{C} + {}^7\mathrm{G}$ of lexicographical combination classes arranged in increasing order, the general definition of the sum given by Hindenburg implies that $^7\mathrm{D}$, $^7\mathrm{E}$, and $^7\mathrm{F}$ are empty classes, and thus they can be omitted in the expression of the sum. Therefore, the empty class is a neutral element for the sum operation in Hindenburg's algebra of classes.

From this, Hindenburg deduces the following algebraic relations:

$$'A = {}^1A + {}^2A + {}^3A + \cdots,$$

$$'B = {}^2B + {}^3B + {}^4B + \cdots,$$

$$'C = {}^3C + {}^4C + {}^5C + \cdots,$$

$$\text{etc.},$$

and hence:

$$'J = 'A + 'B + 'C + \cdots = {}^1A + \left({}^2A + {}^2B\right) + \left({}^3A + {}^3B + {}^3C\right) + \cdots = {}^1J + {}^2J + {}^3J + \cdots.$$

He also points out that it is possible to define compound combination classes of partitions ${}^p_nA$, ${}^{qp}_nB$, ${}^{rqp}_nC$, etc., based on a compound substitution rule like rule 4.33, by assigning different interpretations to digits in a similar manner as in compound permutation classes of partitions, but he does not elaborate further on this topic.

Hindenburg also draws attention to the fact that classes of permutations can be expressed in terms of classes of combinations as follows:

$$'A = \mathfrak{a}'A = \mathfrak{a}\left({}^1A\right) + \mathfrak{a}\left({}^2A\right) + \mathfrak{a}\left({}^3A\right) + \mathfrak{a}\left({}^4A\right) + \cdots$$

$$'B = \mathfrak{b}'B = \mathfrak{b}\left({}^2B\right) + \mathfrak{b}\left({}^3B\right) + \mathfrak{b}\left({}^4B\right) + \cdots$$

$$'C = \mathfrak{c}'C = \mathfrak{c}\left({}^3C\right) + \mathfrak{c}\left({}^4C\right) + \cdots$$

$$\vdots$$

In these expressions, Gothic letters represent a general multinomial coefficient, that is to say that each element of the corresponding class of combinations must be preceded by its respective multinomial coefficient. Since these elements are understood to be uninterpreted symbols, their respective multinomial coefficients cannot represent an arithmetical product. Instead, they represent a sort of combinatorial multiplicity that indicates the number of permutations that can be obtained by rearranging the terms of the combination, and so the general coefficient that precedes a class of combinations stands for the multiplicity of its elements with respect to the corresponding class of permutations.

Although it is rather straightforward to assume that sums of classes of partitions, whether of permutations or combinations, are equivalent to sums of lexicographical involutions, this result was only quickly mentioned by Hindenburg without emphasizing its role in his abstract algebra (Hindenburg 1796a, 267). In 1797, Pfaff gives expressly the equation:

$$^nJ = {}^n\mathbb{J},$$

but he uses the notation "$^n\mathrm{I}$" instead of "$^n\mathrm{J}$", and "$^n\mathrm{J}$" instead of "$^n\mathbb{J}$", probably due to typographical limitations, and he places it at the center of the theory (Pfaff 1797b, 278 ff.). This is an important equation for the combinatorial analysis since all mathematical results of this theory achieved before 1795 have been formulated on the basis of classes of partitions, and this equation ensures that those results can be also formulated on the basis of lexicographical involutions. Thus, Hindenburg's algebra of classes is in fact an algebra of involutions whose aim is to provide the mathematical foundation of the combinatorial school's system of science.

4.2.3.2 Involutions and Reversion of Series

The theorem on reversion of series is one of the topics to be dealt with in order to push ahead with plans to transform the theory of series into a mathematical theory based on the doctrine of combinations. Although Eschenbach's dissertation studied in Sect. 4.1.3.1 was a great forward step in this direction, Pfaff was the first member of the combinatorial school to publish an entire treatise devoted to the study of this theorem from an analytic-combinatorial point of view (Pfaff 1797b). Pfaff's interest in this question originated with his successful attempt to deduce Eschenbach's formula for the reversion of series from Lagrange's inversion theorem, but at that time he did not really intend to commit himself to Hindenburg's project. With the passage of time, he felt more and more positive about the combinatorial analysis, as he put it himself in 1796 (Pfaff 1796a, 152):

> It seems to me that this opens up a vast field of analytic inquiry, whose interest lies at least in its difficulty and innovation.[25]

In 1797, this feeling evolved from the intellectual curiosity expressed in this statement toward the solid conviction that the combinatorial analysis was a promising mathematical theory, which is reflected throughout his treatise on the reversion of series.

When reading Pfaff's treatise, however, one cannot fail to be struck by its lack of unity. It is composed of three chapters. Chapters one and three present Lagrange's inversion theorem and its applications to the solution of equations by series respectively, whereas chapter two offers an in-depth look at the combinatorial analysis with due emphasis on the theorem on reversion of series. Chapter two is almost twice as long as the combined length of chapters one and three, and could have been singled out and been granted its own treatise. It seems as if it were a very long digression artificially inserted between chapters one and three. This is clearly a consequence of Pfaff's struggle with the difficulty of integrating general mathematical results into the specialized theory of the combinatorial analysis. Although he tries on two occasions to apply the method of involutions and classes of partitions to the problems treated in chapter three (Pfaff 1797b, 338, 345–347), the truth is that these two brief applications are rather unnecessary in the sense that a coherent

---

[25] Es eröfnet sich hier, wie es mir vorkommt, ein weites Feld für analytische Speculationen, die wenigstens durch ihre Schwierigkeit und Neuheit Interesse zu haben scheinen.

exposition of the subject can be achieved without them. In fact, even if Pfaff was one of the members of the combinatorial school, he did not always use the methods of the combinatorial analysis in his writings. For instance, in (Pfaff 1797a) he addresses several problems using series and trigonometric functions, but no attempt is made to express these functions and series by means of the combinatorial analysis, despite the fact that the study of these functions by means of analytic-combinatorial methods was included in the scientific programmatic agenda of the combinatorial school. This lack of systematicity in the work of some members of the combinatorial school will become an impediment to the reform of mathematical analysis and to the establishment of the theory of series on the doctrine of combinations.

On the other hand, if chapter two is taken on its own, it offers a coherent account of the combinatorial analysis, including the most recent improvements brought about by the use of Hindenburg's algebra of involutions. In fact, the chapter is structured around the notion of combinatorial involution and the algebraic operations defined for involutions. The chapter begins with an introduction to the doctrine of combinations, but this doctrine is framed in terms of Hindenburg's algebra of involutions. The choice of this approach has grown out of Pfaff's increasing intellectual commitment to the combinatorial analysis and, particularly, to the view that this theory has reached an important turning point with regard to the doctrine of combinations. A long time ago, when Hindenburg's combinatorial tables of partitions were but a tool to prove the multinomial theorem, which was nothing more than a modest corollary of the theory of series, the doctrine of combination did not reduce to the use of mathematical tables or the theoretical ideas supporting these tables. On the contrary, for Hindenburg, the nascent algebra of involutions is nothing but the very essence of the doctrine of combinations. In his 1796 paper, the section devoted to the algebra of involutions has been entitled "*Die Combinationslehre ist eine selbstständige Grundwissenschaft*". It is plain that, for Hindenburg, this autonomous, foundational property of the doctrine of combinations comes from the possibility of formulating an abstract algebra of classes. Thus, this algebra should not be considered to be merely a simple approach but the true expression of the doctrine of combinations. In other words, certainly this doctrine does not reduce to the use of mathematical tables, but it does reduce to this abstract algebra. Pfaff not only agrees with Hindenburg, but even contributes to the development of the algebra of involutions, as seen at the end of Sect. 4.2.3.1.

Once Pfaff has described and summarized this algebra of involutions, he proceeds to apply it to the multinomial theorem. Its application consists in showing that Hindenburg's non-recursive formula:

$$^m\mathfrak{A}a^{m-1}\mathfrak{a}\left(^n\mathrm{A}\right) + {}^m\mathfrak{B}a^{m-2}\mathfrak{b}\left(^n\mathrm{B}\right) + {}^m\mathfrak{C}a^{m-3}\mathfrak{c}\left(^n\mathrm{C}\right) + \cdots + {}^m\mathfrak{N}a^{m-n}\mathfrak{n}\left(^n\mathrm{N}\right) \qquad (4.34)$$

for calculating the general coefficient of the multinomial expansion of $(a+bx+cx^2+\ldots)^m$, where m is any rational (or even real) number, can be justified on the basis of this algebra. As seen in Sect. 3.2.1.2, Hindenburg originally obtained his formula by associating a given combination class of partitions to a specific term of the multinomial expansion, and, as a consequence, the combinatorial properties of his method of proof are not independent of the power series expansion itself, since the method is useless if such an association cannot be made. In this sense, the sum 4.34 of the relevant combination classes of partitions that serve to determine the coefficient of the term x^n of the multinomial expansion arises from the process of constructing the power series expansion itself. It is worth noting that, back then, the sum 4.34 does not really represent a sum of classes but a sum of monomials of the form $a^p b^q c^r \cdots$, since the elements of each combination class of partitions have been previously interpreted as monomials, so that the sum 4.34 is in fact a common algebraic expression, a sum of real numbers. Moreover, without this interpretation, the mathematical expression given by formula 4.34 had no meaning at that time. On the contrary, the creation of the abstract algebra of involutions allows to understand formula 4.34 as a sum of classes, which is independent of any interpretation of the elements of the classes, that is to say, this sum is a combinatorial object whose meaning does not depend on the coefficients of a given polynomial or any other mathematical object. Hence, the construction of the object expressed by formula 4.34 does no longer depend upon the construction of any power series expansion, and furthermore the coefficients of the multinomial expansion can be seen as a particular example of a more general combinatorial object. Pfaff's justification of Hindenburg's formula by means of the algebra of involutions must be understood according to this idea. Indeed, Pfaff also explains, in terms of combinatorial involutions, the proofs of the multinomial theorem given respectively by Moivre and Boscovich, closely following the reconstruction of these proofs given by Hindenburg (see Sects. 4.2.1.2 and 4.2.1.3). Thus, in full accordance with the most recent ideas of the combinatorial school, Pfaff presents the multinomial theorem as a mathematical result of the theory of series that is based on the doctrine of combinations in a strong foundational sense.

Pfaff then introduces the theorem on reversion of series. He presents first the non-recursive formula discovered by Eschenbach and reworked by Rothe, and justifies this formula by means of the multinomial theorem. As Eschenbach, Pfaff also translates Moivre's recursive solution for finding the coefficients of a reversed series into the language of Hindenburg's combination classes of partitions. Thus, no innovation has been fielded regarding the theorem itself. However, Pfaff's presentation pinpoints a general, innovative shift in perspective, at least regarding his own work. As seen in Sect. 4.1.3.5, in 1794 Pfaff thought that Eschenbach-Rothe's formula can be conceptualized as a consequence of Lagrange's inversion theorem. In his 1797 treatise on the reversion of series, as pointed out above, this theorem of Lagrange constitutes the main subject matter of chapter one, but he decided this time not to deduce Eschenbach-Rothe's formula from Lagrange's theorem, he deduced it instead from the multinomial theorem which was, in its turn, derived from the algebra of involutions. This reflects an interesting conceptual shift from the idea of investigating the relation of two theorems belonging to two different mathematical fields toward the purpose of organizing mathematical knowledge in a scientific system. In this sense, the principal aim of Pfaff's treatise is not to improve the techniques involved in the process of reversing a series, but to place the theorem on reversion of series in the mathematical landscape that the combinatorial school is depicting. In doing so, Pfaff incorporates not only Eschenbach-Rothe's and Moivre's formulae into a mathematical scientific system based on the doctrine of combinations, but also Lagrange's inversion theorem since Rothe proved the equivalence between this theorem and Eschenbach-Rothe's formula in 1795. This equivalence ensures that all results obtained by applying Lagrange's inversion theorem, like the solution of equations by means of series, are translatable into the language of this combinatorial mathematical system, even if Pfaff struggles with the difficulty of adapting analytic-combinatorial methods to the differential techniques of Lagrange's theorem throughout the entire book. Pfaff's treatise is, then, a major contribution to the attempt at systematizing mathematical knowledge by means of the doctrine of combinations.

4.2.3.3 Product, Power, and Division of Series

Products, powers and divisions of series are topics included in the scientific work program presented in the *Novi systematis*. The challenge now is to revisit and systematize them into a comprehensive account based on the new theoretical background of the combinatorial school. Again, it is Hindenburg who shows the way forward to reformulate these questions in terms of the algebra of involutions. In his article (Hindenburg 1796a, 228 ff.), Hindenburg deduces from this algebra some formulae for calculating the general term of products, powers, and divisions of series. Let's summarize the main results established by Hindenburg.

Hindenburg points out that, given the compound substitution rule 4.33, the terms of the sequences p, q, r, etc. can be associated to the coefficients of series in the following way:

$$
\begin{array}{ccccccccc}
1 & & 2 & & 3 & & 4 & & \cdots \\
a & + & bz & + & cz^2 & + & dz^3 & + \cdots & = p \\
A & + & Bz & + & Cz^2 & + & Dz^3 & + \cdots & = q \\
\mathfrak{a} & + & \mathfrak{b}z & + & \mathfrak{c}z^2 & + & \mathfrak{d}z^3 & + \cdots & = r \\
\end{array}
\qquad (4.35)
$$

etc.,

where the first row of digits just recalls substitution rule 4.33. Products of series can be calculated using compound permutation classes of partitions. The general term of the product of two series p and q is given by:

$$(qp)\daleth\,(n+1) = {}^{n+2}\overset{qp}{B}z^n,$$

that of three series p, q, and r by:

$$(rqp)\daleth\,(n+1) = {}^{n+3}\overset{rqp}{C}z^n,$$

and in general that of a product of m series is given by:

$$(\ldots tsrqp)\mathord{\text{\r{1}}}(n+1) = {}^{\substack{\ldots tsrqp \\ n+m}}\mathcal{M}z^n.$$

In particular, if the m series are the same series p, the general term of the positive integer power m of p can be written as follows:

$$p^m \mathord{\text{\r{1}}}(n+1) = \mathfrak{m}\left({}^{n+m}\text{M}\right)z^n,$$

where $\mathfrak{m}$ is a general expression for designating the multinomial coefficient corresponding to each element of the combination class of partitions ${}^{n+m}\text{M}$. Thus, this class can be seen as the following sum of combination classes of partitions:

$$\mathfrak{m}\left({}^{n+m}\text{M}\right) = {}^m\mathfrak{A}a^{m-1}\mathfrak{a}\left({}^n\text{A}\right) + {}^m\mathfrak{B}a^{m-2}\mathfrak{b}\left({}^n\text{B}\right) + {}^m\mathfrak{C}a^{m-3}\mathfrak{c}\left({}^n\text{C}\right) + \cdots$$

$$\begin{pmatrix} 1 & 2 & 3 & \ldots \\ a & b & c & \ldots \end{pmatrix} \qquad \begin{pmatrix} 1 & 2 & 3 & \ldots \\ b & c & d & \ldots \end{pmatrix} \tag{4.36}$$

It is worth insisting that the key idea of Hindenburg's work is to define the product operation for series in terms of a more abstract mathematical structure.

Formula 4.36 is in fact a particular case of formula 4.34 for calculating the coefficient of the term z^n in the power series expansion of p^m, where m can be any rational number. As he had done in 1778, Hindenburg proved again his formula 4.34 by using the binomial theorem (Hindenburg 1796a, 232). It should be noted that this proof does not mean that the multinomial theorem depends on the binomial theorem, since Hindenburg subscribes to Klügel's proof of independence of the multinomial theorem, as seen in Sect. 4.2.2.1. However, the use of the binomial theorem offers a straightforward way to construct formula 4.34. Hindenburg also addresses the particular case of the power P^m for a series $P = a + b + c + \cdots$ and for any rational number m, whose general term and expansion are given by:

$$P^m \mathord{\text{\r{1}}}(n+1) = {}^m\mathfrak{N}a^{m-n}\mathfrak{n}\left('\mathcal{N}\right)$$

$$P^m = a^m + {}^m\mathfrak{A}a^{m-1}\mathfrak{a}\left('\text{A}\right) + {}^m\mathfrak{B}a^{m-2}\mathfrak{b}\left('\text{B}\right) + {}^m\mathfrak{C}a^{m-3}\mathfrak{c}\left('\text{C}\right) + \cdots$$

$$(a \quad b \quad c \quad d \quad \ldots) \tag{4.37}$$

These formulae allow to deal with the problem of raising a series to any rational power.

Hindenburg recalls the reader's attention to the possibility of symbolizing the general coefficient 4.34 by means of local symbols (for the notion of local symbols $p^m x(n)$, see Sect. 4.1.3.4), as proposed by Rothe in formula 4.13, and connects Rothe's combinatorial functions to the new algebra of involutions via the following formula:

$$p^m x\,(n+1) = \left(^{m}\overset{*-1}{\mathfrak{A}}\,\mathfrak{a}\right) a^{m-*}\,(^{n}\mathsf{J}) = \left(^{m}\overset{*-1}{\mathfrak{A}}\,\mathfrak{a}\right) a^{m-*}\,(^{n}\mathbb{J})$$

$$p[a \quad b \quad c \quad \ldots] \qquad \begin{pmatrix} 1 & 2 & 3 & \cdots \\ b & c & d & \cdots \end{pmatrix}$$

In this expression, the symbol "*" actually serves as an index of summation which takes values from 1 to n, so that sums on the right-hand side are equal to formula 4.34 (remember that $^{0}\overset{m}{}\mathfrak{A} = {^{m}\mathfrak{A}}$, $^{1}\overset{m}{}\mathfrak{A} = {^{m}\mathfrak{B}}$, $^{2}\overset{m}{}\mathfrak{A} = {^{m}\mathfrak{C}}$, and so on according to Hindenburg's notations). As a consequence, the analytic-combinatorial functions developed by Rothe in order to provide the field of the combinatorial analysis with its own functions have been redefined as sums of abstract classes instead of ordinary algebraic sums.

Local symbols are used to implement products of powers. Thus, for any rational numbers a, b, c, etc., and for series p, q, r, etc. given in Eqs. 4.35, one can write the following expansions:

$$p^a = p^a x(1) + p^a x(2)z + p^a x(3)z^2 + p^a x(4)z^3 + \cdots,$$
$$q^b = q^b x(1) + q^b x(2)z + q^b x(3)z^2 + q^b x(4)z^3 + \cdots,$$
$$r^c = r^c x(1) + r^c x(2)z + r^c x(3)z^2 + r^c x(4)z^3 + \cdots,$$

etc.,

and Hindenburg can define a compound substitution rule to express the expansion of $q^b p^a$ and its general coefficient in terms of compound permutation classes of partitions as follows:

$$\left(q^b p^a\right) \mathsf{1}(n+1) = \;{}^{q^b p^a}_{n+2}B\; z^n$$

$$q^b p^a = \;{}^{q^b p^a}_{2}B \;+\; {}^{q^b p^a}_{3}B\; z + \;{}^{q^b p^a}_{4}B\; z^2 + \cdots$$

$$\begin{pmatrix} 1 & 2 & 3 & \cdots \\ p^a x(1) & p^a x(2) & p^a x(3) & \cdots \\ q^b x(1) & q^b x(2) & q^b x(3) & \cdots \end{pmatrix}$$

In general, for m series p^a, q^b, r^c, etc., the coefficient of the $(n+1)$th term of their product is given by:

$$\left(\ldots r^c q^b p^a\right) \daleth (n+1) = {}_{n+m}\mathcal{M}^{\ldots r^c q^b p^a} z^n,$$

so that its expansion is:

$$\ldots r^c q^b p^a = {}_m\mathcal{M}^{\ldots r^c q^b p^a} + {}_{m+1}\mathcal{M}^{\ldots r^c q^b p^a} z + {}_{m+2}\mathcal{M}^{\ldots r^c q^b p^a} z^2 + \cdots$$

$$\begin{pmatrix} 1 & 2 & 3 & \cdots \\ p^a x(1) & p^a x(2) & p^a x(3) & \cdots \\ q^b x(1) & q^b x(2) & q^b x(3) & \cdots \\ r^c x(1) & r^c x(2) & r^c x(3) & \cdots \\ \vdots & \vdots & \vdots & \cdots \end{pmatrix}$$

Note that, since local symbols have been redefined as sums of abstract classes, this substitution rule implies that ${}_{n+m}\mathcal{M}^{\ldots r^c q^b p^a}$ is a class of classes, and thus Hindenburg's algebra of involutions allows to operate with families of classes, although he has not explicitly addressed this issue.

In a similar vein, Pfaff uses local symbols to express products and powers of series (Pfaff 1796b). The key distinction between Hindenburg's and Pfaff's approaches lies in the fact that Pfaff's solution to these questions depends on the theorem according to which if the coefficients of two series p and q are such that:

$$p \, x \, (n+1) = \frac{s}{s+nd} q^{s+nd} x \, (n+1),$$

for $n = 1, 2, 3, \ldots$, then the equation:

$$p^m \, x \, (n+1) = \frac{ms}{ms+nd} q^{ms+nd} x \, (n+1) \tag{4.38}$$

is valid for any rational (or even real) number m. Indeed, this theorem is a consequence of the theorem on reversion of series. Pfaff supposes that:

$$q = Ay^{-1} + By^{-1+d} + Cy^{-1+2d} + etc.,$$

where $A = qx(1)$, $B = qx(2)$, $C = qx(3)$, etc., are known coefficients. Eschenbach-Rothe's formula 4.16 for the reversion of series gives $y^s x \, (n+1) = \frac{s}{s+nd} q^{s+nd} x \, (n+1)$, and also $y^{ms} \mathrm{x} (n+1) = \frac{ms}{ms+nd} q^{ms+nd} \mathrm{x} (n+1)$. Thus, Eq. 4.38 follows by putting $p = y^s$.

Given the following relations between general coefficients:

$$p \mathrm{x} (n+1) = \frac{s}{s+nd} q^{s+nd} \mathrm{x} (n+1),$$

$$p' \mathrm{x} (n+1) = \frac{s'}{s'+nd} q^{s'+nd} \mathrm{x} (n+1),$$

$$p'' \mathrm{x} (n+1) = \frac{s''}{s''+nd} q^{s''+nd} \mathrm{x} (n+1),$$

etc.,

where $p' = p^{\frac{s'}{s}}$, $p'' = p^{\frac{s''}{s}}$, $p''' = p^{\frac{s'''}{s}}$, etc., and thus $(p \cdot p' \cdot p'' \cdot \ldots) = p^{\frac{s+s'+s''+\cdots}{s}}$, the general coefficient of the product of series p, p', p'', etc. is given by:

$$(p \cdot p' \cdot p'' \cdot \ldots) \mathrm{x} (n+1) = \frac{s+s'+s''+\cdots}{s+s'+s''+\ldots+nd} q^{s+s'+s''+\ldots+nd} \mathrm{x} (n+1)$$

according to Eq. 4.38.

The expansion of p^m, for a series p that satisfies the conditions required for applying formula 4.38, is given directly by this formula:

$$p^m = \left(\frac{s}{s} q^s \mathrm{x}(1) x^\alpha + \frac{s}{s+d} q^{s+d} \mathrm{x}(2) x^{\alpha+\beta} + \frac{s}{s+2d} q^{s+2d} \mathrm{x}(3) x^{\alpha+2\beta} + \cdots \right)^m$$

$$= \frac{ms}{ms} q^{ms} \mathrm{x}(1) x^{m\alpha} + \frac{ms}{ms+d} q^{ms+d} \mathrm{x}(2) x^{m\alpha+\beta} + \frac{ms}{ms+2d} q^{ms+2d} \mathrm{x}(3) x^{m\alpha+2\beta} + \cdots .$$

Pfaff also addresses other problems concerning product and powers of series, for instance the problem of calculating the general coefficient of the product of powers of series $\left(p^m \cdot (p')^{m'} \cdot (p'')^{m''} \cdot \ldots \right)$, but they depend on the solutions of the two previous problems. It is worth noting that, even if Pfaff does not state it explicitly in his article, his formulae should be interpreted on the basis of the abstract algebra of involutions, since local symbols have been reinterpreted on this basis by Hindenburg and Pfaff himself considers the theorem on reversion of series as a consequence of this algebra, as seen in Sect. 4.2.3.2.

Division of two series $p = a+bx+cx^2+dx^3+\cdots$ and $q = \alpha+\beta x+\gamma x^2+\delta x^3+\cdots$ can also be interpreted in terms of the algebra of involutions. To this end, Hindenburg supposes that the quotient $\frac{p}{q}$ can be expanded into a series $P = \dot{A} + \dot{B}x + \dot{C}x^2 + \dot{D}x^3 + \cdots$, where

$\dot A$, $\dot B$, $\dot C$, etc. are unknown coefficients (Hindenburg 1796a, 289). Then, these coefficients are given by:

$$\dot A = \frac{a}{\alpha},$$

$$\dot B = \frac{b - \overset{qP}{\underset{2}{}}B}{\alpha} = \frac{b - (qP)\mathrm{x}(1)}{\alpha} = \overset{q^{-1}p}{\underset{3}{}}B = (q^{-1}p)\mathrm{x}(2),$$

$$\dot C = \frac{c - \overset{qP}{\underset{3}{}}B}{\alpha} = \frac{c - (qP)\mathrm{x}(2)}{\alpha} = \overset{q^{-1}p}{\underset{4}{}}B = (q^{-1}p)\mathrm{x}(3),$$

etc.,

and, in general, the coefficient of the $(n + 1)$th term of the expansion is:

$$\overset{n}{\dot A} = \frac{\overset{n}{a} - \overset{qP}{\underset{n+1}{}}B}{\alpha} = \frac{\overset{n}{a} - (qP)\mathrm{x}(n)}{\alpha} = \overset{q^{-1}p}{\underset{n+2}{}}B = (q^{-1}p)\mathrm{x}(n + 1),$$

	1	2	3	...			1	2	3	...
P[	$\dot A$	$\dot B$	$\dot C$	...	]	p[	a	b	c	...]
q[	α	β	γ	...	]	q^{-1}[	$q^{-1}\mathrm{x}(1)$	$q^{-1}\mathrm{x}(2)$	$q^{-1}\mathrm{x}(3)$	...]

The tables that accompany this equation are at the same time the scales of the corresponding local symbols and the compound substitution rule for the corresponding compound permutation classes of partitions.

4.2.3.4 A New Analytic-Combinatorial Formula

A gifted scientist, with eclectic intellectual interests, Christian Kramp (1760–1826) initially pursued a career in medicine, but in 1809 became professor of mathematics at the University of Strasbourg, his birth town, after having taught physics, chemistry, and mathematics in Cologne. His first contact with the combinatorial analysis came in 1786, when he undertook the study of Hindenburg's theory, but it was not until about 1795 that his original research on the field began to take shape (Kramp 1796, 96–100). One of his most famous contributions to science took place in the context of consolidation of the German combinatorial analysis and was closely related to Hindenburg's project of developing new mathematical notations to express analytic-combinatorial formulae. Kramp endorsed the view of the combinatorial school concerning the role attributed to the multinomial theorem in the reorganization of mathematical analysis and proposed, in 1796, a combinatorial formula for multinomial coefficients (Kramp 1796), which was at the root of that famous contribution: the creation of a new mathematical function called "the factorial function".

The key idea of Kramp's method to find the coefficients of the multinomial expansion consists in using equations instead of mathematical tables. At first glance it would appear that this idea goes against the original objective of the combinatorial analysis according to which the development of the theory of series must be based on the construction of mathematical tables. However, although tables remain fundamental components of the combinatorial analysis, this theory continues to evolve toward more sophisticated and abstract mathematical concepts, such as an algebra of classes, and Kramp's work seems to point in the same direction. To find the coefficient of the term x^ω in the expansion of $(ax^\alpha + bx^\beta + cx^\gamma + etc.)^m$, where all exponents are positive whole numbers, Kramp puts $A = ax^\alpha$, $B = bx^\beta$, $C = cx^\gamma$, etc. in order to transform the initial problem into the equivalent problem of finding the coefficient of the monomial $A^p B^q C^r \cdots$, where $p+q+r+\cdots = m$. One can then pose the equation $\alpha p + \beta q + \gamma r + \cdots = \omega$, whose positive integer solutions for $p, q, r, \ldots$ serve to find the desired coefficient, which is indeed a sum of monomials of the form $a^p b^q c^r \cdots$. For instance, to find the coefficient of x^{120} in the expansion of $(ax^3 + bx^5 + cx^7 + dx^{10})^{24}$, on poses $3p + 5q + 7r + 10s = 120$ (Kramp 1796, 103–104). Then, one can see that, for example, $p = r = s = 0$ and $q = 24$, and $p = r = 1, q = 22$ and $s = 0$ are solutions of the previous equation. From these solutions, one gets $a^0 b^{24} c^0 d^0$ and $a^1 b^{22} c^1 d^0$ respectively, and by calculating the corresponding multinomial coefficients, one obtains b^{24} and $552ab^{22}c$. Thus, after finding all integer positive solutions of this equation, the result can be written as $(b^{24} + 552ab^{22}c + \cdots)x^{120}$. Kramp writes the general coefficient of the multinomial expansion as follows:

$$\int K a^p b^q c^r \cdots$$

The symbol $\int$ stands for the sum of monomials $a^p b^q c^r \cdots$, and K is defined as:

$$K = \frac{m'}{p'q'r's'etc.},$$

where $m' = m(m-1)(m-2)\cdots 1$, $p' = p(p-1)(p-2)\cdots 1$, and so on, thus K is the multinomial coefficient of each monomial $a^p b^q c^r \cdots$.

As pointed out by Hindenburg, the origins of Kramp's method go back as far as the works of Moivre and Jacob Bernoulli (Kramp 1796, 118–119). However, although this is not a very innovative method, its importance lies rather in the new notations introduced by Kramp, particularly in the introduction of an apostrophe to signify a decreasing product of successive positive integers. In the absence of a theoretical explanation of this symbol in Kramp's article, one could be inclined to believe that the expression m' is a simple conventional abbreviation for the long sequence of products $m(m-1)(m-2)\cdots 1$. Nevertheless, this new symbolic representation meets all the expectations of the programmatic agenda of the combinatorial school, summarized originally in Hindenburg's *Novi systematis*, according to which one of the main purposes of the combinatorial analysis is to express its results by means of the most convenient mathematical notations and to use these notations to create new mathematical functions. From this point of view, the expression $m' = m\,(m-1)\,(m-2)\cdots 1$ is not an abbreviation but a function defined in the field of the combinatorial analysis.

The required theoretical background of this function was presented by Kramp three years later in a treatise devoted to the theory of refraction, where he elaborates further on the meaning of this function (Kramp 1799). Since it is not so common to attend the birth of a mathematical function, let me quote the passage where this happens (Kramp 1799, 46):

> I will refer to products whose factors constitute an arithmetic progression by the name of "faculties", such as:
>
> $$a(a+r)(a+2r)\cdots(a+mr-r).$$
>
> This class of analytic functions was never given a name: we will see that it deserved to be given one.[26]

This seemingly simple fact of calling a mathematical expression by name hides the complex and tortuous historical path leading to the creation of a new mathematical concept. In this case, the path begins with Newton's formula for the binomial expansion and his theory of series, and all the previous pages of this book prove Kramp right when he affirms that no one had realized that this kind of products was a recurrent object in several formulae of the theory of series and could be conceptualized as a function. More precisely, a faculty is defined as the following function:

$$a^{m|r} = a(a+r)(a+2r)\cdots(a+mr-r).$$

In $a^{m|r}$, a is called "the base", m "the exponent" or "the number of factors", and r is "the difference of the faculty". The choice of writing "$m|r$" as an exponent goes along with the idea that the power function can be defined in terms of faculties as $a^{m|0} = a^m$. Thus, a faculty is a more general function than powers. It should be noted that a historical antecedent of Kramp's notion of "faculty" can be found in an article of Vandermonde (Vandermonde 1772). Kramp added a note to his book in which he

---

[26] Je désignerai par le nom de facultés, les produits dont les facteurs constituent une progression arithmétique ; tels que

$$a(a+r)(a+2r)\cdots(a+mr-r).$$

Cette classe de fonctions analytiques n'a jamais eu de dénomination particulière : nous allons voir qu'elle méritoit d'en avoir une.

explains that Vandermonde's article was unknown to him at the time his book was already in press (Kramp 1799, xiii). Thus, Kramp's invention of faculties was an independent result.

Kramp built an incipient mathematical theory around the notion of "faculty". First, he gave without proof some elementary calculation rules:

$$a^{m|-r} = a(a-r)(a-2r)\cdots(a-mr+r)$$
$$= (a-mr+r)^{m|r},$$
$$a^{m|r} = (a+mr-r)^{m|-r},$$
$$a^{m|r} = b^{m|-r} \quad \text{if} \quad b-a = mr-r,$$
$$a^{m|r} = a^m \left(1^{m|\frac{r}{a}}\right) = r^m \left(\frac{a}{r}\right)^{m|1},$$
$$a^{(m+n)|r} = a^{m|r}(a+mr)^{n|r} = a^{n|r}(a+nr)^{m|r}.$$

Kramp assumed that these rules remain valid for rational values of the exponent m, although he did not formulate a definition of a faculty with rational exponents. However, his theory of rational faculties is related to the theory of integration and, more particularly, to the gamma function. Then, Kramp applied his calculation rules to establish some mathematical identities, such as:

$$\frac{\sin(m\pi)}{\sin(n\pi)} = \frac{m\,1^{n|1}(1-m)^{m|1}}{n\,1^{m|1}(1-n)^{n|1}}.$$

But the most relevant result of his incipient theory in the context of the German combinatorial analysis was the formulation of the binomial theorem for faculties, which Kramp qualified as one of the most important theorems of analysis:

$$(a+b)^{n|r} = a^{n|r} + na^{n-1|r}r + \frac{n(n-1)}{1\cdot2}a^{n-2|r}b^{2|r} + \frac{n(n-1)(n-2)}{1\cdot2\cdot3}a^{n-3|r}b^{3|r} + \text{etc.}$$

Kramp's theorem can be considered as a generalization of Newton's binomial theorem, and it has been formulated as a consequence of a new combinatorial function belonging to the combinatorial analysis.

In his *Calcul des dérivations*, the Alsatian mathematician, and friend of Kramp, Louis François Antoine Arbogast (1759–1803) studied Kramp's function making explicit reference to Kramp's *Analyse des réfractions astronomiques*, but he called it "factorial" (*factorielle*) instead of "faculty" (Arbogast 1800, 364 ff.). In 1808, Kramp decided to honor the memory of his friend by adopting the term "factorial" to refer to his function, considering that it was a more convenient denomination and was more in tune with the French language identity (Kramp 1808, XI–XII). On the contrary, his notation $a^{m|r}$ for his general factorial function remains unchanged. However, in his *Élémens d'arithmétique universelle*, Kramp revived his original idea of using a simple symbol to denote a decreasing product of successive positive whole numbers, but replaced the apostrophe by an exclamation mark, so that $m! = m\,(m-1)\,(m-2)\cdots 1$ for any positive integer m. Although he presented this formula in a book on arithmetic, he clarified that this function belongs to the doctrine of combinations and gives the number of permutations of a given number m of objects (Kramp 1808, 219). This particular factorial can be expressed by means of Kramp's general factorial function as $1^{m|1} = m!$ for any positive integer m.

On the other hand, Kramp's method for calculating the general coefficient of a multinomial raised to a positive integer power was included in both his *Analyse des réfractions astronomiques* and his *Élémens d'arithmétique universelle*, but in this latter book Kramp acknowledges that Hindenburg's method of tables or combinatorial involutions has three advantages over his own method based on the solution of equations. First, Hindenburg's method does not depend on the heavy task of solving thorny problems of equations. Second, Hindenburg's method is more efficient, since even if one succeeds in finding all the solutions of a given equation, the set of solutions can include more information than necessary. Third, the lexicographical order of involutions offers a better representation of the monomials required for determining the coefficient of a given term. However, even though Hindenburg's mathematical techniques were considered as a better alternative by Kramp himself, he provided the combinatorial analysis with a function and a mathematical notation that have been able to find a place in mathematics and remain in force until today.

4.2.3.5 Other Contributions to the Combinatorial Analysis

Products, powers, and reversion of series were the most extensively studied topics during the period of consolidation of the combinatorial analysis. On the contrary, the problems referred to in the remaining items of the combinatorial school's agenda received comparatively less attention. In the end, there was no comprehensive and systematic examination of those remaining topics but isolated texts, or even short passages of texts, that were somehow connected with them. Some of these will be reviewed below.

Moritz von Prasse (1769–1814) was considered a member of the combinatorial school. As Hindenburg, he was born in Dresden, moved to Leipzig to pursue his mathematical studies at the university, became Privatdozent in 1796, then extraordinary professor in 1798, and finally ordinary professor of mathematics in 1799 at the same university, and died also in Leipzig. In the light of this educational and professional background, it is no wonder that Hindenburg's theories deeply influenced Prasse's thought. Indeed, his *Usus logarithmorum infinitinomii in theoria aequationum*, published in 1796, was the first attempt to summarize in one (brief) volume the results achieved so far by the combinatorial school (von Prasse 1796). Although this booklet was not intended to be a treatise in which an analytic-combinatorial solution to every possible question of the theory of series could be found, it offers an overview of the most recent developments of the combinatorial analysis. It contains a partial presentation of the method of combinatorial involutions and the algebra of involutions, and shows how to apply them to the study of the logarithm function. As indicated in the title, this function is the main subject matter of the booklet and over two-thirds of its pages are devoted to the logarithm (some details of Prasse's work will be addressed in Sect. 5.1.3). In fact, the first applications of Hindenburg's combinatorial methods to the logarithm date from the period prior to the formation of the combinatorial school. Hindenburg gives some examples of possible applications in his 1779 *Infinitinomii* using his method of powers, as seen in Sect. 3.2.2. In 1793, Rothe illustrates the use of formula 4.16 for the reversion of series applying it to the logarithm series expansion, and in 1796, when the combinatorial school already existes, Hindenburg takes up this example and places it in the context of his nascent algebra of involution (Hindenburg 1796a, 299; Rothe 1793, 21–22).

Besides logarithms, other types of transcendental functions considered by the combinatorial school concern trigonometric functions. Examples of possible applications of combinatorial methods to the study of these functions were already discussed in 1779 by Hindenburg, as seen in Sect. 3.2.2. Other isolated examples can be found in some texts of the combinatorial school, for instance in (von Prasse 1796, 34 ff.), where Prasse applies the method of involutions to some examples concerning the sine and cosine functions. However, it was Johann Karl Burckhardt (1773–1825) who tackled this question directly and in a more extensive way in his paper (Burckhardt 1799), where he analyzes trigonometric addition formulae in terms of classes of combinations. Given the angles α, α', α'', ..., $\alpha^{(n)}$ such that $\tan(\alpha) = t$, $\tan(\alpha') = t'$, $\tan(\alpha'') = t''$, ..., $\tan(\alpha^{(n)}) = t^{(n)}$, Burckhardt proposes the following addition formula:

$$\tan\left(\alpha + \alpha' + \cdots + \alpha^{(n-1)}\right) = \frac{\underset{(n)}{A'} - \underset{(n)}{C'} + \underset{(n)}{E'} - \underset{(n)}{G'} + \cdots}{1 - \underset{(n)}{B'} + \underset{(n)}{D'} - \underset{(n)}{F'} + \cdots}$$

where $\underset{(n)}{A'}$, $\underset{(n)}{B'}$, $\underset{(n)}{C'}$, etc. stand respectively for the classes of 1-combinations, 2-combinations, 3-combinations, etc. of the set of n values t, t', ..., $t^{(n-1)}$. He proves the validity of this equation by induction. To illustrate this formula, let's consider the cases where $n = 2$ and 3. For $n = 2$, one gets:

$$\tan\left(\alpha + \alpha'\right) = \frac{t + t'}{1 - tt'} = \frac{\underset{(2)}{A'}}{1 - \underset{(2)}{B'}}$$

since $\underset{(2)}{A'}$ contains the elements t and t' only, and $\underset{(2)}{B'}$ contains the element tt' only (as we know, a class of combinations appearing in an algebraic formula means that the elements of the class must be added). For $n = 3$, one gets:

$$\tan\left(\alpha + \alpha' + \alpha''\right) = \frac{\tan\left(\alpha + \alpha'\right) + \tan\left(\alpha''\right)}{1 - \tan\left(\alpha + \alpha'\right)\tan\left(\alpha''\right)}$$

$$= \frac{\left(t + t' + t''\right) - tt't''}{1 - \left(tt' + tt'' + t't''\right)}$$

$$= \frac{\underset{(3)}{A'} - \underset{(3)}{C'}}{1 - \underset{(3)}{B'}},$$

since $\underset{(3)}{A'}$ contains three elements t, t', and t'', $\underset{(3)}{B'}$ contains three elements tt', tt'', and $t't''$, and $\underset{(3)}{C'}$ contains one single element $tt't''$. If $\alpha = \alpha' = \cdots = \alpha^{(n)}$, and thus $t = t' = \cdots = t^{(n)}$, it follows from this formula that:

$$\tan\left(n\alpha\right) = \frac{{}^n\mathfrak{A}t - {}^n\mathfrak{C}t^3 + {}^n\mathfrak{E}t^5 - {}^n\mathfrak{G}t^7 + \cdots}{1 - {}^n\mathfrak{B}t^2 + {}^n\mathfrak{D}t^4 - {}^n\mathfrak{F}t^6 + \cdots},$$

where ${}^n\mathfrak{A}$, ${}^n\mathfrak{B}$, ${}^n\mathfrak{C}$, etc. are, as we know, the binomial coefficients expressed in Hindenburg's notation.

Burckhardt uses involutions to write the addition formulae for the sine and cosine functions, for $\sin\left(\alpha\right) = s$, $\sin\left(\alpha'\right) = s'$, ..., $\sin(\alpha^{(n)}) = s^{(n)}$, and $\cos\left(\alpha\right) = c$, $\cos\left(\alpha'\right) = c'$, ..., $\cos(\alpha^{(n)}) = c^{(n)}$. In Fig. 4.11, one can see the involutions that serve to calculate the addition formulae for these functions. Gnomons with a single line stand for sine, while gnomons with double line stand for cosine. These involutions must be constructed simultaneously, for they are interdependent. Burckhardt enunciates the following recursive rules to construct these involutions:

1. The first gnomon contains only:
   a. For the involution of sine, the term $+s$.
   b. For the involution of cosine, the term $+c$.
2. For $n \geq 2$, the nth gnomon is given by:
   a. For the involution of sine:
      i. Write the $(n-1)$th gomon of the involution of cosine below the $(n-1)$th gnomon of the involution of sine.
      ii. Write the term $c^{(n)}$ to the right of each row of the $(n-1)$th gnomon of sine, and write the term $s^{(n)}$ to the right of each row of the $(n-1)$th gnomon of cosine.
   b. For the involution of cosine:
      i. Write the $(n-1)$th gomon of the involution of sine below the $(n-1)$th gnomon of the involution of cosine, but change first the sign at the beginning of each row of the $(n-1)$th gomon of the involution of sine to its opposite.

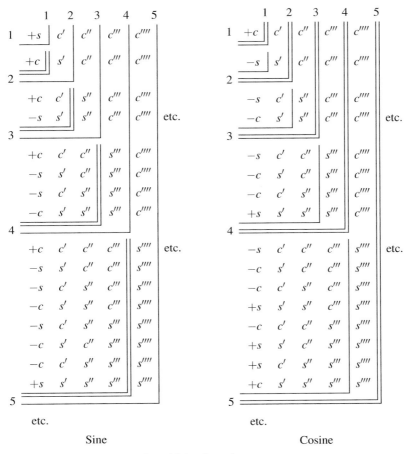

Fig. 4.11 Burckhardt's involutions for addition formulae

ii. Write the term $c^{(n)}$ to the right of each row of the $(n-1)$th gnomon of cosine, and write the term $s^{(n)}$ to the right of each row of the $(n-1)$th gnomon of sine.

Then, the sum of the elements contained in the nth gnomon of these involutions is equal to $\sin\left(\alpha+\alpha'+\cdots+\alpha^{(n-1)}\right)$ and $\cos\left(\alpha+\alpha'+\cdots+\alpha^{(n-1)}\right)$ respectively. For instance, from the third gnomon of the involutions shown in Fig. 4.11, it follows that:

$$\sin\left(\alpha+\alpha'+\alpha''\right)=sc'c''+cs'c''+cc's''-ss's'',$$

and:

$$\cos\left(\alpha+\alpha'+\alpha''\right)=cc'c''-ss'c''-sc's''-cs's''.$$

The corresponding general addition formulae are given, however, in terms of classes of combinations:

$$\sin \left(\alpha + \alpha' + \cdots + \alpha^{(n-1)} \right) = \underset{(n)}{A'}c^* - \underset{(n)}{C'}c^* + \underset{(n)}{E'}c^* - \underset{(n)}{G'}c^* + \cdots ,$$

and:

$$\cos \left(\alpha + \alpha' + \cdots + \alpha^{(n-1)} \right) = cc'c'' \cdots c^{(n-1)} - \underset{(n)}{B'}c^* + \underset{(n)}{D'}c^* - \underset{(n)}{F'}c^* + \cdots ,$$

where $\underset{(n)}{A'}$, $\underset{(n)}{B'}$, $\underset{(n)}{C'}$, etc. stand respectively for the classes of 1-combinations, 2-combinations, 3-combinations, etc. of the set of n values s, s', $\ldots$, $s^{(n-1)}$, and where the symbol c^* means that each element of these classes must be multiplied by some of the cosines c, c', $\ldots$, $c^{(n-1)}$. To determine which of these cosines are the appropriate factors for each element of the classes $\underset{(n)}{A'}$, $\underset{(n)}{B'}$, $\underset{(n)}{C'}$, etc., Burckhardt points out that, given the construction of the involutions of sine and cosine, the components of each element of the nth gnomon are such that they are arranged in increasing order with respect to their indexes, from 0 to $n - 1$, and no index from 0 to $n - 1$ is missing. From this, he correctly infers that, given an element of one of the classes $\underset{(n)}{A'}$, $\underset{(n)}{B'}$, $\underset{(n)}{C'}$, etc., this element must be multiplied by all the cosines among c, c', $\ldots$, $c^{(n-1)}$ whose indexes are different from the indexes of the sines included in the element. For instance, ss' and $s^{(3)}s^{(6)}$ are elements of $\underset{(n)}{B'}$, then the symbol c^* in $\underset{(n)}{B'}c^*$ means that ss' must be multiplied by $c''c''' \cdots c^{(n-1)}$ so that one obtains $ss'c''c''' \cdots c^{(n-1)}$, and $s^{(3)}s^{(6)}$ must be multiplied by $cc'c''c^{(4)}c^{(5)}c^{(7)}c^{(8)} \cdots c^{(n-1)}$ so that one obtains $cc'c''s^{(3)}c^{(4)}c^{(5)}s^{(6)}c^{(7)}c^{(8)} \cdots c^{(n-1)}$. Burckhardt proves the validity of these formulae by induction. From these formulae, he deduces that:

$$\sin(n\alpha) = {}^n\mathfrak{A}sc^{n-1} - {}^n\mathfrak{C}s^3c^{n-3} + {}^n\mathfrak{E}s^5c^{n-5} - {}^n\mathfrak{G}s^7c^{n-7} + \cdots ,$$

and:

$$\cos(n\alpha) = c^n - {}^n\mathfrak{B}s^2c^{n-2} + {}^n\mathfrak{D}s^4c^{n-4} - {}^n\mathfrak{F}s^6c^{n-6} + \cdots ,$$

when $\alpha = \alpha' = \cdots = \alpha^{(n)}$, and where one should be careful not to confuse $s^{(n)}$ and $c^{(n)}$ with s^n and c^n, these latter expressions mean the power n of the sine s and the cosine c respectively.

Other than the topics listed by Hindenburg in his *Novi systematis*, new research subjects were adopted as the work of the combinatorial school progressed. For instance, Klügel's proof of the multinomial theorem analyzed in Sect. 4.2.2.1 paved the way for the study of the combinatorial properties of coefficients appearing in series. As suggested in Sect. 4.2.2.2, Rothe's proof of the binomial theorem was inspired by Klügel's idea, or at least both mathematicians shared the same purpose, which translates for Rothe into showing that the combinatorial properties of the binomial coefficients are independent of the binomial expansion, and that this expansion is, indeed, a consequence of those properties. Thus, Rothe showed that if one can prove a combinatorial property of the binomial coefficients independently of the theory of series, as he did with regard to the Vandermonde's identity, and if this property can be successfully used to prove a theorem of the theory of series, then one can affirm that this theorem has been deduced from the doctrine of combinations. This foundational strategy did not go unnoticed and, in 1797, Karl Heribert Ignatius Buzengeiger (1771–1835) tried to emulate it. Buzengeiger was a German scientist that probably came into contact with the combinatorial school's theories via his former mathematics teacher Christoph Friedrich Pfleiderer (1736–1821) who usually contributed with Hindenburg's *Archiv der reinen und angewandten Mathematik* publishing papers on several topics. In fact, Buzengeiger did not maintain an abiding interest in the combinatorial analysis throughout his life, and he is best known for his professional relationship with Karl Feuerbach (1800–1834) (Fritsch 1979). In any case, following Rothe's argumental strategy, in 1797 Buzengeiger tried to prove a combinatorial property of the binomial coefficients in order to apply it to several problems of analysis (Buzengeiger 1797). His idea consisted in giving a formula for calculating the sum of the squares of the binomial coefficients in order to use this formula for solving certain problems of analysis. More precisely, Lagrange, to whom Buzengeiger makes reference, gave the following formula in 1776 (which is expresed here in our contemporary notations):

$$\sum_{k=0}^{n} \binom{n}{k}^2 = \prod_{k=1}^{n} \frac{4k-2}{k}$$

for calculating the sum of the squares of the elements appearing in a given row of the Pascal's triangle, and remarked that it would not be easy to prove it (Lagrange 1868, 182). Some years later, Euler, to whom Buzengeiger does not make reference, published two papers in which a proof of this formula can be found, though both were presented to the Saint Petersburg Academy of Sciences in 1776 in accordance with the records (Euler 1784; 1785). Buzengeiger intended to prove his formula for the sum of the squares of the binomial coefficients, which is different from the formula given above, by deducing it from Rothe's formula 4.31, i.e. from Rothe's version of the Vandermonde's identity. However, several of Buzengeiger's formulae are wrong, including the main formula for calculating this sum of squares, which is utterly baffling considering that Lagrange wrote correctly this formula in his paper. This is why Buzengeiger's paper will not be reviewed in detail here (and why we have allowed ourselves to use contemporary notations in the formula given above), but what seems important to us is that, despite Buzengeiger's errors, his paper still exemplifies the kind of foundational thinking that guided the research carried out by the combinatorial school in order to deduce the theorems belonging to the field of mathematical analysis from the doctrine of combinations.

Other more peripheral themes were dealt with during this period of effervescence in the history of the combinatorial analysis. Most of them were introduced by Hindenburg himself, although he did not include them explicitly in his programmatic agenda for the development of his theory. For instance, in 1794 Burckhardt wrote a booklet on continued fractions from the perspective of the combinatorial analysis, building on a previous text of Hindenburg published in 1781 (Burckhardt 1794; Hindenburg 1781a). In turn, the same year of 1794, Hindenburg reworked Burckhardt's results by means of the method of combinatorial involutions (Hindenburg 1794a; Bullynck 2006, 264–268). Occasionally, Hindenburg also explored possible applications of the combinatorial analysis to cryptography, with particular emphasis on the cipher technique known as turning grilles, about which he wrote two articles in 1795 and 1796 (Hindenburg 1795b; 1796b). Three years later, as part of his inaugural address as ordinary professor of mathematics at the University of Leipzig in 1799, Prasse elaborated further on the theme of turning grilles (von Prasse 1799; Bauer 2000, 97–98).

*

In 1796, Hindenburg shares the following thoughts on combinatorics (Hindenburg 1796a):

> Here the doctrine of combinations appears to be, especially as far as the representation and development of its forms are concerned, which is of great importance in analysis, as an independent fundamental science, a science that stands by itself and needs no external assistance. It is certainly possible to apply arithmetical concepts and theorems to combinatorial ones, as has been done many times here with great results, hence one can say:
> *alterius sic altera poscit opem res et coniurat amice.*
> But it is important to distinguish purely combinatorial methods, as I usually call them, from composite methods. The first ones are, if that were possible, even simpler than the second ones, and their outcomes, which usually lead to involutions, are based (1) on putting or annexing, (2) on removing or separating, (3) on certain changes or exchanges, as well as on certain arrangements of the remaining elements.[27]

---

[27] Die Combinationslehre tritt hierbey, besonders was die Darstellung und Entwickelung ihrer Formen, worauf in der Analysis doch so viel ankommt, anbetrifft, als selbstständige Grundwissenschaft auf, die, sich allein genügend, fremder Hülfe nicht bedarf. Man kann zwar arithmetische Begriffe und Sätze auf combinatorische anwenden, und hat solches bereits häufig und mit großem Vortheile gethan; so daß man auch hier sagen kann *alterius sic altera poscit opem res et coniurat amice* aber es ist wichtig, die rein-combinatorischen Verfahren, wie ich sie zu nennen pflege, von den gemischten zu unterscheiden. Jene sind, wenn es möglich ist, noch einfacher als diese, und die dabey vorkommenden Veränderungen, die gewöhnlich geradezu auf Involutionen führen, beruhen (1) auf Ansetzen oder Beyfügen (2) auf Wegnehmen oder Absondern (3) auf Aus- oder Umtauschung gewisser, so wie auf bestimmter Anordnung der übrigen Elemente.

This passage sums up well the final stage reached by the combinatorial analysis. The doctrine of combinations is conceived as a science that has pure and simple procedures, such as those enumerated above by Hindenburg, and these procedures cannot be expressed in terms of concepts belonging to other fields of mathematics. This is because the doctrine of combinations relies on a certain law of Nature, discovered first by Moivre. Although the notion of this law remains a vague concept, it drives the development of new mathematical ideas, particularly the development of the algebra of involutions, which is a promising theory based on the study of abstract mathematical structures called "classes". On the basis of this abstract algebra, some power series expansions are conceptualized as merely consequences of combinatorial properties. Among them, the power series expansion expressed by the multinomial theorem is characterized as the most elementary of them all, and thus this theorem becomes the most important theorem in the whole of analysis for the combinatorial school.

The Decline of the German Combinatorial Analysis 5

5.1 Second Thoughts About the Foundation of Analysis

From 1794 to 1798, the German combinatorial analysis enters into a phase of growth characterized by major changes in the theoretical organization of the discipline. Not only is there an increasing attention to Hindenburg's ideas, reflected in the increasing number of articles devoted to them, but there is also a clear improvement in the combinatorial analysis with respect to, at least, the following aspects of this theory: (a) combinatorial methods based on mathematical tables become more efficient thanks to the introduction of involution tables, (b) there is an active interest in creating new mathematical notations and this part of Hindenburg's project proves its worth by showing that the creation of mathematical notations leads to the discovery of new mathematical functions, as the functions called "local symbols" by Rothe or the factorial function of Kramp, (c) the reinterpretation of the theory of series in terms of the concepts and methods of the combinatorial analysis is intensified during this period, (d) the theoretical bases of this new field of mathematics is clarified in the course of the combinatorial school's research, identifying its combinatorial basis with the algebra of involutions and the analytical basis of the theory of series with the mutinomial theorem. However, this positive momentum toward making significant progress in the combinatorial analysis will experience a fast deceleration at the turn of the century. During the first decade of the nineteenth century, there will still be research studies that can be categorized as research carried out within the framework of the combinatorial school's scientific project, but these studies will not contribute anymore to the development of this project.

In particular, the theoretical foundation of the combinatorial analysis was not properly strengthened. On the one hand, no one was curious enough to dig deeper and examine in detail the incipient algebra of involutions in order to produce a coherent body of knowledge about these new abstract mathematical entities called "classes", not even Hindenburg, who gave us a rough outline of this abstract algebra but, unfortunately, was not able to continue his original scientific work toward the end of his life. On the other hand, the multinomial theorem ran into serious problems relating to proving its independence from the binomial theorem. Klügel intended to justify its independence by giving a proof of the multinomial theorem for rational exponents in which there was no application of the binomial theorem. However, in this section, the criticism addressed to Klügel's argument based on analogy will be reviewed, and it will be shown that Klügel's proof was discredited in the eyes of the German mathematical community. Then it will be argued that, since such independence cannot be guaranteed, it is rather difficult to conceptualize the multinomial theorem as the foundation of the theory of series. As a consequence, the viability of reorganizing the field of mathematical analysis on the basis of the multinomial theorem is compromised. Furthermore, it will be seen that the conception of the combinatorial analysis as a system of science is not reflected in the first textbooks on the combinatorial analysis appeared in 1800. In these textbooks, the multinomial theorem is presented as a proposition among others, and the combinatorial analysis seems to come down to the classical doctrine of combinations of the eighteenth century. The misleading conception of these manuals seems to put an end to the long-awaited project of constructing the German combinatorial analysis as a system of science.

5.1.1 The Foundation of Analysis Crumbles

The historical and conceptual evolution of the combinatorial analysis describes an ascensional trajectory that begins in 1794 with the formation of the combinatorial school, reaches its apex in 1796 with the publication of several articles related to the multinomial theorem by different members of the combinatorial school, and goes into decline around 1798, when the publications linked to the combinatorial analysis do not seem to respond any longer to the central question about the possibility of organizing all mathematical knowledge on the basis of the algebra of involutions. The apex reached in 1796 has its antecedent in Hindenburg's announcement made in 1795 that the multinomial theorem was the most fundamental proposition of analysis. This characterization of the multinomial theorem allows the combinatorial school to articulate the full scope of its work and to engage in scientific research based on one single principle. However, this characterization depends, at least in part, on Klügel's proof of the multinomial theorem discussed in Sect. 4.2.2.1, in which Klügel attempts to formulate a purely combinatorial justification without using the binomial theorem. Of course, the success of Klügel's justification was essential to the success of founding mathematical analysis on only this single ultimate principle. Perhaps this is why Hindenburg openly supported and applauded Klügel's proof and no member of the combinatorial school criticized it, despite its evident lack of rigor. Nevertheless, Klügel's proof came under fire from beyond the combinatorial school's frontiers. In the same year of its publication, an anonymous reviewer called into question the validity of Klügel's arguments (Book review 1796a, 586–587):

> To extend the validity of the [multinomial] theorem to any exponent, on the whole Mr. Klügel
> has advanced the same argument, though more complete, on which he based the universal
> validity of the binomial theorem in the Appendix of his excellent *Analytical Trigonometry*,
> namely, that since analytical operations (multiplication, division, power, etc.) determine the
> *form* of that which is sought exclusively by means of the form of that which is given, but leave
> quantities undetermined, the different quantity and nature of the exponents have no influence
> on the form of the expansion of the power [of a multinomial]. This reasoning surprises by
> its sagacity and leaves nothing to be desired, unless it be that the last theorem should be
> proved with absolute mathematical rigor, and we would have made sure that no leap in the
> proof had been dissimulated by the very obscure concept of form. In that case, it is possible
> to say that the binomial theorem appears to be a particular case and a simple corollary of
> the multinomial theorem; but the question is to know whether this generality would entail
> sacrificing the elegance of Hindenburg's new proof.[1]

Thus, the anonymous reviewer's criticism targeted the following weak points of
Klügel's reasoning, listed here in order of increasing seriousness: (1) from a mathematical
point of view, Klügel's proof is not very elegant, (2) there are some values of the
coefficients in the power series expansion that have not been explicitly determined, (3) the
mathematical rigor of the proof is questionable because of the obscure concept of "form",
foreign to mathematics, and (4) from this proof, it is not possible to conclude that the
binomial theorem is a corollary of the multinomial theorem.

The criterion of mathematical elegance to judge the quality of Klügel's proof is far
from ideal because of the relative subjectivity involved in the notion of elegance, but
it is interesting to note here that the anonymous reviewer considers Hindenburg's "new
proof" as an appropriate point of comparison for mathematical elegance. This new
proof concerns Hindenburg's formulae 4.34 and 4.37 (Hindenburg 1796a, 231–234). As
seen in Sect. 4.2.3.3, Hindenburg presents these formulae as possible applications of his
new algebra of involutions to the theory of series. Although they express the general
coefficients of multinomial expansions, in Hindenburg's paper the emphasis is not placed

---

[1] Um die Gültigkeit des Satzes auf jeden beliebigen Exponenten auszudehnen, bedient Hr. *Kl.* sich
im Ganzen desselben, (hier jedoch noch vervollständigten) Raisonnements, auf das er im Anhange
seiner trefflichen analytischen Trigonometrie die allgemeine Gültigkeit des Binomialsatzes stützte,
dass nämlich, weil die analytischen Operationen, (Multiplication, Division, Potenzirung etc.) nur
allein die *Form* des zu Suchenden aus der Form des Gegebenen bestimmen, die Grössen aber
unbestimmt lassen, die verschiedene Grösse und Beschaffenheit der Exponenten auf die Form der
entwickelten Potenz keinen Einfluss habe. Dieses Raisonnement überrascht durch seinen Scharfsinn,
und lasst nichts zu wünschen übrig, als höchstens, dass der letzte Satz mathematisch in völliger
Strenge dargethan, und wir gehörig gesichert würden, dass nicht etwa mit dem so dunklen Begriff
von Form ein Sprung im Beweise verdeckt werde. Artig ist es auf diesem Wege den binomischen
Lehrsatz als einzelnen Fall und blosses Corollarium aus dem polynomischen hervortreten zu sehen;
doch ware es die Frage, ob diese Allgemeinheit nicht der Eleganz des neuen Hindenburgischen
Beweises anfzuopfern seyn möchte.

on the multinomial theorem as such, but on solving a specific problem (*Aufgabe*) by means of the algebra of involutions. Hindenburg methodically shows how the techniques and concepts of his abstract algebra can be successively used to construct the general coefficient of the expansion of a multinomial raised to a positive whole number, then to a negative whole number, and finally to a rational number. In this sense, Hindenburg's text proposes an exercise of the algebra of involutions concerning powers of series. In this exercise, he relies on the binomial theorem to move from integer to rational exponents. Particularly, it is this step in the proof that the anonymous reviewer qualifies as elegant, since it is this step which was omitted by Klügel in his proof of the multinomial theorem. Instead of applying the binomial theorem, Klügel introduced the notion of form, and, according to the anonymous reviewer, this compromises the correct structure of the proof, leaving open a deductive gap. If the critic is right, there is then a correct structure for the proof of the multinomial theorem, in which this theorem must be deduced from the binomial theorem. Thus, what we have here is a conception according to which there is a theoretical hierarchy between these two theorems and the binomial theorem is considered as a more fundamental proposition. As seen in Chap. 2, such a conception has existed from the beginning of the history of the multinomial theorem, but at the end of the eighteenth century it becomes a serious obstacle for a proof that aims to reverse this hierarchy.

In 1800, Klügel published an article in which he responded to those criticisms and attempted to defend the validity and rigor of his proof (Klügel 1800). In order to remove the objections raised in (2) and (3), he endeavored to clarify the meaning of "form", at least the meaning he attributed to this term, on three separate fronts. Then, the refutation of objections (2) and (3) would entail the refutation of objection (4) as well. Klügel did not address objection (1) directly, but this objection alone is completely irrelevant from a mathematical point of view. Let's examine the three arguments advanced by Klügel to defend his demonstration of the multinomial theorem.

The first argument might be called the argument of the conceptual perfection of sciences. Klügel argues that scientific disciplines go through successive stages of development on which the clarity and precision of their concepts depend. The more a scientific discipline goes on to mature, the more clarity and precision its concepts acquire. On the other hand, there are transversal concepts in science in the sense that they belong to more than one discipline, like the concept of "form". The clarity and precision of a transversal concept vary depending on the discipline in which it is used. As a result, the same concept can be considered obscure in the context of philosophy, but clear and precise in the context of mathematics. According to Klügel, this is the case concerning the transversal concept of "form". Since philosophy has not reached yet its highest stage of development, conceptual obscurity does not seem to dissipate in this discipline. On the contrary, the conceptual perfection achieved in mathematics allows us to precisely characterize the notion of form: a form is determined by "the method of composition

of a quantity by another quantity."[2]   A form is then a specific symbolic expression of
a quantity obtained by applying a mathematical method to another given quantity. This
characterization comes very close to being the very notion of "function" presupposed by
the combinatorial school. In particular, as Hindenburg pointed out on several occasions,
one of the main objectives of the combinatorial analysis is to create useful mathematical
notations to express the quantities studied in analysis as specific functions or formulae.
Klügel echoed this objective, stating (Klügel 1800, 146):

> The whole of mathematics [...] is the science of the form of quantity. Form is the concept
> thanks to which mathematics is possible. Formless quantity belongs to metaphysics.[3]

Thus, the idea that a quantity can be univocally determined by using an appropriate
mathematical symbolism constitutes Klügel's first attempt to clarify his concept of "form".

The second argument proposed by Klügel is an argument from authority, inasmuch as
its force depends on the undeniable genius and renown of Euler. Klügel maintains that
the acclaimed mathematician of Basel had based his 1774 functional proof of the binomial
theorem, which has been analyzed in Sect. 2.1.3.2, on a similar mathematical notion of
"form", hence this should be a valid notion in mathematics. He writes the following
comment about Euler's proof (Klügel 1800, 147–148):

> Euler calls *ratiocinium non vulgare* the fact that, in the multiplication of $1 + \frac{m}{1}x + \frac{m(m-1)}{2}x^2 +$
> *etc.* by $1 + \frac{n}{1}x + \frac{n(n-1)}{2}x^2 + etc.$, the *ratio compositionis* (i.e. the form) of the coefficients
> belonging to the product does not depend on the nature of the quantities m, n, which can
> designate a whole number or any other number.[4]

For Klügel, the *ratio compositionis* to which Euler refers corresponds to the symbolic
representation of a quantity, and this symbolic representation can be algebraically
manipulated no matter what the symbols may represent, since the relationships that emerge
from these algebraic manipulations must be true as they are consequences of a sort of
implicit rule contained in the symbolism of a formula. In other words, Klügel reads Euler's
text as if Euler had suggested that it is possible to symbolically manipulate formulae
without taking any care at all of the initial assumptions that are expressed in mathematical
propositions about the specific properties that the objects represented by symbols must
satisfied to fall within the scope of a given formula.

---

[2] Klügel (1800, 146): Art der Zusammensetzung einer Größe aus andern Größen.

[3] Ja die ganze Mathematik [...] ist Wissenschaft der Formen der Größen. Form ist der Begriff,
wodurch Mathematik möglich ist. Die formelose Größe gehört der Metaphysik zu.

[4] Euler nennt es schon ein *ratiocinium non vulgare*, daß bey der Multiplication von $1 + \frac{m}{1}x +$
$\frac{m(m-1)}{2}x^2 + etc.$ durch $1 + \frac{n}{1}x + \frac{n(n-1)}{2}x^2 + etc.$, die *ratio compositionis* (d.i. Form) der
Coefficienten in dem Producte nicht von der Beschaffenheit der Größen, m, n, abhange, sie mögen
ganze oder irgend andere Zahlen bedeuten.

The third argument consists of taking advantage of philosophical notions in order to justify the use of the concept of "form" in the proof of the multinomial theorem. Klügel admits to having tacitly introduced philosophical notions in the last part of his proof concerning rational exponents. In particular, the philosophical notion of "similitude" (*Aehnlichkeit*), or more precisely "the principle of similitude" (*Princip der Aehnlichkeit*), has been used in the proof to jump from integer to rational exponents. In his text, Klügel specifies neither the philosophical currents to which his thinking is related, nor the content of the principle of similitude, probably considering that this was a well-established and well-known principle of philosophy. He confines himself to saying that, from a philosophical point of view, similitude differs from equality (*Gleichheit*) and points out that this idea is in harmony with "the spirit of geometry". Nevertheless, in what follows, we will try to trace back the specific sources of his principle of similitude in hopes of reconstructing its theoretical content.

In the first volume of the *Mathematisches Wörterbuch*, one can find an article by Klügel devoted to the notion of "mathematical similitude" (Klügel 1803, 15–21). In a final historical remark, Klügel lets us know that Wolff, inspired by his reading of Leibniz, was the first to introduce "the general concept of similitude" in the field of mathematics. It seems that Klügel suggests that Wolff intended, for the first time, to establish the study of problems involving similar geometric figures on more general metaphysical bases. Indeed, this idea can be supported by the mathematical work of Wolff. For instance, definition 12 of Wolff's *Elementa Arithmetica*, book quoted by Klügel, characterizes the notion of "similitude" in the most general way, that is to say that this definition concerns all things in a metaphysical sense, and so it is not restricted to mathematical objects (Wolff 1732, 18): "Things that are identical with respect to what they should be mutually different are *similar*."[5] In his article (Wolff 1715, 73), which Klügel had read, Wolff captures in a scholium the idea of similitude that Leibniz had transmitted to him and from which he derived his own general definition: "Things that cannot be distinguished, except by copresence, are similar."[6] A more elaborate explanation of this Leibnizian definition was given by Leibniz himself in a letter written in 1677, addressed to Galloys, but which was never sent (Leibniz 1677, 179–180):

For instance, no one has ever defined what it is to be similar, and however without first having defined it, we would not succeed in giving natural demonstrations of several important propositions of metaphysics and mathematics. After a lot of searching, I came to the conclusion that two things are exactly similar when, except on *copresence*, they are indiscernible; for example, two unequal circles made of the same material would be indiscernible unless they are observed together, since in that case one can clearly see that one is bigger than the other. You will say to me: I will measure one of them today and the other tomorrow, and so I will be able to distinguish them without having put them together. I

---

[5] *Similia* sunt, in quibus ea eadem sunt, per quae a se invicem discerni debebant.

[6] Similia, quae non possunt distingui, nisi per compraesentiam.

affirm that the distinction was still made not *per memoriam, sed per compraesentiam*, since you have retained the first measure not in memory, magnitudes not being susceptible of being stored in the mind, but in a material record engraved on a ruler or another thing. Since if all the things around us were proportionally shrunk, it is clear that no one could tell the difference. By means of this definition I have proved very general and very beautiful propositions; for example, the proposition that if two things are similar with respect to a given operation or consideration, they are similar with respect to any other, for example, if two towns are unequal in magnitude but exactly similar when viewed from the eastern side, I affirm that they will also be similar when viewed from the western side, provided that one can contemplate the entire town at each glance. This proposition is just as important in Metaphysics and even in Geometry and Analysis as the proposition that the whole is bigger than a part. And nevertheless, as far as I know, nobody has ever enunciated it.[7]

According to this explanation, two things are then similar if they are indiscernible when considered separately. But this individual evaluation occurs at a rather perceptual level: it is only thanks to a simultaneous perception of both things that we are able to distinguish them, since magnitude is not a property susceptible to be apprehended by the mind independently of the faculty of perception. In this sense, Leibniz proposes a phenomenological concept of "similitude". Leibniz's phenomenological approach to this concept has been emphasized, for instance, by Heidegger, which is certainly not very surprising (Heidegger 1925, 322–325). However, more neutral specialists of Leibniz's work agree with this phenomenological interpretation, for instance see (De Risi 2007, ch. 2). It is worth to note that this Leibnizian definition of "similitude" was the expression of a deeply-pondered conviction about which Leibniz reflected at different points of his life (Leibniz 1679a; 1679b, 162; 1683, 514; 1696, 868).

---

[7] Par exemple il n'y a personne qui ait bien defini ce que c'est que semblable, et cependant avant que de l'avoir defini, on ne sçauroit donner des demonstrations naturelles de plusieurs propositions importantes de metaphysique et de mathematique. Apres avoir bien cherché, j'ay trouvé que deux choses sont parfaitement semblables, lorsqu'on ne les sçauroit discerner que *per compraesentiam*, par exemple, deux cercles inegaux de même matiere ne se sçauroient discerner qu'en les voyant ensemble, car alors on voit bien que l'un est plus grand que l'autre. Vous me direz : je mesureray aujourdhuy l'un, demain l'autre ; et ainsi je les discerneray bien sans les avoir ensemble. Je dis que c'est encore les discerner non *per memoriam, sed per compraesentiam* : parce que vous avez la mesure du premier presente, non pas dans la memoire, car on ne sçauroit retenir les grandeurs, mais dans une mesure materielle gravée sur une regle, ou autre chose. Car si toutes les choses du monde qui nous regardent, estoient diminuées en même proportion, il est manifeste, que pas un ne pourroit remarquer le changement. Par cette definition je demonstre aisement des propositions tres belles et tres generales, par exemple que deux choses estant semblables selon une operation ou consideration, le sont selon toutes les autres ; par exemple soyent deux villes inegales en grandeur, mais qui paroissent semblables parfaitement, lorsqu'on les regarde au costé oriental, je dis qu'elles paroistront aussi semblables, quand on les regardera du costé occidental, pourveu que à chaque veue on découvre toute la ville. Cette proposition est aussi importante en Metaphysique et même en Geometrie et en Analyse, que celle du tout plus grand que sa partie. Et neantmoins personne que je sçache l'a enoncée.

Some years later, Leibniz extends this phenomenological interpretation to the concept of "magnitude", while the concept of "quality" is defined negatively in relation to the first (Leibniz 1715, 18–19):

> Quantity or magnitude is what cannot be known in things except through copresence (or through simultaneous perception). [...]
> On the other hand, quality is what can be known in things when they are individually observed, without copresence being necessary.[8]

It is worth noting that Leibniz also uses the term of "form" to designate the concept of "quality". For instance, in Leibniz (1693, 179) one can find the expression "*qualitatem seu formam*" in the same context of discussion about the concepts of "similitude", "quantity", and "quality". Thus, since form (or quality) does not depend on any kind of copresence, it would be possible to try to give a metaphysical definition of "form", insofar as the form of a thing should be determined by its intrinsic properties. Such a definition would enable us in its turn to metaphysically define the notion of "similitude", so that its previous phenomenological definition would become just a secondary characterization. However, Leibniz never succeeded in formulating such a metaphysical definition of "form", and he was also well aware that this would render fruitless any attempt to define the concept of "similitude" on the basis of the concept of "form" (Leibniz 1693, 180): "Thus, it is not enough to say that the things whose form is the same are similar, if we do not already have a general concept of *form*."[9] Failing this, all we have left is the perceptual definition of "similitude".

It should be mentioned here that these latter reflections of Leibniz were unpublished at the time when Wolff and Klügel wrote. In particular, without being able to rely on Leibniz's guidance, Wolff ventures to formulate a metaphysical definition of "similitude" without having previously formulated a general definition of "form" (Wolff 1747, 10–13). In the Wolffian tradition, Alexander Gottlieb Baumgarten (1714–1762), for instance, transcribes Wolff's definition in his *Metaphysik*, work that had an enormous influence on the German philosophy of the second half of the eighteenth century (Baumgarten 1766, 20–21). Baumgarten's *Metaphysik* was printed for the first time in 1732 in a Latin version, then it was translated into German in 1766, and was reprinted seven times between 1732 and 1779. Thus, it can be considered as a successful book at the time, whose influence on the German philosophy has been studied, for instance, in Reiss (1994). Baumgarten's transcription of the Wolffian definition states that two things are said to be similar if they

---

[8] Quantitas seu Magnitudo est, quod in rebus sola compraesentia (seu perceptione simultanea) cognosci potest. [...] Qualitas autem est, quod in rebus cognosci potest cum singulatim observantur, neque opus est compraesentia.

[9] Itaque non sufficit similia dicere, quorum eadem forma est, nisi *formae* rursus generalis notio habeatur.

are equal (*einerley*) with respect to its quality (*Beschaffenheit* or *qualitas*), but no attempt was made to previously define the notion of "form" or "quality".

Since Klügel's sources for the concept of "similitude" are the mathematical and philosophical works of Wolff, we can say with confidence that he inherits from the Wolffian tradition this metaphysical definition of "similitude", which can be identified with what he called the principle of similitude. Thus, when Klügel claims that his argument from analogy can correctly justify the multinomial theorem for a positive fractional exponent, he means that his argument is supported in turn by the Wolffian definition of "similitude". According to this definition, since $(a + bz + cz^2 + dz^3 + \ldots)^m$ and $(a + bz + cz^2 + dz^3 + \ldots)^{\frac{1}{m}}$ are of the same form, where m is a positive integer, their respective series expansions must be similar to each other in whatever respects concerning their forms (i.e., their expression by a formula), hence the multinomial theorem for a positive fractional exponent $\frac{1}{m}$ has been justified on the basis of the principle of similitude. For Klügel, the similitude between these series expansions would be a sufficient condition to prove the validity of the multinomial theorem for fractional exponents. However, by appealing to the Wolffian concept of "similitude", Klügel introduces a metaphysical dimension into his proof, not realizing that it would be necessary to rigorously establish first a general definition of "form", as Leibniz had quite well understood a century before. Unfortunately, as Klügel points out himself, philosophy has not yet reached that stage of progress where such a definition of "form" can be formulated. In the absence of this definition, Klügel's reasoning collapses under the weight of the Wolffian metaphysics.

In sum, far from being a refutation of the original objections, Klügel's article reveals substantial problems in the structure of his proof, causing severe detriment to the interests of the combinatorial analysis given that: (a) without this proof, it is not clear that the multinomial theorem rests exclusively on the doctrine of combinations, since its independence from the binomial theorem remains unjustified; and (b) if philosophical presuppositions were necessary to support the combinatorial analysis, these presuppositions should be carefully examined to avoid the introduction of fallacies. Hence Klügel's article fails in its main goal of convincing German mathematicians in general about the correctness of the proof, but it is more worrying that the members of the combinatorial school do not seem concerned with this failure and seem to be satisfied with such an unsound proof. In fact, as will be seen in Sect. 6.2.1, it will take 20 years before someone else tackles again the problem of proving the mutinomial theorem for rational exponents by means of combinatorial methods and without using the binomial theorem, at a time when the German combinatorial school no longer exists.

It is interesting to note that, despite the lack of acceptance among certain German mathematical communities, it is possible that Klügel's conception about the principle of similitude played some non negligible role in the development of the work on algebra of George Peacock (1791–1858), professor of mathematics at the University of Cambridge. As can be seen in his 1833 *Report on the recent progress and present state of certain branches of analysis* to the British Association for the Advancement of Science, Peacock had a significant knowledge of the combinatorial analysis and, more importantly, expressed sincere admiration for this theory, though he was not very enthusiastic about all the innovative notations proposed by the combinatorial school (Peacock 1834, 288). In this report, the so-called principle of the permanence of equivalent forms was set forth in the following way (Peacock 1834, 198–199):

> Whatever form is algebraically equivalent to another when expressed in general symbols, must continue to be equivalent, whatever those symbols denote.

And:

> Whatever equivalent form is discoverable in arithmetical algebra considered as the science of suggestion, when the symbols are general in their form, though specific in their value, will continue to be an equivalent form when the symbols are general in their nature as well as in their form.

Peacock conceived algebra as a field of mathematics divided into two areas according to the nature of the objects represented by algebraic symbols. He called "arithmetical algebra" the area of this discipline concerned with symbolical representation and manipulation of positive integers, whereas "symbolic algebra" deals with the symbolic manipulation of any number by following the rules of the arithmetical algebra. Thus, a proposition established in arithmetical algebra can be viewed "as suggesting" a more general proposition whose validity should be demonstrated in the field of symbolic algebra. The principle of the permanence of equivalent forms provides the theoretical background needed to justified the generalization of propositions about positive integers, which belong to arithmetical algebra, to propositions about any number, which belong to symbolic algebra. According to this principle, a formula deduced in arithmetical algebra is equivalent in form to the same formula enunciated in symbolic algebra if the symbols of the formula in arithmetical algebra are abstract symbols. In that case, the nature of the mathematical objects represented by those abstract symbols does not matter and it is a legitimate procedure to make this kind of generalizations on the basis of the notion of "equivalent form", which is the central idea of Klügel's principle of similitude. Peacock's principle conveys exactly the same idea as the principle of similitude if the latter is restricted to mathematics.

Besides this coincidence of content between both principles, one of the main examples of application of the principle of the permanence of equivalent forms is the generalization of what Peacock calls the most important theorem in analysis, namely, the binomial theorem (Peacock 1834, 204). As Klügel, Peacock claims that Euler had used an analogous principle and, as Klügel, Peacock argues that Euler's functional proof of the binomial theorem for rational exponents given in 1774 is based on the notion of equivalent or similar forms (Peacock 1834, 204–206). Peacock makes no reference to Klügel in his works, but it is difficult to believe that these fundamental resemblances, at both a theoretical and an argumentative level, are simply the result of chance. Perhaps he does not mention Klügel's principle of similitude because of its controversial reception in German mathematics.

Indeed, although Peacock's algebra was appreciated in England, the introduction and use of his principle was not free from controversy. Even his former brilliant student Arthur Cayley (1821–1895) refused to use it and, in 1869, wrote an article on the binomial theorem in which he carefully explains why it is false to say that the functional proof of Euler is based on Peacock's principle (Cayley 1869).

5.1.2   First Textbooks

While Klügel strives to save his proof of the multinomial theorem, the first textbooks devoted to Hindenburg's and the combinatorial school's ideas appear in print. At first sight, this seems to be a genuine contribution to the diffusion of the German combinatorial analysis. However, a distinction should be made between transmitting a group of mathematical techniques related by their dependence on combinatorial concepts, rather than on the structural aspect of a scientific system, and transmitting the objectives and values of the combinatorial school's project concerning the organization of mathematics as a hierarchical structure of knowledge. The main question to be answered in this section is which of these two different images of the combinatorial analysis has been conveyed by those two textbooks. If they did not spread the second image related to the construction of a system of science, then it is necessary to conclude that they rather constitute an obstacle to the development of the project, and so to the development of the combinatorial analysis itself, since in such a case they transmit a misleading conception of this discipline.

5.1.2.1 Stahl's *Grundriss*

Conrad Diedrich Martin Stahl was born in Brunswick in 1771 and died in Munich in 1833. In his youth, he wanted to pursue a law carrier in Helmstedt, but his vocation led him to science, graduating in mathematics and physics. In 1795, he starts working as Privatdozent at the University of Jena, and he is very soon promoted to be extraordinary professor of mathematics as a result of the recommendation made in December, 1798, by Johann Wolfgang von Goethe (1749–1832) (von Goethe 1798). During this period of professional academic growth, Stahl gives more focused attention to combinatorics, which has spread a little bit everywhere in Germany, with a more pronounced presence in the north, but he surely acquired his first knowledge of the combinatorial analysis through the teaching of Pfaff, who had earlier been his university teacher. His interest in this discipline will leave an indelible mark on the memory that remains of him. An obituary tells us that "he was one of the first expert educators (*Fortbildner*) of the theory of combinations created by Hindenburg."[10] This reputation was certainly forged by the relative notoriety achieved by

---

[10] Anonym (1833): Er war einer der ersten Fortbildner der von Hindenburg erfundenen Combinationslehre.

his book *Grundriss der Combinationslehre nebst Anwendung derselben auf die Analysis* at the beginning of the nineteenth century (Stahl 1800).

This textbook will be favorably received not only by Hindenburg and the combinatorial school but also by the German intellectual elite in general, to the point of becoming an unavoidable reference source for any scholar interested in the combinatorial analysis. Moreover, thanks in part to this textbook, Stahl got a new academic promotion in 1804 to ordinary professor of mathematics at the University of Würzburg. On this occasion, Friedrich Wilhelm Joseph von Schelling (1775–1854) was part of the tenure committee and gave three reasons why Stahl deserved to be appointed to the vacant chair. Here is the third (Schelling's text can be found in Fabbianelli (2009, 308)):

> Regarding his active contributions to the progress of mathematics as a whole and of its methods, the most important of them is, without doubt, that related to the invention of Hindenburg's *theory of combinations*, which has provided the whole of analysis with a new, simpler method, and which its author has presented only in fragments, but which Stahl, thanks to this latter, has systematically reworked, considerably enlarging its range of applications.[11]

This is an example of how the book was perceived by the German academic community, namely, as a book that finally completes the unfinished theory of Hindenburg. However, although Schelling became acquainted firsthand with this theory by attending Hindenburg's lessons from 1796 to 1798 during his sojourn in Leipzig, his opinion about the fragmentary state of Hindenburg's theory and its completion by Stahl is not quite accurate, as will be seen below (for Schelling's sojourn in Leipzig, see Durner 1990).

The thematic organization of Stahl's textbook follows a simple clear division in three parts:

1. Part one: Introduction to the theory of combinations.
   (a) Characterization of the theory of combinations.
   (b) Distinction between "pure combinatorial representation" and "arithmetical combinatorial representation".
2. Part two: Theory of combinations.
   (a) Study of combinations.
   (b) Study of permutations.
3. Part three: Applications of the theory of combinations to mathematical analysis.
   (a) Application to products of series.
   (b) Application to powers of series.

---

[11] Seinen thätigen Antheil an den Fortschritten der gesammten Mathematik und ihrer Methode betreffend, so ist der bedeutendste dieser Art ohne Zweifel der, welcher durch Erfindung der *Combinations-Lehre* von Hindenburg gemacht wurde, als welche der gesammten Analysis eine neue, einfachere Methode giebt, und die von ihrem Urheber nur fragmentarisch dargestellt, von Stahl aber zu großem Dank des letzteren, systematisch bearbeitet und in der Anwendung viel weiter geführt wurde.

(c) Application to powers of products of series.

(d) Application to divisions of series.

(e) Application to the reversion of series.

Let's examine the theoretical presuppositions on which each of these parts rests.

In the first part, one can find a general characterization of what should be regarded as the theory of combinations. Stahl's characterization of this scientific domain starts from the idea that it has unrestricted scope, as can be seen in the following passage (Stahl 1800, 1):

> The theory of combinations deals with the composition of given things into several wholes. The given things are called elements and are generally denoted with letters, and the composition of the elements is denoted by simply writing them adjacent to each other.[12]

This simple definition summarizes Stahl's vision of the theorems, methods, and problems addressed by the combinatorial school. To better understand this definition, it is convenient to have a reference point on how these compositions or combinations, which are the subject matter of this discipline, were conceived at the time. In the *Encyclopédie*, D'Alembert gives the following definition (d'Alembert 1753, 663):

> COMBINATION should be applied, strictly speaking, only to two by two aggregates of several things, but this term is applied in mathematics to all possible ways of taking a certain number of given quantities.[13]

As opposed to this conception, in which combinations are aggregates of quantities as far as mathematics is concerned, Stahl claims that the theory of combinations is not restricted to the study of compositions of quantities, but it deals with compositions of things in general, no matter to which sector of reality these things belong. In other words, the theoretical principles of the theory of combinations govern not only the possible combinatorial relations among mathematical objects, but also among any kind of objects. This assertion should not be understood in the trivial sense that combinatorics can be applied to the study of different phenomena of the world, but in the metaphysical sense that reality is structured according to the laws of combinations. This conception is consistent with Hindenburg's idea according to which there exists a combinatorial law of Nature. When only quantities are involved in combinatorial compositions, or more precisely when the elements are numbers, they are denoted by digits and the composition is said to be

---

[12] Die Combinationslehre beschäftigt sich mit des Zusammensetzung gegebener Dinge zu mehrern Ganzen. Die gegebenen Dinge nennt sie Elemente, bezeichnet sie im allgemeinen mit Buchstaben und zeigt das Zusammensetzen derselben durch ein blosses Schreiben neben einander an.

[13] COMBINAISON ne devroit se dire proprement que de l'assemblage de plusieurs choses deux à deux ; mais on l'applique dans les Mathématiques à toutes les manieres possibles de prendre un nombre de quantités données.

"an arithmetical combinatorial representation". On the contrary, when the nature of the elements is not specified, they are denoted by letters and the composition is said to be "a pure combinatorial representation".

All throughout the first part, the distinction between arithmetical combinatorial representations and pure combinatorial representations is further explained by means of examples. In particular, the combinations and permutations of digits that were studied by Hindenburg in order to create his method of partitions fall into the category of "arithmetical combinatorial representation". In fact, this distinction is not new. Hindenburg had already pointed out that a difference should be made between the domain of pure combinatorics and the domain of arithmetic combinatorics, but Stahl misinterprets the sense of this distinction. As seen in Chap. 4, after reworking Moivre's original proof of the multinomial theorem in 1795, Hindenburg becomes convinced that, contrary to what he had previously believed, the doctrine of combinations does not rest on arithmetical principles. Moivre's proof causes him to understand that what he had thought to be an arithmetical basis of this doctrine was in fact a combinatorial law expressed through the combinatorial involutions of integer partitions. Then he acts on this new conviction and draws the first lines of a new, unfinished theory that we have called in Chap. 4 "the algebra of involutions". Hence the specificity of Hindenburg's doctrine of combinations, in contrast to other theories of combinations, is supposed to come from the analysis of integer partitions as being the most elementary among combinatorial operations. Thus, the combinatorial study of integer partitions not only belongs to the domain of pure cominatorics, but it is also the most important part of this domain. Stahl deviates from Hindenburg's conception by classifying the study of integer partitions under arithmetic combinatorics. This deviation is probably due to the fact that, unlike Hindenburg, Stahl does not regard digits as uninterpreted symbols when used in Hindenburg's combinatorial involutions. On the contrary, he considers that digits denote always numbers, and thus their study by means of the theory of combinations falls into the domain of a sort of applied combinatorics, called arithmetical combinatorics. Furthermore, it should be pointed out here that the term "combinatorial analysis", which designates the theory endorsed by the combinatorial school, appears nowhere in Stahl's book. It is as if, in the eyes of Stahl, Hindenburg and the combinatorial school had never tried to create and develop a new mathematical theory, but had instead simply tried to apply the existing theory of combinations to other mathematical fields, notably to mathematical analysis, which is reflected even in the title of Stahl's textbook. As seen in Chap. 4, the combinatorial school's scientific project consisted in creating a mathematical theory whose foundation is both the algebra of involutions and the multinomial theorem. However, in his book, Stahl does not seem to endorse this project, even if he presents the combinatorial methods developed by the combinatorial school.

In the second part of his textbook, Stahl presents the theoretical elements of the theory of combinations. He introduces the operations of combinations with and without repetition, permutations with and without repetition, and integer partitions. He defines Hindenburg's classes of combinations and explains the method of combinatorial involutions. In short, this part contains the rudiments of the theory of combinations that Hindenburg had studied and published since 1778.

In the third part, Stahl shows how these rudiments can be applied to find a solution for the problems (a)–(e) of point (3) listed above. The solutions offered in the text agree with those previously given by the combinatorial school. However, there is a subtle but important difference regarding the solution proposed by Stahl to problem (b) of point (3). In the combinatorial school's writings, the approach to this problem began with supposing we are given a possibly infinite multinomial $az + bz^2 + cz^2 + \ldots$, which should be raised to a rational power, and the question consists in finding out whether there is a general formula to determine, directly or recursively, the coefficients of the corresponding series expansion, formula that was given by the multinomial theorem. Stahl modifies this argument scheme. For him, the main question consists in determining a formula for the power series expansion of a binomial raised to a rational power, and then this result is used to justify the power series expansion of any multinomial raised to a rational power. As a consequence, for Stahl, the multinomial theorem depends on the binomial theorem. Ironically, Stahl's textbook appears in the same year that Klügel seeks to defend his proof of the multinomial theorem, whose aim was to justify the independence of this theorem from the binomial theorem. Thus, probably knowing nothing of Klügel's response to his critics, Stahl confirms the objections already stated with regard to the unjustified independence of the multinomial theorem, which was one of the principal obstacles to the combinatorial school's reconstruction of mathematical analysis as a new system of science.

Besides its mathematical content, Stahl's book sketches out a very brief history of the process that led to the theoretical construction of a pure and fundamental theory of combinations (Stahl 1800, V–VI). According to Stahl's narrative, this historical process began with Leibniz, who set out the first attempts to apply the theory of combinations to mathematical analysis. However, because of the imperfection of his knowledge about combinatorial operations, particularly about combinatorial involutions, Leibniz did not succeed in establishing the ultimate foundations of this theory. Still in accordance with Stahl's narrative, Cramer, Boscovich, and Castillon worked hard to achieve this goal, but it was Hindenburg who finally discovered these ultimate foundations. Hence, for Stahl, Hindenburg's theory is nothing else than the improvement of Leibniz's theory of combinations. Perhaps Stahl came to this conclusion on the basis of Hindenburg's rhetorical use of Leibniz's image, consisting in the convention of placing an epigraph by Leibniz at the head of some sections of the 1779 *Infinitinomii* and at the head of the *Novi systematis*, as seen in Chap. 3. However, as seen in Chaps. 3 and 4, Hindenburg's combinatorial analysis was born in the Newtonian tradition of the theory of series and the Lambertian tradition of numerical tables, and later Moivre's implicit work on combinatorial involutions, contained in his formulation of the multinomial theorem, was explicitly designated by Hindenburg as the historical and theoretical antecedent of his algebra of involutions. Furthermore, Leibniz's application of the theory of combinations to the theorem on reversion of series was severely criticized not only by Hindenburg, but also by other members of the combinatorial school. Thus, Stahl's narrative does not fit the historical evidence on the development of the combinatorial analysis, and will convey a false picture according to which the work of the combinatorial school belongs to the Leibnizian tradition.

In 1801 Stahl publishes another textbook on the combinatorial school's theory, whose content is more or less the same, except that it does not contain a section devoted to applications of the theory of combinations to analysis; instead of this, a further section on combinatorial involutions was included (Stahl 1801). Thrice in the Introduction, mention is finally made of the term "combinatorial analysis", but always in contrast to the term "Hindenburg's theory of combinations". This distinction between both terms suggests that, for Stahl, the combinatorial analysis is not the original and fundamental theory developed and endorsed by the combinatorial school. His second textbook deals with the theory of combinations exclusively.

5.1.2.2 Weingärtner's *Lehrbuch*

Johann Christoph Weingärtner was born in Erfurt, Germany, in 1771 and died in 1833 in his native town. He dedicated his life to theology and mathematics. In this latter domain, he authored a textbook on the combinatorial analysis, which was published in two volumes in 1800 and 1801 under the title of *Lehrbuch der combinatorischen Analysis nach der Theorie des Herrn Professor Hindenburg* (Weingärtner 1800; 1801). This is the major contribution of Weingärtner to the theory of Hindenburg, and it will be examined in what follows.

A New Epistemological Substrate for the Combinatorial Analysis

The title of Weingärtner's textbook advertises that the reader will find a comprehensive exposition of the combinatorial analysis, written in full accordance with the principles established by Hindenburg. Unlike Stahl, not only does Weingärtner proposes to provide unity to the different texts where the combinatorial analysis has been presented, but he also proposes to do so by closely following the ideas of its creator. In this sense, Weingärtner's textbook seemed to promise the fulfillment of the combinatorial school's research project, at least at the introductory level. However, Weingärtner's characterizations of the theory of combinations and the combinatorial analysis appear to tug in a different direction from Hindenburg's project (Weingärtner 1800, 1):

> The theory of combinations is the science of connecting given things according to certain laws.
> It can be applied to the most diverse purposes of all sciences and arts. However, when it is applied in particular to mathematical analysis, it is called combinatorial analysis.[14]

---

[14] Die Combinationslehre ist die Wissenschaft von den Verbindungen gegebener Dinge nach gewissen Gesetzen.

And he adds further (Weingärtner 1800, 6):

> The combinatorial analysis consists of a theoretical part and a practical one. The theoretical part deals with the combinatorial operations themselves; the practical part contains applications of those operations to the theorems and problems of analysis.[15]

Hence, for Weingärtner, the combinatorial analysis is the result of applying the theory of combinations to mathematical analysis. Its theoretical content, which does not go beyond the scope of the classical combinatorial operations, strictly belongs to the theory of combinations. From a practical point of view, the combinatorial analysis can be seen as an applied science, and thus no theorem of analysis belongs to this applied science. Containing no theorems of its own and having no theoretical content of its own, the combinatorial analysis could hardly aspire someday to enter the elite group of scientific theories. The term "combinatorial analysis" is understood by Weingärtner to have to do with the way some combinatorial methods can be used for the benefit of other mathematical domains, but, in some sense, these methods are not indispensable to the scientific justification of those domains. This is the opposite idea to Hindenburg's, which considers that the combinatorial analysis constitutes a science in its own right, and even more than that, a scientific system within which any mathematical knowledge can be fully justified.

Thus, Weingärtner's general vision of the theory developed by the combinatorial school concurs with that of Stahl. Weingärtner's textbook is composed of two parts. The first volume corresponds to the first part and deals with the theoretical aspects, that is, with the theory of combinations. The basic combinatorial operations, combinations with and without repetition, permutations with and without repetition, and integer partitions, are studied in this part by means of the method of involutions. The second volume contains the practical part of the theory, that is, the application of those combinatorial operations to the solution of problems belonging to the theory of series. The second volume will be discussed in the next paragraph.

Putting aside the fact that Weingärtner's conception of the combinatorial analysis deeply differs from that of Hindenburg, and from that of the combinatorial school in general, his textbook stands out for its careful exposition of the topics covered in both volumes. On the other hand, no innovation has been introduced regarding the theory of combinations and its basic operations. On the contrary, the introduction to the first volume offers, as compared to Stahl's historical narrative, a more comprehensive and interesting

---

Sie hat auf alle Wissenschaften und Künste die mannigfaltigsten Anwendungen. In so ferne sie aber insbesondere auf die mathematische Analysis angewendet wird, heißt sie combinatorische Analysis.

[15] Die combinatorische Analysis hat einen theoretischen und einen praktischen Theil. Der theoretische Theil behandelt die combinatorischen Operationen selbst; der praktische enthält Anwendungen derselben auf Lehrsätze und Aufgaben der Analysis.

reconstruction of what Weingärtner takes to be the past and present of Hindenburg's theory. Mostly focused on the practical aspects of applications, Weingärtner's historical narration of the evolution of the theory of combinations traces its origins back to the *Ars magna* of Ramon Llull (fl. 1233–1315), whose main goal was, according to this narrative, to put combinatorial operations at the service of inventive logic (*logische Invention*). Then, Weingärtner makes the connection between Llull's treatise and the work on combinatorics by Athanasius Kircher (1602–1680), who *grosso modo* repeated, according to Weingärtner, what Llull had already said. These two authors have not been chosen at random by Weingärtner. Both of them are quoted in the *De arte combinatoria* by Leibniz, and it is well know the influence exerted by Llull and Kircher upon the combinatorial research of Leibniz (Brauen 1982; Friedländer 1937). In Weingärtner's narrative, Leibniz is the thinker who takes up the torch of the future development of combinatorics and opens "a new path of applications of combinations to logic", exerting a profound influence on Jacob Bernoulli. The publication of the *De arte combinatoria* marked the beginning of a new era for the theory of combinations, since Leibniz brilliantly showed its universality by applying it to the most diverse areas of science, from chemistry to natural history. Furthermore, and this is the crucial question at issue, Leibniz's most important innovation in the field concerns, continues Weingärtner, the application of the theory of combinations to the solution of problems belonging to mathematical analysis, as the problem of reversing series, and this innovative idea led Leibniz to recognize the possibility of formulating a superior algebra or a universal characteristic. "And as a result", writes Weingärtner, "particular applications of the theory of combinations have been occasionally proposed by great mathematicians."[16] These great mathematicians, which must then necessarily be seen as intellectual heirs of Leibniz, are: Jacob Bernoulli because of the application of combinations to the binomial and multinomial theorems; Johann Bernoulli, who used combinations to solve certain problems of tangents; Euler, Lambert, and Daniel Bernoulli because of the applications of the theory of combinations to the study of continued fractions; Moivre and Boscovich, who applied this theory to the multinomial theorem.

However none of these mathematicians, including Leibniz himself, succeeded, according to Weingärtner, in bringing the theory of combinations to its most perfect formulation. For Weingärtner, all these works on combinatorics suffer from two main defects (*Mängel*). First, it was necessary to find an efficient method to enumerate all possible combinations of a given number of objects, particularly an efficient method to enumerate all partitions of a given positive integer. Weingärtner points out that Leibniz had addressed this question in 1666 and later Euler had made sincere efforts to solve it, but those attempts were inconclusive (Weingärtner 1800, XXV–XXVI). Second, not one of these mathematicians had tried to provide the theory of combinations with a suitable scientific notation for

---

[16] Weingärtner (1800, XXII): Auch in der Folge sind hin und wieder einzelne Anwendungen der Combinationslehre von großen Mathematikern gemacht worden.

describing its own concepts. The resolution of these two problems ushered in a new epoch in the history of mathematical thinking, one in which "the reinvention of the theory of combinations and its applications to analytical problems belongs to Professor Hindenburg."[17]   After the foundations of the new theory of combinations were laid, Eschenbach, Rothe, Burckhardt, Hauber, Klügel, Kramp, Pfaff, Prasse, and Töpfer cooperated with Hindenburg on the further construction of the theory. Weingärtner's textbook is supposed to be a compendium of this collaborative work.

As can be seen, both Weingärtner's historical narrative about the evolution of the combinatorial analysis and that of Stahl converge on the same conclusion: the intent of Hindenburg and the combinatorial school was to refine the theory of combinations, whose origin went back to Leibniz and evolved from there. On the other hand, Weingärtner and Stahl disagree about the possibility of creating a universal characteristic on the basis of the theory of combinations. Weingärtner thought that Leibniz conceived the universal characteristic as the last stage in the development of the theory of combinations, but he did not believe that even the most perfect formulation of this theory could produce such an impressive system of human thoughts as the Leibnizian characteristic. This is, in fact, a point of coincidence with Hindenburg, who was rather skeptical about Leibniz's universal characteristic, as seen in Sects. 3.2.2 and 4.2.1.1. However, Weingärtner goes a little further than rejecting this Leibnizian project when he philosophically argues that the theory of combinations cannot yield a universal characteristic (Weingärtner 1800, XII–XIII):

Leibniz's dissertation [*De arte combinatoria*] already contains the first ideas about a theory of combinations at the service of philosophy, which were connected with his ideas of introducing a language of symbols into philosophy and the metaphysical calculus based on it, ideas that gave rise to his extraordinarily high conception of such an *Arte Combinatoria Characteristica*, to the point of believing that it could afford us as much certainty in philosophy as in mathematics. [...] We know too well now that the benefits Leibniz hoped would result from a philosophical art of symbols and the use of the combinatorial calculus are unattainable, since the essential difference between mathematical and philosophical knowledge, and between its respective methods, has been revealed by the critical philosophy in the most satisfactory way.[18]

---

[17] Weingärtner (1800, XVII): die neue Erfindung der Combinationslehre, und ihrer Anwendung auf analytische Aufgaben ist ein Eigenthum des Herrn Professor Hindenburg.

[18] Die angeführte Dissertation Leibnitzens enthält schon die ersten Ideen einer Combinationslehre zum Behuf der Philosophie, womit sich hernach seine Ideen von einer in der Philosophie einzuführenden Zeichensprache und dem dadurch zu begründenden metaphysischen Calcul verbanden, und ihm eine ausserordentlich hohe Vorstellung von einer solchen *Arte Combinatoria Characteristica* beybrachten, so daß er dadurch in der Philosophie eine eben so große Evidenz als in der Mathematik hervorzubringen glaubte. [...] Daß diese Vortheile, die sich Leibnitz von einer philosophischen Zeichenkunst und dem damit anzustellenden combinatorischen Calcul versprach, nie erreichbar sind, siehet man jetzt wohl ein, da endlich durch die critische Philosophie der wesentliche Unterschied

In short, for Weingärtner, the Leibnizian project of a universal characteristic turns out to be an unreasonable project in the light of the Kantian critical philosophy.

Weingärtner's criticism of the Leibnizian project points to the difficulty of extending the reach of the universal characteristic beyond the limits of mathematics, since such an extension presupposes an illegitimate dogmatic use of reason, in the sense that the combinatorial principles of a metaphysical calculus based on the Leibnizian characteristic would not be supported by pure intuition, thus violating the conditions of possibility of knowledge. Therefore, although the principles and methods of the theory of combinations can be considered as scientific truths, their application to the domain of philosophy cannot give certainty to philosophical propositions. On the contrary, this kind of application would make the results so obtained empty and meaningless. Weingärtner interprets Hindenburg's choice of restricting the scope of applications of the theory of combinations to mathematics as a conscious choice of avoiding the exercise of an illegitimate dogmatic use of reason. As a result, Kant's critical philosophy set out, according to Weingärtner's interpretation, a major theoretical framework for Hindenburg's scientific project, and thus became the ultimate source of epistemological authority for Hindenburg's mathematical theories.

It should be noted that this Kantian interpretation of Hindenburg's thought cannot be confirmed by data. There is absolutely no textual evidence linking the writings of Hindenburg on the combinatorial analysis, or for that matter even the writings of any other member of the combinatorial school, to the critical philosophy. Weingärtner's Kantian interpretation of the combinatorial analysis shows the way Weingärtner understood this mathematical theory in the context of the intellectual climate of the turn of the century, but it is important to realize that he offers an inaccurate, idealized, historical narrative that obscures the comprehension of the actual historical evolution of the combinatorial analysis. Sometimes one effect of this is to mislead scholars in their historical comprehension of the period. For instance, Otte generalizes Weingärtner's historic inaccuracies about Hindenburg and the development of the combinatorial analysis to include all the members of the combinatorial school, as if all of them had supported the idea that the Kantian critical philosophy invalidates the Leibnizian project concerning a universal characteristic (Otte 1989, 32, footnote 6). In fact, other mathematicians of the first half of the nineteenth century that used the methods developed by Hindenburg attached the combinatorial analysis to other philosophies of mathematics. For instance, Jahnke has shown that the work of Spehr, who used the methods of the combinatorial analysis, was based on a philosophy of mathematics different from that of Kant (Jahnke 1992). However, among these philosophical interpretations developed in the nineteenth century, Weingärtner's inaccurate historical narrative seems to have exerted the strongest influence on the way the historical evolution of the combinatorial analysis is perceived.

---

der mathematischen und philosophischen Erkenntniß und ihrer beiderseitigen Methoden auf das befriedigendste ist aufgedeckt worden.

Applications of the Theory of Combinations

The second volume of Weingärtner's textbook contains a survey of the possible applications of the theory of combinations to mathematical analysis. The problems addressed in this volume are:

1. Products of binomials.
2. The binomial theorem for positive integer, negative integer, and fractional exponents.
3. Kramp's factorial.
4. Products and division of powers of a binomial.
5. Products of multinomials.
6. The multinomial theorem for a fractional exponent: presentation of both the recursive formula and the non-recursive formula for the general coefficient of the multinomial expansion.
7. Composition of series.
8. Division of series.
9. Reversion of series.
10. Logarithms.
11. Some trigonometric applications.

In general, these are the same questions addressed by the combinatorial school, and this volume intends to summarize the results of the combinatorial school.

   This list also shows the order in which the selected problems of the theory of series are dealt with in Weingärtner's book. Thus, questions related to binomials take the first place in the thematic organization chosen for the volume. In particular, the binomial theorem is the first important result to be established in Weingärtner's second volume, and this choice obeys to the fact that this particular thematic organization corresponds to a deductive scheme. Indeed, the recursive formula presented in the book for calculating the power series expansion of a multinomial raised to a fractional power is deduced from the binomial theorem, and the same applies to the non-recursive formula for directly determining the general coefficient of the multinomial expansion (Weingärtner 1801, 141, 146, 150). Therefore, the multinomial theorem depends on the binomial theorem. As in the case of Stahl's textbook, this lends credence to the objections raised against the possibility of proving the multinomial theorem independently of the binomial theorem, putting the combinatorial school's project in check again. Paradoxically, Weingärtner shares the combinatorial school's view that the multinomial theorem is the most important theorem of analysis (Weingärtner 1800, XXIX). However, his treatment of the question suggests otherwise by putting the binomial theorem at the top of the deductive hierarchy in the theory of series, and this assumed hierarchy will end up depriving the slogan "the multinomial theorem is the most important theorem of analysis" of sense.

Another relevant aspect of the volume concerns the theorem on reversion of series, whose proof was modified by Weingärtner. As seen in Sect. 4.1.3.4, Rothe demonstrated in 1793 Eschenbach's formula for directly calculating the general coefficient of a reversed series, but the proof was based on differential calculus since formula 4.14, on which the proof depends, was established using basic methods of differentiation. Weingärtner reworked this part of Rothe's proof in order to eliminate its dependence on differential calculus, and proposed instead a combinatorial justification for formula 4.14. This Weingärtner's original modification of the proof can be seen, then, as a positive contribution to the combinatorial school's project. Weingärtner's proof of formula 4.14 depends on two preliminary results about the sum of an arithmetic progression.

The proof of the first preliminary result goes as follows (Weingärtner 1801, 189–190). Suppose one is given a complexion or combination $1^\alpha 2^\beta 3^\gamma 4^\delta \cdots$, where the exponents are positive integers and stand for the multiplicity of the respective elements in the complexion, that is, this complexion is a finite list $\underbrace{1 \cdots 1}_{\alpha \text{ times}} \underbrace{2 \cdots 2}_{\beta \text{ times}} \cdots$, so that $\alpha + \beta + \gamma + \delta + \cdots = m$ is the length of the list and $1\alpha + 2\beta + 3\gamma + 4\delta + \cdots = n$ is the sum of the elements contained in the complexion. Then Weingärtner asserts that:

$$\alpha s + \beta (s + d) + \gamma (s + 2d) + \cdots = sm + (n - m) d, \tag{5.1}$$

for any arithmetic progression $s, s+d, s+2d$, and so on. By noting that $\beta + 2\gamma + 3\delta + \cdots = n - m$, the proof of this equation is immediate since:

$$\alpha s + \beta (s + d) + \gamma (s + 2d) + \cdots = s (\alpha + \beta + \gamma + \cdots) + d (\beta + 2\gamma + \cdots)$$
$$= sm + (n - m) d.$$

Equation 5.1 serves to derive the second preliminary result which states that:

$$\frac{sm + (n - m) d}{m} \text{m} \left({}^n \text{M} \right) = \tag{5.2}$$

$$s a m \left({}^{n-1} \overset{-1}{\text{M}} \right) + (s + d) \, b m \left({}^{n-2} \overset{-1}{\text{M}} \right) + \cdots + (s + (n - m) d) \, {}^{n-m} a \, m \left({}^{m-1} \overset{-1}{\text{M}} \right)$$

$$\begin{pmatrix} 1 & 2 & 3 & \cdots & n \\ a & b & c & \cdots & {}^{n-1}_{} a \end{pmatrix}$$

where, as we know, capital letters with a left superscript are Hindenburg's classes of combinations, and s, $s+d$, $s+2d$, ... is an arithmetic progression for any values of s and d. To this end, Weingärtner took up a corollary established by Prasse and illustrated through the following example (von Prasse 1796, 22). Suppose one is given the combination (2 2 2 3 3), whose length is equal to $m = 5$ and the sum of its elements is equal to $n = 12$. The permutations of this combination can be arranged in a rectangular pattern as follows:

$$\begin{pmatrix} 2\,2\,2\,3\,3 \\ 2\,2\,3\,2\,3 \\ 2\,2\,3\,3\,2 \\ 2\,3\,2\,2\,3 \\ 2\,3\,2\,3\,2 \\ 2\,3\,3\,2\,2 \\ 3\,2\,2\,2\,3 \\ 3\,2\,2\,3\,2 \\ 3\,2\,3\,2\,2 \\ 3\,3\,2\,2\,2 \end{pmatrix}$$

The number of permutations can be calculated by applying the classical formula for multinomial coefficients to the given combination, and so, in this case, the multimnomial coefficient m is equal to m = 10. Prasse established that the sum of any column of this rectangular pattern is equal to $\frac{n}{m}$m. On the basis of this corollary, Weingärtner considers a combination $1^\alpha\ 2^\beta\ 3^\gamma\ 4^\delta \cdots p^\rho$ of length $\alpha + \beta + \gamma + \delta + \cdots + \rho = m$ and the sum of its elements is equal to $1\alpha + 2\beta + 3\gamma + 4\delta + \cdots + p\rho = n$. The permutations of $1^\alpha\ 2^\beta\ 3^\gamma\ 4^\delta \cdots p^\rho$ can then be arranged in a rectangular pattern, similar to that of (2 2 2 3 3), and the number of permutations in the pattern is given by the multinomial coefficient m of $1^\alpha\ 2^\beta\ 3^\gamma\ 4^\delta \cdots p^\rho$. The first column of this pattern can be seen, in turn, as a combination $1^{\alpha'}\ 2^{\beta'}\ 3^{\gamma'}\ 4^{\delta'} \cdots p^{\rho'}$, whose length is given by the multinomial coefficient m of $1^\alpha\ 2^\beta\ 3^\gamma\ 4^\delta \cdots p^\rho$, hence one has $\alpha' + \beta' + \gamma' + \delta' + \cdots + \rho' = m$ for some positive integers α', β', γ', δ', ..., ρ', and the sum of its elements is, according to Prasse's corollary, equal to $\frac{n}{m}$m. Given a (finite) arithmetic progression s, $s + d$, $s + 2d$, ..., $s + (p - 1)d$, it follows from Eq. 5.1 that:

$$\alpha's + \beta'(s+d) + \cdots + \rho'(s+(p-1)d) = sm + \left(\frac{n}{m}m - m\right)d = \frac{sm + (n-m)d}{m}m.$$

This sum can be interpreted by the rectangular pattern of permutations as expressing the following relation:

$$\frac{sm + (n - m)\,d}{m}\,\mathrm{m}\left(1^{\alpha}\,2^{\beta}\,3^{\gamma}\,4^{\delta}\cdots p^{\rho}\right) =$$

$$\begin{pmatrix}
s & 1 & \cdots & 1 & 2 & \cdots & 2 & 3 & \cdots & 3 & \cdots \\
\vdots & \vdots & \vdots & \vdots & \vdots & \vdots & \vdots & \vdots & \vdots & \vdots & \vdots \\
s & 1 & p & \cdots & \cdots & \cdots & \cdots & \cdots & \cdots & \cdots & \cdots \\
s + d & 2 & 1 & \cdots & 1 & \cdots & \cdots & \cdots & \cdots & \cdots & \cdots \\
\vdots & \vdots & \vdots & \vdots & \vdots & \vdots & \vdots & \vdots & \vdots & \vdots & \vdots \\
s + d & 2 & p & \cdots & \cdots & \cdots & \cdots & \cdots & \cdots & \cdots & \cdots \\
\vdots & \vdots & \vdots & \vdots & \vdots & \vdots & \vdots & \vdots & \vdots & \vdots & \vdots \\
s + (p - 1)d & p & \cdots & p & \cdots & \cdots & \cdots & \cdots & \cdots & \cdots & \cdots
\end{pmatrix}$$

This kind of relation is also a corollary established by Prasse, which Weingärtner applied to arithmetic progressions (von Prasse 1796, 22). It could be said that, in this relation, the coefficient $\frac{sm+(n-m)d}{m}$ m expresses the multiplicity of the combination $1^{\alpha}\,2^{\beta}\,3^{\gamma}\,4^{\delta}\cdots p^{\rho}$ in the rectangular pattern according to the coefficients $s, s + d, s + 2d, \ldots, s + (p - 1)d$ of the permutations. In view of this relation, it follows immediately that:

$$\frac{sm + (n - m)\,d}{m}\,\mathrm{m}\left({}^{n}\mathrm{M}\right) =$$

$$s\mathrm{a}\left({}^{n-1}\overset{-1}{M}\right) + (s + d)\,\mathrm{b}\left({}^{n-2}\overset{-1}{M}\right) + \cdots + (s + (n - m)\,d)\,{}^{n-m}\mathrm{a}\left({}^{m-1}\overset{-1}{M}\right)$$

$$\begin{pmatrix}
1 & 2 & 3 & \cdots & n \\
\mathrm{a} & \mathrm{b} & \mathrm{c} & \cdots & {}^{n-1}\mathrm{a}
\end{pmatrix}$$

where capital letters in italic font with a left superscript represent the classes of permutations corresponding to the class of combinations $({}^{n}\mathrm{M})$, and this equality is simply another way of writing Eq. 5.2.

Now, let's examine the combinatorial proof of Rothe's formula 4.14 proposed by Weingärtner. The proof begins with the expression in local symbols of the general coefficient of the multinomial expansion:

$$p^{m+1} \text{x} (n+1) = {}^{m+1}\mathfrak{A} \mathrm{a}^m \mathfrak{a} \left({}^n A\right) + {}^{m+1}\mathfrak{B} \mathrm{a}^{m-1} \mathfrak{b} \left({}^n B\right) + \cdots + {}^{m+1}\mathfrak{N} \mathrm{a}^{m-n+1} \mathfrak{n} \left({}^n \mathcal{N}\right),$$

$$p[a, b, c, d, \ldots] \qquad \begin{pmatrix} 1 & 2 & 3 & \cdots & n \\ b & c & d & \cdots & \overset{n}{a} \end{pmatrix}$$

where m is a rational (or even real) number. And thus, one gets:

$$\frac{p^{m+1} \text{x} (n+1)}{m+1} = \mathrm{a}^m \mathfrak{a} \left({}^n A\right) + \frac{{}^m \mathfrak{A}}{2} \mathrm{a}^{m-1} \mathfrak{b} \left({}^n B\right) + \cdots + \frac{{}^m \overset{-1}{\mathfrak{N}}}{n} \mathrm{a}^{m-n+1} \mathfrak{n} \left({}^n \mathcal{N}\right), \qquad (5.3)$$

and also:

(1) $\mathrm{a} p^m \text{x} (n+1) = {}^m \mathfrak{A} \mathrm{a}^m \mathfrak{a} \left({}^n A\right) + {}^m \mathfrak{B} \mathrm{a}^{m-1} \mathfrak{b} \left({}^n B\right) + \cdots + {}^m \mathfrak{N} \mathrm{a}^{m-n+1} \mathfrak{n} \left({}^n \mathcal{N}\right),$

(2) $\mathrm{b} p^m \text{x} (n) =$

$$ {}^m \mathfrak{A} \mathrm{a}^{m-1} \mathfrak{b} \mathrm{a} \left({}^{n-1} A\right) + {}^m \mathfrak{B} \mathrm{a}^{m-2} \mathfrak{b} \mathfrak{b} \left({}^{n-1} B\right) + \cdots + {}^m \overset{-1}{\mathfrak{N}} \mathrm{a}^{m-n+1} \mathfrak{b} \mathfrak{n} \left({}^{n-1} \overset{-1}{\mathcal{N}}\right),$$

(3) $\mathrm{c} p^m \text{x} (n-1) =$

$$ {}^m \mathfrak{A} \mathrm{a}^{m-1} \mathfrak{c} \mathrm{a} \left({}^{n-2} A\right) + {}^m \mathfrak{B} \mathrm{a}^{m-2} \mathfrak{c} \mathfrak{b} \left({}^{n-2} B\right) + \cdots + {}^m \overset{-2}{\mathfrak{N}} \mathrm{a}^{m-n+2} \mathfrak{c} \mathfrak{n} \left({}^{n-2} \overset{-2}{\mathcal{N}}\right),$$

$$\vdots$$

(n) $\overset{n-1}{\mathrm{a}} p^m \text{x} (2) = {}^m \mathfrak{A} \mathrm{a}^{m-1} \overset{n-1}{\mathrm{a}} \mathfrak{a} \left({}^1 A\right),$

(n + 1) $\overset{n}{\mathrm{a}} p^m \text{x} (1) = \mathrm{a}^m \overset{n}{\mathfrak{a}}.$

Then Weingärtner multiplies the equalities from (1) to $(n+1)$ by $s, s+d, s+2d, s+3d, \ldots, s+nd$ respectively, which gives in particular for the fist equality:

$$s \mathrm{a} p^m \text{x} (n+1) = s^m \mathfrak{A} \mathrm{a}^m \mathfrak{a} \left({}^n A\right) + s^m \mathfrak{B} \mathrm{a}^{m-1} \mathfrak{b} \left({}^n B\right) + \cdots + s^m \mathfrak{N} \mathrm{a}^{m-n+1} \mathfrak{n} \left({}^n \mathcal{N}\right)$$

$$= sm \mathrm{a}^m \mathfrak{a} \left({}^n A\right) + s (m-1) \frac{{}^m \mathfrak{A}}{2} \mathrm{a}^{m-1} \mathfrak{b} \left({}^n B\right) + \cdots \qquad (5.4)$$

$$+ s (m-n+1) \frac{{}^m \overset{-1}{\mathfrak{N}}}{n} \mathrm{a}^{m-n+1} \mathfrak{n} \left({}^n \mathcal{N}\right),$$

and for the last equality:

$$(s + nd)\,\overset{n}{\text{a}}\,p^m x(1) = (s + nd)\,\text{a}^m\overset{n}{\text{a}} = (s + nd)\,\text{a}^m\left(^n\text{A}\right). \tag{5.5}$$

Adding the equalities obtained after multiplication from (2) to (n), one gets:

$$(s + d)\,\mathfrak{b}p^m x\,(n) + \cdots + (s + (n - 1)d)\,\overset{n-1}{\text{a}}\,p^m x(2) =$$

$$^m\mathfrak{A}\text{a}^{m-1}\left[(s + d)\,\mathfrak{b}\text{a}\left(^{n-1}\text{A}\right) + (s + 2d)\,\text{ca}\left(^{n-2}\text{A}\right) + \cdots + (s + (n - 1)\,d)\,\overset{n-1}{\text{a}}\,\text{a}\left(^1\text{A}\right)\right]$$

$$+^m\mathfrak{B}\text{a}^{m-2}\left[(s + d)\,\mathfrak{b}\mathfrak{b}\left(^{n-1}\text{B}\right) + (s + 2d)\,\text{cb}\left(^{n-2}\text{B}\right) + \cdots + (s + (n - 2)\,d)\,\overset{n-2}{\text{a}}\,\mathfrak{b}\left(^2\text{B}\right)\right]$$

$$+ \cdots +^m\mathfrak{N}\text{a}^{m-n+1}\left[(s + d)\,\mathfrak{b}\text{n}\left(^{n-1}\overset{-1}{\mathcal{N}}\right)\right].$$

From Eq. 5.2, it follows that:

$$\frac{2s + nd}{2}\,\mathfrak{b}\left(^n\text{B}\right) =$$

$$\left[(s + d)\,\mathfrak{b}\text{a}\left(^{n-1}\text{A}\right) + (s + 2d)\,\text{ca}\left(^{n-2}\text{A}\right) + \cdots + (s + (n - 1)\,d)\,\overset{n-1}{\text{a}}\,\text{a}\left(^1\text{A}\right)\right],$$

$$\frac{3s + nd}{3}\,\mathfrak{c}\left(^n\text{C}\right) =$$

$$\left[(s + d)\,\mathfrak{b}\mathfrak{b}\left(^{n-1}\text{B}\right) + (s + 2d)\,\text{cb}\left(^{n-2}\text{B}\right) + \cdots + (s + (n - 2)\,d)\,\overset{n-2}{\text{a}}\,\mathfrak{b}\left(^2\text{B}\right)\right],$$

$$\vdots$$

$$\frac{ns + nd}{n}\,\mathfrak{n}\left(^n\mathcal{N}\right) = \left[(s + d)\,\mathfrak{b}\text{n}\left(^{n-1}\overset{-1}{\mathcal{N}}\right)\right],$$

and therefore:

$$(s + d)\,\mathfrak{b}p^m x\,(n) + \cdots + (s + (n - 1)d)\,\overset{n-1}{\text{a}}\,p^m x(2) =$$

$$^m\mathfrak{A}\text{a}^{m-1}\frac{2s + nd}{2}\,\mathfrak{b}\left(^n\text{B}\right) + \cdots +^m\mathfrak{N}\text{a}^{m-n+1}\frac{ns + nd}{n}\,\mathfrak{n}\left(^n\mathcal{N}\right).$$

Adding 5.4 and 5.5 to this latter equality, one obtains:

$$sap^m \mathbf{x}\, (n+1) + (s+d)\, bp^m \mathbf{x}\, (n) + \cdots + (s+nd)\, \overset{n}{\mathrm{a}} p^m \mathbf{x}(1) =$$

$$(s\,(m+1)+nd)\, \mathrm{a}^m \mathfrak{a}\, \left(^n \mathrm{A}\right) + (s\,(m+1)+nd)\, \frac{^m \mathfrak{A}}{2}\mathrm{a}^{m-1}\mathfrak{b}\left(^n \mathrm{B}\right) + \cdots$$

$$+ (s\,(m+1)+nd)\, \frac{^m \overset{-1}{\mathfrak{N}}}{n}\mathrm{a}^{m-n+1}\mathfrak{n}\left(^n \mathcal{N}\right).$$

One obtains Rothe's formula 4.14 from this latter equality and from the product of Eq. 5.3 by $(s\,(m+1)+nd)$.

Later in his book, Weingärtner transcribes the rest of the proof of the formula for calculating the general coefficient of a reversed series given by Rothe in 1793 (Weingärtner 1801, 353–362). Thanks to Weingärtner's proof of formula 4.14, Eschenbach-Rothe's formula 4.16 concerning the reversion of series was for the first time demonstrated by using combinatorial methods exclusively.

*

Before the publication of Stahl's and Weingärtner's textbooks, there was no comprehensive work that covered all of the problems and methods studied by the combinatorial school. Pfaff and Prasse had published extensive monographs related to the combinatorial analysis, but these were restricted to specific topics, as the study of the reversion of series for Pfaff and that of logarithms for Prasse. As a consequence, it was hard to get an overall view of the combinatorial school's ideas without reading most of their research articles. In this sense, Stahl's and Weingärtner's works meant a great advantage for those interested in the combinatorial analysis, but these didactic books failed to capture correctly the foundational ideas that guided the development of the combinatorial analysis. However, it was through these textbooks that many people, beyond the circle of the combinatorial school, became acquainted with "the theory of Hindenburg" during the nineteenth century. For instance, in the *Conversations-Lexicon*, a general encyclopedia whose first edition appeared in 1796 and which continued to be published until our times, the definition of "combinatorial analysis" was taken from Weingärtner's textbook, characterizing the combinatorial analysis as an application of the theory of combinations to mathematical analysis (Anonym 1820, 715). In 1829, the *Encyclopaedia Americana* was launched in the United States and its articles were based on the *Conversations-Lexicon*, favoring the dissemination of Weingärtner's vision of the combinatorial analysis (Lieber 1830, 355). But, mostly, this vision permeated large sectors of the German culture, and not only through general encyclopediae, but also through mathematical books. In his highly successful book *Sammlung aus der Buchstabenrechnung und Algebra*, reprinted twenty times between 1804 and 1890 and translated into English and French, Meier Hirsch (1765–1851), professor of mathematics in Berlin, included a brief section on combinations, and there the only texts recommended for reading in connection with "the theory of combinations as a self-founded science" are the textbooks of Stahl and Weingärtner (Hirsch 1804, 80–94). It is worth noting that, in the French translation of Hirsch's book, the section devoted to the theory of combinations was omitted, which reflects the general disinterest among French scholars in the combinatorial analysis (Hirsch 1832, IV).

While these textbooks find their way into the general culture and the German education through publications of wide circulation, they are also very well received in the combinatorial school. Quite soon after their publication, Hindenburg rushed to make the following very flattering comment (Hindenburg 1800, VIII):

> The wish expressed several times to have a textbook on the combinatorial analysis in which its first and simplest foundations, as well as its scientifically ordered theorems and applications, were presented in a clear and intelligible way; a textbook that brings together everything that has been said upon the subject, in which nothing is assumed as known beforehand, in which all references to other writings, except those of a historical and literary nature, are therefore completely omitted; that wish is probably coming true at the moment I write this. Professor Stahl in Jena and vice rector Weingärtner in Erfurt have been working for some time now on the composition of such a valuable work, according to my theory and my symbolic notations.[19]

In the circle of the combinatorial school, these textbooks filled the gap left by the missing treatise on the combinatorial analysis that Hindenburg never wrote. For instance, as Hindenburg, Bürmann recommended the texts of Stahl and Weingärtner to those who wished to learn what the combinatorial analysis is (Bürmann 1801c, 129). Thus, the most favorable conditions for the successful dissemination of an inaccurate image of the combinatorial analysis were met at the turn of the century, an image that does not match very well the goals of the project the combinatorial school had worked on over the six last years of the eighteenth century.

---

[19] Der Wunsch, den man bereits mehrmals geäussert hat, ein eigenes Lehrbuch der combinatorischen Analysis zu haben, welches von den ersten und einfachsten Gründen an, die Hauptsätze und ihre Anwendung wissenschaftlich geordnet, faßlich vorgetragen und deutlich entwickelt enthielte; ein Lehrbuch, in welchem alles, was zur Sache gehört, beisammen anzutreffen wäre, worin nichts, als anderswoher bekannt, vorausgesetzt würde, wo also Berufungen und Verweisungen auf andre Schriften, ausgenommen in historischer und litterarischer Rücksicht, ganz wegfielen —dieser Wunsch ist wahrscheinlich bereits itzt, da ich dieses schreibe, in Erfüllung gegangen. Es haben nähmlich, Herr Professor Stahl in Jena und Herr Conrector Weingärtner in Erfurt, sich schon seit einiger Zeit mit Ausfertigung eines solchen Werts, nach meiner Theorie und meinen Zeichen, beschäftigt.

5.1.3 A Last Combinatorial Proof of the Binomial Theorem

At the beginning of the nineteenth century, the orientation and goals of some important writings about the combinatorial analysis undergo significant changes with respect to the seminal ideas of the combinatorial school developed between 1794 and 1800. This was a time marked by profound transformations regarding the original objectives and the general conception of the combinatorial analysis. Perhaps without wishing to do so, Prasse took part in this transformation by trying to prove the binomial theorem using combinatorial methods exclusively, which led him to forget about the fact that the combinatorial analysis was meant to be a system of science based on the multinomial theorem. This is the last combinatorial proof of the binomial theorem proposed by one of the members of the combinatorial school, and was originally published in von Prasse (1803b), and later reprinted in von Prasse (1804, 39–54). In Panza (1992, vol. 2, 680–689), on can find a study of this proof in the broader context of the historical evolution of mathematical analysis in the eighteenth century. In what follows, Prasse's proof will be examined on the basis of the 1803 original booklet.

Prasse's proof depends on the method of combinatorial involutions. In particular, he uses Hindenburg's lexicographical permutation involutions, which has been studied in Sect. 4.2.3.1, but they are denoted by the symbol $^n\mathcal{J}$ instead of Hindenburg's notation $^n\mathbb{J}$. Besides, Prasse introduces the symbol $\mathfrak{j}\ ^n\mathcal{J}$, in which the Gothic letter $\mathfrak{j}$ means that every row of the involution $^n\mathcal{J}$ should be preceded by 1 if the first element of the row (from left to right) is α, or 2 if the first element of the row (from left to right) is β, or 3 if the first element of the row (from left to right) is γ, and so on. Figure 5.1 shows an example of this kind of involutions. This poor choice of notation can cause confusion, since Gothic lowercase letters usually denote multinomial coefficients in the writings of the combinatorial school, but this explanation should be enough to avoid any problem.

Prasse derives the binomial coefficients from two theorems. The first states that:

$$\mathfrak{j}^n\,\mathcal{J} = \mathfrak{j}^{n-1}\,\mathcal{J}\alpha + \mathfrak{j}^{n-2}\,\mathcal{J}\beta + \cdots + \mathfrak{j}^2\,\mathcal{J}\overset{-2}{\upsilon} + \mathfrak{j}^1\,\mathcal{J}\overset{-1}{\upsilon} + n\upsilon, \tag{5.6}$$

$$(\alpha, \beta, \gamma, \ldots \upsilon) \qquad \left(\alpha, \beta, \gamma, \ldots \overset{-1}{\upsilon}\right)$$

where n is a positive integer. Prasse does not bother to prove it, but it is quite obvious from the construction of gnomons, as can be easily seen in the example of Fig. 5.1. The second theorem consists of the following logarithmic-combinatorial identity:

$$\log\left(1 - \left(\alpha x + \beta x^2 + \gamma x^3 + \ldots\right)\right) = -\left(\mathfrak{j}^1\mathcal{J}\frac{x}{1} + \mathfrak{j}^2\mathcal{J}\frac{x^2}{2} + \cdots + \mathfrak{j}^n\mathcal{J}\frac{x^n}{n} + \cdots\right),$$

$$\begin{pmatrix} 1 & 2 & 3 & \cdots \\ \alpha & \beta & \gamma & \cdots \end{pmatrix} \tag{5.7}$$

Fig. 5.1 Prasse's
lexicographical permutation
involution

$$
\begin{array}{r}
1\cdot\ \alpha\ |\alpha|\alpha|\alpha|\alpha \\
2\cdot\ \beta\ |\alpha|\alpha|\alpha \\
1\cdot\ \alpha\ \ \beta\ |\alpha|\alpha \\
3\cdot\ \gamma\ |\alpha|\alpha \\
1\cdot\ \alpha\ \ \alpha\ \ \beta\ |\alpha \\
2\ \ \beta\ \ \beta\ |\alpha \\
1\cdot\ \alpha\ \ \gamma\ |\alpha \\
4\cdot\ \delta\ |\alpha \\
\hline
1\ \ \alpha\ \ \alpha\ \ \alpha\ \ \beta \\
2\ \ \beta\ \ \alpha\ \ \beta \\
1\cdot\ \alpha\ \ \beta\ \ \beta \\
\gamma\ \ \beta \\
1\cdot\ \alpha\ \ \alpha\ \ \gamma \\
2\cdot\ \beta\ \ \gamma \\
1\cdot\ \alpha\ \ \delta \\
5\cdot\ \varepsilon
\end{array}
\quad = \quad
\begin{array}{c}
\mathbf{j}^{5}\ \mathscr{J} \\
{\scriptstyle (\alpha,\beta,\gamma,\delta,\varepsilon)}
\end{array}
$$

where $z = \alpha x + \beta x^2 + \gamma x^3 + \cdots$ is any (possibly infinite) multinomial, and where "log" stands for natural logarithm. By noting that:

$$
\log\,(1 - z) = -\left(\frac{z}{1} + \frac{z^2}{2} + \cdots + \frac{z^n}{n} + \cdots\right)
$$

and that:

$$
z = {}^{1}Ax + {}^{2}Ax^2 + {}^{3}Ax^3 + \cdots ,
$$

$$
z^2 = {}^{2}Bx^2 + {}^{3}Bx^3 + {}^{4}Bx^4 + \cdots ,
$$

$$
z^3 = {}^{3}Cx^3 + {}^{4}Cx^4 + {}^{5}Cx^5 + \cdots ,
$$

$$
\vdots
$$

$$
z^m = {}^{m}Mx^m + {}^{m+1}Mx^{m+1} + {}^{m+2}Mx^{m+2} + \cdots ,
$$

where capital letters are permutation classes of partitions, it follows that:

$$\log(1-z) = \log\left(1 - \left(\alpha x + \beta x^2 + \gamma x^3 + \cdots\right)\right)$$

$$= -\left(\begin{array}{l} \frac{1}{1}\left(^1A\right)x + \left(\frac{1}{1}\left(^2A\right) + \frac{1}{2}\left(^2B\right)\right)x^2 + \cdots \\ \cdots + \left(\frac{1}{1}\left(^nA\right) + \frac{1}{2}\left(^nB\right) + \ldots + \frac{1}{n}\left(^nN\right)\right)x^n + \cdots \end{array}\right)$$

$$= -\left(\begin{array}{l} \frac{1}{1}\left(^1A\right)\frac{x}{1} + \left(\frac{2}{1}\left(^2A\right) + \frac{2}{2}\left(^2B\right)\right)\frac{x^2}{2} + \cdots \\ \cdots + \left(\frac{n}{1}\left(^nA\right) + \frac{n}{2}\left(^nB\right) + \cdots + \frac{n}{n}\left(^nN\right)\right)\frac{x^n}{n} + \cdots \end{array}\right).$$

Then Prasse points out that, for any positive integer n:

$$j^n \mathcal{J} = n\left(\frac{^nA}{1} + \frac{^nB}{2} + \cdots + \frac{^nN}{n}\right), \tag{5.8}$$

which concludes the proof of Eq. 5.7.

Let's explain formula 5.8 with an example, as Prasse does (although our example is different from his). Suppose one is given the permutation class of partitions:

$$^5C = \begin{pmatrix} \alpha & \alpha & \gamma \\ \alpha & \beta & \beta \\ \alpha & \gamma & \alpha \\ \beta & \alpha & \beta \\ \beta & \beta & \alpha \\ \gamma & \alpha & \alpha \end{pmatrix} = m(\alpha\ \alpha\ \gamma) + n(\alpha\ \beta\ \beta)$$

where m and n are the multinomial coefficients of $(\alpha\ \alpha\ \gamma)$ and $(\alpha\ \beta\ \beta)$ respectively, and thus the expression $m(\alpha\ \alpha\ \gamma) + n(\alpha\ \beta\ \beta)$ means that ^5C is composed of the permutations of the combination $(\alpha\ \alpha\ \gamma)$ and of those of the combination $(\alpha\ \beta\ \beta)$. In the previous section we have seen that Prasse established the following result:

$$\begin{pmatrix} 1 & \alpha & \alpha & \gamma \\ 1 & \alpha & \gamma & \alpha \\ 3 & \gamma & \alpha & \alpha \end{pmatrix} = \frac{n}{m}m(\alpha\ \alpha\ \gamma) = \frac{5}{3} \times 3(\alpha\ \alpha\ \gamma),$$

where $m = 3$ is the length of the combination $(\alpha\ \alpha\ \gamma)$, $n = 5$ is the sum of its elements (since this is a partition of 5: α is associated with 1 and γ with 3), and $m = 3$ is its multinomial coefficient, and where the first column of numerical values has been determined by adding 1 if the first element of the row (from left to right) is α, or 2 if the first element of the row (from left to right) is β, or 3 if the first element of the row (from left to right) is γ. On the other hand, the permutations $(1 \cdot \alpha \; \alpha \; \gamma)$, $(1 \cdot \alpha \; \gamma \; \alpha)$, and $(3 \cdot \gamma \; \alpha \; \alpha)$ are included in the involution shown in Fig. 5.1, hence this involution contains $\frac{5}{3} \times 3(\alpha \; \alpha \; \gamma)$. The same applies to the combination $(\alpha \; \beta \; \beta)$. Therefore, $\frac{5}{3}({}^5C) = \frac{5}{3}m(\alpha \; \alpha \; \gamma) + \frac{5}{3}n(\alpha \; \beta \; \beta)$ is contained in the involution. A similar point can be made about the classes ${}^5A$, ${}^5B$, ${}^5D$, ${}^5E$, and therefore:

$$\mathfrak{j}^5 \mathcal{J} = 5\left(\frac{{}^5A}{1} + \frac{{}^5B}{2} + \frac{{}^5C}{3} + \frac{{}^5D}{4} + \frac{{}^5E}{5}\right)$$

Prasse remarks that this argument can be generalized, and thus Eq. 5.8 holds for any positive integer n. From this, one can derive Eq. 5.7.

To prove the binomial theorem, Prasse supposes that any power of a binomial can be expanded in an infinite series, as was usual at the time:

$$(1 - x)^u = 1 - \left(\alpha x + \beta x^2 + \gamma x^3 + \cdots\right),$$

where α, β, γ, etc. are undetermined coefficients. At the end of his booklet, Prasse claims that u can take imaginary values, but this statement has not really been justified, not even for irrational numbers as pointed out by Panza (1992, vol. 2, 688). Let's assume, then, that u can be any rational number, so that the power series expansion of the logarithm is well established, by the standards prevailing at the time. Thus, taking the logarithm of both sides of the equation, one gets:

$$\log\left(1 - \left(\alpha x + \beta x^2 + \gamma x^3 + \cdots\right)\right) = \log(1 - x)^u = u \log(1 - x)$$

$$= -u\left(\frac{x}{1} + \frac{x^2}{2} + \cdots + \frac{x^n}{n} + \cdots\right).$$

From this expansion and from that shown in Eq. 5.7, he obtains the following identities:

$$u = \mathfrak{j}^1 \mathcal{J} = \mathfrak{j}^2 \mathcal{J} = \cdots = \mathfrak{j}^n \mathcal{J}.$$

$$(\alpha, \beta, \gamma, \ldots v)$$

Applying formula 5.6 to each of these identities, it follows that:

$$u = \mathfrak{j}^1 \mathcal{J} = 1\alpha,$$

$$u = \mathfrak{j}^2 \mathcal{J} = \mathfrak{j}^1 \mathcal{J}\alpha + 2\beta = u\alpha + 2\beta,$$

$$u = \mathfrak{j}^3 \mathcal{J} = \mathfrak{j}^2 \mathcal{J}\alpha + \mathfrak{j}^1 \mathcal{J}\beta + 3\gamma = u\alpha + u\beta + 3\gamma,$$

$$\vdots$$

$$u \;=\; j^n \mathcal{J} \;=\; j^{n-1} \mathcal{J}\alpha + j^{n-2} \mathcal{J}\beta + \cdots + j^1 \mathcal{J}\overset{-1}{v} + nv$$

$$= \; u\alpha + u\beta + \cdots + u\overset{-1}{v} + nv.$$

$$(\alpha, \beta, \gamma, \ldots v) \qquad \left(\alpha, \beta, \gamma, \ldots \overset{-1}{v}\right)$$

From these, Prasse deduces the following equation system:

$$u = \alpha,$$

$$u = u\alpha + 2\beta,$$

$$u = u\alpha + u\beta + 3\gamma,$$

$$\vdots$$

$$u = u\alpha + u\beta + u\gamma + \cdots + u\overset{-1}{v} + uv,$$

whose solution gives the binomial coefficients:

$$\alpha = u = {}^n\mathfrak{A},$$

$$\beta = -\frac{u\,(u-1)}{1.2} = -{}^n\mathfrak{B},$$

$$\gamma = \frac{u\,(u-1)\,(u-2)}{1.2.3} = {}^n\mathfrak{C},$$

$$\vdots$$

$$v = \pm\frac{u\,(u-1)\,(u-2)\cdots(u-n+1)}{1.2.3\ldots n} = \pm{}^n\mathfrak{N}.$$

This completes the proof of the binomial theorem.

However, Prasse draws attention to the fact that, in his work (von Prasse 1803a), he had previously obtained the power series expansion of the logarithm by applying the binomial theorem, result that was later reprinted in von Prasse (1804, 25–38). It might seem, then, that there is a vicious circle involved here. In order to avoid this logical flaw, he proposes that the successive steps in the proof of the binomial theorem should follow a strict deductive order, namely:

1. The binomial theorem for a positive integer exponent.
2. The binomial theorem for a negative integer exponent.

3. The multinomial theorem for a positive and a negative exponent.
4. Power series expansion of $\log(1 \pm x)$.
5. Power series expansion of $\log\left(1 - \left(\alpha x + \beta x^2 + \gamma x^3 + \cdots\right)\right)$.
6. The binomial theorem for any exponent.
7. The multinomial theorem for any exponent.

Sticking to this deductive order would avoid any logical problem.

At the time, point (1) is a well-known result that can be justified by elementary methods. In von Prasse (1803b, 12–14), Prasse proves point (2) by induction using point (1). As seen in Chap. 2, there is a long-standing tradition of proofs of the multinomial theorem for an integer exponent based on the binomial theorem for an integer exponent. In addition, Prasse himself has demonstrated point (4) on the basis of point (1). The proof of point (5) corresponds to that of Eq. 5.7, which depends on points (3) and (4). As shown above, the proof of point (6) rests on all the previous points, particularly on point (4). Finally, the multinomial theorem for any exponent can be justified by means of the binomial theorem. Therefore, no logical error has been committed in Prasse's reasoning.

From the point of view of the combinatorial analysis, Prasse's text on the binomial theorem is very interesting because of its paradoxical theoretical assumptions. On the one hand, in his proof, Prasse not only makes use of the method of involutions, but more importantly he uses Hindenburg's algebra of involutions. Prasse himself published treatises related to this method and to the algebra of involutions and contributed to the progress of the combinatorial school's project. Indeed, his 1803 proof of the binomial theorem is supported on his research about combinatorial methods carried out in 1796. After all, it should be remembered that the discovery of the method of involutions is deeply tied to the discovery of the ultimate foundation of the combinatorial analysis, which aroused the interest of several members of the combinatorial school, as Rothe, in the binomial theorem. On the other hand, the deductive order listed above shows that, despite the use of the method and the algebra of involutions, Prasse's proof has not been guided by a concern for the way that this order could impact on the combinatorial analysis conceived as a system of science. For instance, in Rothe's proof of the binomial theorem studied in Sect. 4.2.2.2, it is possible to glimpse the idea that the multinomial theorem should occupy the first place in the deductive structure of mathematical analysis. On the contrary, in Prasse's work, the multinomial theorem has been relegated to the lowest rung on his deductive stratification order, lagging even further behind other power series expansions, as that of the logarithm. Although Prasse does not clarify how point (7) is to be demonstrated on the basis of the previous points, there is no doubt that combinatorial methods should be employed, as well as the binomial theorem. Thus, it is highly likely that Hindenburg's proof of the multinomial theorem published in 1778 in the *Methodus nova* can be seen as the proof that should be inserted in point (7). However, placing Hindenburg's proof at the end of the list in 1803 is tantamount to denying the combinatorial school's project of reorganizing mathematical knowledge into a new scientific system called "combinatorial analysis", since that Hindenburg's old proof does not reflect the evolution of Hindenburg's ideas, particularly Prasse's list does not take into account that the scientific system proposed by the combinatorial school must be based on the algebra of involutions and that the bridge connecting this algebra to mathematical analysis is the multinomial theorem inasmuch as this theorem, one might say, contains the analytical translation of the combinatorial principles of the algebra of involutions.

In any event, there is no doubt that all these discordant opinions about what kind of role the binomial theorem plays or should play in the combinatorial analysis sowed confusion among the scientific community. For example, this is the case of the anonymous reviewer who criticized Klügel's proof of the multinomial theorem by pointing out that this theorem should be derived from the binomial theorem, since the reviewer did not grasp the importance of the multinomial theorem for the combinatorial school's project. But he was not the only one to get confused by the discordant opinions of the combinatorial school. The textbooks of Stahl and Weingärtner, in which the binomial theorem takes the first place, can be seen, at the same time, as a consequence of this and as decisive texts that powerfully crystallized the idea of the binomial theorem as the central theorem in analysis for the combinatorial school. In the first half of the nineteenth century, there are scientists convinced that that indeed was the case. For instance, in the first few sentences of his book on the binomial theorem published in 1816, Bernard Bolzano (1781–1848) affirms that this theorem can be considered as the most important theorem of the whole of analysis, since it serves to derive the multinomial theorem, as well as the power series expansions of logarithms and exponential functions (Bolzano 1816, III). To understand the implication of these sentences in the context of the combinatorial analysis, suffice it to say that Bolzano had studied Prasse's proof described above. A few lines later, Bolzano gives a relatively long list of mathematicians "who have attempted to find a completely strict proof of the [binomial] theorem", including several members of the combinatorial school. In particular, Hindenburg is mentioned in the list, which is merely a product of Bolzano's imagination since Hindenburg never attempted to find a proof of the binomial theorem at all. This kind of miscomprehension of the combiantorial analysis increases during the first decades of the nineteenth century.

5.2   Against the Current

As time goes by, the theoretical demarcation of the combinatorial school's project becomes more and more blurred, and the initial cohesion of the group tends to ebb away. In fact, these two aspects of the decline of the combinatorial analysis are two sides of the same problem. The combinatorial school was not a scientific team in the sense understood today, but a fragile intellectual alliance of mathematicians linked to each other by a common conception of mathematics. If this common conception becomes fragmented or begins to fade, the alliance falls apart and the group disappears. In this sense, the existence of the combinatorial school and its scientific research project depended on preserving the unity of the ideas that led to the birth of the combinatorial analysis. Without this unity, only a handful of combinatorial methods are left and dispersed mathematicians that sometimes applied them to unrelated questions. It is possible to identify some fundamental theoretical divergences that delivered the final blow to this unity, ending all hope of reorganizing mathematical knowledge as a unique theoretical system. Two of the first important divergences appeared early in the history of the combinatorial analysis, just at the crucial moment of consolidation of the central ideas of the group, between 1796 and 1798. Another relevant divergence appeared around the turn of the century. These events will be analyzed in this section.

5.2.1   A Non-Combinatorial Approach to the Multinomial Theorem

Johann Nicolai Tetens was born in 1736 in Tetenbüll, in the Duchy of Holstein. He was a prolific scientist with a broad range of interests. During his years as a teacher at the Academy of Bützow in the 1760s, his research interests were focused on the comprehension of some physical phenomena, as the explanation for the color of the visible sky, and on reflection on traditional problems of philosophy, as that of the existence of God. During this period of his life, mathematics did not seem to pique his curiosity. Only one single paper on the application of the method of *maxima* and *minima* to the study of curves came out of his pen in 1763 (Tetens 1763). Instead, he deepened his knowledge of philosophy by discovering and reading the work of David Hume (1711–1776), and later helped to introduce it in the Germanic world. Indeed, Tetens is better known today for his work to promote Hume's philosophical thought among German academic circles (Cassirer 1907; Keach 1997; Watkins 2005). His obvious philosophical talents and his knowledge on the subject permitted him to be appointed to the chair of philosophy at the University of Kiel in 1776, where he also taught mathematics. It was not until 1785 when he published a book related to mathematical questions under the title *Einleitung zur Berechnung der Leibrenten und Anwartschaften*, and a second volume appeared the next year (Tetens 1785; 1786). By the title of the book, one can make out that he abandoned his studies on mathematical curves and reoriented his interests toward the theory of probability under the form of the study of life annuities. This work has attracted the attention of the specialists in the history of actuarial mathematics because of its detailed mortality tables and its innovative analysis of the mathematical notion of "risk" (Broch 1967, 432–434; Haberman & Sibbett 1995; Keiding 1987, 7–10; Pradier 1998, ch. 1, 2003, 147–151).

In his *Einleitung*, Tetens's approach to the mathematical treatment of this notion consists of applying some techniques based on polynomials which require calculation of the series expansion of powers of polynomials. He remarked that, given a polynomial function:

$$(a + bx + cx^2 + \ldots + qx^n)^m$$

the larger the exponents m and n, the harder it becomes to calculate the expansion. For this reason, he considered Hindenburg's formula for directly determining any term of the multinomial expansion to be of great value (Tetens 1786, 138). About 10 years later, he sent a paper on the multinomial theorem to Hindenburg, who published it in 1796 in one of his scientific journals devoted entirely to the combinatorial analysis (Tetens 1796). The goal of Tetens's article is to propose and validate a new formula and a new method that allow the direct calculation of any term of the multinomial expansion. Because of its subject and its objective, the article seems consistent with the combinatorial school's practice and project. Let's examine in detail Tetens's formula and method.

First of all, it is necessary to make some explanatory points with respect to Tetens's mathematical notations. An expression like $|n|$ stands for the nth coefficient of a possibly infinite multinomial:

$$a + bx + cx^2 + \cdots + |n|x^{n-1} + \cdots ,$$

and the expression $T\,(a + b + c + \cdots + |n|)^m$ denotes the nth coefficient in the series expansion of that multinomial raised to the power m. It is worth saying that Hindenburg went through the entire text and made comments comparing the assertions and techniques of Tetens to those of the combinatorial school. In this case, he underlined the fact that Tetens's symbolic representation of a given coefficient in the multinomial expansion can be expressed more succinctly by means of local symbols, so that $T\,(a + b + c + \cdots + |n|)^m = p^m x(n)$.

Before broaching the topic of powers of multinomials, Tetens addresses the question of calculating the general term of the product of two possibly infinite multinomials $a + bx + cx^2 + \cdots + |n|x^{n-1}$ and $\alpha + \beta x + \gamma x^2 + \cdots + |v|x^{n-1}$, which is given by the expression $(\alpha|n| + \beta|n - 1| + \cdots + |v - 1|b + |v|n)x^{n-1}$. To find this coefficient of the term x^{n-1} of the product, he proposes an approach which consists in constructing a table containing the coefficients of the original multinomials as follows:

a	b	c	$\dots$	$\lvert n - 2\rvert$	$\lvert n - 1\rvert$	$\lvert n\rvert$
$\lvert v\rvert$	$\lvert v - 1\rvert$	$\lvert v - 2\rvert$	$\dots$	γ	β	α

Thus, to calculate the nth coefficient of a multinomial product, one row contains the coefficients of the first factor from the first to the nth in increasing order, whilst the other contains the coefficients of the second factor from the nth to the first in decreasing order. Then, for each column, one takes the product of the elements in the first and second rows, and the sum of these products gives the desired coefficient of the product of two multinomials. For instance, to calculate the eighth coefficient of:

$$(a + bx + cx^2 + dx^3)(\alpha + \beta x + \gamma x^2 + \delta x^3 + \varepsilon x^4)$$

Tetens writes:

$$
\begin{array}{cccccccc}
a & b & c & d & 0 & 0 & 0 & 0 \\
0 & 0 & 0 & \varepsilon & \delta & \gamma & \beta & \alpha
\end{array}
$$

and after multiplication, one gets the term $(d\varepsilon)\, x^7$ of that product. This method based on tables was called "method of substitution" by Tetens.

The question of raising a multinomial to a positive integer power was treated by Tetens as a particular case of multinomial product problems. He used his method of substitution to demonstrate the validity of the following formula:

$$
T(a + \cdots + |n|)^m = m a^{m-1} |n| + \frac{m\,(m-1)}{1.2} a^{m-2} T\,(b + \cdots + |n-1|)^2 \tag{5.9}
$$
$$
+ \frac{m\,(m-1)\,(m-2)}{1.2.3} a^{m-3} T\,(b + \cdots + |n-2|)^3 + \cdots
$$
$$
+ \frac{m\,(m-1)\,(m-2)\ldots(m-(h-1))}{1.2.3\ldots m} a^{m-h} T\,(b + \cdots + |n-(h-1)\,|)^h ,
$$

which serves to directly calculate the nth coefficient of the power series expansion of $(a + bx + cx^2 + \cdots)^m$ for any fractional number m. Hindenburg did not miss out on this opportunity to remark in a footnote that, disregarding the notational differences, this is an identical reproduction of his formula for the multinomial theorem, made public as early as the appearance of the *Infinitinomii*, and reprinted several times since then.

Tetens's proof proceeds by stages. First, he establishes formula 5.9 for a positive integer m by reasoning by induction. His method of substitution allows to corroborate that:

$$
T\,(a + b + \cdots + |n|)^2 = 2a|n| + 2b|n-1| + 2c|n-2| + \cdots ,
$$

and:

$$
T\,(b + \cdots + |n-1|)^2 = 2b|n-1| + 2c|n-2| + \cdots ,
$$

therefore:

$$
T\,(a + b + \cdots + |n|)^2 = 2a|n| + T\,(b + \cdots + |n-1|)^2 .
$$

By assuming the validity of formula 5.9 for a given positive integer m, Tetens presupposes that the following coefficients are known:

$$
T\,(a + \cdots + |n|)^m ,\quad T\,(a + \cdots + |n-1|)^m ,\ldots,\quad T\,(a)^m \tag{5.10}
$$

According to formula 5.9, each of these coefficients is a finite sum. Thus, taking the fist term of each of these sums, Tetens arranges them in the first row of the following table:

$$
\begin{array}{ccccc}
ma^{m-1}|n| & ma^{m-1}|n-1| & ma^{m-1}|n-2| & \cdots & ma^{m-1}b \quad a^m \\
a & b & c & \cdots & |n-1| \quad |n|
\end{array}
$$

By applying the method of substitution, one obtains the sum:

$$
R = ma^m|n| + ma^{m-1}b|n-1| + ma^{m-1}c|n-2| + \ldots + ma^{m-1}b|n-1| + a^m|n|.
$$

But since:

$$
ma^{m-1}b|n-1| + ma^{m-1}c|n-2| + \cdots + ma^{m-1}b|n-1| = ma^{m-1}T\ (b + \cdots + |n-1|)^2,
$$

one gets:

$$
R = (m+1)\,a^m|n| + ma^{m-1}T\ (b + \cdots + |n-1|)^2.
$$

A similar table can be constructed by taking the second terms of the expansions of coefficients 5.10, so that:

$$
\begin{array}{cccc}
\frac{m(m-1)}{1.2}a^{m-2}T(b+\cdots+|n-1|)^2 & \frac{m(m-1)}{1.2}a^{m-2}T(b+\cdots+|n-2|)^2 & \cdots & \frac{m(m-1)}{1.2}a^{m-2}T(b)^2 \\
a & b & \cdots & |n-2|
\end{array}
$$

By applying the method of substitution, one obtains the sum:

$$
\begin{aligned}
S =\ & \frac{m\,(m-1)}{1.2}a^{m-1}T\ (b+\cdots+|n-1|)^2 \\
& + \frac{m\,(m-1)}{1.2}a^{m-2}bT\ (b+\cdots+|n-2|)^2 + \cdots \\
& + \frac{m\,(m-1)}{1.2}a^{m-2}|n-2|T\ (b)^2.
\end{aligned}
$$

But since:

$$
bT\ (b+\cdots+|n-2|)^2 + \cdots + |n-2|T\ (b)^2 = T\ (b+\cdots+|n-2|)^3,
$$

one gets:

$$
S = \frac{m\,(m-1)}{1.2}a^{m-1}T\ (b+\cdots+|n-1|)^2 + \frac{m\,(m-1)}{1.2}a^{m-2}T\ (b+\cdots+|n-2|)^3.
$$

By adding R and S, one obtains:

$$R + S = (m + 1) a^m |n| + \left(m + \frac{m\,(m-1)}{1.2} \right) a^{m-1} T\, (b + \cdots + |n - 1|)^2$$

$$+ \frac{m\,(m-1)}{1.2} a^{m-2} T\, (b + \ldots + |n - 2|)^3 .$$

In general, given:

$$P = \frac{m\,(m-1)\,(m-2)\ldots(m-(h-1))}{1.2.3\ldots m} ,$$

Tetens constructs the table:

$Pa^{m-h} T\,(b + \cdots + |n - (h-1)\,|)^h$	$Pa^{m-h} T\,(b + \cdots + |n - h|)^h$	$\ldots$	$Pa^{m-h} T\,(b)^h$
a	b	$\ldots$	$|n - 2|$

in order to apply the method of substitution, from which he deduces the following expression:

$$Pa^{m-h+1} T\,(b + \cdots + |n - (h-1)\,|)^h + Pa^{m-h} bT\,(b + \cdots + |n - h|)^h + \cdots$$

$$+ Pa^{m-h} T\,(b)^h ,$$

which can be rewritten as:

$$Pa^{m-h+1} T\,(b + \cdots + |n - (h-1)\,|)^h + Pa^{m-h} T\,(b + \cdots + |n - h - 1|)^{h+1} ,$$

by noting that:

$$Pa^{m-h} T\,(b + \cdots + |n - h - 1|)^{h+1} =$$

$$Pa^{m-h} bT\,(b + \cdots + |n - h|)^h + \cdots + Pa^{m-h} |n - 2| T\,(b)^h .$$

This concludes the proof by induction.

To generalize formula 5.9 to any fractional number m, Tetens uses the basic technique of grouping terms in the multinomial in order to obtain a binomial. Since the binomial theorem for rational exponents has been justified by other mathematicians, formula 5.9 for a rational number m can be justified by this theorem and by the validity of this formula for a positive integer. Thus, the method of substitution, as Hindenburg's combinatorial method for proving the multinomial theorem, can only be applied to positive integers. The method of substitution is limited to determine the literal coefficients and the multinomial coefficients of the power series expansion, just as it was the case of

Hindenburg's combinatorial method of proof, and the key idea of the generalization consists in noting that the exponent m only appears in the binomial coefficients, as pointed out by Hindenburg in his original proof of the multinomial theorem for a rational exponent.

Hindenburg was right when he described Tetens's formula and proof as an emulation of his, but Tetens acknowledged an influence of Hindenburg on his work. However, Tetens disagreed with Hindenburg on the crucial question of the importance of combinatorial methods for the development of mathematical analysis (Tetens 1796, 1–3):

> Mr. Hindenburg has taught us how to find a general [coefficient] independently of its order by means of the method of combinations. But then the general formula for calculating the coefficients does not allow to obtain them analytically, that is to say, it does not allow to obtain them in such a way as to exclusively use known substitutions and usual analytic operations in calculations. [...] The combinatorial method must be therefore known in order to use that formula for determining the desired coefficients. It is true that the method is quite simple, once its principles and operations have been established. Thus, it could be included in analytic methods in the same sense that differentiation and integration were. These are recognized as useful methods regarding other analytic problems. However, their application requires a specific and very different work with respect to other analytic operations, hence it should be avoided if possible. So that is why I thought it might be worth a try to find another purely analytic formula for the powers [of a multinomial], and that it would even be an extension of analysis, so to speak, and thus the combinatorial method would no longer be necessary.[20]

For Tetens, Hindenburg's combinatorial method has the same controversial status as that of differential techniques in analysis. As a consequence, it could be tolerated by virtue of its practical utility, but it cannot be considered as a pure or fundamental method. However, pure analytic methods should be preferred when available, that is, methods based on elementary arithmetic procedures. This is why Tetens rewrites Hindenburg's formula and proof, some sixteenth years after its original publication. Hindenburg's remarkable formula should be deprived of the spurious combinatorial method on which it depended. Nevertheless, Tetens does not seem to be well versed in the combinatorial analysis, his only quoted source being the 1779 *Infinitinomii* of Hindenburg. No reference is made

---

[20] Herr Hindenburg hat solche ganz allgemein außer ihrer Folge zu finden gelehrt, mittelst der Combinationsmethode. Aber da giebt alsdann die allgemeine Formel für die Coefficienten diese nicht analytisch an, nehmlich nicht so, daß nichts mehr als bekannte Substitutionen und die gewöhnlichen analytischen Operationen erfordert würden, um sie zu erhalten. [...] Es muß also die Combinationsmethode dem bekannt seyn, der nach einer solchen Formel die Coefficienten herausbringen will. Nun ist diese Methode freylich auf ziemlich einfache. Grundsätze und Operationen gebracht. Man könnte sie daher eben sowohl unter die anatytischen Methoden aufnehmen, als das Differentiiren und Integriren. Ihre Brauchbarkeit bey verschiedenen andern analytischen Problemen ist auch anerkannt. Aber dennoch erfordert sie besondere, von den übrigen analytischen Operationen ganz verschiedene Arbeiten, denen man lieber entgeht, wenn sich ihnen entgehen läßt. Ich habe daher geglaubt, es verlohne sich der Mühe, und es sey gewissermaßen noch eine Erweiterung der Analysis, eine andere blos analytische Formel für die Potenzen zu suchen, wobei man die Combinationsmethode nicht nöthig habe.

in his works to the method of involutions or to any article of other members of the combinatorial school. No reference is made to Hindenburg's claim according to which the theory of combinations is a pure and self-founded mathematical science, even more fundamental than arithmetic. In sum, Tetens's article seems to belong to a bygone period, and matches the outdated conception presented in the *Infinitinomii* and the *Novi systematis*, when Hindenburg himself believed that the theory of combinations rested on arithmetic. On the contrary, if Tetens's article is placed in its chronological context, after the breaking point marked by Hindenburg's rereading of Moivre's work in 1795, it runs counter to the trend in the combinatorial school toward the development of a more abstract theory of mathematical structures based on the study of combinatorial involutions.

Despite this irreconcilable opposition between Tetens's and Hindenburg's points of view, it is not uncommon to find historical writings in which Tetens is presented as a member of the combinatorial school, especially during the twentieth century. For instance, in Pradier (2003), Tetens's arithmetical approach to the study of polynomials is considered to be the result of the combinatorial school's influence on Tetens's work, which belongs then, at least in this respect, to this current of thought. This opinion is untenable for various reasons, particularly because Tetens's arithmetical approach implies the cancellation of the combinatorial school's project. In contrast, the contemporaries of Tetens had a more accurate evaluation of his article. In Book review (1796b, 1797), one can find some reviews that draw attention to the fact that the adoption of the method of substitution involves the elimination of the combinatorial method, and so Tetens's and Hindenburg's approaches exclude each other, as can be seen in this English translation of the second review originally appeared in the *Jenaische Intelligenzblatt der Allgemeinen Literatur-Zeitung* (Book review 1797): "Mr. Tetens appears here as it's antagonist [of Hindenburg's method], but evidently without being sufficiently acquainted with it." What is important here is not so much to "demonstrate" that Tetens did not belong to the combinatorial school, but to understand that, despite some superficial coincidences of Tetens's thought with the early ideas of Hindenburg, his method of substitution is an example of the conceptions that raised under the influence of the combinatorial school's works, and thus seem to be linked with them, but they represent in fact a theoretical alternative that presupposes the removal of the combinatorial analysis from mathematics.

5.2.2   The Exponent Calculus

Another important contribution to changing the fate of the German combinatorial analysis appeared in Hindenburg's scientific journal *Archiv der reinen und angewandten Mathematik* in 1798 (Pasquich 1798a). It was authored by Johann Pasquich (1754–1829), an Austro-Hungarian mathematician born in Senj. Unlike Tetens's, Pasquich's article does not make a direct attack on Hindenburg's combinatorial methods, but it is far from being a contribution to the combinatorial analysis. His study was conducted in the context of an open debate with his fellow mathematicians about differential calculus and its questionable concepts. While differential calculus continues to grow and gradually gains more and more weight as a significant field of mathematics, some attempts are made to clarify its theoretical bases or even to prevent its use by proposing new mathematical techniques. Pasquich's article falls within the second category. It contains a new algebraic method to obtain the results of differential calculus, and the theory behind this method was called "exponent calculus"[21] (*Exponentialrechnung*) by Pasquich. It should be noted that, in a display of modesty, Pasquich claims that his new calculus might be considered unnecessary, given the actual calculation power of differential calculus. From a theoretical point of view, however, it is clear that, despite the practical advantages of differential calculus, Pasquich is worried about the lack of conceptual justification for differential calculus. A brief overview of his exponent calculus was presented in a book published the same year, in which Pasquich advises his readers to consult his more complete presentation of the subject in Hindenburg's *Archiv* (Pasquich 1798c, 42–49). The following discussion is based on the more complete version of exponent calculus presented in Pasquich's paper.

This new calculus rests on the explicit assumption that any function y of x can be expressed as a power series of the form $y = Ax^a + Bx^b + Cx^c + \cdots$, where a, b, c, etc. are rational numbers (Pasquich 1798a, 386), and its central idea consists in emulating basic derivative rules by means of elementary algebraic manipulations. More precisely, Pasquich defines the algebraic operation of "exponentiation" for any function y of x, which is symbolized by the exponentiation operator ϵ, as the result of multiplying each term in the power series expansion of y by its respective exponent, so that the "exponent function" ϵy is given by:

$$\epsilon y = aAx^a + bBx^b + cCx^c + \cdots .$$

As a consequence of this definition and of Pasquich's general assumption, it follows that the exponent function of any constant function $y = C$ is null:

$$\epsilon y = \epsilon(Cx^0) = 0,$$

and that the exponent function of a sum $U + V + \cdots + Z$, where $U, V, \ldots, Z$ are functions of x, is given by:

$$\epsilon(U + V + \cdots + Z) = \epsilon U + \epsilon V + \cdots + \epsilon Z.$$

---

[21] The choice of translating "*Exponential*" as "exponent" better captures the idea of Pasquich's method, as will be seen below, and avoids confusion with the unrelated notions of "exponential function" and "exponentiation". For the same reason, the term "exponentiation", instead of "exponentiation", will be employed below.

Then he shows that the exponentation of a product UV obeys the rule:

$$\epsilon(UV) = U\epsilon V + V\epsilon U.$$

A simple induction based on this product rule shows that:

$$\epsilon(Z^m) = mZ^{m-1}\epsilon Z,$$

for any natural number m. This power rule can be generalized to a rational number $n = \frac{p}{q}$ as follows. Let $y = Z^n = Z^{\frac{p}{q}}$ be a function of x. Then $y^v = Z^u$, and by applying the power rule for natural numbers, one gets $vy^{v-1}\epsilon y = uZ^{u-1}\epsilon Z$. The power rule for (positive) rational exponents:

$$\epsilon(Z^n) = nZ^{n-1}\epsilon Z$$

follows immediately from the last identity. Pasquich shows that the power rule is also valid for negative exponents. The quotient rule for exponentation can be derived from the product rule. If $y = \frac{U}{V}$ is a function of x, one gets the identity $Vy = U$. The application of the product rule to this identity and a simple algebraic manipulation yields the quotient rule:

$$\epsilon\frac{U}{V} = \frac{V\epsilon U - U\epsilon V}{V^2}.$$

These are the five elementary rules of exponent calculus.

Pasquich does not only stress how the simple operation of exponentation might pave the way for an algebraic calculus that includes algebraic rules similar to the fundamental derivative rules. In talking of the theoretical structure of exponent calculus, he attaches great importance to the conceptual integrity and the practical utility of his theory. Its conceptual integrity is supported by well-established concepts and principles of common algebra. Its practical utility depends on whether exponent calculus is about as rich as differential calculus. In this order of ideas, Pasquich also defines the notion of an exponent function of order n, drawing a clear parallel with the notion of a derivative function of order n. Given a function y of x, its exponent function of order $n+1$, for $n \geq 1$, is defined recursively by:

$$\epsilon^{n+1}y = \epsilon\frac{\epsilon^n y}{x}.$$

As a final important conceptual element of his calculus, Pasquich enunciates what can be seen as a chain rule for exponentation (Pasquich 1798a, 395–396). Although he does not write this rule in mathematical symbols and just illustrates this idea by two concrete examples, it is clear that, according to Pasquich, the exponentation of a function V

composed with a function U can be expressed as the product of the exponent function of U by the exponent function of V composed with U.

This theoretical framework seems to be particularly apt for the study of questions usually addressed by differential calculus. Pasquich's bet is that the powerful and undeniable practical advantages of differential calculus in contrast to finite analysis can be duplicated by his algebraic theory. To dispel any doubts, the first application of exponent calculus concerns one of the most robust tools available in the eighteenth century for developing mathematical analysis. Pasquich proposes to calculate the coefficients of the power series expansion of $(a + b)^m$, where m is a rational number, by using the operation of exponentation exclusively, and thus the binomial theorem for rational exponents and its consequences would belong to finite analysis. In fact, Pasquich does not precise the nature of the exponent m, but since his calculations depend on the power rule, m is certainly a rational number. He first reduces the problem to finding the coefficients of:

$$y = (1 + x)^m$$
$$= 1 + Ax + Bx^2 + Cx^3 + \cdots + Px^r + Qx^{r+1} + \cdots,$$

given that $(a + b)^m = a^m(1 + x)^m$ for $x = \frac{b}{a}$, and where A, B, C, etc. are undetermined coefficients. By applying the power rule, he obtains the exponent function of y:

$$\epsilon y = mx(1 + x)^{m-1} = \frac{mx(1 + x)^m}{1 + x} = \frac{mxy}{1 + x},$$

and then:

$$(1 + x)\epsilon y = mxy.$$

On the other hand, the definition of exponentation gives the following alternative expression of the exponent function of y:

$$\epsilon y = Ax + 2Bx^2 + 3Cx^3 + \cdots + rPx^r + (r + 1)Qx^{r+1} + \cdots,$$

and so:

$$mxy = (1 + x)(Ax + 2Bx^2 + 3Cx^3 + \cdots + rPx^r + (r + 1)Qx^{r+1} + \cdots),$$

Then Pasquich says that the binomial coefficients can be easily determined from this last equation by the "usual method". It is impossible to know what he meant by that exactly, but it is highly likely that he meant the method of undetermined coefficients. Indeed, by applying this method to the last equation, one gets the binomial coefficients:

$$A = m, \qquad B = \frac{mA - A}{2} = \frac{m(m-1)}{2}, \qquad C = \frac{mB - 2B}{3} = \frac{m(m-1)(m-2)}{2 \cdot 3}, \cdots$$

Thus the binomial theorem for rational exponents belongs to finite analysis.

With the binomial theorem in hand, Pasquich embarks on a brief but relevant exhibition of what can be done in mathematical analysis by using his exponent calculus. On the basis of the binomial theorem for rational exponents and the exponentiation operation, he calculates the power series expansions of the natural logarithmic function, of the exponential function, of the sine and cosine functions. Besides, he shows that the derivative of all these functions, and of the secant, cosecant, tangent, arcsine, arccosine, arctangent, arccotangent, arcsecant and arcosecant functions, can be formulated in terms of exponentations. In sum, this selected group of examples serves as a rather strong argument for confirming that exponent calculus provides a valuable alternative for developing the theory of series and the theory of transcendental functions.

Not devoid of interest is the fact that Pasquich's article appeared in Hindenburg's *Archiv*. As seen in Chap. 4, this scientific journal became one of the most important means of dissemination for the combinatorial school's ideas; and the reconstruction of the theory of series and the study of transcendental functions without using differential calculus was one of the main objectives of this school. Thus, Pasquich's article seems very close to, at least, the combinatorial school's general conceptual approach to mathematical analysis. All this could give the false impression that Pasquich was somehow related to the combinatorial school. For instance, Phili describes Pasquich as being "skilled at manipulating infinite series because of his 'learning' in Hindenburg's school" (Phili 1990, 108). Perhaps Phili is right in suggesting that Pasquich developed a substantial interest in the theory of series under the intellectual sway of the combinatorial school. However, care must be taken not to jump to the conclusion that this plausible influence reveals a sort of collaborative work with the combinatorial school. While the main target of Pasquich's theory was differential calculus, the first paragraph of his paper makes it clear that, because of its conceptual simplicity, its theoretical rigor, and the universality of its foundations, exponent calculus deserves more praise than many other methods of calculation intended to achieve differential calculus's results (Pasquich 1798a, 385). Notable among these many other methods is the combinatorial analysis, which is focused on the theory of series and whose programmatic agenda also includes the study of transcendental functions. In addition, as can be seen above, exponent calculus excludes the use of mathematical techniques other than those based on pure algebraic principles, including combinatorial methods. In this sense, Pasquich's article points in the same direction as that of Tetens analyzed in Sect. 5.2.1, but it does so in a more radical way, for it is not limited to the confines of a single theorem: on the contrary, it provides a theoretical framework that removes the entire combinatorial analysis from mathematical analysis itself.

5.2.3   The Combinatorial Characteristic

Not much is known about the life of Heinrich Bürmann. He died in the summer of 1817 in Mannheim, where he worked as a business teacher, but no record of his place and date of birth survives. One can tell from his writings that he exercised both his scientific and literary talents. As an example of his poetic work, one can consult for instance (Bürmann 1807a). In fact, as will be seen below, he thought that literary creation follows scientific combinatorial laws. According to his claims that appear at various places in his texts, he was self-taught and had no formal training in mathematics. By the end of the eighteenth century, his work on differential and integral calculus led him to the discovery of the combinatorial analysis, and he imagined that there was a powerful connection among these apparently unrelated disciplines. This connection and his ambition of entirely renewing algebraic notations awakened in him the dream of an unrealistic scientific project. In regard to formal sciences, this project was referred to as "combinatorial mathematics", but in its broadest sense concerning all human knowledge it was referred to as "the combinatorial characteristic" or "the syntactic characteristic". In what follows, it will be analyzed the impact that Bürmann's project had on the combinatorial analysis.

5.2.3.1 The Analytic Tachygraphy

Some particular results of Bürmann's general project were set down in a manuscript entitled *Essai de calcul fonctionnaire aux constantes ad libitum* (Bürmann 1798b). In this section, its content will be reviewed in some detail, emphasizing three formulae that were considered as fundamental in analysis by Bürmann. The aim of the next Sect. 5.2.3.2 is to present a historical reconstruction of the manuscript in order to get a better understanding of it and its relation with the combinatorial analysis. Section 5.2.3.3 deals with Bürmann's general project of the combinatorial characteristic.

One main concern of the manuscript is to develop new algebraic notations, since Bürmann considered that mathematics had indiscriminately adopted all symbols used by mathematicians in their works, which yielded an ambiguous algebraic notation. Specifically, he identified three major deficiencies concerning mathematical notations:

1. Ambiguity: a given symbol has different meanings in different texts, or even in a same text.
2. Notational diversity: there exists no unified notation system that could be put at the service of all mathematicians. On the contrary, each time a mathematical paper is written, it is necessary to explain notations in words.
3. Mathematical prolixity: this deficiency concerns series particularly, and is related to the practice of expressing a series by writing its first terms given that there is no conventional way to express it by means of a one single concise formula.

The analytic tachygraphy aims to supply these deficiencies by developing a unique symbolic system, which should be used to rebuild algebra and, as a consequence, to improve mathematical analysis. For Bürmann, algebra is indeed only a coherent system of symbols intended to be used in scientific research, and thus algebra is primarily the mathematical language that should be employed in analysis (Bürmann 1798b, 10, footnote). The more precise algebra or mathematical language is, the more powerful mathematical analysis becomes.

Nevertheless, in his *Essai*, Bürmann remains prudent in terms of his notational reform of mathematics, stressing the fact that, as far as languages are concerned, changes should affect as few components as possible. His innovations were indeed still quite conservative in relation to his final notation system, as will be seen later, and are mostly restricted to the introduction of indexes for variables, functions, and mathematical operators.

Functions are denoted by Greek letters, for example φ and ψ, and indexes placed over the letter $\overset{1}{\varphi}$, $\overset{2}{\varphi}$, $\overset{3}{\varphi}$, ... can be used to denote as many functions as needed. Function composition is represented by simple juxtaposition: $\varphi\psi$ means φ composed with ψ. But the iteration of a function is denoted by an exponent: $\varphi^1 = \varphi$, $\varphi^2 = \varphi\varphi$, $\varphi^3 = \varphi\varphi\varphi$ and so on. The corresponding inverse functions are φ^{-1}, φ^{-2}, φ^{-3}, and so on. Some usual functions are denoted by particular symbols. For instance, sin, tang, Asin, and Atang correspond to the sine, the tangent, and their inverse functions. The natural logarithm is represented by $\mathcal{L}$, and the exponential by $\mathcal{N}$.

There are three pairs of operators for representing the basic operations in analysis and their opposites: sum $\sum$ and finite difference Δ, integration $\int$ and differentiation ∂, product $\prod$ and quotient $\underset{\curlywedge}{\Delta}$. Except for $\underset{\curlywedge}{\Delta}$, which is a creation of Bürmann, all these symbols existed already in the literature; in particular, Bürmann probably took the sum operator $\sum$ from Lagrange's works. For Bürmann, common mathematical practice should be modified in order to clarify its concepts. The fist step in this conceptual clarification has to be to create an appropriate terminology. Thus, finite differences should be clearly distinguished from infinitesimal differences by establishing two different disciplines: a discipline dealing with finite differences exclusively, and differential calculus. The operator $\underset{\curlywedge}{\Delta}$ was invented by Bürmann to maintain equilibrium among operations, so that, for each operation, there is an opposite operation. Here is an example of how this symbol should be interpreted:

$$\underset{\curlywedge}{\Delta} = \frac{(z + \Delta z)}{\varphi z}$$

Operators can be accompanied by indexes, but Bürmann's treatment of indexes can get confusing sometimes, partly due to his ambiguous use of operators: these are sometimes used as operators, sometimes as the operations themselves. In general, operators have only one index which indicates the final value, while the initial value should be guessed from context. For instance, given a positive integer n, its factorial is represented as:

$$\prod_n (n + \mathrm{i}) = 1 \cdot 2 \cdot 3 \cdots n,$$

where $\dot{\text{i}}$ indicates that the index increases by one unit each time from 1 to n. Symbols ∞, $\overset{0}{\infty}$, $\overset{1}{\infty}$, ... are general indexes that take the successive values 0, 1, 2, 3, ... and can be used in expressions such as:

$$\sum \overset{\infty}{Z} y^{\infty} = \overset{0}{Z} + \overset{1}{Z} y + \overset{2}{Z} y^2 + \cdots + \overset{n}{Z} y^n + \cdots,$$

where $\overset{n}{Z}$ is the nth coefficient of this series. The following examples show the power series expansions of four well-known function written with Bürmann's symbolism:

$$\mathcal{L}(x+1) = \sum \frac{z\,(-z)^{\infty}}{\infty+1},$$

$$\mathcal{N}z = \sum \frac{z^{\infty}}{\prod(\infty+1)},$$

$$\sin z = \sum \frac{z\,(-z^2)^{\infty}}{\prod(2\infty+\dot{\text{i}})},$$

$$\cos z = \sum \frac{(-z^2)^{\infty}}{\prod(2\infty+\dot{2})}.$$

The right-hand side of these identities is what Bürmann calls a tachygraphic analytic expression.

It is worth mentioning that Bürmann also invented a new notation for continued fractions. Instead of giving a general pattern, he offers some examples. For instance, the following is the expansion of the tangent function in continued fraction:

$$\text{tang } x = \cfrac{x}{1-\cfrac{x^2}{3-\cfrac{x^2}{5-\frac{x^2}{7-\cdots}}}} = \cfrac{x}{1-}\,\cfrac{x^2}{3-}\,\cfrac{x^2}{5-}\,\cfrac{x^2}{7-\cdots}$$

which is close to some recent notations, as that of Pringsheim.

On the other hand, it is rather difficult to clarify the meaning of the notion "constants ad libitum", which was considered to be one of the main contributions of the manuscript to mathematical notations. Although a full section of Bürmann's *Essai* is devoted to this notion, there is no definition anywhere in the text as to what that term means. Bürmann simply translates it as "arbitrary constant", but gives no explicit characterization of what kind of object an arbitrary constant can be. In section 3 of his manuscript, when he uses the arbitrary constant a in one of the formulae transcribed below, he just says that "a can be anything we want it to be, even a function of z; that is what I call *CONSTANT*

AD LIBITUM."[22] Clearly, this lack of precision does not help us to better understand this notion. Indeed, a concern for concise writing in mathematics motivated the introduction of Bürmann's constants *ad libitum*, rather than a concern for developing new mathematical concepts. Bürmann planned to write a mathematical encyclopedia containing all theorems and results of this science, and he believed that his notation reform would allow him to summarize all (present and future) mathematical knowledge in about ten volumes. Constants *ad libitum* were part of the solution for achieving this goal: one single symbol could replace long mathematical expressions. However, Bürmann does not seem to have taken into account how these arbitrary substitutions could affect readability and legitimacy of mathematical formulae.

These notations are used to formulate three main theorems in the manuscript in terms of the analytic tachygraphy. They have been transcribed here with their respective names, and will be discussed in the next section.

Theorem of expansion. Given $\varphi z = \varphi \psi^{-1} (\psi a + (\psi z - \psi a))$, the series expansion of φz in terms of ψz is given by:

$$\varphi z = \sum \frac{d^\infty \varphi a \, (\psi z - \psi a)^\infty}{d\psi a^\infty \cdot \prod (\infty + 1)} = \varphi a + \frac{d\varphi a}{d\psi a} \frac{(\psi z - \psi a)^1}{1} + \frac{d^2 \varphi a}{d\psi a^2} \frac{(\psi z - \psi a)^1}{1.2} + \&$$

(5.11)

where *a* is a constant *ad libitum*.

Theorem on reversion (or Lagrange formula of reversion). This is a corollary of the previous theorem, taking $\psi z = 0$:

$$\varphi z = \sum \frac{d^\infty \varphi a \, (-\psi a)^\infty}{d\psi a^\infty \cdot \prod (\infty + 1)} = \varphi a - \frac{d\varphi a}{d\psi a} \frac{\psi a}{1} + \frac{d^2 \varphi a}{d\psi a^2} \frac{\psi a^2}{1.2} - \&, \qquad (5.12)$$

where *a* is a constant *ad libitum*.

Theorem of integration:

$$\sum_x{}^n X \Delta x^n = \left(n \sum_x{}^n \Delta x^n \right) \left(\sum \frac{\prod (-x^{\infty+1}) \cdot \Delta^\infty X}{\prod (\infty + 1) \cdot (n + \infty) \Delta x^\infty} \right), \qquad (5.13)$$

which can be expressed in common algebra, says Bürmann, as:

$$\sum{}^n X \Delta x^n = n \sum{}^n \Delta x^n \left[\frac{X}{n} - \frac{x + \Delta x}{1(n+1)} \frac{\Delta X}{\Delta x} + \frac{(x + \Delta x)(x + 2\Delta x)}{1.2(n+2)} \frac{\Delta^2 X}{\Delta x^2} - \frac{(x + \Delta x)(x + 2\Delta x)(x + 3\Delta x)}{1.2.3(n+3)} \frac{\Delta^3 X}{\Delta x^3} + \& \right]$$

---

[22] *a* est tout ce qu'on veut, même fonction de *z* ; c'est ce que je nomme *CONSTANTE AD LIBITUM.*

where:

$$n\sum_{x}^{n}\Delta x^{n} = \frac{x\,(x-\Delta x)\,(x-2\Delta x)\ldots(x-n\Delta x+\Delta x)}{1.2.3\ldots(n-1)} \tag{5.14}$$

A recent review of this formula can be seen in Lubet (2010).

5.2.3.2 Ducit in Vitium Culpae Fuga

Bürmann's manuscript is dated February 18, 1798, and was written with the intention to be published in the *Mémoires* of the French Academy of Sciences. On the 8th of April, Bürmann addressed a letter to Jérôme de Lalande (1732–1807), director of the Observatory of Paris, asking for advice about how to submit his manuscript to the scrutiny of the Academy (Bürmann 1798f). Lalande formally notifies his colleagues of the Academy about Bürmann's intention on 20 April 1798, pointing out that he gave the necessary instructions to Bürmann for the submission of the manuscript, which was sent to the Institut de France on 23 May (Anonym 1798a, 377; Bürmann 1798e). Adrien-Marie Legendre (1752–1833) and Lagrange are appointed by the Academy on 9 June to deliver an opinion on the manuscript (Anonym 1798b, 403). A few days later, on 13 June, Bürmann addresses another letter to Lalande in which he gives some additional explanations of the formulae used in his *Essai* and adds some corrections, but more importantly he asks Lalande to eliminate section 5 of the manuscript, which was devoted to the main notion of "constant *ad libitum*" (Bürmann 1798g). Bürmann does not elaborate on his decision to delete this section, invoking his reading of Lagrange's *Théorie des fonctions analytiques* as the only motivation. However, it is impossible to identify precisely which aspects of Lagrange's treatise made him change his mind and led him to abandon that notion, supposed to be one of the main contributions of the manuscript. Bürmann speaks now simply of "constants" instead of "constants *ad libitum*". Yet, this unexpected turn of events does not discourage him from pursuing his plan of completely renewing algebra; he feels, on the contrary, more motivated to unveil a more realistic description of his plan, and promises to send a new document next autumn.

The promise is kept and Bürmann sends a new manuscript entitled *Supplément à l'Essai de calcul fonctionnaire aux ad-libitum* to the Academy (Bürmann 1798i). Unlike his *Essai*, the *Supplément* does not propose a conservative reform of algebraic notations, but a genuine renewal that implies the exclusion of practically every known mathematical symbol. This renewal is considered to be a new mathematical notation system called "analytic ideography" (*Idéographie analytique*). Bürmann claims that his analytic ideography provides the solution to scientific problems linked to the diversity of languages used to communicate scientific results, including confusions due to the diversity of mathematical symbols. The analytic ideography is conceived as a "characteristic system", that is, a symbolic language based on characteristics, which are simple, intelligible, universal symbols that can be understood by "any nation" since these characteristics would be explained in a brief dictionary, some only a few pages long (Bürmann 1798i, 1).

Concerning algebraic symbols, they should be replaced with combinations of lines and points arranged in a specific way. The following are the ideographic expressions proposed by Bürmann for representing some mathematical objects:

1. Symbols like:

$$\ulcorner\, , \quad \ulcorner. \, , \quad \ulcorner: \, , \quad \ulcorner\vdots \, , \quad \dots \, ,$$

   denote mathematical functions. The different number of points serves to signify that these are, in principle, different functions. For instance, functions φ and ψ can be represented as $\ulcorner$ and $\ulcorner.$ using the analytic ideography.
2. Iterations of a function are expressed as follows:

$$\ulcorner^{n} x = \ulcorner\ulcorner\ulcorner \dots \ulcorner x.$$

   One should be careful not to confuse the iteration of a function with the nth power of a function, which is:

$$\ulcorner x \urcorner^{n}.$$

3. Inverse functions are expressed as:

$$\urcorner\, , \quad .\urcorner\, , \quad :\urcorner\, , \quad \dots \, ,$$

   with respect to the functions in point (1).
4. The differential operator d becomes a left subscript added to the symbol of a function. Thus, one has:

$$_{1}\ulcorner x = d\varphi x, \quad _{2}\ulcorner x = d^{2}\varphi x, \quad \dots, \quad _{n}\ulcorner x = d^{n}\varphi x.$$

Although this renewal is quite radical, Arabic numerals have still been conserved in the notation system, perhaps so as not to excessively disconcert the French mathematicians of the Academy.

   Using these symbols, except for the differential operator, formulae 5.11, 5.12, and 5.13 given in the *Essai* become:

   Theorem of expansion:

$$\ulcorner.x = \ulcorner.v + \frac{d\ulcorner.v}{d\ulcorner v}\frac{\left(\ulcorner x - \ulcorner v\right)}{1} + \frac{d^{2}\ulcorner.v}{d\ulcorner v^{2}}\frac{\left(\ulcorner x - \ulcorner v\right)^{2}}{1.2} + \frac{d^{3}\ulcorner.v}{d\ulcorner v^{3}}\frac{\left(\ulcorner x - \ulcorner v\right)^{3}}{1.2.3} + \&, \qquad (5.15)$$

whose general coefficient, which does not appear in the *Essai*, is given by:

$$\frac{d^n\overline{.v}}{d\overline{v}^n} = \frac{d^{n-1}\left(\frac{d\overline{.v}}{d\overline{v}}\left(\frac{\bigwedge - v}{\overline{\bigwedge}-\overline{v}}\right)\right)}{dv^{n-1}},\tag{5.16}$$

where $\bigwedge$ is "the final value of v", according to Bürmann.

Theorem on reversion:

$$\overline{.x} = \overline{.v} - \frac{d\overline{.v}}{d\overline{v}}\frac{\overline{v}}{1} + \frac{d^2\overline{.v}}{d\overline{v}^2}\frac{\overline{v}^2}{1.2} - \frac{d^3\overline{.v}}{d\overline{v}^3}\frac{\overline{v}^3}{1.2.3} + \&.\tag{5.17}$$

Theorem of integration, which Bürmann deduces from formula 5.15 by multiplying by $\Delta\overline{x}^m$ and adding m times:

$$\sum^m\overline{.X}^m\Delta\overline{.x}^m = \overline{.v}^m\sum^m\Delta\overline{.x}^m + \frac{d\overline{.v}}{d\overline{v}}\frac{\sum^m\left(\overline{x} - \overline{v}\right)\Delta\overline{.x}^m}{1} + \&,\tag{5.18}$$

This formula replaces formula 5.13 given in the *Essai*.

In the meanwhile, Bürmann addresses a copy of the *Supplément* and of a part of the *Essai* to Hindenburg on 17 August and on 15 September 1798, according to two letters partially published in the *Archiv der reinen und angewandten Mathematik* without distinguishing the texts of each one (Bürmann 1798a). In these letters, Bürmann claims not to be very familiar with Hindenburg's theory, but he emphasizes that his analytic ideography could benefit from the combinatorial analysis. A similar remark was included in the *Essai*, where Bürmann describes Hindenburg's combinatorial analysis as a very clever notational system (Bürmann 1798b, Préface). It seems as if Bürmann wanted to convince Hindenburg of the relationship between both theories. In any case, an excerpt from the *Supplément* appears in German in the same issue of the *Archiv* (Bürmann 1798j). Since articles are normally published in their original language in the *Archiv*, it is possible that Bürmann wrote an original German version of his texts, now lost. In a brief introductory comment to the excerpt of the *Supplément*, Hindenburg lets it be known that he will discuss the details of Bürmann's works in the second volume of the *Sammlung*, but no commitment is made to publish them integrally. We will return attention to Hindenburg's remarks a bit later, but now let's see what is happening in the French Academy of Sciences.

Toward the end of August, Bürmann was somewhat worried at not having received an answer concerning his manuscripts. Then he wrote to the Academy again, this time to Legendre, in order to get information about the views of the academic commission on his work (Bürmann 1798h). On 23 August, Legendre signs the letter in which he informs

Bürmann of the rather unfavorable opinion of both manuscripts (Legendre 1798a). He points out that the delay has been basically caused by the new notations used by Bürmann, which hinders the understanding of the texts. Furthermore, Legendre severely questions the viability of Bürmann's general project. On the one hand, he affirms that, even though notational improvements in algebra are possible, some mathematical works, as those of Euler and Lagrange, have shown that the current algebraic symbolism is accurate, simple, and elegant, hence a radical renewing as that proposed by Bürmann is not needed. On the other hand, Legendre thinks that the analytic tachygraphy is in fact impracticable, since there are mathematical series that cannot be expressed by means of a concise formula, either because no such a formula is known, or because the series is too complicated to be reduced to a formula without sacrificing conceptual clarity. Regarding Bürmann's theorems, only formulae 5.13 and 5.16 are selected as deserving further examination and careful study, the rest of the mathematical results presented by Bürmann being nothing more, according to Legendre, than obvious corollaries deduced from Taylor's theorem.

In the face of Legendre's severe criticism, Bürmann seems to withdraw his notation reforms of algebra and recant his deviation of the standard mathematical practice, as can be seen in a letter dated on 11 October 1798 and addressed to the French Academy (Bürmann 1798d):

> This spring I had the honor of submitting my *Essai de Calcul Fonctionnaire* for your consideration, but I made the big mistake of deviating from standard notation. *Ducit in vitium culpae fuga.*[23]

Horace's Latin verse makes this situation dramatically clear. Bürmann seems intent on rejecting his own project, adopting a self-critical attitude. He even declares that both manuscripts are completely worthless, and they should be replaced with a third manuscript entitled *Formules de développement, de retour et d'intégration* (Bürmann 1798c). It is this third manuscript that was finally reviewed by Lagrange and Legendre in order to deliver an opinion to the Academy. Lagrange's and Legendre's review was read at the meeting held on 26 December 1798 (Anonym 1798c). In accordance with the positive review, the members of the Academy decided to publish Bürmann's third manuscript in the collection of foreign scholars and the review in the *Mémoires* of the Academy. Lagrange's and Legendre's review was effectively published the next year, but the publication of Bürmann's manuscript was delayed (Lagrange and Legendre 1799). In 1801, a brief notice confirms that the manuscript will be printed under the auspices of the Institut de France, but it remains unpublished to this day (Anonym 1801).

In *Formules de développement, de retour et d'intégration*, there is no trace left of Bürmann's exuberant project plans, whether under the name of analytic tachygraphy or of analytic ideography. The manuscript is written in a terse and sober style, excluding

---

[23] En vous faisant, le printems dernier, l'hommage de mon *Essai de Calcul Fonctionnaire*, j'ai eu le grand tort de m'écarter de la notation reçue. *Ducit in vitum culpae fuga.*

all personal reveries about the future form of mathematical knowledge and even avoiding using a single concise formula to express a series instead of writing its first terms. The content of Bürmann's manuscript is articulated around the following problem. Suppose that X and ξ are functions of the same variable x and independent of each other. The problem is to express X as a power series expansion of ξ by supposing that one is given the relation:

$$X = T + \overset{1}{T}\frac{\xi}{1} + \overset{2}{T}\frac{\xi^2}{1.2} + \cdots + \overset{n}{T}\frac{\xi^n}{1.2.3\ldots n} + \&. \tag{5.19}$$

Thus, the question is to determine the general coefficient $\overset{n}{T}$, which becomes of the form, as Bürmann points out:

$$\overset{n}{T} = \frac{d^n X}{d\xi^n},$$

by differentiating n times Eq. 5.19 and by taking $\xi = 0$ after differentiation.

To determine the value of this coefficient, Bürmann first establishes two basic lemmas. The first one states that:

$$\frac{d^n \xi^r \left(\frac{z}{\xi}\right)^n}{dz^n} = n(n-1)(n-2)\cdots(n-r+1)\frac{d^{n-r}\left(\frac{z}{\xi}\right)^{n-r}}{dz^{n-r}}, \tag{5.20}$$

where z is another function of x independent of X and ξ, and where one takes $z = 0$ after differentiation. The second lemma states that:

$$\frac{d^{n-1}\xi^r d\left(\frac{z}{\xi}\right)^n}{dz^n} = n(n-1)(n-2)\cdots(n-r+1)\frac{d^{n-r}\left(\frac{z}{\xi}\right)^{n-r}}{dz^{n-r}}, \tag{5.21}$$

for $r < n$, as remarked by Legendre in his analysis of Bürmann's manuscript (Legendre 1798b).

By applying Eq. 5.19 to the product $\frac{d^n X\left(\frac{z}{\xi}\right)^n}{dz^n}$, Bürmann gets the following expansion:

$$\frac{d^n X\left(\frac{z}{\xi}\right)^n}{dz^n} = T\frac{d^n\left(\frac{z}{\xi}\right)^n}{dz^n} + \frac{\overset{1}{T}}{1}\frac{d^n \xi\left(\frac{z}{\xi}\right)^n}{dz^n} + \frac{\overset{2}{T}}{1.2}\frac{d^n \xi^2\left(\frac{z}{\xi}\right)^n}{dz^n} + \&,$$

from which he obtains by formula 5.20:

$$\frac{d^n X \left(\frac{z}{\xi}\right)^n}{dz^n} = T \frac{d^n \left(\frac{z}{\xi}\right)^n}{dz^n} + \frac{n}{1} \frac{1}{T} \frac{d^{n-1} \xi \left(\frac{z}{\xi}\right)^{n-1}}{dz^{n-1}}$$

$$+ \frac{n(n-1)}{1.2} \frac{2}{T} \frac{d^{n-2} \xi^2 \left(\frac{z}{\xi}\right)^{n-2}}{dz^{n-2}} + \cdots + \frac{n}{1} \frac{n-1}{T} \frac{d \left(\frac{z}{\xi}\right)^{n-1}}{dz} + \frac{n}{T}.$$

In the same way, by using formulae 5.19 and 5.21, he obtains the expansion:

$$\frac{d^{n-1} X d \left(\frac{z}{\xi}\right)^n}{dz^n} = T \frac{d^n \left(\frac{z}{\xi}\right)^n}{dz^n} + \frac{n}{1} \frac{1}{T} \frac{d^{n-1} \left(\frac{z}{\xi}\right)^{n-1}}{dz^{n-1}}$$

$$+ \frac{n(n-1)}{1.2} \frac{2}{T} \frac{d^{n-2} \left(\frac{z}{\xi}\right)^{n-2}}{dz^{n-2}} + \cdots + \frac{n}{1} \frac{n-1}{T} \frac{d \left(\frac{z}{\xi}\right)}{dz}.$$

From these two expansions, it follows that:

$$\frac{n}{T} = \frac{d^n X \left(\frac{z}{\xi}\right)^n}{dz^n} - \frac{d^{n-1} X d \left(\frac{z}{\xi}\right)^n}{dz^n} = \frac{d^n X}{d\xi^n},$$

but given that:

$$d^n X \left(\frac{z}{\xi}\right)^n = d^{n-1} \left(dX \left(\frac{z}{\xi}\right)^n\right) + d^{n-1} X d \left(\frac{z}{\xi}\right)^n,$$

one obtains finally:

$$\frac{n}{T} = \frac{d^n X}{d\xi^n} = \frac{d^{n-1} \left(\frac{dX}{dz} \left(\frac{z}{\xi}\right)^n\right)}{dz^{n-1}}, \tag{5.22}$$

where one takes $z = \xi = 0$ after differentiation. Formula 5.22 is a generalization of formula 5.16, which was presented in Bürmann's *Supplément* without proof. Bürmann goes to the trouble of demonstrating it now because Legendre had deplored the fact that it was presented without proof in the *Supplément* (Legendre 1798a).

By setting $X = \varphi x$ and $\xi = \psi x - \psi t$, Bürmann derives the following equality from Eq. 5.19:

$$\varphi x = \varphi t + T \frac{1}{1} \frac{(\psi x - \psi t)^1}{1} + T \frac{2}{1.2} \frac{(\psi x - \psi t)^2}{1.2} + \&,$$

where the coefficient $\overset{n}{T}$ is given by formula 5.22. This equality correspond to formula 5.11 of the *Essai* and to formula 5.15 of the *Supplément*.

Formulae 5.12 and 5.17 for the reversion of series appear now in the form of:

$$\varphi x = \varphi t + \overset{1}{T}\frac{(\psi t)^1}{1} + \overset{2}{T}\frac{(\psi t)^2}{1.2} + \&,$$

where the general coefficient $\overset{n}{T}$ is given by formula 5.22. Besides, Bürmann remarks in his *Formules de développement, de retour et d'intégration* that Lagrange's inversion theorem, which can be consulted at the end of our Sect. 4.1.2.1, can be seen as a particular case of this formula for reversing a series. Considering, as Lagrange, an initial equation $x = t + u\psi x$, and hence $u = \frac{x-t}{\psi x}$, Bürmann writes the main formula given by Lagrange's inversion theorem as:

$$\varphi x = \varphi t + \overset{1}{T}\frac{u}{1} + \overset{2}{T}\frac{u^2}{1.2} + \&,$$

where, according to formula 5.22, the general coefficient is:

$$\overset{n}{T} = \frac{d^{n-1}\left(\frac{d\varphi t}{dt}\psi t^n\right)}{dt^{n-1}}.$$

At the end of his manuscript, Bürmann also translates his formulae 5.14 and 5.18 of integration into standard algebraic notations. Despite his observation that formula 5.18 was the best expression of his ideas, Lagrange and Legendre never paid attention to this formula and only took into account formula 5.14, which was included without proof in their review of Bürmann manuscript, since this formula was not proved in *Formules de développement, de retour et d'intégration*. However, in the *Essai*, Bürmann offers a proof by induction. In France, Bürmann's formula of integration will be disseminated through Lagrange's and Legendre's review in the form of formula 5.14. In Germany, the publication of an excerpt of the *Supplément* in Hindenburg's scientific journal enables the dissemination of Bürmann's formula of integration in the form of formula 5.18.

Although Bürmann's *Formules de développement, de retour et d'intégration* was never published, its mathematical results spread to the scientific community indirectly through Legendre's work principally. In 1817, they were included in Legendre's *Exercices de calcul intégral*, together with a more detailed explanation than that given in his review of the manuscript published by the French Academy (Legendre 1817, 224 ff.). For instance, 2 years later, Silvestre-François Lacroix reworked Bürmann's results on the basis of Legendre's account given in the *Exercices* (Lacroix 1819, 623 ff.). Naturally, Lagrange's and Legendre's review contributed to the dissemination too. For instance, in 1814, François-Joseph Servois (1767–1847) made reference to this review when dealing with Bürmann's results (Servois 1814, 42). Over the years, what will remain of Bürmann's manuscript is his reversion formula that generalizes Lagrange's inversion theorem, which will be frequently included in textbooks on differential calculus. In these textbooks, however, the treatment of Bürmann's formula is in general subordinated to Lagrange's work, as can be seen, for instance, in Bertrand (1864, 312–324). On the contrary, the ambitious project of renewing algebraic language, program that met with opposition from the French Academy of Sciences and with Bürmann's refusal to uphold its relevant aspects of improving mathematical symbolism, will not be conveyed by Bürmann's theorems of expansion, reversion, and integration, which were supposed to be the main examples of the advantages gained by using the analytic ideography.

5.2.3.3 A Universal Notation System

As seen in the previous section, Bürmann felt compelled to expunge his innovative notations from his manuscripts in order to persuade the members of the French Academy to pay attention to his work. The modification of the texts led to the elimination of the analytic ideography and the adoption of the standard mathematical notation. Current algebraic notations were defended by Legendre on the basis of some epistemic values, such as simplicity, precision, and elegance. But these values rest on a specific mathematical practice evidenced by the work of renown mathematicians, as Euler or Lagrange. However, Bürmann's renewing of algebraic notation was motivated by similar epistemic values, as universality or concision, according to Bürmann's presentation of his project in the *Essai* and the *Supplément*. The problem is that there is no mathematical practice to support his notation system. Bürmann realizes that the future of his project depends on the training of a new generation of young mathematicians committed to both the understanding of the analytic ideography and its successful application to solve mathematical problems. In this sense, he alludes to Fontanelle who had said with respect to Leibniz's universal characteristic that, apart from the ability to discover it, it would be essential to make everyone willing to adopt it (Bürmann 1801b, 2). But, continues Bürmann in his text published in 1803 by Hindenburg though written in 1801 as can be inferred from the reference made to Stahl's and Weingärtner's textbooks printed "the past year", the job of convincing the entire scientific community to modify the intellectual habit of using the current algebraic symbolism can be an extremely tough task. However, the existence of evolved languages as that of arithmetic and algebra itself shows, for Bürmann, that this task is not insurmountable, but to succeed it requires to focus on young people that, unlike experts, are not anchored in the limitations imposed by previously acquired knowledge. There is a very clear allusion here to the members of the French Academy of Sciences who denied the analytic ideography the opportunity to grow and to compete with other scientific theories. This opportunity cannot be created, for Bürmann, in the ossified thinking of Academies, but in the open mind of someone eager to learn about a new mathematical language and willing to work hard to make it prosper, just as the current experts made prosper algebraic symbolism in the past, leaving behind the norms and standards of their teachers. Thus, Bürmann did not abandon his project when writing his *Formules de développement, de retour et d'intégration* in accordance with the current notations, he just conformed to the standards. To promote his ideas about the analytic ideography, he had to seek support from Hindenburg, who seemed to be more receptive to innovations.

At the end of the eighteenth century, Hindenburg had won considerable intellectual influence in German academic circles, which certainly motivated Bürmann to try to gain Hindenburg's sympathy. Indeed, Bürmann was determined to find a way to reach a large audience for his ideas and Hindenburg's influence could be an excellent channel to introduce his combinatorial characteristic in the academic world. His desire to reach large audiences could explain why he wrote in French his article on the combinatorial characteristic, despite the fact that it was intended to be publish in Hindenburg's German journal *Sammlung*, whose main target audiences were German scholars. In fact, Bürmann's article was finally included in an anthology because of the demise of the *Sammlung* (Bürmann 1801b). Approximately half of the articles contained in the anthology are authored by Bürmann, who pointed out that these articles formed part of a treatise on the combinatorial characteristic, whose composition was still in progress at the time of the anthology's publication. It is not known if this treatise was ever completed. In 1807, 4 years after the anthology's release, Bürmann continued to work on it and encouraged whosoever was interested in understanding the functioning of his universal writing system to pay him a visit at home (Bürmann 1807b, 28).

Besides the desire to promote his ideas, it seems that Bürmann had a genuine interest in the combinatorial analysis, although this is difficult to confirm. In his correspondence with Legendre concerning the submission of the *Essai*, Bürmann affirms that he has the intention of using Hindenburg's combinatorial analysis to further develop his analytic tachygraphy (Bürmann 1798h). In the same year of 1798, he addresses a letter to Hindenburg to inform him of his intention to establish a theoretical link between the combinatorial analysis and the analytic ideography (Bürmann 1798a, 510). In this letter, he tells the story of his project. During his youth he was struck by the "marvelous thought" of the great Leibniz about the existence of a universal language, able to put an end to sterile philosophical discussions, and dreamed of discovering the Leibnizian language. After several failed attempts, he acknowledged himself defeated in the face of the monumental enterprise of classifying human thoughts, considered by Leibniz as the basic elements of the universal language. Then renewed hope arose when he found that the *Pasigraphie* of Joseph de Maimieux (1753–1820) came close to what he was looking for, and from there he was able to sketch the key ideas of the analytic ideography. Soon after that, he began to study the combinatorial analysis which provided him with insights into the direction to follow, namely, basing the analytic ideography on the number system. Thus, three different conceptions converge in Bürmann's project. The background to the analytic ideography appears to be the universal language conceived by Leibniz. But instead of basing it on a system of human thoughts, it should be based on a system of symbols, as suggested by Maimieux. The *Pasigraphie* to which Bürmann makes reference is a booklet printed in 1797 in which Maimieux proposed to reduce all of human natural languages to a universal alphabet composed of twelve symbols and a few syntactic rules (de Maimieux 1797). In fact, Maimieux's *Pasigraphie* is more concerned with translation issues than with an epistemological program. Its aim is to provide an encoding method for human natural languages, so that a message expressed in a particular natural language could be encrypted by means of the pasigraphy and then deciphered in any natural language, preserving the sense of the original message. While the Leibnizian background and the symbolic reduction of language are plain to see in the analytic ideography, Bürmann gives no explanation of the relation between his project and Hindenburg's combinatorial analysis, other than the remark about the number system.

The name used by Bürmann to designate his project changes from one epoch to another, or, at the same period, from one correspondent to another. In 1798, he speaks of "analytic ideography" when addressing Hindenburg, and of "analytic tachygraphy" when addressing Legendre. In 1801, he replaces both of them by "combinatorial characteristic", making explicit reference to the "algebra of thought" about which Leibniz "never wrote a single word" even though he spoke of it frequently (Bürmann 1801b, 1). Thus, the combinatorial characteristic should be seen as the incarnation of Leibniz's dreamy speculations, although Bürmann claims that his combinatorial characteristic differs from Leibniz's speculations in that it is not based on metaphysics, probably due to the fact that Bürmann abandoned Leibniz's idea of basing the universal language on some system of human thoughts. On the other hand, he also claims that his syntactic system is better than that of Maimieux because of its mathematical simplicity and its concision. In 1801, his system is composed of the following nine symbols:

In 1807, the number of symbols has increased to twelve, the same number of symbols used by Maimieux. Here are Bürmann's symbols (Bürmann 1807b, 1):

In 1807, Bürmann refers to his project as "syntactic characteristic".

The main difficulty in judging the merits of Bürmann's symbolic system is that the usage rules (or the grammar) of his twelve-letter alphabet appear nowhere in Bürmann's works (by the way, this is why he invited people to pay him a visit). There are isolated examples throughout his writings showing partial applications of these symbols, as the mathematical formulae included in the *Supplément*, which contain, however, foreign symbols, like numerals. But again, the general rules are not explained, thereby foreclosing all possibility of an understanding of the combinatorial characteristic itself. On the contrary, one can attempt to reconstruct the general outlines of Bürmann's project with the help of the articles published in Hindenburg's anthology. For the purposes of our study, three different aspects of Bürmann's project of the combinatorial characteristic will be distinguished: his notation reform, the theoretical assumptions of the project, and the general intended application areas.

Bürmann's notation reform concerns the whole of mathematics, including the combinatorial analysis. Therefore, a successful implementation of the reform implies the elimination of the combinatorial notations invented by Hindenburg since 1778 and complemented by the subsequent notations created by other members of the combinatorial school. According to Hindenburg, among the manuscripts sent by Bürmann, two articles were devoted to the renewing of mathematical notations (Hindenburg 1800, XI). These articles were not included in Hindenburg's anthology and remained unpublished. It seems that, in these articles, Bürmann addressed directly the question of improving the combinatorial school's notations. At least this is what can be inferred from a relatively long response given by Hindenburg in the introduction to the second volume of his journal *Sammlung*, in which he argues in favor of the mathematical notations developed by him and his school. On the basis of Hindenburg's response, one can understand that Bürmann suggested to Hindenburg that the copious amount of alphabetic symbols used by the combinatorial school could be advantageously replaced by the nine symbols of the combinatorial characteristic. Hindenburg provides the following three reasons in favor of the combinatorial school's notations:

1. Regarding local symbols. Rothe created local symbols with the intention of providing mathematics with a suitable, brief expression to represent the particular terms of a given series. For Hindenburg, there is no mathematical expression in Bürmann's combinatorial characteristic equivalent to local symbols (Hindenburg 1800, XXI).
2. Regarding simplicity and concision. For Bürmann, a good notation system should contain the smallest possible number of symbols. Thus, a mathematical system based on the whole of the letters of the alphabet, as the combinatorial analysis, should be excluded by definition. Hindenburg argues that the combinatorial analysis offers the possibility of using as few alphabetic symbols as one wants. For instance, according to Bürmann, a combination could be symbolized as $\lfloor$, while classes of combination are represented by the letters A, B, C, etc. in the combinatorial school's notation. Hindenburg points out that one can take, for example, the letter C and add indexes, so that $\overset{1}{C} = A$, $\overset{2}{C} = B$, $\overset{3}{C} = C$, etc., and the general expression $\overset{r}{C}$ is as simple and concise as that of Bürmann.
3. Regarding Gothic letters. Bürmann was particularly critical of the use of Gothic letters in the combinatorial analysis, and suggested that this was one of the main causes that the combinatorial analysis remained confined to German-speaking territories. Hindenburg replies that Euler made use of this kind of typography in some of his writings, and thus foreign mathematicians should be acquainted with it (Hindenburg 1800, X). However, Hindenburg is not categorical on this point, and concedes that the use of Gothic letters has been a source of problems for the dissemination of his theory.

Hindenburg concludes that the use of Bürmann's symbols does not represent a real advantage over the use of the combinatorial school's notations and that, at best, Bürmann's symbolism may result in a more accessible mathematical representation for foreigners and uncultured men (Hindenburg 1800, XIX). It should be mentioned that Hindenburg was aware of the difficulties posed by the use of Gothic letters, since a foreign colleague warned him in 1800 against the strong hostility to these symbols in other countries (Hindenburg 1803a, 135). Indeed, this problematic affected German mathematical circles too, despite the undeniable authority and influence of Euler. For instance, while some mathematicians external to the combinatorial school, as Hennert, adopted Gothic capital letters to represent the binomial coefficients in accordance with the combinatorial school's practice, these letters were entirely excluded from the books of others, as in the case of Johann Albert Eytelwein (1764–1848), even from chapters devoted to the exposition of the combinatorial analysis (Eytelwein 1824a, 25 ff.; Hennert 1799).

For Bürmann, these arguments lack the power to compel Hindenburg's conclusion. But he preferred to give no answer, since Hindenburg's support for diffusing the combinatorial characteristic was more valuable than refuting these arguments (Bürmann 1801b, 7–8):

> The circumspection required in the correspondence among men of letters has prevented me from giving all the force I could to my arguments, hence I am absolutely not surprised that they failed to convince HINDENBURG.[24]

Nevertheless, Hindenburg showed a great deal of tolerance and open-mindedness for a project whose implementation would erase the product of more than 20 years of work on mathematical notations. In the interest of science, Hindenburg remained as impartial as he could (Hindenburg 1800, 21):

> What does the future hold for these symbols [those of Bürmann and those of Hindenburg]? Will any of them manage to prevail over the other in practice? This has little to do with the authors that created them [...] when science alone can benefit from it; and it certainly will if Mr. Bürmann continues with his project and presents his combinatorial analysis soon.[25]

In this passage, Hindenburg makes clear that Bürmann's is a rival project to that of the combinatorial school and does not meet the general ideas of the combinatorial analysis that he created long time ago. This has been even more emphasized by means of the

---

[24] Une circonspection nécessaire dans les correspondances des gens de lettres m'ayant empeché de donner à mes raisons tout le poids que j'aurais pû, je ne m'étonne aucunement qu'elles n'ayent pas convaincu HINDENBURG.

[25] Was das künftige Schicksal dieser Symbole seyn, und ob die eine Art derselben vor der andern ein merkliches Uebergewicht im Gebrauche bekommen werde? das kann die Urheber derselben wenig kümmern [...] wenn nur die Wissenschaft dadurch gewinnt; und das wird sie gewiß, wenn Herr B. seinen Plan verfolgt, und das Publikum recht bald mit seiner Analyse combinatoire beschenkt.

possessive in the expression "his combinatorial analysis" at the end of the quote, putting prudent distance between both projects.

On the other hand, according to Bürmann, from a theoretical point of view the theory of combinations is composed of two distinct branches: mathematical combinatorics and metaphysical combinatorics (Bürmann 1801b, 14). About this latter branch, Bürmann gives no information, except that it forms part of his project of combinatorial characteristic and he shall expose the principles behind this discipline in a future treatise, which was never written though. Mathematical combinatorics is also called combinatorial analysis. According to Bürmann, the systematic study of this area of knowledge was initiated by Leibniz, and improved by Bernoulli, Moivre, and Boscovich. But it was Hindenburg who established the precise technical rules of the discipline, and so he is "the real father of the combinatorial analysis". Bürmann underlines the fact that Hindenburg's combinatorial analysis is focused on the theory of series, particularly on the study of power series expansions. However, for Bürmann, the most fundamental theorems of power series expansions were the theorems he presented in his *Essai de calcul fonctionnaire*, that is, formulae 5.11 and 5.12. These theorems belong, for Bürmann, to differential calculus. Since combinatorial analysis deals with power series expansions and the fundamental theorems on this kind of expansions belong to differential calculus, Bürmann claims that these branches of science are more than complementary disciplines. To make them progress, they should be merged into a single new discipline (Bürmann 1801b, 15). Bürmann's ideas on differential calculus were presented in an article entitled *Développement général aux fonctions arbitraires*, whose content corresponds to that of the *Essai* approximately (Bürmann 1801a). In his article *Polynôme combinatoire*, Bürmann deals with some of the results of the combinatorial school on the theory of combinations: problems of enumerating combinations and permutations, combinatorial involutions, and the multinomial theorem (Bürmann 1801c). On the other hand, Bürmann points out that the single new discipline obtained by fusing combinatorial analysis into differential calculus should be complemented by incorporating another theory, namely, Arbogast's calculus of derivations. For Bürmann, a discipline composed of differential calculus, combinatorial analysis, and the calculus of derivations constitutes what he calls the "functional calculus" (*calcul fonctionnaire*) (Bürmann 1801a, 29). It is impossible to know how Bürmann understood Arbogast's theory and its eventual mergence with the other two disciplines, since one can find no account of the calculus of derivations or of its role in the functional calculus in Bürmann's article. This is a delicate issue because of the evident incompatibility of the general conception on which each of these disciplines rests. Arbogast's calculus of derivations was conceived as an attempt to provide the concept of "differential" or "derivative" with an algebraic basis (Arbogast 1800; Friedelmeyer 1994), and Hindenburg's combinatorial analysis was created as an alternative to the differential methods used in the study of series expansions. Therefore, these disciplines seem to be rather mutually exclusive competing theories than complementary ones. Bürmann does not seem to be aware of this theoretical incompatibility which presupposes three different ways of conceiving what mathematical analysis should be, or at least he gives no clue as to what meaning he might have attached to this mergence.

Bürmann's intended application areas of the combinatorial characteristic cover almost every field of knowledge: mathematical analysis, geometry, trigonometry, astronomy, medicine, chemistry, crystallography, cryptography, universal languages, politics, jurisprudence, economics, and even fine arts. But, perhaps, the most astonishing application comes in the area of metaphysics because of the millenary challenge he solved, or thought he solved, by means of the theory of combinations. He gave a combinatorial proof for the existence of God. His proof is in fact a mixture of an argument from beauty and an argument from design (Bürmann 1801c, 67):

> [S]ince everything is consequence of necessary laws, phenomena that better express an intention, such as the structure of organic bodies or the conservation of the species, are really nothing else but the amazing result obtained by loaded dice. One cannot forget that loaded dice are the poof of an intellect who has loaded them and that our reason judges according to probability instead of according to facts. Besides, despite all objections, the design manifested in an eye, a foot, a wing, etc., demonstrates the existence of a supreme being. This proof, without question the best, is based on the theory of combinations.[26]

In this argument, the thesis that the design of the world reveals the existence of God is opposed by the antithesis that the order of the world can be explained by pure chance. Bürmann's combinatorial proof consists in saying that necessary laws of Nature are indeed the result of chance, but chance can be explained by the principles of the theory of combinations, hence chance does obey rational laws and there is no antithesis at all. God designed the world not mathematically but combinatorially, and one should not forget that combinatorics is a much more general discipline than mathematics for Bürmann. Since reality has been organized combinatorially, all human knowledge should be susceptible of being expressed by means of a universal notation system whose syntax is shaped by combinatorial laws. Thus, Bürmann's universal notation system is neither a universal language nor a comprehensive mathematical system, but the synthesis of all possible knowledge.

In sum, regarding notations and their role in the system, and the theoretical assumptions, and the intended application areas, Bürmann's project drastically differs from Hindenburg's combinatorial analysis. These differences, some of which were outlined by Hindenburg himself, are significant enough not to be ignored, and they clearly indicate that Bürmann's system and the combinatorial school's theory should not be confused with each

---

[26] [T]out étant conséquence de loix nécessaires, les phénomènes qui annoncent le plus de dessein, tels que la structure des corps organiques, la conservation des espèces, ne sont au fond qu'un résultat surprenant obtenu par des dés pipés. L'auteur oublie que des dés pipés prouvent une intelligence qui les a pipés et que notre raison juge d'aprés la probabilité et non d'après l'événement. Aussi, malgré toutes les objections, le dessin manifeste que présente un œil, un pied, une aîle, etc., démontre un être suprême. Cette preuve, sans contredit la meilleure de toutes, est due à la théorie des combinaisons.

other. However, Bürmann's ideas had considerable influence on those who were interested in Hindenburg's combinatorial analysis during the first decades of the nineteenth century. There are some books belonging to this period that present the combinatorial analysis as Bürmann described it, that is, as a syntactic system, as will be seen in Sect. 6.2.3. From the point of view of historiography, especially that of our time, it is not uncommon to find the mistake of confusing Bürmann's project with that of the combinatorial school. For instance, in Martin (2005, 58), one can read that Hindenburg, Kramp, and Bürmann endorsed Leibniz's project of the universal characteristic, which is absolutely false since neither Hindenburg's nor Kramp's works were related to Leibniz's thought. At the beginning of the nineteenth century, the misleading assimilation of the combinatorial analysis to Bürmann's combinatorial characteristic further distorts comprehension of the actual mathematical project of the combinatorial school.

*

Since early in the evolution of the German combinatorial analysis, some voices raise against the dominant combinatorial current impulsed by Hindenburg and his school. These voices call for a return to the values of finite analysis, in which classic algebraic methods have been correctly established, and dismiss the power of new approaches, even of those that have the right to be considered part of "finite mathematics", as Hindenburg's combinatorial approach to mathematical analysis. The clamor for algebraic purity is intensified by the fact that it has become widespread by means of Hindenburg's scientific journals mainly devoted to the combinatorial analysis, making apparent the opposition between both mathematical conceptions. Furthermore, the pages of Hindenburg's journals and anthologies also expose a new combinatorial analysis, which is profoundly different from that of the combinatorial school. Despite these differences, Bürmann's combinatorial analysis is dressed in the combinatorial school's terminology and fitted to meet every challenge faced by the German combinatorial analysis. Thus, Bürmann's theory jeopardizes the survival of the combinatorial school's project.

A Combinatorial Current of Thought

6

6.1   Isolated Research

At the close of the eighteenth century, the internal cohesion of the German combinatorial school shows increasingly signs of erosion. The dynamic collaborative research environment of its emergence and growth periods not only stagnates but also becomes fragmented due to polarized positions on several delicate issues about what the German combinatorial analysis should be. From a contemporary perspective, it may seem a little bit silly to profusely argue the order of theoretical priority between the binomial and multinomial theorems. However, it should be understood that this was not a trivial topic at the time: what was at stake in the debate was the determination of the last principles of mathematical knowledge. Thus, when in 1800 Klügel engages in battle to defend the theoretical priority of the multinomial theorem, whereas in 1803 Prasse engages in battle to defend the theoretical priority of the binomial theorem, the disastrous panorama unfolds itself in such a way that anyone can appreciate the alarming schism at the heart of the combinatorial school. Moreover, the disagreement also stems from a breakdown in the general conception of the discipline: while the combinatorial school has always considered the combinatorial analysis as part of pure mathematics, Stahl's and Weingärtner's textbooks rather convey the image of an applied science. Certainly, this critical situation does not improve much if the new combinatorial analysis of Bürmann bursts onto the scene to rival the German combinatorial analysis, and, to top it all off, it does so in books and journals edited by Hindenburg.

All these events, plus the incipient concern about combinatorial methods as compared with pure algebraic methods, will have catastrophic consequences for the German combinatorial analysis, entailing its extinction, but at the same time they will have positive consequences for the development of German mathematics in the nineteenth century, as will be seen in Sect. 6.2, because the ideas of the combinatorial school motivated a large amount of mathematicians to explore new ways of thinking mathematics. The internal disagreement among the members of the combinatorial school gave the final push for the definitive dissolution of their tacit intellectual collaboration. It should be remembered that this collaboration surrounding a common research project was the only reason that brought the combinatorial school into existence. However, no trace remains of such collaboration in the nineteenth century, which implies that the combinatorial school is no more. It is not hard to corroborate that their interest in a common cause vanished at the dawn of the century. Töpfer, whose fight for Hindenburg's combinatorial analysis was decisive in bringing together the first members of the combinatorial school, abandoned Hindenburg's theory and wrote no more on the subject during the nineteenth century. Pfaff, who tried to translate some results of differential calculus into the combinatorial analysis, ended up working on differential calculus exclusively, ignoring combinatorial methods. Prasse, well we have been over that already and, in Sect. 6.2.4, we will talk about his new conception of combinatorics at the beginning of the nineteenth century. Hindenburg himself seems to have given up developing his theory; during the last 8 years of his life, he published no further original papers on the combinatorial analysis. Probably, the absence of Hindenburg's leadership during the first decade of the nineteenth century was a major element in the disintegration of the combinatorial school.

Another important element was education. The combinatorial school emerged first and foremost from Hindenburg's classrooms. His teaching endeavor helped form a generation of mathematicians actively engaged in improving and organizing Hindenburg's combinatorial theory. At the turn of the century, those mathematicians became the new educators of new generations of scientists, but they failed to transmit the enthusiasm they brought from Hindenburg's lessons to their students. This failure was shared by the other members of the combinatorial school, who were professional mathematicians working on a specific research project, but they did not transmit the values of that project to new generations. As a consequence, there was no renewal of the combinatorial school other than occasional affiliations of some mathematicians that contributed to the project with one or two papers. This ephemeral structure of the combinatorial school could not last forever and seemed destined to end as it did, as an early example in history of a scientific team that did not understand how important the transmission not only of the theoretical content but also of the values of a scientific project was for the existence of the team. Thus, when the members of the combinatorial school began to reorient their professional interests, there was no one left to resume the scientific project of the team.

As seen in previous chapters, Rothe was one of the most active members of the combinatorial school. His teaching, however, is an example of what has been said in the previous paragraph. In 1804, Rothe moved to Erlangen where he was appointed ordinary professor of mathematics at the university. As expected, Hindenburg's combinatorial methods were part of his lectures. This specific content had a profound influence on Martin Ohm, who studied at the University of Erlangen. Ohm possessed an exceptional aptitude for mathematics, which Karl Christian Langsdorf (1757–1834) soon noticed and he made sure that Ohm's talent was not wasted, encouraging him to pursue a career in mathematics and later becoming his doctoral supervisor. In 1811, Ohm presented his doctoral dissertation, entitled *De elevatione serierum infinitarum secundi ordinis ad potestatem exponentis indeterminati* (Ohm 1811). Although Langsdorf was his adviser, his dissertation shows that Ohm was definitely under the influence of Rothe's teachings. Let's examine his dissertation in some detail in order to understand how Rothe's teaching is an example of what has been said in the previous paragraph.

Ohm's dissertation can be described as a specialized study in the mathematical field of the German combinatorial analysis, despite the very limited amount of bibliographical sources used in the text. Hindenburg's 1779 *Infinitinomii* and *Problema solutum maxime universale ad serierum reversionem* are mentioned in the dissertation, but no result of Hindenburg's books is actually discussed. Kramp's *Analyse des réfractions astronomiques et terrestres* and Rothe's *Formulae de serierum reversione* are quickly mentioned too. On the contrary, Ohm makes constant reference to Weingärtner's textbook to justify most of his mathematical notations, and his formulae, and his combinatorial concepts, and his mathematical propositions. Indeed, each time Ohm mentions Hindenburg, Kramp, or Rothe, he does not forget to add a complementary reference to Weingärtner's textbook, in which an exposition of the subject can be found. These five texts represent the totality of Ohm's bibliographical sources on the combinatorial analysis, but it is clear that Weingärtner's textbook is his primary source.

Ohm's admiration and respect for his teacher Rothe are not only reflected in some kind words addressed to him in the preface of the dissertation, but also in his choice of theme. The main topic of his dissertation is the notion of "local symbol", created by Rothe in 1793 to provide Hindenburg's combinatorial analysis with a new kind of mathematical function (see Sect. 4.1.3.4). The central objective of Ohm's dissertation is to generalize the notion of "local symbol", which applies to (infinite) polynomials in one variable, in such a way that it can be applied to certain (infinite) polynomials in two variables.

The first part of the dissertation presents some elementary combinatorial notions, including Kramp's generalized factorial (see Sect. 4.2.3.4), which is called "faculty" by Ohm even if Kramp preferred the term "factorial", but it is not really employed later in the text. Ohm uses Rothe's notations:

$$\underset{(0,1,2,...)}{M_V^n} \quad \text{and} \quad \underset{(0,1,2,...)}{M_C^n} \, ,$$

for Hindenburg's classes of variations and combinations with repetition respectively. The numbers in parentheses are the elements to be permuted or combined, n is the number of elements to be taken at a time from $(0, 1, 2, \ldots)$, the letter M is a general designation for "class" (a practice started by Hindenburg, and largely adopted by Weingärtner), and thus $\underset{(0,1,2,...)}{M_V^n}$ and $\underset{(0,1,2,...)}{M_C^n}$ denote, according to Hindenburg's terminology, the nth class of variations and combinations with repetition respectively, i.e. the set of n-variations and n-combinations with repetition of the elements $(0, 1, 2, \ldots)$. Probably, Ohm learned firsthand these notations during Rothe's lectures. Ohm also employs Rothe's notation for the $(n + 1)$th binomial coefficient m_n, where m is the power to which a binomial has been raised, so:

$$m_0 = 1, \quad m_1 = \frac{m}{1}, \quad m_2 = \frac{m(m-1)}{1 \cdot 2}, \quad m_3 = \frac{m(m-1)(m-2)}{1 \cdot 2 \cdot 3}, \quad \text{etc.}$$

This notation for the binomial coefficients does not appear in any work published by Rothe until before 1820, when it is finally included in a book of Rothe, but the meticulous details about Rothe's notations provided by Ohm allows us to know that Rothe invented his notation since at least 1811 (Ohm 1811, 7; Rothe 1820, 44). For the class of n-variations with repetition in which the sum of the terms of each n-variation is equal to a given number m, Ohm uses the symbol $\overset{m}{\underset{(0,1,2,\ldots)}{V}}\overset{n}{}$, which has been probably taken from Weingärtner's textbook, but Ohm lists the elements of the class to the right of the symbol. In Fig. 6.1, the first two tables are examples of this kind of classes. Ohm also introduces the notion of "conjugate variation". Let $a\ b\ c\ \cdots\ n$ and $A\ B\ C\ \cdots\ N$ be two variations, then their conjugate variation is given by $aA\ bB\ cC\ \cdots\ nN$. Thus, a conjugate class is the class obtained by conjugating each variation of a given class with each variation of another class. For instance, the third table in Fig. 6.1 is the conjugate class of the classes $\overset{3}{\underset{(0,1,2,\ldots)}{2V}}$ and $\overset{3}{\underset{(0,1,2,\ldots)}{1V}}$.

Ohm calls "infinite series of second order" or "of dimension two" an infinite series in two variables x and y arranged in a two dimensional array as follows:

a	$+ bx$	$+ cx^2$	$+ \text{etc.}$
$+ \,'ay$	$+ \,'bxy$	$+ \,'cx^2y$	$+ \text{etc.}$
$+ \,''ay^2$	$+ \,''bxy^2$	$+ \,''cx^2y^2$	$+ \text{etc.}$
$+ \text{etc.}$	$+ \text{etc.}$	$+ \text{etc.}$	$+ \text{etc.}$

array that can be denoted by P. He learned this concept of series from Rothe, who used it since about 1800 according to the preface of a book of Rothe published in 1820 (Rothe 1820). This arrangement in two dimensions must be considered as a sort of infinite matrix with rows and columns, on which depends Ohm's generalization of Rothe's notion of "local symbol". For Rothe, a local symbol is a function (i.e. a group of mathematical symbols) that expresses a coefficient of an infinite series in one variable, which is called by Ohm "infinite series of first order" or "of dimension one". Ohm's local symbol of dimension two is a function:

$$Px(m + 1, n + 1),$$

$$
{}_{2}V^{3}_{(0,1,2,\dots)}
\begin{vmatrix}
0 & 0 & 2 \\
0 & 1 & 1 \\
0 & 2 & 0 \\
1 & 0 & 1 \\
1 & 1 & 0 \\
2 & 0 & 0
\end{vmatrix}
\qquad
{}_{1}V^{3}_{(0,1,2,\dots)}
\begin{vmatrix}
0 & 0 & 1 \\
0 & 1 & 0 \\
1 & 0 & 0
\end{vmatrix}
$$

00	00	21
00	01	20
01	00	20
00	10	11
00	11	10
01	10	10
00	20	01
00	21	00
01	20	00
10	00	11
10	01	10
11	00	10
10	10	01
10	11	00
11	10	00
20	00	01
20	01	00
21	00	00

Fig. 6.1 Ohm's variations classes and conjugate class

where m and n are natural numbers (including 0) and whose values are the coefficients of the infinite series P of dimension two. The value of $Px(m+1, n+1)$ is the coefficient placed in column $m+1$ and in row $n+1$; for instance, $Px(1,2) = {}'a$. It is worth noting that Ohm's notion of "function" is different from that of Rothe in this case. Ohm's function $Px(m+1, n+1)$ is not just a group of symbols that stands for a quantity, but an association between the couples of positive integers and the real numbers given by the coefficients of P. In this sense, Ohm's notion of "function" is closer to ours than to that of the eighteenth century. Using a local symbol of dimension two, the following formula:

$$
Px(m+1, n+1)x^{m}y^{n}
$$

expresses the general term of the two-dimensional series P.

Local symbols of two dimensions allow calculation of the general term of the product PQ, where P and Q are two-dimensional series of general term $Px(m+1, n+1)x^{m}y^{n}$ and $Qx(\overset{1}{m}+1, \overset{1}{n}+1)x^{\overset{1}{m}}y^{\overset{1}{n}}$. Here, Ohm follows Hindenburg's typical notations $\overset{1}{m}$, where the digit over the letter is an index. To determine the general term $(PQ)x(M+1, N+1)x^{M}y^{N}$ of PQ, where $M = m + \overset{1}{m}$ and $N = n + \overset{1}{n}$, Ohm calculates the conjugate class of ${}_{M}V^{2}_{(0,1,2,\dots)}$

and $N\overset{2}{V}_{(0,1,2,...)}$, then interprets the table so obtained as the values for $m, n, \overset{1}{m}$ and $\overset{1}{n}$ in $Px(m+$

$1, n+1)$ and $Qx(\overset{1}{m}+1, \overset{1}{n}+1)$: each row gives a product $Px(m+1, n+1)Qx(\overset{1}{m}+1, \overset{1}{n}+1)$ and the sum of all these products is equal to $(PQ)x(M+1, N+1)$. For instance, to determine $(PQ)x(2, 3)$ (thus $M = 1$ and $N = 2$), one constructs the classes:

$$
1\overset{2}{V}_{(0,1,2,...)}\begin{vmatrix} 0 & 1 \\ 1 & 0 \end{vmatrix} \qquad \text{and} \qquad 2\overset{2}{V}_{(0,1,2,...)}\begin{vmatrix} 0 & 2 \\ 1 & 1 \\ 2 & 0 \end{vmatrix},
$$

and so, one gets the conjugate class:

m	n	$\overset{1}{m}$	$\overset{1}{n}$
0	0	1	2
0	1	1	1
0	2	1	0
1	0	0	2
1	1	0	1
1	2	0	0

from which it follows that:

$$
(PQ)x(2, 3) = Px(1, 1)Qx(2, 3) + Px(1, 2)Qx(2, 2) + Px(1, 3)Qx(2, 1)
$$
$$
+ Px(2, 1)Qx(1, 3) + Px(2, 2)Qx(1, 2) + Px(2, 3)Qx(1, 1).
$$

This method can be generalized to a product $PQR \cdots X$ of p two-dimensional series of general term $Px(m + 1, n + 1)x^m y^n$, $Qx(\overset{1}{m} + 1, \overset{1}{n} + 1)x^{\overset{1}{m}} y^{\overset{1}{n}}$, $Rx(\overset{2}{m} + 1, \overset{2}{n} + 1)x^{\overset{2}{m}} y^{\overset{2}{n}}$, $\ldots$, $Xx(\overset{p-1}{m} + 1, \overset{p-1}{n} + 1)x^{\overset{p-1}{m}} y^{\overset{p-1}{n}}$. In this case, one puts $M = m + \overset{1}{m} + \overset{2}{m} + \cdots + \overset{p-1}{m}$ and $N = n + \overset{1}{n} + \overset{2}{n} + \cdots + \overset{p-1}{n}$, and then one constructs the classes $M\overset{p}{V}_{(0,1,2,...)}$ and $N\overset{p}{V}_{(0,1,2,...)}$ to determine their conjugate class, from which one derives the general coefficient $(PQR \cdots X)x(M+1, N+1)$: it is a sum of products of the form $Px(m+1, n+1)Qx(\overset{1}{m}+1, \overset{1}{n} + 1)Rx(\overset{2}{m} + 1, \overset{2}{n} + 1) \cdots Xx(\overset{p-1}{m} + 1, \overset{p-1}{n} + 1)$.

Naturally, the power P^p, for a positive whole number p, is a particular case of a product, and its general term can be obtained by the method explained above. Besides this method, Ohm offers another way to calculate the general coefficient of P^p, which consists in constructing an involution of combinations. Both methods are in fact equivalent, the only difference being that the first method is based on variations, whereas the second is based on combinations. Ohm generalizes this result to any number p as follows. Let

$$X = \int P\mathbf{x}(m+1, n+1)x^m y^n$$
$$\scriptstyle (m,n)$$

be a two-dimensional series, where the symbol $\int$ stands for "sum" and the expression (m, n) under the general term represents the summation indices, and let

$$S = \int Q\mathbf{x}(m+1, n+1)x^m y^n$$
$$\scriptstyle (m,n)$$

be another two-dimensional series such that $Q\mathbf{x}(m+1, n+1) = P\mathbf{x}(m+1, n+1)$ whenever $m+n > 0$, and $Q\mathbf{x}(m+1, n+1) = 0$ if $m = n = 0$. Thus, one gets:

$$X = P\mathbf{x}(1, 1) + S,$$

and by the binomial theorem:

$$X^p = p_0[P\mathbf{x}(1, 1)]^p S^0 + p_1[P\mathbf{x}(1, 1)]^{p-1}S^1 + p_2[P\mathbf{x}(1, 1)]^{p-2}S^2 + \text{etc.},$$

where p_0, p_1, p_2 are binomial coefficients written in Rothe's notation. From this, it is possible to infer that the nature of p depends on the binomial theorem, so p can be a rational number. This result can be seen as a generalization of the multinomial theorem for rational exponents to two-dimensional series. This generalized theorem and two-dimensional local symbols are then applied by Ohm to problems in reversion of series, including Eschenbach-Rothe's formula 4.16 discussed in Sect. 4.1.3.4.

Ohm's dissertation contains, as can be seen above, two results that might be particularly attractive to the combinatorial school, two results that largely conform to its research project. Local symbols, as Hindenburg pointed out on several occasions, were fundamental functions defined in the context of the combinatorial analysis. Therefore, a generalization of this kind of functions should be considered as a substantial contribution to the development of the combinatorial analysis. A generalization of the multinomial theorem should be a matter of equal notoriety, even if it rests on the binomial theorem. According to all the rules of the art, Ohm's generalized multinomial theorem has been justified by means of an elegant combinatorial method, notable for its practical simplicity and its use of elementary combinatorial tables. It could even be said that Ohm's dissertation contributes to enlarge the production of mathematical notations, one of the main objectives of the combinatorial school's project. Indeed, Ohm publishes for the first time Rothe's mathematical notations for binomial coefficients and combinations classes, and he even allows himself to make minor modifications to Weingärtner's notations. But how did the (former) members of the combinatorial school marshal all these strengths to improve the German combinatorial analysis? Well, they did not. No (former) member of the combinatorial school reworked Ohm's results, or included them in his own works, or at least casually mentioned them down in a remote corner of a page. In 1820, Rothe just mentioned Ohm's dissertation in a footnote, but saying nothing about Ohm's results (Rothe 1820, vi, footnote). Some time ago, in 1814, Rothe and Ohm published a booklet, which they wrote together, on some questions of polygons (Rothe and Ohm 1814). In the booklet, they make use of the binomial theorem, calculate the coefficients of a multinomial raised to a given power, and make quick reference to reversions of series (Rothe and Ohm 1814, 32–35). This could have been a wonderful opportunity for them to mention Ohm's results, or, even better, to use them, but they let the opportunity slip by. Nonetheless, it is true that a brief reference to Ohm's dissertation is given in the booklet, but only in relation to the notion of "series of different orders", which just means here a series in several variables according to the context of the booklet (Rothe and Ohm 1814, 36). It should be stressed that, despite the use of the binomial theorem, the calculation of the coefficients of a multinomial raised to a given power, and the reference to reversions of series, no combinatorial method was applied, but rather algebraic methods, and no mathematical notation of the combinatorial school was employed to write the binomial or multinomial coefficients, which is rather disturbing given the authors of the booklet. All this strongly suggests that the combinatorial school does not exist any longer and that, in particular, Rothe failed to transmit the values of the combinatorial school's project to Ohm, or he did not care anymore about it, even though he successfully taught Ohm some combinatorial methods. Ohm's dexterity in these methods is not followed by an actual interest in the German combinatorial analysis in itself, that is to say, in the German combinatorial analysis considered as a particular conception of mathematics. Ohm's 1814 booklet serves as a minor example of his detached attitude toward this conception of mathematics, but we will return to this in Sect. 6.2.4. Thus, Ohm's dissertation is in fact a piece of isolated research in two senses: first, the dissertation is not taken into account in developing new scientific results; second, although it is a display of dexterity in combinatorial methods, it did not improve the field to which it belongs because no one even tried to relate its content to the existing results of the German combinatorial analysis. Indeed, this isolation is more problematic and widespread than it looks, since, as far as we know, no mathematician foreign to the extinct combinatorial school resumed the combinatorial results that Ohm presented in his dissertation.

In 1820, Rothe published his book *Theorie der combinatorischen Integrale* (Rothe 1820). However, this work went through a long process of composition. The final manuscript was signed on 28 September 1819, according to the date recorded at the end of the preface. Rothe affirms, in the preface, that he started working on it 18 years ago, expecting that, after such a long time, the theory he presents would be as complete as possible. Thus, the first ideas for his theory of combinatorial integrals appeared near the end of the year 1801 or in early 1802. This date is corroborated by Hindenburg, who dedicated a brief section of an article published in 1803 to make a comment on combinatorial integrals, underlying the advantages of combinatorial methods over Arbogast's calculus (Hindenburg 1803b, 251–268). In this article, Hindenburg says that Rothe wrote him on 15 January 1803 to inform him that he was seriously working toward the development of a comprehensive theory of combinatorial integrals (Hindenburg 1803b, 266). Between the years 1803 and 1820, Rothe published nothing regarding his work on this theory. However, it seems that he talked about it in his lectures, or at least he shared his ideas with some of his talented students. For instance, Ohm tells in 1811 that he was taught the theory of combinatorial integrals by Rothe (Ohm 1811, 2–3). As pointed out above, Rothe's notation for binomial coefficients and his notion of "series of dimension two" were created in the context of his work on the theory of combinatorial integrals, which both appear in Rothe's 1820 book, though he does not elaborate on the notion of "series of dimension two", but just mentions it in passing. This long gestation period of the book makes the comprehension of its historic role in the development of the German combinatorial analysis more complex, but let's start with a brief overview of its content.

The term "integral" could be a little misleading, even in Rothe's time. Rothe's theory is rather a theory of sums, and not a theory of integrals, as he pointed out in the preface of his book. According to Rothe, the difference lies in the notion of infinity. While common integrals deals with, or are related to, infinitesimal quantities, combinatorial integrals avoid the use of such dubious quantities. It is unclear whether Rothe thought that a theory of combinatorial integrals could displace the theory of integrals related to infinitesimals, or he conceived such a theory just as an interesting mathematical alternative. In both cases, Rothe's theory aligns itself with the general conceptual framework of the German combinatorial analysis not only because of its combinatorial nature, but also because of its concern to provide mathematics with "finite" methods of calculation. Rothe does not offer an exact definition of what a combinatorial integral is, since his definition is based on a particular example. However, the general idea consists in selecting the terms of a sum through a combinatorial pattern. For instance, let γ, ϵ, and η be positive whole numbers, including 0, and let n be a positive whole number. All the solutions of the equation $\gamma + \epsilon + \eta = n$ are seen by Rothe as variations with repetition. If $n = 2$, then the following table of variations with repetition contains all the solutions of the given equation:

γ	ϵ	η
0	0	2
0	1	1
0	2	0
1	0	1
1	1	0
2	0	0

Now, consider the function $(a^\gamma + \epsilon z)^\eta$. Using the values of our variation table, one gets the following table for our function:

$$(a^0 + 0z)^2$$
$$(a^0 + 1z)^1$$
$$(a^0 + 2z)^0$$
$$(a^1 + 0z)^1$$
$$(a^1 + 1z)^0$$
$$(a^2 + 0z)^0$$

The sum of the elements of this table is a combinatorial integral, which Rothe writes as follows:

$$\int\limits_{\substack{(\gamma,\epsilon,\eta) \\ \gamma+\epsilon+\eta=2}} \overline{(a^\gamma + \epsilon z)^\eta}$$

Thus, γ, ϵ, and η are summation indices, but subject to the condition $\gamma + \epsilon + \eta = 2$, which is considered as a combinatorial condition.

The lack of explicit conditions in a combinatorial integral should be understood as signifying that the variations with repetition on which the indices depend are defined on the totality of positive whole numbers and zero. For example:

$$\int \overline{x^\alpha} = x^0 + x^1 + x^2 + x^3 + \cdots,$$

and:

$$\int \overline{x^\alpha y^\beta} = \quad \begin{array}{llll} x^0 y^0 & +x^0 y^1 & +x^0 y^2 & + \text{etc.} \\ +x^1 y^0 & +x^1 y^1 & +x^1 y^2 & + \text{etc.} \\ +x^2 y^0 & +x^2 y^1 & +x^2 y^2 & + \text{etc.} \\ + \text{etc.} & + \text{etc.} & + \text{etc.} & + \text{etc.} \end{array}$$

are combinatorial integrals. Thus, series of dimension one, as the first, and series of dimension two, as the second, are combinatorial integrals. More generally, any infinite series is a combinatorial integral. Therefore, the notion of "combinatorial integral" enables Rothe to conceptualize infinite series in combinatorial terms, that is to say, through this notion his theory should offer a theoretical framework for interpreting infinite series as combinatorial objects, as objects that belong to the realm of the theory of combinations.

However, despite his stated intent to formulate a theory of combinatorial integrals as complete as possible, he does not develop his theoretical framework for infinite series. The entire book is restricted to the discussion of finite sums, except the last four pages. Indeed, in the final pages of the book, an attempt was made to justify the binomial theorem for rational exponents by using the principles of the new combinatorial integral calculus. Rothe's justification is based on a functional relation, following the general idea of the functional proofs discussed in Sect. 2.1.3.2, the major difference being that Rothe insists on writing the binomial expansion in terms of Kramp's generalized factorials. Using the binomial theorem and Kramp's generalized factorials, Rothe also obtains the expansion of natural logarithms $\log(1 - x)$ and $\log(1 + x)$.

Indeed, Rothe's research continues Kramp's work on combinatorial integrals. Rothe credits Kramp for having coined the term "combinatorial integral (or aggregate)", and says that it appears for the first time in print in an article by Kramp published in 1800 (Kramp 1800; Rothe 1820, iv). In fact, Kramp uses the term "*aggrégat* (or *intégrale*) *combinatoire*" in his *Analyse des réfractions astronomiques et terrestres* published in 1799 (Kramp 1799, 73). Rothe also credits Kramp for having further explored this subject, which was previously addressed, according to Rothe, by Lagrange and the Bernoulli brothers. Kramp's work motivated Arbogast to devote two paragraphs of his *Calucul des dérivations* to this topic (Arbogast 1800). However, according to Rothe, none of them elaborated a theory of combinatorial integrals, and so Rothe's mathematical contribution to the field consisted in formulating such a theory. More precisely, since Rothe's research rests on Kramp's conception of combinatorial integrals exclusively, it is a contribution to, or rather a prolongation of, Kramp's work. The delicate question here is whether Rothe's book is also a contribution to the German combinatorial analysis. There is no doubt that Rothe's research, when it began back in 1801, was part of the collaborative work of the combinatorial school: Rothe's work further develops that of Kramp, and Hindenburg makes a comment on the advantages of these works over Arbogast's approach, even though

Rothe's results had not yet been published. However, the book published in 1820 gets lost in the void that the extinction of the combinatorial school left between the German combinatorial analysis and German mathematics in general. No one further develops the theory of combinatorial integrals and no one seems concerned enough to figure out a way to integrate this theory appeared in 1820 into the German combinatorial analysis. For instance, a book of Franz Ferdinand Schweins (1780–1856) published in 1820 includes a quick mention of Rothe's *Theorie der combinatorischen Integrale*, in which Schweins underlines the fact that he had not yet had the opportunity to read it, but that he was convinced that Rothe's book contained many important mathematical developments that might have been inserted in his own work (Schweins 1820, XXXII). Five years later, in another book of Schweins partially devoted to the combinatorial analysis, the theory of combinatorial integrals was neither mentioned nor used, although Schweins had certainly read Rothe's text by then (Schweins 1825). Rothe's book becomes in 1820 a piece of isolated research in the same twofold sense as Ohm's dissertation.

It is not possible to state categorically that Ohm's dissertation and Rothe's book on the theory of combinatorial integrals are the only cases of original research published after the first decade of the nineteenth century and closely related to the German combinatorial analysis. Perhaps there are a few more. If some day other examples are found, it would probably be prudent for historical research to question whether they are cases of isolated research. Another difficulty arises from the fact that the theory of combinations, as a field of mathematics, is larger than the German combinatorial analysis: not everything that is produced by those working in this field belongs to the German combinatorial analysis. For instance, Carl Gustav Wunder (1793–1850), a professor of mathematics and physics in a Lyceum in Wittenberg, published in 1826 a booklet entitled *Ueber Kombinationen des zweiten Grades oder Kombinationen von Kombinationen* (Wunder 1826). Wunder's main idea consists in using Hindenburg's combinatorial involutions to construct a sort of block involution, that is to say, an involution whose elements are not only digits (as in the case of Hindenburg's involutions), but can also be blocks of combinations. However, for Wunder, the theory of combinations is larger than the combinatorial school's theory. For him, the theory of combinations includes the works of Leibniz, Bernoulli (Wunder does not give the full name), Hindenburg, Weingärtner, Fischer and Krause (Wunder 1826, 1). Thus, since among these authors only Hindenburg belongs to the combinatorial school and only Weingärtner presents the theory of combinations from the perspective of the German combinatorial analysis, it seems inappropriate to say that Wunder's booklet was intended to be a contribution to the German combinatorial analysis, even if Hindenburg's combinatorial involutions are employed. Wunder's booklet is an example of a work influenced by the combinatorial school's ideas, but, as will be seen Sect. 6.2.4, Wunder's conception of mathematical analysis was different from that of the combinatorial school. In any case, this kind of original research studies is rare in the nineteenth century as well as those that are more closely related to the German combinatorial analysis.

6.2 Combinatorics and Mathematics

The isolated research discussed in the previous section gives historical evidence that the collaborative work among the former members of the combinatorial school stopped somewhere around 1803 and 1810, and, as a consequence, it can be stated that the combinatorial school disappeared in the first decade of the nineteenth century. Given the informal nature of the school, a more precise date cannot be assigned to this event. On the other hand, the dissolution of the combinatorial school does not automatically imply the dissolution of its research project. In this section, it will be seen that, without the combinatorial school, the development of the German combinatorial analysis stopped too, at the same time. This does not mean that the ideas of the combinatorial school were forgotten or disregarded, but simply that the German combinatorial analysis as a project that offers a specific conception of mathematics was relegated from the vanguard of the German mathematical thought to the honorable place of historical background of new conceptions of mathematics. All these new conceptions have in common the fact that they consider that combinatorics should play a certain role in our understanding of mathematics, and *this* combinatorics is specifically related to Hindenburg's school. This is what we call a combinatorial current of thought in the nineteenth century inspired by the German combinatorial analysis. More than a hundred textbooks attests of the existence of such a current in German-speaking countries. Four main branches of this current have been identified, depending on the role combinatorics plays in explaining what mathematics is, and they will be discussed below. It should be noted that this combinatorial current of thought is not a homogeneous movement, not even within each of the four branches. Naturally, this multiplicity of mathematical conceptions, several of which are incompatible with one another, shows that the German combinatorial analysis disappeared along with the combinatorial school.

6.2.1 Foundationalist Current

The thought that the theory of combinations contains the most fundamental principles of mathematics became one of the guiding ideas of the German combinatorial analysis. As seen in previous chapters, this thought was not just a matter of mere blind conviction, a Hindenburg's personal obstinacy, it was a matter of concrete mathematical results. For instance, Hindenburg's method of combinatorial involutions had in a sense shown that some delicate issues of the theory of series could be reformulated in terms of combinations. On the other hand, the combinatorial school's foundationalist project, which started with the publication of Hindenburg's *Novi systematis* in 1781, remained unfinished after 20 years of join efforts. The dissolution of the combinatorial school in the first decade of the nineteenth century seemed to condemn the foundation of mathematics on combinatorics to failure. However, the combinatorial school's partial results concerning its foundational project encouraged others to attempt similar studies of mathematics from a foundationalist perspective, although they did not pursue the goal of completing the German combinatorial analysis. They pursued personal projects, as will be seen in this section.

Jakob Friedrich Fries (1773–1843) was a German philosopher born in Barby. He received his philosophical formation both at the University of Leipzig and at the University of Jena, where he attended Fichte's lectures for a short time, which were such a disappointment to him because of Fichte's lack of rigor (Beiser 2014, 39). At a time when German idealism was in full swing, Fries criticized Fichte's and Schelling's philosophies, and tried to steer philosophy away from idealistic speculation and back in the tracks laid down by Kant. This facet of Fries's thought took shape in a book called *Neue Kritik der Vernunft*, appeared in 1807 (Fries 1807). In this work, Fries attempts to reformulate the Kantian epistemology based on categories and pure intuition, by postulating that, besides these Kantian principles of knowledge, there exists a primordial knowledge hidden in human reason. This primordial knowledge, which human reason possesses and is evidenced by the various functions of the mind, cannot be explained by categories or intuition, whether or not pure; on the contrary, synthetic a priori judgments depend on it.

This kind of epistemology, which explains knowledge as the result of an obscure psychological process, could not escape accusations of psychologism (Beiser 2014, 23–88). Concerning mathematical knowledge, Fries's psychologism is partially true. Fries considered that the classical division of pure mathematics into geometry and arithmetic had been incorrectly understood by contemporary mathematicians (Fries 1807, vol. 2: 112 ff.). He pointed out that this division had been wrongly attributed, on the one hand, to the difference between continuous and discrete magnitudes by Kästner and others, and, on the other hand, to the difference between space and time by Gottlob Ernst Schulze (1761–1833) and others. The problem with viewing pure mathematics as a field divided according to these interpretations is that it puts us in a situation where it becomes vividly clear that mathematics lacks unity, since the division rests on the irreconcilable nature of continuous and discrete magnitudes, or of space and time. Here space and time are understood as pure intuitions, and thus Fries's remark is a criticism of Kant's philosophy. However, this does not mean that Kant's conception of space and time proves inadequate, according to Fries, for explaining the nature of mathematics. Kant's space and time alone are just not sufficient.

What is lacking in Kant's philosophy that has failed to guarantee the unity of mathematics is a principle of the mind on which the production of mathematical knowledge rests. Fries identifies this principle with the irreducible and fundamental function of combining (Fries 1807, vol. 2: 113). This function of the mind allows the faculty of productive imagination to perform its basic operation of constructing new representations. Mathematics is a constructive science in that it incorporates new knowledge into its field by means of two types of construction based on the function of combining. The first type is a schematic construction, which consists in combining elements of the same kind. Schematic construction is not restricted to mathematics, but when the productive imagination used it in this science, the result is arithmetic. The second type is called figurative construction (*bildliche Konstruktion*), and geometry depends on it. According to Fries, who follows Kant's philosophy here, arithmetic is related to time, and geometry to space, but these relations are only possible because of the function of combining. This interpretation of arithmetic and geometry provides, for Fries, a solid justification for Kant's doctrine of combination of motions, known as phoronomy. More generally, this interpretation provides a justification for the unity of mathematics: there is a unique function of the mind that gives rise to the whole of mathematics.

While mathematical knowledge is explained as a consequence of a function of the mind, Fries's conception of mathematics does not only originate from Fries's psychologism, but arises also from the combinatorial school's conception of mathematics. Fries went to the University of Leipzig in the autumn of 1795 and spent a year in the city, before moving to Jena. Fries's main intellectual interest was philosophy, but he also studied natural sciences and mathematics at the University of Leipzig, where Hindenburg had become a prominent professor.   Thus, although Fries does not mention Hindenburg or the combinatorial school in his *Neue Kritik der Vernunft*, he certainly came in contact with Hindenburg's combinatorial analysis at the University of Leipzig, becoming highly influenced by the general idea of founding mathematics on the theory of combinations. In fact, in other book, Fries mentions Kramp, uses Kramp's mathematical notation for the factorial, as well as Hindenburg's notation for binomial coefficients, and writes the expansion of $(a + b + c + d + \cdots)^m$ in terms of Hindenburg's classes of combinations (Fries 1842, 36, 49, 61). Thus, his philosophical conception of mathematics is based in his knowledge of the German combinatorial analysis. Fries does not make reference to Weingärtner, but perhaps he knew about Weingärtner's Kantian interpretation of Hindenburg's theory, according to which the German combinatorial analysis must be framed within the limits of the critical philosophy (see Sect. 5.1.2.2). Putting Weingärtner's Kantian interpretation together with Fries's psychologistic reformulation of the critical philosophy, one gets a simple explanation of the origin of Fries's conception of mathematics described above. However, even if Fries knew nothing about Weingärtner's work, it is clear that his foundationalist conception of mathematics derives from that of the combinatorial school, but it is different from the German combinatorial analysis.

Another foundationalist approach guided by a philosophical doctrine was proposed by Ferdinand von Sommer (ca. 1802–1849), who published, in 1823, a booklet inspired by Hindenburg's ideas (von Sommer 1823). This inspiration borders on delirium: the first statement of the text affirms that the progress made in mathematics since Descartes, including all contributions of the great Newton, cannot compete with Hindenburg who grasped the canon of the understanding for the organon of reason (whatever that means). Sommer's philosophy seems to be a fusion of Kant's transcendental doctrine and Hegel's dialectic of the unity. He explains arithmetic as the result of constructing *the* number in an act of primary perception in which multiplicity becomes unity. This primary perception produces what Sommer calls a "primary unity". Then, in a sort of dialectic unfoldment, multiplicity can arise again from primary unity by linking together the successive apprehensions of this primary unity in a temporal sequence. The knowledge obtained by means of this act of linking apprehensions is called "theory of temporal sequence apprehension" (*Apprehensions-Zeitfolgelehre*) or, in worldly terms, "theory of combinations in a broad sense" (*Combinationslehre im weitesten Sinne*). Thus, the theory of combinations is the foundation of mathematics because of a philosophical explanation of knowledge. In Sommer's booklet, one cannot recognize Hindenburg's combinatorial analysis, except for the use of some of Hindenburg's mathematical notations.

It can be said that those of Fries and Sommer are external approaches to the foundation of mathematics on combinatorics, inasmuch as they depend on elements that come from outside the field of mathematics. The first internal foundationalist approaches appeared later in the 1820s. The first was proposed by Schweins, a professor of mathematics at the University of Heidelberg. In 1820, he published a book with the laconic title *Analysis* (Schweins 1820). This is a book focused on the theory of series, whose study is based on some combinatorial methods developed by the combinatorial school. Its most important result is a proof of the multinomial theorem for rational exponents, which does not depend on the binomial theorem and relies on combinatorial techniques of proof (Schweins 1820, 9–47). Since this was a major issue of concern to the combinatorial school, let's examine this proof.

The first part of the proof deals with the case of positive integer exponents, which is treated as a consequence of the operation of polynomial multiplication. Schweins establishes the equation:

$$(\overset{1}{a} + \overset{2}{a} + \cdots)(\overset{1}{b} + \overset{2}{b} + \cdots)(\overset{1}{c} + \overset{2}{c} + \cdots) \cdots = {}^{n}\overset{n}{C} + {}^{n+1}\overset{n}{C} + {}^{n+2}\overset{n}{C} + {}^{n+3}\overset{n}{C} + \cdots \atop \left(\overset{1}{a}, \overset{2}{a}, ..., \overset{1}{b}, \overset{2}{b}, ..., \overset{1}{c}, \overset{2}{c}, ... \right)$$

for the product of n possibly infinite polynomials. As we know, the symbol ${}^{m}\overset{n}{C}$ stands for the class of n-variations with repetition such that the sum of the terms of each n-variation is equal to m, and a digit n over a letter $\overset{n}{a}$ is an index, following Hindenburg's notations. For instance, ${}^{5}\overset{3}{C}$ is the class shown in Fig. 6.2, which has to be interpreted as the sum $\overset{1}{a}\overset{1}{b}\overset{3}{c} + \overset{1}{a}\overset{2}{b}\overset{2}{c} + \overset{1}{a}\overset{3}{b}\overset{1}{c} + \overset{2}{a}\overset{1}{b}\overset{2}{c} + \overset{2}{a}\overset{2}{b}\overset{1}{c} + \overset{3}{a}\overset{1}{b}\overset{1}{c}$ when appearing in an equation such as the equation given above. Indeed, Schweins does not demonstrate this equation, but illustrates it with

Fig. 6.2 Schweins's variations class

1	1	3
1	2	2
1	3	1
2	1	2
2	2	1
3	1	1
a	*b*	*c*

the example $(\overset{1}{a}+\overset{2}{a}+\overset{3}{a})(\overset{1}{b}+\overset{2}{b}+\overset{3}{b})(\overset{1}{c}+\overset{2}{c}+\overset{3}{c})$, maybe thinking that the general case follows by a simple induction. From this, one gets immediately:

$$(\overset{1}{a}x^1 + \overset{2}{a}x^2 + \cdots)(\overset{1}{b}x^1 + \overset{2}{b}x^2 + \cdots)(\overset{1}{c}x^1 + \overset{2}{c}x^2 + \cdots)\cdots =$$

$$\overset{n}{\underset{(\overset{1}{a},\,\overset{2}{a},\,\ldots,\,\overset{1}{b},\,\overset{2}{b},\,\ldots,\,\overset{1}{c},\,\overset{2}{c},\,\ldots)}{{}^{n}C}}x^n + {}^{n+1}\overset{n}{C}x^{n+1} + {}^{n+2}\overset{n}{C}x^{n+2} + \cdots.$$

The multinomial theorem for a positive integer exponent n is just a particular case of this:

$$(\overset{1}{a}x^1 + \overset{2}{a}x^2 + \cdots)^n = \underset{(\overset{1}{a},\,\overset{2}{a},\,\ldots)}{{}^{n}\overset{n}{C}}x^n + {}^{n+1}\overset{n}{C}x^{n+1} + {}^{n+2}\overset{n}{C}x^{n+2} + \cdots,$$

where the general coefficient ${}^{n+m}\overset{n}{C}$ is expressed as a Hindenburg's class of partitions and can be calculated by constructing a simple table. Noting that, for $m \geq 1$:

$$\underset{(\overset{1}{a},\,\overset{2}{a},\,\ldots)}{{}^{n+m}\overset{n}{C}} =$$

$${}^{n}\mathfrak{A}\left(\frac{1}{a}\right)^{n}\,{}^{0}_{m}C + {}^{n}\mathfrak{A}\left(\frac{1}{a}\right)^{n-1}\,{}_{m+1}^{1}C + {}^{n}\mathfrak{A}\left(\frac{1}{a}\right)^{n-2}\,{}_{m+2}^{2}C + \cdots + {}^{n}\,\underset{(\overset{2}{a},\,\overset{3}{a},\,\ldots)}{{}^{m}\mathfrak{A}}\left(\frac{1}{a}\right)^{n-m}\,{}_{m+m}^{m}C,$$

where:

$${}^{k}_{s}\mathfrak{A} = \frac{s(s - 1)(s - 2)\cdots(s - k + 1)}{1 \cdot 2 \cdots k}$$

is a binomial coefficient, written in a mixture of Hindenburg's notations, and where the first term is always equal to zero since ${}^{0}_{m}C$ is empty, the multinomial expansion can be written as:

$$(\overset{1}{a}x^1 + \overset{2}{a}x^2 + \cdots)^n = \tag{6.1}$$

$$\left(\frac{1}{a}\right)^{n}x^n + {}^{n}\mathfrak{A}\left(\frac{1}{a}\right)^{n-1}\,{}^{2}C x^{n+1} + \left({}^{n}\mathfrak{A}\left(\frac{1}{a}\right)^{n-1}\,{}^{3}C + {}^{n}\mathfrak{A}\left(\frac{1}{a}\right)^{n-2}\,{}^{4}C\right)\underset{(\overset{1}{a},\,\overset{2}{a},\,\ldots)}{}x^{n+2} + \cdots,$$

This last expansion displays the structure of a class ${}^{n+m}\overset{n}{C}$ more explicitly.

For a negative integer exponent, the proof rests on the operation of polynomial division. Dividing the constant polynomial 1 by the (infinite) polynomial $1 - \overset{1}{a}x^1 - \overset{2}{a}x^2 - \overset{3}{a}x^3 - \cdots$ according to the increasing powers of x, Schweins obtains the following expansion:

$$\frac{1}{1 - \overset{1}{a}x^1 - \overset{2}{a}x^2 - \overset{3}{a}x^3 - \cdots} = \overset{0}{D}x^0 + \overset{1}{D}x^1 + \overset{2}{D}x^2 + \overset{3}{D}x^3 + \cdots, \tag{6.2}$$

where the coefficients on the right-hand side of this equation are then given by:

$$\overset{0}{D} = 1,$$

$$\overset{1}{D} = \overset{1}{a}$$

$$\overset{2}{D} = \overset{2}{a} + \overset{1}{D}\overset{1}{a}$$

$$\overset{3}{D} = \overset{3}{a} + \overset{1}{D}\overset{2}{a} + \overset{2}{D}\overset{1}{a}$$

$$\vdots$$

$$\overset{p}{D} = \overset{p}{a} + \overset{1}{D}\overset{p-1}{a} + \overset{2}{D}\overset{p-2}{a} + \overset{3}{D}\overset{p-3}{a} + \cdots + \overset{p-1}{D}\overset{1}{a}.$$
$$\left(\overset{1}{a},\ \overset{2}{a},\ \overset{3}{a},\ \ldots \right)$$

It is easy to see that these coefficients can be written in terms of Hindenburg's classes of partitions by the formula:

$$\overset{p}{D} = \overset{1}{{}^pC} + \overset{2}{{}^pC} + \overset{3}{{}^pC} + \cdots + \overset{p}{{}^pC}.$$
$$\left(\overset{1}{a}, \overset{2}{a}, \overset{3}{a}, \ldots \right)$$

Thus, while this expansion rests on an algebraic method, it has been clad in combinatorial terminology, so that the general coefficient $\overset{p}{D}$ is a Hindenburg's class of partitions. Now, dividing the (infinite) polynomial $1 + \overset{1}{b}x^1 + \overset{2}{b}x^2 + \cdots$ by $1 - \overset{1}{a}x^1 - \overset{2}{a}x^2 - \cdots$ according to the increasing powers of x, Schweins obtains the expansion:

$$\frac{1 + \overset{1}{b}x^1 + \overset{2}{b}x^2 + \cdots}{1 - \overset{1}{a}x^1 - \overset{2}{a}x^2 - \cdots} = 1 + \begin{vmatrix} \overset{1}{b} \\ \overset{1}{D} \end{vmatrix}x^1 + \begin{vmatrix} \overset{2}{b} \\ \overset{1}{b}\ \overset{1}{D} \\ \overset{2}{D} \end{vmatrix}x^2 + \begin{vmatrix} \overset{3}{b} \\ \overset{2}{b}\ \overset{1}{D} \\ \overset{1}{b}\ \overset{2}{D} \\ \overset{3}{D} \end{vmatrix}x^3 + \cdots, \tag{6.3}$$

where the coefficients of this expression should be interpreted as sums of rows. For example, the coefficient of x^2 is $\overset{2}{b} + \overset{1}{b}\overset{1}{D} + \overset{2}{D}$. By setting $\overset{p}{b} = \overset{p}{D}$ in this equation, it follows from Eq. 6.2 that:

$$\frac{1}{(1 - \overset{1}{a}x^1 - \overset{2}{a}x^2 - \cdots)^2} = 1 + \begin{vmatrix} \overset{1}{D} \\ \overset{1}{D} \end{vmatrix} x^1 + \begin{vmatrix} \overset{2}{D} \\ \overset{1}{D}\,\overset{1}{D} \\ \overset{2}{D} \end{vmatrix} x^2 + \begin{vmatrix} \overset{3}{D} \\ \overset{2}{D}\,\overset{1}{D} \\ \overset{1}{D}\,\overset{2}{D} \\ \overset{3}{D} \end{vmatrix} x^3 + \cdots,$$

Next, Schweins repeats the procedure by setting:

$$\overset{1}{b} = \begin{vmatrix} \overset{1}{D} \\ \overset{1}{D} \end{vmatrix}, \quad \overset{2}{b} = \begin{vmatrix} \overset{2}{D} \\ \overset{1}{D}\,\overset{1}{D} \\ \overset{2}{D} \end{vmatrix}, \quad \overset{3}{b} = \begin{vmatrix} \overset{3}{D} \\ \overset{2}{D}\,\overset{1}{D} \\ \overset{1}{D}\,\overset{2}{D} \\ \overset{3}{D} \end{vmatrix}, \quad \text{etc.}$$

in Eq. 6.3, which gives:

$$\frac{1}{(1 - \overset{1}{a}x^1 - \overset{2}{a}x^2 - \cdots)^3} = 1 + \begin{vmatrix} \begin{vmatrix} \overset{1}{D} \\ \overset{1}{D} \end{vmatrix} \\ \begin{vmatrix} \overset{1}{D} \\ \overset{1}{D} \end{vmatrix} \end{vmatrix} x^1 + \begin{vmatrix} \begin{vmatrix} \overset{2}{D} \\ \overset{1}{D}\,\overset{1}{D} \\ \overset{2}{D} \end{vmatrix} \\ \begin{vmatrix} \overset{1}{D} \\ \overset{1}{D} \end{vmatrix} \\ \begin{vmatrix} \overset{1}{D}\,\overset{1}{D} \\ \overset{2}{D} \end{vmatrix} \end{vmatrix} x^2 + \begin{vmatrix} \begin{vmatrix} \overset{3}{D} \\ \overset{2}{D}\,\overset{1}{D} \\ \overset{1}{D}\,\overset{2}{D} \\ \overset{3}{D} \end{vmatrix} \\ \begin{vmatrix} \overset{2}{D} \\ \overset{1}{D}\,\overset{1}{D} \\ \overset{2}{D} \end{vmatrix}\,\begin{vmatrix} \overset{1}{D} \\ \overset{1}{D} \end{vmatrix} \\ \begin{vmatrix} \overset{1}{D} \\ \overset{1}{D} \end{vmatrix}\,\begin{vmatrix} \overset{2}{D} \\ \overset{1}{D}\,\overset{1}{D} \\ \overset{2}{D} \end{vmatrix} \\ \begin{vmatrix} \overset{1}{D}\,\overset{2}{D} \\ \overset{3}{D} \end{vmatrix} \end{vmatrix} x^3 + \cdots,$$

The coefficients of this expression should be interpreted as sums of rows. For example, the coefficient of the term in x^2 is the sum $\overset{2}{D} + \overset{1}{D}\overset{1}{D} + \overset{2}{D} + \overset{1}{D}\overset{1}{D} + \overset{1}{D}\overset{1}{D} + \overset{2}{D}$. A successive iteration of this procedure yields the expansion:

$$\frac{1}{(1 - \overset{1}{a}x^1 - \overset{2}{a}x^2 - \cdots)^n} = 1 + \overset{n}{{}^1G}x^1 + \overset{n}{{}^2G}x^2 + \cdots + \overset{n}{{}^pG}x^p + \cdots$$
$$\left(\overset{1}{a}, \overset{2}{a}, \overset{3}{a}, \ldots\right)$$

where the coefficients on the right-hand side of this equation are abbreviations for the corresponding coefficients obtained by n iterations of this procedure. However, in order to prevent this iterative calculation, Schweins proves that the general coefficient $\overset{n}{{}^pG}$ satisfies the identity:

$$\overset{n}{{}^pG} = -\begin{pmatrix}-n\\1\end{pmatrix}\mathfrak{A}\begin{pmatrix}p\\1\end{pmatrix}C + \cdots (-1)^q \begin{pmatrix}-n\\q\end{pmatrix}\mathfrak{A}\begin{pmatrix}p\\q\end{pmatrix}C + \cdots (-1)^p \begin{pmatrix}-n\\p\end{pmatrix}\mathfrak{A}\begin{pmatrix}p\\p\end{pmatrix}C,$$
$$\left(\overset{1}{a}, \overset{2}{a}, \ldots, \overset{p}{a}\right)$$

where $\overset{k}{{}^s\mathfrak{A}}$ is a binomial coefficient, as indicated above. By replacing $\overset{1}{a}, \overset{2}{a}, \ldots$ by $-\frac{\overset{1}{a}}{a}, -\frac{\overset{2}{a}}{a}$, …respectively, it follows from the expansion of $(1 - \overset{1}{a}x^1 - \overset{2}{a}x^2 - \cdots)^{-n}$ that:

$$\frac{1}{(a + \overset{1}{a}x^1 + \overset{2}{a}x^2 + \cdots)^n} = a^{-n} + \overset{n}{{}^1G}x^1 a^{-n} + \overset{n}{{}^2G}x^2 a^{-n} + \cdots + \overset{n}{{}^pG}x^p a^{-n} + \cdots$$
$$\left(-\frac{\overset{1}{a}}{a}, -\frac{\overset{2}{a}}{a}, \ldots\right)$$

for any positive whole number n. Thus, although Schweins employs the algebraic method of polynomial division, the coefficients of the multinomial expansion are expressed by combinatorial tables whose elements are Hindenburg's classes of partitions. Ultimately, Schweins's procedure shows that this algebraic method is actually a combinatorial method, since polynomial division is nothing but properly combining the coefficients of the given polynomials in order to determine the coefficients of the quotient.

Schweins was aware that his proof was the first to be independent of the binomial theorem, but its most important part, namely, the part in which he deals with a positive rational exponent, presents only a very general sketch. In his book, he successively demonstrates the binomial theorem for positive integer exponents, the multinomial theorem for the same kind of exponents; then the binomial theorem for negative integer exponents, and the multinomial theorem for the same exponents; next, the binomial theorem for positive rational exponents, and finally the multinomial theorem for the same exponents. The proofs for the binomial and multinomial theorems follow the same pattern, but they are independent of each other. Schweins's proof of the binomial theorem for rational exponents is clearly inspired by that of Rothe analyzed in Sect. 4.2.2.2, even if Schweins does not say it. As Rothe, Schweins proposes a proof based on the Vandermonde's identity and pursues the goal of showing that, as a consequence, the binomial theorem is a purely combinatorial result. In his sketch of proof, he indicates that the multinomial theorem for positive rational exponents rests on a generalization of the Vandermonde's identity. The entire sketch consists in establishing this generalization, but its application to the proof of the multinomial theorem for rational exponents has been deliberately neglected. Thus, we will fill in some blanks to make the idea of the proof more accessible, but without trying to give a complete proof of our own, which goes beyond the acceptable limits of a historical study.

Schweins defines the quantity $f+g\overset{m}{A}$ as follows:

$$f+g\overset{m}{A} = f+g\overset{0}{\mathfrak{A}} \, {}^{m}_{}\overset{0}{C}a^{f+g} + f+g\overset{1}{\mathfrak{A}} \, {}^{m}\overset{1}{C}a^{f+g-1} + f+g\overset{2}{\mathfrak{A}} \, {}^{m}\overset{2}{C}a^{f+g-2} + \cdots + f+g\overset{m}{\mathfrak{A}} \, {}^{m}\overset{m}{C}a^{f+g-m}$$
$$\left(\overset{1}{a}, \overset{2}{a}, \overset{3}{a}, \ldots\right)$$

where m, f and g are positive whole numbers, and it should be noted that the first term of this sum is always equal to zero, since ${}^{m}\overset{0}{C}$ is empty. His idea consists in showing that this quantity can be decomposed into a sum of products in a similar manner as the Vandermonde's identity, which establishes the following relation for binomial coefficients:

$$f+g\overset{h}{\mathfrak{A}} = f\overset{h}{\mathfrak{A}} \, g\overset{0}{\mathfrak{A}} + f\overset{h-1}{\mathfrak{A}} \, g\overset{1}{\mathfrak{A}} + f\overset{h-2}{\mathfrak{A}} \, g\overset{2}{\mathfrak{A}} + \cdots + f\overset{0}{\mathfrak{A}} \, g\overset{h}{\mathfrak{A}}.$$

In the sketch of proof, Schweins's generalization of this formula involves elementary algebraic manipulations and some combinatorial identities that depend on the Vandermonde's identity, by means of which Schweins manages to establish the following formula:

$$f+g\overset{m}{A} = f\overset{m}{A} \, g\overset{0}{A} + f\overset{m-1}{A} \, g\overset{1}{A} + f\overset{m-2}{A} \, g\overset{2}{A} + \cdots + f\overset{0}{A} \, g\overset{m}{A}. \tag{6.4}$$

In a certain way, this formula generalizes that of Vandermonde to Hindenburg's classes of partitions. Since the coefficients of the multinomial expansion are expressed in terms of these classes of partitions, Schweins affirms that the justification of the multinomial theorem for positive rational exponents follows from his formula. This is where his sketch of proof ends.

To apply Schweins's formula to the multinomial theorem, it should be noted that the expansion of $(a + \overset{1}{a}x^1 + \overset{2}{a}x^2 + \cdots)^n$ for a positive whole number n can be derived from Eq. 6.1, and that the coefficients of this expansion can be expressed by $f+g\overset{m}{A}$, for $m \geq 1$, putting $^{f+g}\overset{0}{A} = a^{f+g}$, and setting $f + g = n$. Indeed, using Schweins's notations, this expansion can be written as follows:

$$(a + \overset{1}{a}x^1 + \overset{2}{a}x^2 + \cdots)^n = {}^n\mathfrak{A}\overset{0}{}a^n + {}^n\mathfrak{A}\overset{1}{}\overset{1}{a}a^{n-1}x^1 + \left({}^n\mathfrak{A}\overset{1}{}\overset{2}{a}a^{n-1} + {}^n\mathfrak{A}\overset{2}{}\left(\overset{1}{a}\right)^2 a^{n-2}\right)x^2 +$$

$$\left({}^n\mathfrak{A}\overset{1}{}\overset{3}{a}a^{n-1} + {}^n\mathfrak{A}\overset{2}{}\,{}^2\mathfrak{A}\overset{1}{}\,{}^2\overset{1}{C}\overset{2}{a}\overset{1}{a}a^{n-2} + {}^n\mathfrak{A}\overset{3}{}\left(\overset{1}{a}\right)^3 a^{n-3}\right)x^3 + \cdots$$

$$= {}^n\mathfrak{A}\overset{0}{}a^n + {}^n\mathfrak{A}\overset{1}{}\,{}^1C a^{n-1}x^1 + \left({}^n\mathfrak{A}\overset{1}{}\,{}^2C a^{n-1} + {}^n\mathfrak{A}\overset{2}{}\,{}^2C a^{n-2}\right)x^2 +$$

$$\left({}^n\mathfrak{A}\overset{1}{}\,{}^3C a^{n-1} + {}^n\mathfrak{A}\overset{2}{}\,{}^3C a^{n-2} + {}^n\mathfrak{A}\overset{3}{}\,{}^3C a^{n-3}\right)x^3 + \cdots$$

$$= {}^n\overset{0}{A} + {}^n\overset{1}{A}x^1 + {}^n\overset{2}{A}x^2 + {}^n\overset{3}{A}x^3 + \cdots,$$

where the classes of partitions are composed of elements belonging to $\left(\overset{1}{a}, \overset{2}{a}, \overset{3}{a}, \ldots\right)$. Then, the idea would be to prove by induction, and by formula 6.4, the following identity for any positive whole number s:

$$^{a_1+a_2+\cdots+a_s}\overset{0}{A} + {}^{a_1+a_2+\cdots+a_s}\overset{1}{A}x^1 + {}^{a_1+a_2+\cdots+a_s}\overset{2}{A}x^2 + \cdots =$$

$$\left({}^{a_1}\overset{0}{A} + {}^{a_1}\overset{1}{A}x^1 + \cdots\right)\left({}^{a_2}\overset{0}{A} + {}^{a_2}\overset{1}{A}x^1 + \cdots\right)\cdots\left({}^{a_s}\overset{0}{A} + {}^{a_s}\overset{1}{A}x^1 + \cdots\right)$$

where $a_1, a_2, \ldots, a_s$ are positive rational numbers. Indeed, Schweins establishes a similar property for the series:

$$^{a_1+a_2+\cdots+a_s}\mathfrak{A}\overset{0}{}x^{a_1+a_2+\cdots+a_s} + {}^{a_1+a_2+\cdots+a_s}\mathfrak{A}\overset{1}{}x^{a_1+a_2+\cdots+a_s-1} + \cdots$$

in his proof of the binomial theorem for positive rational exponents. Here there is a gap in Schweins's reasoning, since these properties depend on the Vandermonde's identity, which Schweins establishes for positive whole numbers only. However, Rothe demonstrated that the Vandermonde's identity holds for two real numbers f and g, as seen in Sect. 4.2.2.2. Since Schweins's scheme of proof was inspired by Rothe's proof of the binomial theorem, this gap can be disregarded. Thus, by putting $a_1 = a_2 = \cdots = a_s = \frac{n}{s}$, it follows that:

$$^n\overset{0}{A} + {}^n\overset{1}{A}x^1 + {}^n\overset{2}{A}x^2 + \cdots = \left({}^{\frac{n}{s}}\overset{0}{A} + {}^{\frac{n}{s}}\overset{1}{A}x^1 + {}^{\frac{n}{s}}\overset{2}{A}x^2 + \cdots\right)^s,$$

and thus:

$$(a + \overset{1}{a}x^1 + \overset{2}{a}x^2 + \cdots)^{\frac{n}{s}} = \overset{n}{\underset{s}{A}}^0 + \overset{n}{\underset{s}{A}}^1 x^1 + \overset{n}{\underset{s}{A}}^2 x^2 + \cdots,$$

which completes the proof of the multinomial theorem for positive rational exponents.

Schweins does not address the issue of negative rational exponents, but this is a minor point. He was the first to achieve one of the most desired objectives of the combinatorial school. By means of the same combinatorial method inherited from Rothe, the independence of the multinomial theorem has been finally justified, and, almost as a bonus, Rothe's foundational proof of the binomial theorem for rational exponents has been reworked. The two theorems, which were designated in the eighteenth century as the most important theorems of the whole of analysis by talented mathematicians, like Hindenburg and Euler respectively, the two theorems belong to the theory of combinations: this is the general theoretical conclusion that can be drawn from Schweins's proofs. His book, entitled *Analysis*, can thus be seen as an attempt to found mathematical analysis on the theory of combinations. All the results of his book depend, in one way or another, either on the multinomial theorem or on the binomial theorem: reversions of series, series expansions of exponential and logarithmic functions, all the series expansions treated in the book, even the theory of faculties or generalized factorials of Kramp. However, Schweins's foundationalist conception of mathematics does not coincide with the combinatorial school's project. Looking just at this book, one could get the wrong impression that this is actually a contribution to the German combinatorial analysis. In 1825, Schweins published another more comprehensive book on analysis (Schweins 1825). One can find there several sections focused on the theory of combinations and its applications to the theory of series, a brief overview of the theory of functions, a brief discussion of finite differences, and differential calculus takes up a large part of the book. In 1825, Schweins's foundationalist conception is found to be narrower than what the title of his 1820 book suggests. Indeed, the theory of combinations is the theoretical basis of the theory of series, but not of the whole of analysis. According to Schweins's more comprehensive work, mathematical analysis is a mixture of the theory of series based, in a foundationalist sense, on the theory of combinations, of the calculus of finite differences, and of differential calculus. This is the way that Bürmann conceived mathematical analysis, as a mixture (called "functional calculus") of the combinatorial analysis, of Arbogast's calculus of derivations, and of differential calculus. Instead of Arbogast's calculus, Schweins introduces the theory of finite differences. Thus, Schweins's conception of mathematical analysis looks more like Bürmann's functional calculus than a contribution to the German combinatorial analysis. In Sect. 5.2.3, it has been explained that the conceptual differences between Bürmann's project and that of the combinatorial school are significant enough not to be overlooked. The same precaution principle should apply to Schweins here, especially considering that differential calculus and the German combinatorial analysis were competing theories. Another significant difference between the German combinatorial analysis and Schweins's approach to the theory of combinations and its application to the theory of series is the complete lack of attention given to Hindenburg's theory of involutions in Schweins's works. As seen in Chap. 4, Hindenburg's incipient algebra of involutions became the most important ingredient of the German combinatorial analysis; Schweins does not even use the word "involution" in his texts.

Schweins was not the only one to advocate an internal combinatorial foundationalist view. In 1823, a young mathematician named Friedrich Wilhelm Spehr (1799–1833) showed first signs of what would 1 year later reveal itself to be a radical foundationalist position. Spehr offered a proof of the multinomial theorem for rational exponents, which was, as that of Schweins, independent of the binomial theorem and based on combinatorial methods (Spehr 1823, 46–60). Apparently, he knew nothing about Schweins's work. According to him, there were two kinds of proofs: differential proofs and those that are based on the binomial theorem. The first kind is unacceptable because of the poor theoretical foundation of differential calculus; and the second, because general propositions should not be deduced from particular ones. Spehr's proof was intended to redress the wrongs of mathematics in this respect by using elementary mathematics, whose principles rest, as will be seen below, on the theory of combinations.

Spehr claims that it is well known that the multinomial expansion for a positive whole number n is given by:

$$(\overset{0}{a}x^{\alpha} + \overset{1}{a}x^{\alpha+\delta} + \cdots + \overset{r}{a}x^{\alpha+r\delta} + \cdots)^n = \overset{0}{_p}\overset{n}{C}x^{n\alpha} + \overset{1}{_p}\overset{n}{C}x^{n\alpha+\delta} + \cdots + \overset{r}{_p}\overset{n}{C}x^{n\alpha+r\delta} + \cdots,$$

where $\overset{n}{_p}C$ is a class of n-combinations with repetition of the elements $(\overset{0}{a}, \overset{1}{a}, \ldots)$ in which each n-combination is preceded by the number of permutations with repetition of its terms, number represented by p, and in which the sum of the (indices of the) terms of each n-combination is equal to r. For instance:

$$\overset{4}{_p}\overset{3}{C}(\overset{0}{a}, \ldots, \overset{5}{a}) = \begin{array}{c|ccc} 3 & 0 & 0 & 4 \\ 6 & 0 & 1 & 3 \\ 3 & 0 & 2 & 2 \\ 3 & 1 & 1 & 2 \\ \hline & a & a & a \end{array} = 3\left(\overset{0}{a}\right)^2 \overset{4}{a} + 6\overset{0}{a}\overset{1}{a}\overset{3}{a} + 3\overset{0}{a}\left(\overset{2}{a}\right)^2 + 3\left(\overset{1}{a}\right)^2\overset{2}{a}.$$

The last member of this identity does not represent the class $\overset{3}{_p}\overset{4}{C}$, but this is the way of interpreting it when appearing in an equation such as that of the multinomial expansion. Although Spehr does not justify this multinomial expansion, he establishes the following recursive formula for the general coefficient, for $r \geq 1$:

$$\overset{r}{\underset{p}{}}C = \frac{(n-r+1)\ \overset{r-1}{\underset{p}{}}C\overset{n}{a}^{1} + \cdots + (hn-r+h)\ \overset{r-h}{\underset{p}{}}C\overset{n}{a}^{h} + \cdots + (rn)\ \overset{0}{\underset{p}{}}C\overset{n}{a}^{r}}{r\overset{0}{a}}.$$

It seems that this kind of recursive formulae came to be regarded by Spehr as more scientific than the direct expression of the class itself, as if a recursive construction provided some sort of scientific explanation for the multinomial expansion. This scientific explanation should be understood as giving a general recursive form for the coefficients of the expansion no matter what the actual coefficients of the expansion are, so if the multinomial expansion is written as:

$$(\overset{0}{a}x^{\alpha} + \cdots + \overset{r}{a}x^{\alpha+r\delta} + \cdots)^{n} = \overset{0}{A}x^{n\alpha} + \cdots + \overset{r}{A}x^{n\alpha+r\delta} + \cdots,$$

the coefficients $\overset{0}{A}$, ..., $\overset{r}{A}$ are formal symbols, i.e. they are neither undetermined nor determined coefficients, but symbols that formally indicate what the coefficients should be, and these formal coefficients are supposed to verify, according to Spehr, the recursive formula given above, so:

$$\overset{r}{A} = \frac{(n-r+1)\overset{1}{a}\overset{r-1}{A} + \cdots + (hn-r+h)\overset{h}{a}\overset{r-h}{A} + \cdots + rn\overset{r}{a}\overset{0}{A}}{r\overset{0}{a}}.$$

This is clearly a very questionable idea, in our time as well as in Spehr's.

Spehr's proof of the multinomial theorem for rational exponents depends on this questionable idea. In fact, this idea is nothing new, it goes back to 1796, when Klügel sought to justify the independence of the multinomial theorem by giving a proof based on the notion of "form" (see Sect. 4.2.2.1). The formal coefficients $\overset{0}{A}$, ..., $\overset{r}{A}$ can be seen as Spehr's attempt to provide Klügel's philosophical notion of "form" with mathematical content. However, Spehr got lost in the labyrinth of an empty formalism, without realizing that mathematical content cannot arise from just writing down a symbol. As Klügel, Spehr claims that the form of the multinomial expansion (*forma evolutionis*) should be the same regardless the nature of the exponent. He assumes that this form is expressed by his formal coefficients, so that:

$$(\overset{0}{a}x^{\alpha} + \cdots + \overset{r}{a}x^{\alpha+r\delta} + \cdots)^{\frac{1}{m}} = \overset{0}{A}x^{\frac{\alpha}{m}} + \cdots + \overset{r}{A}x^{\frac{\alpha}{m}+r\delta} + \cdots,$$

or equivalently:

$$\overset{0}{a}x^{\alpha} + \cdots + \overset{r}{a}x^{\alpha+r\delta} + \cdots = (\overset{0}{A}x^{\frac{\alpha}{m}} + \cdots + \overset{r}{A}x^{\frac{\alpha}{m}+r\delta} + \cdots)^{m}.$$

By the multinomial theorem for positive integer exponents, the power $(\overset{0}{A}x^{\frac{\alpha}{m}} + \cdots + \overset{r}{A}x^{\frac{\alpha}{m}+r\delta} + \cdots)^m$ can be expanded as a formal series:

$$\overset{0}{B}x^{m\frac{\alpha}{m}} + \cdots + \overset{r}{B}x^{m\frac{\alpha}{m}+r\delta} + \cdots,$$

where the coefficients are formal symbols. This dangerous game of empty symbols allows him to get, by the method of undetermined coefficients, the following identity:

$$\overset{r}{a} = \overset{r}{B} = \frac{(m-r+1)\overset{1}{A}\overset{r-1}{a} + \cdots + (hm-r+h)\overset{h}{A}\overset{r-h}{a} + \cdots + rm\overset{r}{A}\overset{0}{a}}{r\overset{0}{A}},$$

from which, by simple algebraic manipulations, he deduces the general form of the coefficient $\overset{r}{A}$ in the expansion of $(\overset{0}{a}x^{\alpha} + \cdots + \overset{r}{a}x^{\alpha+r\delta} + \cdots)^{\frac{1}{m}}$:

$$\overset{r}{A} = \frac{(\frac{1}{m}-r+1)\overset{1}{a}\overset{r-1}{A} + \cdots + (h\frac{1}{m}-r+h)\overset{h}{a}\overset{r-h}{A} + \cdots + r\frac{1}{m}\overset{r}{a}\overset{0}{A}}{r\overset{0}{a}},$$

which has the same form as the general coefficient of the multinomial expansion for positive integer exponents. This imperfect proof lends itself relatively easy to attack, to similar objections to those urged against Klügel's proof and examined in Sect. 5.1.1. As Klügel, Spehr concludes his presentation of the multinomial theorem by stating the binomial theorem as a corollary.

This proof was included next year in another book of Spehr (1824, 207–223), which is probably the most radical example of a combinatorial foundationalist approach to mathematical analysis related to the combinatorial school's ideas. The book opens with the statement that the theory of combinations is the foundation of mathematical analysis, but this theory has hitherto found no scientific presentation in the previous literature. Later, in the introduction, Spehr elaborates further on the second part of this statement. There can be found a historical overview of the theory of combinations. The works of Lull, Kircher, Jacob Bernoulli, Leibniz, Lambert, Euler constitute the remote past of the theory of combinations, while the works of Hindenburg, Rothe, Kramp, Töpfer, Prasse, Eschenbach, and a "few more" constitute its recent past. According to Spehr, the authors of the remote past missed the fact that combinatorics is the most general mathematical science and can, therefore, be used to rebuild mathematical analysis, even though Leibniz, Bernoulli, and Euler made "unscientific attempts" to resolve some analytic problems by means of combinatorial methods. Hindenburg was the first to recognize the significance of the theory of combinations for mathematical analysis, which led him to seek the means to increase understanding of combinatorial processes. The members of the combinatorial school joined him in his quest. Together, they created the basis for a new mathematical science "over the span of just a few years" (*in wenigen Jahren*). Spehr's historical narrative makes clear his belief that the combinatorial school existed only a few years and had already disappeared by his time. It also makes clear that Spehr does not perceive himself as a member of the combinatorial school. Moreover, Spehr does not consider his theory of combinations to be part of the combinatorial school's quest. While he recognizes the value of the combinatorial school's work, it must be considered relative to its groundbreaking features in the history of combinatorics. Indeed, after his historical overview, Spehr makes two criticisms of the German combinatorial analysis. First, he criticizes its overabundance of mathematical symbols, particularly this overabundance would be one of the principal reasons why foreign countries rejected the German combinatorial analysis. Second, Spehr holds that the German combinatorial analysis is rather an art than a science. According to him, Hindenburg and his school formulated some unexplained rules for combinations and their application to analysis, without pursuing the goal of assembling these rules and integrating them into a scientific system. This lack of scientificity would be another main reason why a large number of mathematicians, mostly foreigners, rejected the German combinatorial analysis.

Maybe Bürmann's criticism of the combinatorial school's mathematical notations, and also maybe the perpetual unfinished state of the German combinatorial analysis, motivated Spehr's views. In any case, the aim of Spehr's book was to bring the theory of combinations directly into the foreground, by distilling it into a scientific discipline. This discipline should be able to explain the principles and objects of elementary mathematics, on which higher mathematics depends. Spehr's reductionist approach is plain to see in his definition of mathematical analysis (Spehr 1824, 141):

> Analysis is the science of the laws which regulate the combinations of compound numbers, and function is its main object of study.[1]

A compound number is a quantity that depends on several quantities to be determined, and the manner of putting together all these quantities in order to form another quantity is expressed by a function. The act of putting together is nothing but the act of combining, and thus, because of its very nature, functions and mathematical analysis can be reduced to the theory of combinations. More than half the book deals with "pure combinatorics", i.e. with the notions of "combination", "permutation", and "variation". The scientificity of Spehr's pure theory of combinations consists, according to Spehr, in its recursive formulae, similar to that given above for the general coefficient of the multinomial expansion, which translate into mathematical terms the laws of combinations. However, this part of the book does not cover a wide breadth of topics. For instance, the entire theory of combinatorial involutions has been excluded. Thus, Spehr's scientific approach turns out

---

[1] Die Analysis ist die Wissenschaft von den Gesetzen der Verknüpfungen zusammengesetzter Zahlen, und ihr Haupt-Object ist die Function.

to be more restrictive than the artisanal approach of Hindenburg's school, eliminating at a single stroke the most important methods developed by this school, which, as artisanal as they were, proved to be very efficient. The range of applications of pure combinatorics to analysis is not much wider than the range of topics treated in the first half of the book: Kramp's generalized factorials, product and division of series, the binomial and multinomial theorems, the series expansion of the exponential function, and some few pages devoted to the theory of probability. In sum, Spehr's foundationalist conception is more radical than that of the combinatorial school according to the eloquent image of pure combinatorics portrayed by Spehr in some paragraphs containing his reflections on the nature of mathematics, but the implementation of his plan falls short of achieving what is expected of such a cogent image.

A second edition of Spehr's book was published posthumously 7 years after Spehr died (Spehr 1840). No improvement was made in this edition. It contains the same faulty proof of the multinomial theorem for rational exponents, the same topics, and the same applications of the theory of combinations to analysis. Thus, Spehr's plan of founding mathematical analysis on combinatorics did not evolve further. It should be mentioned that, as a part of his plan, Spehr intended to reform differential calculus as well. His first attempts in that direction took place in his 1823 book, in which he originally proposed his combinatorial proof of the multinomial theorem. Further and more comprehensive attempts were made in a book on differential calculus published in 1826, in which Spehr aimed to reformulate differential calculus in algebraic terms, avoiding the use of infinitesimals and evanescent quantities (Spehr 1826). Since algebra can be reduced to the theory of combinations according to Spehr's views discussed above, his algebrization of differential calculus belongs to his larger plan of founding mathematics on combinatorics.

As has been seen above, Fries thought that all mathematical knowledge derives from combinations, but this view was supported by a philosophical theory of the mind. As far as we know, Spehr was the only one that professed to accept such an extreme position without subscribing to any external philosophical doctrine, that is to say, the only one that adopted this extreme position as a result of the influence of the combinatorial school's ideas alone. Indeed, there were few who were moved to embrace a foundationalist approach with characteristics similar to those of the German combinatorial analysis. Schweins's and Spehr's conceptions of mathematics were close enough to certain main ideas of the combinatorial school to give the impression of being part of the German combinatorial analysis. In both cases, a more detailed study reveals some differences that could justify the theoretical distinction of these three conceptions, but it is clear that both Schweins and Spehr held a foundationalist position. There is another case that is, as those of Schweins and Spehr, very close to the German combinatorial analysis and that could be linked to this foundationalist current of thought, namely, the case of Andreas von Ettingshausen (1796–1878), a professor of mathematics and physics at the University of Vienna.

Ettingshausen wrote a book entitled *Die combinatorische Analysis*, which had a non negligible impact on the transmission of this discipline, but whose foundationalist tendency does not jump out at its readers right away (von Ettingshausen 1826). He claims that learning combinatorial methods could help to develop in students the necessary skills to meet the needs of their education in higher mathematics (von Ettingshausen 1826, v), but that does not imply that higher mathematics is based on the theory of combinations. One result of the textbook suggests, however, that Ettingshausen was inclined to think in this way. He offered a combinatorial proof of the multinomial theorem for rational exponents that does not depend on the binomial theorem (von Ettingshausen 1826, 107 ff.). This is the last proof of this kind that appeared in the large framework of the combinatorial current of thought in the nineteenth century (of which the foundationalist current forms part). The proof follows the same general lines as that of Schweins, that is, it depends on the Vandermonde's identity, although it is not based on Schweins's work. It is highly likely that Ettingshausen did not have access to Schweins's texts, if not actually unaware of their existence. In the preface, Ettingshausen indicates that his presentation of the combinatorial analysis relies on the few original available writings concerning this discipline, and he carefully makes reference to the works of the combinatorial school throughout the book, but Schweins is not mentioned. Their similar approach to the proof of the multinomial theorem can be explained by the fact that they followed Rothe's scheme of proof for the binomial theorem and generalized it in such a way that it becomes suitable for the study of the multinomial expansion. As indicated before, the key idea behind this kind of proof is that the multinomial expansion can be constructed as a combinatorial object, which points to the reconstruction of the theory of series by means of this kind of objects. This unveils a foundationalist conception in Ettingshausen's work, but it is certainly not as radical as that of Spehr.

Ettingshausen's textbook is, indeed, a partial presentation of the German combinatorial analysis. One of Ettingshausen's main contributions to mathematics is the invention of our contemporary symbol for binomial coefficients, and he also writes the value of a binomial coefficient using the factorial symbol invented by Kramp, as has become customary (von Ettingshausen 1826, 30, 102):

$$\binom{n}{k} = \frac{n!}{k!(n-k)!}.$$

Besides this contribution, Ettingshausen applies the multinomial theorem to solve problems concerning reversions of series, to justify the binomial theorem, and to the study of the exponential function. The textbook also addresses some questions of trigonometric functions and Kramp's generalized factorials. Despite some five examples shown in the book, Ettingshausen does not present Hindenburg's theory of involutions. Thus, his book offers a general overview of the German combinatorial analysis, even if it contains a rather small number of applications of the theory of combinations to mathematical analysis and some important theoretical topics are not treated fully.

Three, and no more, are the authors that pursued mathematical projects so closely related to the German combinatorial analysis that it becomes difficult to tell them apart, and this is so not only with respect to the foundationalist current, but also to the totality of authors of the combinatorial current of thought in the nineteenth century. First, Ettingshausen's book shows no conceptual differences compared to the German combinatorial analysis, so it can be said that it partially reproduces the theory of Hindenburg's school. Second, Spehr has been deeply influenced by the combinatorial school, but he himself differentiates quite distinctively his project from the German combinatorial analysis. And finally, Schweins resumes some ideas of the combinatorial school but develops them in a similar direction to that of Bürmann's functional calculus, which should be enough to distinguish Schweins's project from that of Hindenburg's school. The historian of mathematics Arthur Arneth (1802–1858) pointed out, when telling the story of the theory of combinations, that Schweins "gave a different form to this branch [of mathematics, i.e. the German combinatorial analysis] and further developed it in his Analysis" (Arneth 1852, 275). Thus, some contemporaries of Schweins were able to see the differences between Schweins's combinatorial conception of mathematics and that of the combinatorial school.

Despite the conceptual differences of these three authors, their works are linked by a common interest in showing that the multinomial theorem in its most general expression belongs to the field of combinatorics and is independent of the binomial theorem. In 1796, the correct proofs of Schweins and Ettingshausen would have led to a cataclysm in German mathematics by proving Hindenburg right about the multinomial theorem and the theory of combinations: those proofs would have provided evidence that mathematical analysis is essentially nothing but a subbranch of the theory of combinations. But those proofs arrived too late, 30 years too late. In 1826, nobody cares about that anymore. Spehr's proof has to be discarded because of its argumentative deficiencies, and no mathematician made reference to it. Ettingshausen's book had an influence on other textbooks. For instance, Jakob Philipp Kulik (1793–1863) explicitly mentions the name of Ettingshausen and uses Ettingshausen's combinatorial notations, as that of binomial coefficients, but he deduces the multinomial theorem (for positive integer exponents) from the binomial theorem, so he deliberately ignores Ettingshausen's proof (Kulik 1831). Joseph Salomon (1793–1856) makes explicit reference to Ettingshausen's book as an important source for his own textbook, proves the binomial theorem for rational exponents, and deduces from it the multinomial theorem, ignoring Ettingshausen's proof deliberately (Salomon 1831). Carl Gustav Reuschle (1812–1875), Leopold Carl Schulz von Straßnitzki (1803–1852), and Wunder use Ettingshausen's combinatorial notations, deal with the binomial theorem, and do not even mention the multinomial theorem (Reuschle 1850; von Straßnitzki 1844; Wunder 1844). Some traces of Ettingshausen's combinatorial notations can be detected in the works of Mathias Hartmann von Franzens-Huld (1807–1866) and Friedrich Reidt (1834–1894), but they say nothing about the multinomial theorem (Hartmann von Franzens-Huld 1843; Reidt 1880). In 1822, Anton Müller (1799–1860) published a brief study on the history of the binomial and multinomial theorems (Müller 1822). The study focuses mostly on the part of this history that is related to the combinatorial school. In this part of the history, Schweins's combinatorial proof of the multinomial theorem should be considered as a major historical event, but it was not included in Müller's narration. Therefore, one of the most important problems of the German combinatorial analysis loses all interest for the German-speaking mathematical community in the nineteenth century, which is another piece of evidence to be taken into consideration concerning the end of the German combinatorial analysis in the first decade of the nineteenth century.

The fragmentation of the combinatorial foundationalist current into personal research projects can also be illustrated by Fischer's conception of mathematics. As seen in Chap. 4, Fischer was among the first to show his appreciation for Hindenburg's ideas, even before the combinatorial school took form, but he did so in a very reprehensible manner. Instead of recognizing the authorship of Hindenburg, he decided to plagiarize Hindenburg's ideas and presented them under the name of "theory of dimension symbols". More than two decades later, his faith in combinatorial methods remained intact, as can be seen in a comment he wrote on his own textbook of elementary mathematics. This comment is composed of several volumes that appeared since 1820. In the third volume, one can find an interesting reflection on the theory of combinations (Fischer 1824, 5 ff.). According to Fischer, this theory is more general than the traditional parts of mathematics because its scope is not restricted to magnitudes, whether continuous or discrete, and its principles hold for any kind of objects. On the other hand, he classifies mathematics into two categories: calculatory mathematics and non-calculatory mathematics. Roughly speaking, non-calculatory mathematics corresponds to geometry, and calculatory mathematics to the rest. Calculatory mathematics is divided into three disciplines: arithmetic, algebra (or the theory of equations), and analysis (or the theory of functions). Fischer claims that the act of counting, which is the most elementary operation of arithmetic, consists in combining discrete magnitudes, given the generality of objects with which the theory of combinations can deal. As a consequence, the four elementary arithmetical operations can be explained by combinations, and arithmetic rests on combinatorics. Then, algebra should be based on arithmetical principles, and analysis on algebraic ones. Therefore, calculatory mathematics depends on the theory of combinations in a foundationalist sense. However, arithmetic "still ought to find its Euclid" since, according to Fischer, there is no satisfactory formulation of this science. Since arithmetic depends on the theory of combinations, Fischer's remark implies that the Euclid of combinatorics has not yet been born. To keep up appearances, Fischer makes a nod to Hindenburg and Hindenburg's former students (who exposed Fischer's plagiarism) in saying that they have the merit of having elaborated a complete scientific exposition of the theory of combinations, including its applications to calculatory mathematics, and this scientific exposition is far superior to his theory of dimension symbols. However, despite its great scientific superiority, immediately afterwards he throws away Hindenburg's combinatorial analysis by claiming that it is not suitable for the purposes of his presentation of arithmetic, which will be based on his own version of combinatorics. Obviously, with all proper courtesy, Fischer recommends the books of Stahl, Weingärtner, and Lorenz to those interested anyway in Hindenburg's combinatorial analysis. Thus, Fischer's combinatorial foundationalist conception of mathematics is really interesting from a historical point of view. It involves the paradox of being a copy of the German combinatorial analysis, because of the plagiarism, and of rejecting the German combinatorial analysis at the same time. On the other hand, as seen in Sect. 5.1.2, Stahl's and Weingärtner's textbooks transmit an idea of the German combinatorial analysis that seems to point in a different direction than that of the combinatorial school's project, and thus Fischer's recommendation helps blur even further the boundaries of the extinct German combinatorial analysis.

Far from this troubled passionate relationship between Fischer and the combinatorial school, it is possible to identify some influence of Hindenburg's school in the work of Johann Friedrich Schaffer (1776–1844). Schaffer, like so many other mathematicians, was unsatisfied with the current theory of differential calculus which introduces infinitesimal increments into the mathematical world of magnitudes. He thought he had to stop this nonsense by rethinking the ancient method of exhaustion, which was the aim of his book on differential calculus published in 1824 (Schaffer 1824). His idea consisted in using a calculus of finite differences to develop differential calculus, avoiding the introduction of things that are not magnitudes into the field of magnitudes. In other words, it was necessary to develop an appropriate theory of magnitudes. Such a theory should be able to explain all kind of compositions of magnitudes in order to justify mathematical calculations, but these compositions are seen by Schaffer as combinations of magnitudes (*Combination der Größen*) (Schaffer 1824, x). Then he points out that the theory of series (*Theorie der Zahlenreihen*) is already based on the principles of combinations of magnitudes. Here Schaffer makes no reference at all to the combinatorial school, but he states this idea as if it were a mathematical fact, firmly established and accepted by anyone. The lack of reference to the combinatorial school in the entire book allows to understand, on the one hand, how thoroughly the combinatorial school's ideas had permeated German mathematics, to such an extent that even those who show no particular bond with this school, like Schaffer, can fall under its sway, and, on the other hand, how carefully one must approach the subject of what counts as a contribution to the German combinatorial analysis. Schaffer is convinced that the theory of series rests on the theory of combinations, and this is why his books opens with a section devoted to combinations, but Schaffer shows no signs of attachment to Hindenburg's school. However, this belief led him to theorize that differential calculus rests on the theory of combinations too: the theory of functions depends on the theory of series, the theory of finite differences should be based on the theory of functions, and differential calculus should be developed from calculus of finite differences. Schaffer holds a combinatorial foundationalist position which, whether consciously or unconsciously, points in a similar direction to that of Bürmann's functional calculus.

6.2.2   Countercurrent: Algebraic Purity

The combinatorial current of thought that is being analyzed in this chapter can be characterized, naturally, by the application of combinatorial methods to other areas of mathematics, but it is indispensable that this commitment to a combinatorial perspective should be related to some extent to the combinatorial school, otherwise it is just a mathematician using the theory of combinations, which is not the exclusive property of this school. The relation with this school is expressed by the role the combinatorial school's ideas play in the assumed perspective. This role can be reduced to nothing with regard to the theory of combinations, which means that the combinatorial school's project of founding mathematical analysis on the theory of combinations has been explicitly rejected because of its combinatorial component. In this sense, part of the historical consequences of Hindenburg's school has to do, at least partially, with the rise of voices calling for algebraic purity in mathematics. These voices can be considered as a countercurrent inside the combinatorial current of thought that is being analyzed here. This countercurrent belongs to the general combinatorial current because of its relation with the combinatorial school's ideas: the authors of this countercurrent appreciate the work on the theory of series developed by the combinatorial school, but think that mathematical analysis should rest on pure algebraic principles.

As seen in Chap. 5, this countercurrent took shape in the last years of the eighteenth century and was born in the pages of Hindenburg's scientific journals, mostly devoted to the dissemination of the German combinatorial analysis. Tetens openly manifested his admiration for Hindenburg's work, especially for Hindenburg's non-recursive formula for calculating the general coefficient of the multinomial expansion, but he also openly manifested his opposition to combinatorial methods from a theoretical point of view, and tried to emulate Hindenburg's formula, as pointed out by Hindenburg himself, by using algebraic techniques. Pasquich created his exponent calculus with the explicit goal of improving differential calculus, but the implementation of his calculus renders combinatorial methods unnecessary to develop the theory of series. Although Pasquich does not say explicitly, perhaps as a mark of respect to Hindenburg, that his calculus is an algebraic alternative to the German combinatorial analysis, the scope of problems treated by means of his calculus leaves little room for doubt. As Tetens pointed out in 1796, mathematical analysis should be based on analytic principles, even if other methods, as differential or combinatorial ones, have been used in practice because of their calculatory advantages. Seeds of intellectual dissidence were thus sown among the pages in which Hindenburg's theory grew up. Seeds of intellectual dissidence had begun to sprout beneath the ground of the German combinatorial analysis, and the consequences were harvested during the nineteenth century.

For Hindenburg's theory, the nineteenth century began with the publication of another mathematical theory that challenged it. Jean Philippe Gruson (1768–1857) invented his "exposition calculus" (*calcul d'exposition*), as he called it, toward the end of the eighteenth century, and it appeared in print in 1802 (Gruson 1802). Exposition calculus was a mathematical theory designed to explain differential calculus algebraically. Among its advantages, Gruson underlines that, unlike differential calculus, exposition calculus needs to make no disputable metaphysical assumptions about mathematical objects, and it is as simple, general, and efficient as Hindenburg's combinatorial analysis. Furthermore,

Gruson affirms that he probably would not have achieved his objective without his prior knowledge of the German combinatorial analysis. This suggests that, as Tetens and probably Pasquich, Gruson found in Hindenburg's combinatorial analysis what must be the central idea of any acceptable theory on mathematical analysis. However, as Tetens and Pasquich, Gruson opposes the use of combinatorial methods to justify analytic results, even though he does not set out to diminish the merit of Hindenburg's theory. Thus, Gruson not only showed respect for Hindenburg, but he conceived his calculus in such a way that it fed off the German combinatorial analysis neglecting, ironically, all combinatorial elements. Despite its Hindenburgian roots, exposition calculus arrived in the nineteenth century to compete, assuredly, with differential calculus, but also with the German combinatorial analysis.

Gruson's exposition calculus is identical to Pasquich's exponent calculus discussed in Sect. 5.2.2. It depends both on the assumption that any function can be expressed as a power series, though Gruson, unlike Pasquich, does not explicitly state this, and on the definition of an exponent function identical to that of Pasquich. As Pasquich, Gruson formulates then some rules that mirror those of differential calculus, including the necessary stipulations for the formation of exponent functions of higher order, i.e. algebraic correlates of higher order derivatives. On 24 June 1798, Pasquich wrote a note concerning Gruson's plan of elaborating an algebraic theory similar to his in which he claims to have begun his research on exponent calculus 9 years ago, without knowing anything about Gruson's calculus (Pasquich 1798b). Pasquich learned about Gruson's plan from a remark in this respect included in Gruson's translation of Lagrange's *Théorie des fonctions analytiques*. In his note, Pasquich points out that his exponent calculus could be consulted in his recently published book *Unterricht in der mathematischen Analysis* and in his paper published in Hindenburg's *Archiv*, and, before its publication, exponent calculus circulated in a manuscript he shared with some German scholars. Gruson read his *Mémoire* on exposition calculus before the Berlin Academy on 14 June 1798, in which he said nothing about Pasquich. It is then difficult to say whether Gruson knew Pasquich's theory or he developed his own calculus independently. The latter is not impossible. After all, the way they defined an "exponent function", on which their entire theories depend, makes explicit an idea that had been latent for a long time concerning the algebraic hidden nature of derivatives. For instance, as has been seen in Sect. 3.3.1, in 1769 Borz explained the process of derivation as an algebraic calculation similar to what Pasquich and Gruson called "exponent function". In any case, there was no accusation of plagiarism against Gruson. On the other hand, Gruson took the analogy with differential calculus further by distinguishing between the "direct analysis of exposition calculus" and the "inverse analysis of exposition calculus" (Gruson 1803). The first corresponds to the algebraic theory of derivation explained above, and the second is the analogous algebraic theory of integration for exponent functions, but the general rules for this theory of integration are missing, since Gruson only gave some examples of application.

In 1812, Pasquich included his exponent calculus in a comprehensive work on mathematics (Pasquich 1812). In this general context, exponent calculus is used to explain Pasquich's theory of analytic functions (Pasquich 1812, II. Theil, 96 ff.). The textbook is organized to help students understand what mathematical analysis is about and how it can be constructed on an adequate theory of magnitudes. The theory of magnitudes is approached from the standpoint of arithmetic, and is developed on the basis of the relation of "identity" and the order relations of numbers ("greater than" and "less than"). From there, Pasquich defines the four elementary arithmetical operations. The arithmetical principles of this theory of magnitudes will be generalized to algebra. These generalized principles and those of Pasquich's exponent calculus lay the foundations for mathematical analysis. Here we find Pasquich reaffirming, 14 years later, his basic tenet of algebraic purity in mathematics, which excludes the combinatorial approach of Hindenburg's school.

Pasquich's exponent calculus was used, in 1803, to argue in support of differential calculus by Johann Schultz (1739–1805), a professor of mathematics at the University of Königsberg (Schultz 1803, xii ff.). Schultz claims that the Leibnizian theory of differential calculus does not need to be reformulated in order to avoid conceptual problems. His main argument rests on the practical utility of this theory. For him, reformulations such as that of Lagrange's *Théorie des fonctions analytiques* and Lacroix's *Traité du calcul différentiel* are too complicated to be of any use, and they miss the point of differential calculus: simplicity and efficacy in practice. Schultz acknowledges, however, that if the theoretical framework of differential calculus were to be questioned, its mathematical status would be doubtful; but he thinks that this theoretical framework has been misunderstood. Instead of explaining the right way to understand it, Schultz affirms that there is only one theory that can conform to the simplicity and efficacy of differential calculus: Pasquich's exponent calculus. Since this algebraic calculus avoids any concept of "infinity" and leads to the same results of differential calculus, Pasquich's exponent calculus is "the most beautiful and evident confirmation" of the rightness of differential calculus. In the end, although reluctantly, Schultz must admit that the rightness of analysis depends on the algebraic nature of its methods. Schultz also makes reference to Gruson's exposition calculus, underlining its troubling resemblance with Pasquich's calculus. Because of their identical features, exposition calculus is another confirmation of the algebraic soundness of differential calculus, and thus of mathematical analysis in general.

In his book, Schultz addresses the problem of raising a polynomial:

$$p = 1 + \overset{1}{c}z + \overset{2}{c}z^2 + \cdots$$

to a positive integer power n:

$$p^n = 1 + \overset{1}{x}z + \overset{2}{x}z^2 + \cdots$$

where $\overset{1}{x}, \overset{2}{x}$, etc. are undetermined coefficients. Schultz's solution is interesting because the procedure for determining the coefficients in the expansion of p^n can be seen as an attempt to algebraize Hindenburg's combinatorial methods. First, Schultz calculates the product pp^n as follows:

$$p^{n+1} = 1 + (\overset{1}{x} + \overset{1}{c})z + (\overset{2}{x} + \overset{11}{cx} + \overset{2}{c})z^2 + \cdots .$$

Then, without giving any justification, he assumes that the undetermined coefficients satisfy the following equations:

$$\overset{1}{x} = n\overset{1}{c}$$

$$\overset{2}{x} = n\overset{2}{c} + \frac{n(n-1)}{1 \cdot 2} {}^{2}\overset{1}{X}$$

$$\vdots$$

$$\overset{r}{x} = n\overset{r}{c} + \frac{n(n-1)}{1 \cdot 2} {}^{2}\overset{r-1}{X} + \cdots + \frac{n(n-1)\cdots(n-m)}{1 \cdot 2 \cdots (m+1)} {}^{m+1}\overset{r-m}{X} + \cdots +$$

$$\frac{n(n-1)\cdots(n-r+1)}{1 \cdot 2 \cdots r} {}^{r}\overset{1}{X}$$

where, for $m \geq 1$ and $m + 1 \leq r$, ${}^{m+1}\overset{r-m}{X}$ is an unknown number. To solve the original problem, it suffices to calculate these numbers. These equations allow him to express the coefficients of p^{n+1} in two different ways. First, by replacing n by $n + 1$, he gets:

$$\overset{1}{x} + \overset{1}{c} = (n+1)\overset{1}{c}$$

$$\overset{2}{x} + \overset{11}{cx} + \overset{2}{c} = (n+1)\overset{2}{c} + \frac{(n+1)n}{1 \cdot 2} {}^{2}\overset{1}{X}$$

$$\overset{3}{x} + \overset{12}{cx} + \overset{21}{cx} + \overset{3}{c} = (n+1)\overset{3}{c} + \frac{(n+1)n}{1 \cdot 2} {}^{2}\overset{2}{X} + \frac{(n+1)n(n-1)}{1 \cdot 2 \cdot 3} {}^{3}\overset{1}{X}$$

$$\vdots$$

Second, by adding the missing terms on both sides:

$$\overset{1}{x} + \overset{1}{c} = (n+1)\overset{1}{c}$$

$$\begin{array}{c} \overset{2}{x} + \overset{2}{c} \\ + \overset{11}{cx} \end{array} = \begin{array}{c} (n+1)\overset{2}{c} + \frac{n(n-1)}{1\cdot2} {}^{2}\overset{1}{X} \\ + n\overset{11}{cc} \end{array}$$

$$\begin{aligned}\overset{3}{x}+\overset{3}{c}\quad&=\quad(n+1)\overset{3}{c}\ +\tfrac{n(n-1)}{1\cdot2}\,2\overset{2}{X}\ +\tfrac{n(n-1)(n-2)}{1\cdot2\cdot3}\,3\overset{1}{X}\\[4pt]+\overset{12}{cx}+\overset{21}{cx}\quad&\qquad\qquad +n(\overset{12}{cc}+\overset{21}{cc})\ +\tfrac{n(n-1)}{1\cdot2}\,\overset{1}{c}\,2\overset{1}{X}\end{aligned}$$

$$\vdots$$

And thus:

$$\tfrac{(n+1)n}{1\cdot2}\,2\overset{1}{X}\ =\ \tfrac{n(n-1)}{1\cdot2}\,2\overset{1}{X}$$
$$+n\overset{11}{cc}$$

$$\tfrac{(n+1)n}{1\cdot2}\,2\overset{2}{X}+\tfrac{(n+1)n(n-1)}{1\cdot2\cdot3}\,3\overset{1}{X}\ =\ \tfrac{n(n-1)}{1\cdot2}\,2\overset{2}{X}\ +\tfrac{n(n-1)(n-2)}{1\cdot2\cdot3}\,3\overset{1}{X}$$
$$+n(\overset{12}{cc}+\overset{21}{cc})\ +\tfrac{n(n-1)}{1\cdot2}\,\overset{1}{c}\,2\overset{1}{X}$$

$$\vdots$$

Then, Schultz identifies the terms on both sides of these identities by columns, i.e. he gets:

$$\frac{(n+1)n}{1\cdot2}\,2\overset{1}{X}=\frac{n(n-1)}{1\cdot2}\,2\overset{1}{X}+n\overset{11}{cc}$$

$$\frac{(n+1)n}{1\cdot2}\,2\overset{2}{X}=\frac{n(n-1)}{1\cdot2}\,2\overset{2}{X}+n(\overset{12}{cc}+\overset{21}{cc})$$

$$\frac{(n+1)n(n-1)}{1\cdot2\cdot3}\,3\overset{1}{X}=\frac{n(n-1)(n-2)}{1\cdot2\cdot3}\,3\overset{1}{X}+\frac{n(n-1)}{1\cdot2}\,\overset{1}{c}\,2\overset{1}{X}$$

$$\vdots$$

without justifying his reasoning. By noting that:

$$\frac{(n+1)n}{1\cdot2}\,2\overset{1}{X}=\frac{n(n-1)}{1\cdot2}\,2\overset{1}{X}+n\,2\overset{1}{X}$$

$$\frac{(n+1)n}{1\cdot2}\,2\overset{2}{X}=\frac{n(n-1)}{1\cdot2}\,2\overset{2}{X}+n\,2\overset{2}{X}$$

$$\frac{(n+1)n(n-1)}{1\cdot2\cdot3}\,3\overset{1}{X}=\frac{n(n-1)(n-2)}{1\cdot2\cdot3}\,3\overset{1}{X}+\frac{n(n-1)}{1\cdot2}\,3\overset{1}{X}$$

$$\vdots$$

Schultz obtains the values of the unknown numbers:

$$^2\overset{1}{X} = \overset{11}{cc}, \quad ^2\overset{2}{X} = \overset{12}{cc} + \overset{21}{cc}, \quad ^3\overset{1}{X} = \overset{1}{c}\,{}^2\overset{1}{X}, \quad \ldots,$$

which completes his solution of the problem. Schultz points out that the binomial theorem for positive integer exponents is a particular case of this problem, but his remark is of no interest because, in a subsequent section of his book, he proves first the binomial theorem for rational exponents, and then he deduces the multinomial theorem for rational exponents from it.

Schultz describes this procedure as an "analytic solution". It is interesting to note that the unknown numbers $^{m+1}\overset{r-m}{X}$ are, in fact, Hindenburg's classes of combinations, but, to avoid calling them that, Schultz introduces them as "unknowns" in some unjustified equations in order to stick as closely as possible to an analytic procedure. In a remark to his solution, Schultz explains how Hindenburg's combinatorial methods can be applied to solve this problem, underlining that the key idea of his analytic solution was invented by Hindenburg who used the theory of combinations instead of algebra. Schultz affirms that the only difference between his solution and that of Hindenburg is his use of algebra instead of the theory of combinations. As Gruson, Schultz thinks that Hindenburg's theory is remarkable, and he even hopes that his analytic method will help students to discover it. However, as Tetens, Pasquich, and Gruson, Schultz is trying to algebraize Hindenburg's theory by eliminating its combinatorial methods, although his translation into algebra leaves some gaps in his reasoning.

Other authors were also engaged in this strange cohabitation with Hindenburg's combinatorial theory. For instance, in a textbook published in 1825 by Heinrich Ferdinand Scherk (1798–1885), some topics of mathematical analysis are treated with the help of the combinatorial school's methods (Scherk 1825). Indeed, there is a profuse use of mathematical notations invented by the school, and Scherk makes explicit reference to Hindenburg and Rothe. Moreover, an entire section of the book is devoted to the theory of combinations. In this section, Scherk makes the following interesting statement (Scherk 1825, 94–95):

> The theory of combinations presents a real mathematical interest for us only inasmuch as it can be applied to analysis. An analytical result that has been established by means of the theory of combinations has been determined, at the same time, by analytic and combinatorial operations, which are mixed together [...]. However, since we are dealing here with a pure analytical result, our efforts must focus on reducing combinatorial operations to analytic operations, or at least on hinting at the possibility of a simple reduction, if it has not been actually performed.[2]

---

[2] Eigentlich mathematisches Interesse hat nämlich die Combinationslehre nur in so fern für uns, als sie sich auf die Analysis anwenden lässt. Ein mit Hülfe der Combinationslehre gefundenes analytisches Resultat wird also durch analytische und combinatorische Operationen, die mit einander

In this passage, Scherk expresses well the theoretical necessity of purging mathematical analysis of the efficient combinatorial methods developed by Hindenburg's school, even if those methods can be used in practice because of its calculatory advantages. Indeed, Scherk's book is an example of this kind of mathematical practice: it benefits from the combinatorial school's results, while it promotes the belief that the theoretical justification of mathematical analysis should depend on analytic methods exclusively.

Sometimes, authors choose not to use the combinatorial tools created by Hindenburg's school, even though they are aware of its potential benefits. Johann Karl Friedrich Hauff (1766–1846), professor of mathematics at the University of Marburg, included Hindenburg's *Beschreibung* in his bibliography of relevant works for arithmetic, but he ignored Hindenburg's combinatorial perspective in his own work on arithmetic (Hauff 1807, 406). As pointed out before, Hartmann von Franzens-Huld uses Ettingshausen's combinatorial notations, and thus he was probably familiar with Ettingshausen's presentation of the German combinatorial analysis (Hartmann von Franzens-Huld 1843). However, his book on universal arithmetic contains no section whatever dealing with the theory of combinations, and the German combinatorial analysis is completely overlooked. He just employs Ettingshausen's notation for binomial coefficients in his brief discussion of the binomial theorem (Hartmann von Franzens-Huld 1843, 85). Johann Christian Martin Bartels (1769–1836) was a former student of Pfaff and became a prominent mathematician in the nineteenth century. One of his last works was a textbook on mathematical analysis (Bartels 1833). In this book, some problems of the theory of series are addressed, such as the binomial theorem and some power series expansions, but there is no trace of any use of the German combinatorial analysis in its pages. This is quite surprising, considering that Pfaff was an important member of the combinatorial school. Thus, it is very likely that Bartels's omission of the combinatorial school's ideas was the result of a well-reasoned decision. Although it is not possible to assert categorically in these three cases that Hauff, Hartmann von Franzens-Huld, and Bartels held a theoretical position of algebraic purity in mathematical analysis, their books could be read in that sense in a time marked by the German combinatorial analysis.

Naturally, this kind of interpretation should always be considered with caution when there is no explicit link between an author and the combinatorial school. Hauff's, Hartmann von Franzens-Huld's, and Bartels's works show a clear link, no matter how simple it may be, with Hindenburg's school. On the contrary, the textbooks of such mathematicians as Joseph Beskiba (1792–1863), Franz Eduard Desberger (1786–1843), Johann Bernhard Friederich (1796–1863), Adam Rudolph Jacob König (1787–1868), among many others, deal with arithmetic and mathematical analysis at different levels

---

vermischt sind, bestimmt werden [...]. Unsere Bemühung aber wird dahin gehen müssen, da es sich hier von einem rein analytischen Resultate handelt, die combinatorischen Operationen auf analytische zu reduciren, oder doch wenigstens, die Möglichkeit einer leichten Reduction einzusehen, wenn sie auch nicht ausgeführt wird.

of depth, but none of them says a word about the theory of combinations or the German combinatorial analysis (Beskiba 1847; Desberger 1831, 1847; Friederich 1831; König 1827, 1828). In these cases, it would be premature to speculate as to whether those works should be classified as examples of the countercurrent of algebraic purity that raised against the combinatorial methods of Hindenburg's school. On the other hand, given the broad dissemination of the combinatorial school's ideas in German-speaking countries, it would be naive to assume that those authors had heard nothing about Hindenburg's combinatorial analysis. Perhaps a reasonable conclusion should be that this kind of works exhibits a conception of mathematics according to which the theory of combinations can be neglected in the study of arithmetic and mathematical analysis. This kind of mathematical conception does not necessarily mean that their authors belong to the countercurrent of algebraic purity, but it helps strengthen this countercurrent through its disregard of combinatorial methods. For instance, in his *Lehrbuch der Arithmetik und Algebra*, August Leopold Crelle (1780–1855) states that (Crelle 1825, iii):

> Mathematical analysis or the art of calculating still lacks, to some extent, the clarity, rigor, and consistency that other parts of mathematics possess, qualities that are within the reach of this science and are especially necessary to achieve its main goal of developing and sharpening the faculty of thought.[3]

It can be inferred from this that Crelle was dissatisfied with the way in which the mathematical theories he knew explained mathematical analysis. Hence the problem boils down to the question whether Crelle knew the German combinatorial analysis well enough to affirm that Hindenburg's research project did not provide mathematical analysis with the desirable clarity, rigor, and consistency that mathematical sciences deserve. The sole absence of combinatorial methods in his textbook does not permit to draw the conclusion that Crelle opposes Hindenburg's school on the matter of the combinatorial nature of mathematical analysis, and that, as a consequence, Crelle's conception of mathematics belongs to the countercurrent of algebraic purity. To reach this conclusion, it would be necessary to establish a factual link between Crelle and the combinatorial school. Without such a link, Crelle's book warrants the weaker conclusion that, for him, the theory of combinations plays no role in the theoretical explanation of what mathematical analysis is. This is the conclusion that can also be established with respect to the works of Beskiba, Desberger, Friederich, König, and many others. Naturally, the exclusion of the theory of combinations entails the exclusion of the German combinatorial analysis as a suitable explanation of mathematical analysis, and, in this sense, all those works reinforce externally the countercurrent of algebraic purity.

---

[3] Der mathematischen Analysis oder Rechenkunst fehlt zum Theil noch diejenige Klarheit, Strenge und Folgerichtigkeit, welche andere Theile der Mathematik besitzen, deren diese Wissenschaft fähig ist und welche ihr zu ihrem Hauptzwecke, die Denkkraft zu entwickeln und zu schärfen, vor Allem nothwendig sind.

In the case of Crelle, there exists a factual link between his thought and the combinatorial school's ideas, although it cannot be found in his textbook on arithmetic and algebra. Crelle wrote an entire book on what he called "analytic faculties" (*analytischen Facultäten*) (Crelle 1823). As seen in Sect. 4.2.3.4, Kramp invented a new mathematical function that was originally called "faculty", but later he renamed it as "factorial" (called by us "generalized factorial" to avoid confusion with our contemporary notion of "factorial", which is a particular case of Kramp's generalized factorial). In German-speaking countries, the term "faculty" prevailed over "factorial" in the nineteenth century, and Kramp's function caught the attention of several scholars. Kramp's function was conceived as a combinatorial function and was one of the main achievements of the combinatorial school. By describing it as an "analytic" function in the very title of his book on Kramp's generalized factorials, Crelle's commitment to a theoretical position of analytic purity rings loud and clear. The content of his book confirms his theoretical position. Using analytic methods, Crelle develops the first elements of the theory of logarithms, powers and exponential functions, then he deals with the binomial theorem which is used to further develop the theory of logarithms, powers and exponential functions, and he then establishes Taylor's theorem. Some of these analytic results will be employed to explain Kramp's theory of generalized factorials. Thus, Crelle's book on faculties is an attempt to purify an interesting result of the combinatorial school by eliminating the use of the theory of combinations in mathematical analysis. In fact, it could even be said that Crelle goes further and actually expropriates a result that originally belonged to the German combinatorial analysis in order to give it to mathematical analysis. Therefore, Crelle's conception of mathematical analysis can be considered as an example of the countercurrent that is being studied here.

In their 1814 booklet, as pointed out in Sect. 6.1, Rothe and Ohm missed the opportunity to use or, at least, to promote Ohm's new results in the field of the German combinatorial analysis. Indeed, not only did they miss this opportunity, in general they did not use the methods developed by the combinatorial school (Rothe and Ohm 1814). This major change can be detected in several of Rothe's writings published in the nineteenth century. In 1804, Rothe presented a dissertation on the problem of dividing a circle into seventeen and thirteen equal parts in which he did not use combinatorial methods (Rothe 1804a). Yet, it could be argued that the subject does not lend itself to this kind of approach. In the same year, he published the first volume of a textbook devoted to pure mathematics, and the second volume appeared in 1811 (Rothe 1804b, 1811). None of the two includes any section on the theory of combinations. No problem, proposition, or theorem has been treated from a general combinatorial point of view, still less from a combinatorial point of view related to the techniques of the combinatorial school. The only result that can be associated with the combinatorial school is the following formula:

$$(x + z)(x + az)(x + a^2 z)(x + a^3 z) \cdots (x + a^{n-1} z) =$$
$$x^n + \frac{a^n - 1}{a - 1} z + \frac{(a^n - 1)(a^n - a)}{(a - 1)(a^2 - 1)} z^2 + \frac{(a^n - 1)(a^n - a)(a^n - a^2)}{(a - 1)(a^2 - 1)(a^3 - 1)} z^3 + \cdots,$$

which appears in the preface of the second volume without proof, and which is the earliest historical antecedent of what is today referred to as the "q-binomial theorem" in q-analysis (Johnson 2020, 49–53; Rothe 1811, XXIX). The product on the left-hand side of this identity is similar to the products studied in Kramp's theory of generalized factorials, in which Rothe probably found a source of inspiration, but this cannot be verified in his text since Rothe made no effort to relate this formula to the German combinatorial analysis or to the theory of combinations. The truth is that these two theories have been left out of Rothe's textbook. Two volumes devoted to pure mathematics without a single word of combinatorial elucidation. Two volumes devoted to arithmetic without a single word of explanation about the combinatorial nature of arithmetic. Two volumes that do not reflect any particular concept developed by the school Rothe contributed to create some 18 years earlier. It seems that Rothe's general conception of mathematics underwent dramatic changes during the nineteenth century. His 1820 book on combinatorial integrals proves that Rothe did not definitively renounce all his beliefs about the German combinatorial analysis, but it is clear that something happened to his core beliefs about mathematical analysis. A textbook on pure mathematics authored by a renowned member of the combinatorial school that contains no combinatorics cannot but be a strong message that the German combinatorial analysis does not provide the right answer to the puzzle about the nature of mathematical analysis. In this sense, Rothe himself joins in the nineteenth century the countercurrent of algebraic purity against his own extinct school.

6.2.3   Syntactic Current

At the end of Chap. 5, Bürmann's combinatorial characteristic or syntactic characteristic has been discussed in the context of the German combinatorial analysis. This was one of the main projects of Bürmann aiming to reform the entire edifice of human knowledge. An important part of his reform consists in reorganizing the totality of mathematics by formulating a new mathematical language. This language is characterized by the incorporation of a new symbolism into mathematics, which should in fact replace the current mathematical symbolism, and by a syntax for his new symbolism. Developing new mathematical symbols and manipulating them by establishing combinatorial rules was one of the objectives of the German combinatorial analysis. Bürmann saw in these objectives a weak version of his own syntactic characteristic and insisted that the German combinatorial analysis should be reformulated in such a way that it could fulfill its destiny to lead mathematics to perfection. Hindenburg pointed out that Bürmann's syntactic characteristic was different from the research project of his school, but he published Bürmann's works in his scientific journals and anthologies. These publications under the auspice of Hindenburg and Bürmann's appropriation of the terminology, problems, theorems, and methods of the German combinatorial analysis resulted in the interpretation of the theory of combinations as a syntactic theory. This kind of mathematical interpretation is what we call a "syntactic current of thought", which should not be confused with the German combinatorial analysis since it emerged from two different projects and, as will be seen below, the authors of this current proposed different ways of understanding this syntactic theory.

While it is true that this syntactic current originates from Bürmann's thought, it was Johann Friedrich Lorenz (1737–1807) who spread Bürmann's syntactic point of view: most of the mathematicians of the nineteenth century learned this syntactic theory from Lorenz, and not directly from Bürmann. Toward the end of his life, Lorenz embarked on preparations to undertake the task of presenting mathematics, at least its calculatory parts (i.e., excluding geometry), in the form of a system. According to him, the calculatory parts of mathematics are arithmetic, syntactics, algebra and analysis. When all these parts are articulated in a system, the system is called "logistic" (*Logistik*). Lorenz published two volumes of a handbook on logistic: the first volume deals with arithmetic, and the second with syntactics (Lorenz 1803, 1806). Probably his plan was to publish a third volume on algebra, and a final volume on analysis, but unfortunately he died before achieving his aim. Thus, we only have a partial view of his system. The first volume covers the typical subjects of arithmetic treated in textbooks of the time: arithmetical operations, a basic presentation of logarithms, arithmetic and geometric series, etc.; but it did not really catch the attention of his fellow mathematicians.

On the contrary, the second volume on syntactics was well received by the German mathematical community. This volume deals with the ideas of the combinatorial school. In an appendix, Lorenz gives a relatively long list of works of the combinatorial school, showing that he is well versed in the German combinatorial analysis. However, his list is neither exhaustive nor exclusive of works on the German combinatorial analysis. For instance, Fischer's books on the theory of dimension symbols appear in the list, while works mentioned in the body of the book are not included in the list. In fact, Lorenz does not say that this is a list of works on the German combinatorial analysis, but of works on syntactics or on the theory of combinations (which are synonyms of each other). In other words, this list suggests that Lorenz is not concerned with a formulation of the German combinatorial analysis, but with a reconstruction of the theory of combinations. Indeed, Lorenz explicitly claims that his book addresses the theory of combinations instead of Hindenburg's combinatorial analysis (Lorenz 1806, 20). It should be noted, however, that the theory of combinations reached its current state, for Lorenz, due to the work of Hindenburg's school, and thus Lorenz's book presents mostly the combinatorial school's results in the theory of combinations. Nevertheless, Lorenz's conception of this theory is not exactly the same as that of Hindenburg's school, given that Lorenz intermixes the theory of Hindenburg's school with other conceptions. Among the works that are not included in Lorenz's list but that are mentioned in his book, the most important are Bürmann's *Essai de caractéristique combinatoire* and *Polynome combinatoire*, in which Bürmann discusses his project of syntactic characteristic and its application to the German combinatorial analysis. From these works of Bürmann, Lorenz takes the term "syntactics" to describe the theory of combinations, and he attempts to reinterpret the combinatorial school's works on the theory of combinations from Bürmann's perspective. Thus, as Bürmann, Lorenz presents the theory of combinations through the methods developed by Hindenburg's school, but he theorizes it differently.

The book contains a detailed explanation of combinations, permutations, and variations. Lorenz uses the concept of "Hindenburg's classes" to develop the first elements of the theory of combinations, and then he discusses Hindenburg's theory of combinatorial involutions. Although the book does not aim to show the large range of applications of combinatorics, some of these can be found in the text. In the field of the theory of series, Lorenz gives a proof of the binomial theorem for rational exponents that can be classified as a functional proof, similar to those studied in Sect. 2.1.3.2. Then he applies this theorem to the multinomial theorem in order to obtain Hindenburg's formula for the general coefficient of the multinomial expansion. Lorenz also applies the theory of combinations to some problems of arithmetic, such as the construction of number systems or some problems of cryptography. Therefore, the thematic choices of the book reflect Lorenz's belief that the essential methods of the theory of combinations have been developed by the combinatorial school, avoiding getting into the discussion of some delicate issues of the German combinatorial analysis, as the priority problem of the binomial and multinomial theorems.

On the other hand, from a theoretical point of view, the theory of combinations can be reduced to the elementary operation of "binding" together (*Verbindung*) elements according to a given condition. In Bürmann's syntactic characteristic, this can be interpreted as follows. In 1801, Bürmann proposed the use of nine symbols to codify mathematical knowledge. This codification was supposed to depend on the operation of binding together these symbols according to certain rules. These rules were considered by Bürmann as the syntax of his language composed of nine symbols, but his syntax was of a combinatorial nature in the sense that those rules were nothing but the specification of how to combine the symbols of the language in order to express a precise piece of mathematical knowledge. However, Bürmann failed to define the combinatorial syntax of his language. In his book, Lorenz subscribes to Bürmann's general conception of combinatorial syntax, but he discards Bürmann's symbolism and is compelled to solve the problem of defining the syntactic rules. Lorenz thinks that these syntactic rules are given by the rules that define the combinatorial methods invented by the combinatorial school, such as the rules that define a combinatorial involution, and the symbols governed by these rules are those created by Hindenburg's school. Although possible in principle, this solution is not practical. Instead of having a general, unified syntax, Lorenz's solution has to cope with the existence of a multiplicity of unconnected rules and its subsidiary problems, as the problem of determining the coherence of the rules. Thus, the challenge of assuming that the theory of combinations is a syntactic theory consists in defining the syntactic rules of the theory. Bürmann failed as well as Lorenz, but other scholars attempted to find a solution to this problem.

In the same vein, the mathematician Jacob Struve (1755–1841) wrote a textbook on mathematics in two volumes, the first devoted to arithmetic and the second to the theory of combinations or syntactics (Struve 1808, 1809). Struve was substantially influenced by the work of Lorenz, to such an extent that his general conception of mathematics replicated that of Lorenz. Following Lorenz, Struve divided mathematics into two parts: geometry and logistic, which was also called "general arithmetic" (*Arithmetik im weitern Umfange*) by Struve (1808, 7–8). Logistic is composed of the same four parts that Lorenz had considered in his classification. In particular, Struve credits Lorenz for taking the initiative to rename Hindenburg's combinatorial analysis with the more meaningful term of "syntactics". Thus, it is possible that Struve was unaware of Bürmann's work on the syntactic characteristic. Struve's remark also makes clear that, for him, the German combinatorial analysis is the theory of combinations, which is other name for syntactics. As we know, the German combinatorial analysis cannot be reduced to the theory of combinations, and thus Struve's view drifts away from the combinatorial school's project, moving nearer to an interpretation of the German combinatorial analysis such as that of Weingärtner. In the preface of the second volume devoted to the theory of combinations, Struve points out, in fact, that, besides Hindenburg, his sources are Lorenz's and Weingärtner's textbooks. The content of the second volume is similar to that of Lorenz's book on syntactics, but it contains more applications. For instance, there are applications to Kramp's theory of generalized factorials and to the theory of probability. On the other hand, Struve reworks Rothe's proof of the binomial theorem for rational exponents, and then he used it to deduce the multinomial theorem. Concerning the delicate issue of defining the syntactic rules of the theory, Struve does not go further than Lorenz. There is, however, an interesting terminological change in Struve's textbook. Traditionally, the general term used to designate a combinatorial arrangement (whether combination, permutation or variation) was "complexion", which is employed by both Lorenz and Struve. Lorenz also uses the term "*Verbindung*" instead of "complexion", but Struve goes one step further and claims that a complexion can also be called "*Syntax*", which could be translated as "syntactic group" to avoid confusion with "*Syntax*" as a part of grammar (Struve 1809, 3). This terminological change obeys the idea that the aim of the theory of combinations consists in providing logistic with syntactic groups in order to express mathematical results. For instance, for Struve, the binomial coefficients $\frac{n}{1}$ and $\frac{n(n-1)}{1\cdot 2}$ are expressed, respectively, by the syntactic terms $^{n}\mathfrak{A}$ and $^{n}\mathfrak{B}$ which are associated to syntactic groups (Struve 1809, 33).[4] Thus, Hindenburg's notations for binomial

---

[4] Diese Binomialcoefficienten $\frac{n}{1}$, $\frac{n}{1}\cdot\frac{n-1}{2}$, u.s.w werden syntaktisch durch $^{n}\mathfrak{A}$, $^{n}\mathfrak{B}$, $^{n}\mathfrak{C}$, ..., $^{n}\mathfrak{M}$, ...ausgedrückt [...].

coefficients are conceptualized as symbolic codifications that contain some information because of the manner in which they are codified. The manner in which they are codified can be reduced, in principle, to syntactic rules. The problem is that Struve, as Lorenz, was not able to verbalize these rules as a coherent, unified system.

As the foundationalist current, the syntactic current harbors a philosophical side as well. The German philosopher Karl Christian Friedrich Krause (1781–1832) studied philosophy at the University of Jena under Schelling, Fichte, and Schlegel, but he was also interested in science and attended some lectures of mathematics, including those delivered by Stahl, who worked as Privatdozent from 1795 to 1799, and then as extraordinary professor of mathematics from 1799 to 1802 at that university. For Krause, philosophy and mathematics should be conceived as being complementary domains in a general system of knowledge. In particular, there is no mathematical knowledge without philosophy, that is to say, mathematical sciences depend on philosophical principles. Probably, Krause paid particular attention to the role of the theory of combinations in mathematics because of Stahl's lectures, or perhaps because of the increasing dissemination of the German combinatorial analysis. In any case, the theory of combinations occupied an important place in what he called a "philosophical system of mathematics" (Krause 1804, vii).

In 1804, Krause gave an overview of his philosophical system of mathematics in his *Grundlage der Arithmetik* (Krause 1804, 3 ff.). According to Krause, the key to making mathematics an actual scientific discipline is to answer whether there is an ultimate principle of mathematical knowledge. This question cannot be answered by mathematics itself, but by philosophy since it is just a particular formulation of a more general problem, namely, that of the existence of an ultimate axiom of all knowledge. Krause elaborates a theory of schematism, clearly inspired by Kant's, to explain knowledge, or, more precisely, the formation of concepts. A concept is formed by means of a schema that links the sense data obtained by sensory apprehension to a mental representation of the object called "idea". However, sensory apprehension, or indeed any kind of apprehension, occurs in time, and thus schematism alone cannot explain that a thing is, but just that something has been in a precise moment of time. The identity of the object remains inaccessible to schematism. The identity of objects rests on what Krause calls "the absolute infinite unity" or "the Absolute". An object in its becoming can be recognized as being the same because it is conceived within the Absolute. In speculative philosophy, the Absolute becomes then the ultimate axiom of knowledge, since it allows the identification of characteristics shared by a multiplicity of things and this identification leads to conceptualize such a multiplicity as being a unity. Formation of concepts depends on the possibility of constructing such unities from multiplicities, and thus the Absolute can be considered as an epistemological principle. In the particular case of mathematics, Krause gives, among others, the example of positive whole numbers. He claims that there exists an infinity of positive whole numbers, but this cannot be proved in mathematics. For him, the conceptual representation of the totality of positive integers can only occur within the framework of the Absolute: in this infinite framework, one can place an infinite totality. Once the infinite totality of positive integers has been so constructed, it is possible to apprehend the idea of "positive integer" as a unity of unites, that is to say, this infinite totality is represented as an actual infinite sequence of unities, then the mind forms a schema for a given positive integer as a determined collection of the unities in the sequence, and finally one gets the concept of "positive integer" as being a determined collection of unities. This construction contains, according to Krause, the elements of what should be a correct mathematical reasoning: first, mathematical reasoning should take place in the framework of the Absolute; second, elaboration of a schema; and finally, formation of mathematical concepts as the result of this schematism framed in the Absolute. In this concept of "positive integer" as a determined collection of unities, the term "determined" means that this collection is a finite and particular unity. Finite and particular unities are called "forms" by Krause. Reality is divided into two spheres: nature and reason. The form of nature is space, the form of reason is conceptual unity, and the form that links nature and reason together is time. As a consequence, mathematics is a theory of forms (Krause 1804, 17):

> Mathematics is the systematic and synthetic representation of forms (limits, formal conditions) of being and becoming of particulars.[5]

And mathematics does have its own ultimate axiom: every aspect of reality can be perfectly understood by representation of forms (Krause 1804, 27).

Mathematics can be divided into two main parts, arithmetic and geometry, related to time and space considered as forms. The theory of combinations is a part of arithmetic, and Krause describes it as a general characteristic in the sense that this term has been used by authors like Leibniz or Bürmann (Krause 1804, 68–69). Indeed, Krause points out that Leibniz, Hindenburg, and "a few others" developed the theory of combinations as a general characteristic. In previous chapters, it has been explained that Hindenburg and the combinatorial school never envisaged such a Leibnizian interpretation of the theory of combinations. It has been shown that even Weingärtner differentiates the Leibnizian characteristic from the combinatorial school's project on the basis of a Kantian interpretation of Hindenburg's work. Thus, Krause's interpretation of the theory of combinations as a general characteristic might be based directly on Leibniz's works or on Bürmann's syntactic characteristic. In his *Grundlage der Arithmetik*, Krause does not make any reference to Bürmann. Therefore, the direct influence of Bürmann cannot be established from this book, even if it is far from being improbable. In any case, Krause wrote a very positive review on Lorenz's textbook in 1807, which had a considerable effect on his conception of the theory of combinations (Krause 1807). As indicated above, Krause defined mathematics as the science of forms in 1804 and thought that the theory of combinations was a part of arithmetic. As a consequence of his reading of Lorenz's book, Krause changed his mind and affirmed that the theory of combinations is

---

[5] Die Mathematik ist systematische und synthetische Darstellung der Formen (Grenzen, förmlichen Bedingungen) des Seins und Werdens alles Individuellen.

not subordinated to arithmetic, but it is an independent mathematical discipline, defined as the science of relations or as the science of the category of relation, echoing Kant's categories (Krause 1807, 2099). Indeed, he asserted in his review that arithmetic and the theory of combinations are at the same level of hierarchy in mathematics, inasmuch as the Kantian's categories of quantity and relation are at the same level of hierarchy in the system of pure understanding. Hence, in his philosophical system of mathematics, the theory of combinations is no longer devoted to the construction of forms, but to the regulation of binding forms correctly. This regulation acts as a sort of syntax of forms, at least from a theoretical point of view. In 1812, Krause and Ludwig Joseph Fischer (?–1813) published an elementary textbook on the theory of combinations and arithmetic (Fischer and Krause 1812). Krause says in the preface that, although pedagogical purposes prevented a full systematic presentation, the content of the book is approached from the perspective of his philosophical system of mathematics, which had captivated the attention of his friend Fischer. Lorenz's book, along with those of Stahl and Weingärtner, are listed by Fischer among the important sources for combinatorics, and both Krause and Fischer call "syntactics" the theory of combinations in accordance with Lorenz. It is worth noting that they recommend to study logic as a propaedeutic for mathematics; in particular, Krause recommends his own syllogistic which has been covered "in a complete combinatorial manner" (*mit combinatorischer Vollständigkeit*). It seems that the discovery of Lorenz's textbook offered Krause a new rare glimpse of the theory of combinations as a theory that might govern mathematical relations by a sort of syllogistic syntax.

As seen in Sect. 6.2.1, Fries developed, in 1807, a combinatorial foundationalist conception of mathematics from a philosophical point of view. He held that the faculty of thought (*Denkkraft*) produces mathematical knowledge through productive imagination's fundamental operation of combining. In 1811, Fries keeps loyal to his foundationalist theory, but he fleshes it out further with a new syntactic component. In his *System der Logik*, he describes how a symbolic mode of representation can function to provide knowledge (Fries 1811, 374 ff.). Every symbolic mode of representation consists of a system of symbols, and every system of symbols is composed of two elements: the matter and the form of symbols. Matter of symbols refers to the objects chosen to serve as symbolic codifications, for instance: letters of different alphabets, digits, smoke, sounds, and so on. Form of symbols refers to the way in which symbols are put together. The objects chosen for a symbolic system are considered as basic symbols in contrast to composed symbols which are symbols formed by putting together some basic symbols in a specific way. Composed symbols have meaning because of their form, that is to say, because of the way in which their basic symbols are arranged. This semiotics, as Fries calls it, is not restricted but can be successfully applied to mathematics. Fries claims that arithmetic is nothing but an artificial combination of the ten digits chosen as basic symbols, and the same applies to algebra and analysis with respect to their corresponding basic symbols. Therefore, in 1811, Fries theorizes mathematics as a particular kind of semiotics which is regulated by combinatorial rules. Unfortunately, his theorization is not supported by a further elaboration of the system of symbols to which mathematics could be reduced: a clear identification of the basic symbols of the system is missing, as well as a precise definition of its combinatorial rules. However, despite the lack of an actual development of this system, Fries's foundationalism moves from being a psychologistic to a syntactic approach. Even though the operation of combining remains an irreducible operation of the mind, it can be translated, in principle, into objective semiotic rules.

Besides these philosophical approaches to the theory of combinations, and besides Lorenz's and Struve's extensive reformulations of the German combinatorial analysis in accordance with Bürmann's syntactic characteristic, the syntactic current manifested itself in some texts in which there is no particular discussion of the characterization of the theory of combinations as a syntactics, but just a presentation of this theory under the name of "syntactics". This lack of discussion might be interpreted as indicating that the conception of the theory of combinations as a syntactic theory has been assimilated and accepted by some mathematicians in their professional practice, even though the syntactic rules of the theory still remain rather obscure. In 1817, Friedrich Schmeißer (1785–1869) called "syntactics" the theory of combinations, which is discussed in a section of his *Lehrbuch der Mathesis* (Schmeißer 1817). Although his presentation of this theory is elementary, Schmeißer makes reference to several works of Hindenburg, including some texts edited by Hindenburg that contains some of Bürmann's works. He also mentions Töpfer, Ernst G. Fischer, Stahl, and Weingärtner. In spite of a long philosophical discussion of over a hundred and fifty pages about the nature of mathematics, which constitutes the preface of his textbook and in which Fries's *System der Logik* is quoted, Schmeißer does not examine the philosophical or mathematical implications of considering the theory of combinations as a mathematical syntactics. In 1844, Ludwig Friedrich Ritter includes a brief section on the theory of combinations in a textbook on algebra and analysis, in which there is no sign of the German combinatorial analysis other than the use of Lorenz's term "syntactics" (Lefébure de Fourcy et al. 1844). In 1858, Christoph Ludwig Schoof (1810–?) uses this term as a synonym of the theory of combinations and devotes the third volume of his *Arithmetik und Algebra* to syntactics, without discussing why the theory of combinations can be conceived as a syntactic theory (Schoof 1858). The use of some mathematical notations of the combinatorial school and a brief application of the theory of combinations to the theory of series, including the binomial and multinomial theorems, revels some influence of the German combinatorial analysis on Schoof. In 1869, Hermann Fahland published a paper on the theory of combinations or syntactics, gives a brief discussion of the binomial theorem, and uses some mathematical notations of the combinatorial school (Fahland 1869). A very elementary introduction to the theory of combinations or syntactics is included in a textbook by Gotthelf Weber, who uses Ettingshausen's combinatorial notations (Weber 1869). The extent of the syntactic current in the nineteenth century can be measured by all these manuals, including those of Lorenz, Struve, Krause, and Fries, but it can also be measured by references to them that can be found in other mathematical textbooks in which no combinatorial perspective is assumed. For instance, Johann Tobias Mayer (1752–1830) published a textbook on mathematical analysis in 1818 in which he gives a differential proof of the binomial theorem and suggests that the multinomial theorem can be easily derived from the first (Mayer 1818, 233 ff.). Then, he points out that more complete investigations of those topics have been carried out in the field of the German combinatorial analysis, and he offers three source for this discipline: Hindenburg's scientific journal *Sammlung*, Töpfer's *Combinatorische Analytik*, and Lorenz's *Syntactik*. This gives witness to the fact that the syntactic current extends even further the ramifications of the general combinatorial current of thought to mathematicians that are not interested in combinatorics but that, unconsciously, help transmit this syntactic approach of the theory of combinations, and it also gives witness to the fact that the German combinatorial analysis retreats as the combinatorial current of thought advances in the nineteenth century.

This kind of indirect promotion of the syntactic current also takes place through some writings that offer some theoretical insights into the theory of combinations. For instance, Christian August Semler (1767–1825) wrote several essays on the theory of combinations in which, without giving a mathematical exposition of the theory, he explores the positive effects that this theory has had in improving human culture in general, from mathematics to fine arts. These essays were originally published in the journal *Neues allgemeines Intelligenzblatt* during the year 1809, then they were collected in a single volume in 1811, which was reprinted in 1822 (Semler 1809, 1811, 1822). The essays do not focus on the German combinatorial analysis or on the authors that have been identified in this chapter as belonging to the combinatorial current of thought of the nineteenth century. Hindenburg is quickly mentioned in just two paragraphs, and no other member of the combinatorial school appears in the pages of the essays. Something similar happens with the authors of the combinatorial current of thought, which are completely absent in Semler's discourse. However, in the second edition of 1822, Semler added a footnote in which he recommends five manuals to his readers as books that fully cover the theory of combinations from a mathematical point of view: those of Stahl, Weingärtner, Lorenz, Schmeißer, and that of Krause and Ludwig Joseph Fischer (Semler 1822, 9–10, footnote). It is interesting to note that Semler does not even include a work of Hindenburg. Thus, Semler's book contributes to indirectly disseminating, on the one hand, Stahl's and Weingärtner's views on the theory of combinations, which are different from those of the combinatorial school, as seen in Chap. 5, and, on the other hand, some different syntactic approaches to the theory of combinations.

The syntactic current and the German combinatorial analysis shared a common interest in developing suitable mathematical notations, but that raised the question of how to decide whether a mathematical notation is suitable or not. At the end of the eighteenth century, Hindenburg and Bürmann were locked in a discussion of the matter which turned out to be a dead end, as seen in Chap. 5. Hindenburg defended the combinatorial school's notations on the basis of their practical advantages, while Bürmann argued that his analytic ideography was even more practical. An important part of the problem was the lack of objective criteria to guide decision. As seen in Chap. 5, Bürmann had to apologize to the French Academy for using his innovative notations and, to be taken seriously, he had to recognize that traditional notations were elegant enough to meet the needs of mathematical sciences. But "elegance" is not really an objective criterion to decide. The problem of constructing a symbolic mathematical language seemed to be stuck in a matter of taste. In 1831, Weingärtner gave a lecture on the mathematical notations of the German combinatorial analysis before the Academy of sciences in Erfurt (*Königliche Akademie gemeinnütziger Wissenschaften zu Erfurt*) (Weingärtner 1831). In his speech, he tries to compare some of the mathematical notations related to the German combinatorial analysis in order to determine their respective advantages. Weingärtner recalls the dispute between Hindenburg and Bürmann, and quotes a long passage of a letter in which Hindenburg tells him the reasons why the combinatorial school's notations are better than those of Bürmann, which have been analyzed in Chap. 5. Then, his speech focuses on what Weingärtner considered to be the most outstanding notations so far, mostly comparing notations for Hindenburg's classes of combinations and binomial coefficients. For him, the best notations are those of Hindenburg-Thibaut, Schweins, and Eytelwein. The works of Thibaut and Eytelwein related to the German combinatorial analysis will be discussed in Sect. 6.2.4. Weingärtner thinks that Thibaut made some improvements to Hindenburg's notations, but that these are indeed slight modifications, and thus Hindenburg's and Thibaut's notations can be considered as the same. Although Weingärtner made a fine effort to understand the theoretical implications of modifying mathematical notations, the truth is that he reaches no conclusions with respect to which one is a more "suitable" notation and why. Instead of being a theoretical contribution to the understanding of mathematics, his speech turns into a relentless historical observation that mathematical notations change over time. Of course, this does not call into question Weingärtner's mathematical talents, nor does it question the talent of the mathematicians working in the direction of the syntactic current. The difficulty to solve the problem of understanding what it is to be a suitable mathematical notation, a problem that all these mathematicians had been trying to solve for the first 70 years of the nineteenth century, shows how challenging it can be to transform mathematics into a symbolic language.

In 1872, Robert Grassmann (1815–1901) took an important step toward this transformation of mathematics when he published his *Die Bindelehre oder Combinationslehre* (Grassmann 1872a). As is well known, Grassmann belonged to a family of mathematicians. His father Justus Günther Grassmann (1779–1852) entered the University of Halle with the intention of studying theology in 1799, but his inclination to scientific reasoning impelled him to the study of mathematics and, as he points out, his teacher of mathematics was Klügel (Grassmann, H. G. 2009, 36). Apparently, Klügel's teachings awakened in him a deep interest in the theory of combinations, which he tried to apply to crystallonomy in one of his major works (Grassmann 1829). Probably, his son Robert learned the rudiments of combinatorics from him, and thus Robert's first notion of combinatorics was closely related to the developments of the combinatorial school. Indeed, Robert's intellectual interests followed a path similar to that of his father. Although Robert never abandoned his studies of mathematics, he devoted increasingly attention to theology when he moved to Bonn in 1834 to pursue his formation at the university. In 1838, he returned home to Stettin and, in his father's home, "the old love of mathematical sciences re-ignited" (Grassmann, R. 2009, 100). Then, he took up his studies in mathematics by reading some of Lacroix's, Cauchy's, and Lagrange's works, and also by improving his knowledge of the German combinatorial analysis through the study of Ettingshausen's textbook (Grassmann, R. 2009, 103).

Clearly, Robert Grassmann's knowledge of the German combinatorial analysis went beyond Ettingshausen's presentation of this theory. For instance, Klügel defined mathematics as "the science of forms of magnitudes" (*die Wissenschaft von den Formen der Größen*) (Klügel 1808, 602), and Grassmann gave mathematics the name "theory of forms" (*Formenlehre*). This indicates that Grassmann's general concept of mathematics owes a debt to the German combinatorial analysis, even though the notion of "form" was as obscure in 1872 when he published his *Die Fromenlehre oder Mathematik* as it was in 1796 when Klügel used it to justify the multinomial theorem for rational exponents (see Sect. 4.2.2.1) (Grassmann 1872b). Indeed, Grassmann gave neither a definition nor an explanation of this notion. Instead, he elaborated his views on mathematics in terms of a symbolic mathematical language. In his *Die Fromenlehre oder Mathematik*, Grassmann developed what could be described as an incipient philosophy of language, within which his general concept of mathematics was framed (Grassmann 1872b, 5 ff.). According to Grassmann, natural languages are intrinsically ambiguous. A given term, for example, "love" (*Liebe*), can have different meanings depending on its specific context of use. This intrinsic ambiguity leads to confusion and fallacies when natural languages are used to express scientific results. On the other hand, even if confusion and fallacies are avoided, scientific knowledge cannot be universally formulated in any natural language because natural languages are historically and socially determined by their linguistic communities. Since the theory of forms must be free from ambiguity and universally valid, its correct formulation cannot be guaranteed by any natural language. For Grassmann, the theory of forms requires a universal symbolic language in which every mathematical relation among magnitudes (*Größen*), which are the general object of all mathematics, could be conveniently expressed. He offers a general description of this universal symbolic language. It is composed of five different sets of symbols: (i) letters e_1, e_2, e_3, etc. designate a specific particular magnitude, thus these symbols are used to represent constants, (ii) letters a, b, c, etc. designate a non specific particular magnitude, thus they are used as variables, (iii) parentheses are punctuation marks, (iv) there are symbols for mathematical operators, for example, Grassmann uses the symbol ∘ as a general undefined operator, and (v) symbols = and ⋛ designate the specific relations "equal to" and "not equal to". Mathematics or the theory of forms consists in constructing formulae by combining symbols of these five categories correctly. Thus, in this symbolic language of formulae, natural languages are completely excluded, although Grassmann points out that the use of a natural language is recommended and even necessary for pedagogical reasons and for human communication. It is worth quoting Grassmann's explanation of the mathematical relation "equal to" (Grassmann 1872b, 7):

> Two magnitudes are said to be equal when one of these magnitudes can by replaced by the other in every arrangement of the theory of forms without modifying the value [of the arrangement].[6]

This constitutes a clear historical antecedent for the well-known substitution rule of Gottlob Frege (1848–1925). Here the term "arrangement" (*Knüpfung*) has a strong combinatorial connotation since it makes reference to a formula obtained by putting together some of the symbols of the five groups (i)–(v) in a precise combination. In the combinatorial current of thought in the nineteenth century, it is not uncommon to find this kind of expressions. For instance, in his textbook, Ettingshausen uses the term "*Verknüpfung*" in the similar sense of arranging symbols combinatorially, without making reference, however, to the alphabet of a predetermined mathematical language such as that of Grassmann. Thus, Grassmann's substitution rule evokes a combinatorial component of this symbolic language, and Grassmann elaborates further on this component in his *Die Bindelehre oder Combinationslehre*.

According to Grassmann, mathematics can be divided into a general branch and four specific branches (Grassmann 1872b, 11 ff.). The general branch of mathematics is called "theory of magnitudes" (*Grösenlehre* [sic]), in which magnitudes are studied in abstraction from any further classification of magnitudes. In other words, the theory of magnitudes aims to establish mathematical laws of addition, multiplication, and exponentiation, but these laws should be common to all mathematical objects. Indeed, Grassmann's concept of mathematics depends on the idea that mathematical knowledge can be articulated in a system by defining laws of addition, multiplication, and exponentiation in each branch. Besides these laws, the four specific branches are characterized as "internal" or "external" depending on the nature of the representations related to the arrangements of the theory of forms: if the representation of a given arrangement is considered as merely conceptual, it is an internal representation; on the contrary, if the representation is considered as being linked to the external world, it is an external representation. There are two internal branches: logic and the theory of combinations; and there are two external branches: arithmetic and external theory (*Ausenlehre*). Internal branches deal with concepts strictly. Logic determines mental contents conceptually, while the theory of combinations is responsible for bringing order to our world of concepts. In particular, this means that, for Grassmann, the theory of combinations should play the role of regulating the correct formation of formulae in his symbolic language. This regulation is implemented through the particular laws of addition, multiplication, and exponentiation of the theory of combinations. Grassmann defined these laws in his *Die Bindelehre oder*

---

[6] Gleich heisen zwei Grösen, wenn man in jeder Knüpfung der Formenlehre die eine statt der andern ohne Aenderung des Werthes setzen kann.

Combinationslehre (Grassmann 1872a, 7 ff.). In the theory of combinations, addition is idempotent, so for every element e of this theory, the following rule applies:

$$e + e = e.$$

On the contrary, multiplication is not idempotent, but classical combinatorial operations are characterized by different properties of multiplication. Commutativity serves to distinguish between combinations and variations, that is to say, commutativity does not hold in general, but just for combinations; so if a product is commutative:

$$e_1 e_2 = e_2 e_1,$$

it is a combination, if not, it is a variation. A combination without repetition is defined as a product that is equal to zero if it contains two equal factors, while a combination with repetition is a product different from zero if it contains two equal factors. Similar definitions are given for variation with and without repetition. The notion of classes of combinations and of variations inherited from the combinatorial school is defined by Grassmann by means of the operation of exponentiation. In the theory of combinations, exponentiation is defined only for positive whole exponents and zero, and the result of this operation is a class. In contemporary terms, if n is a positive integer, Grassmann's exponentiation to the power n produces the set of all the n-combinations or n-variations. This kind of sets corresponds to what was referred to as nth classes of combinations or variations by Hindenburg, classes that Grassmann intends to characterize by means of the notion of "exponentiation". The distributive property is expressed in the theory of combinations by the formula:

$$(S_{1,n} e_\mathfrak{a})^m = S_{1,n} \overline{e_\mathfrak{a} e_\mathfrak{b} e_\mathfrak{c} \cdots},$$

where $S_{1,n}$ is the addition operator with summation indices, and $\mathfrak{a}$, $\mathfrak{b}$, $\mathfrak{c}$, etc. are indices that take all values between 1 and n. For instance, if e_1, e_2 and e_3 are elements of the theory of combinations, one gets:

$$(e_1 + e_2 + e_3 + e_1)^2 = S_{1,3} \overline{e_\mathfrak{a} e_\mathfrak{b}}$$

$$= \begin{vmatrix} e_1 e_1 & e_2 e_1 & e_3 e_1 \\ e_1 e_2 & e_2 e_2 & e_3 e_2 \\ e_1 e_3 & e_2 e_3 & e_3 e_3 \end{vmatrix}$$

because of the idempotence property of addition. In fact, if one performs the exponentiation operation on the left-hand side algebraically, one gets:

$$\begin{vmatrix} e_1e_1 & e_2e_1 & e_3e_1 & e_1e_1 \\ e_1e_2 & e_2e_2 & e_3e_2 & e_1e_2 \\ e_1e_3 & e_2e_3 & e_3e_3 & e_1e_3 \\ e_1e_1 & e_2e_1 & e_3e_1 & e_1e_1 \end{vmatrix} = (e_1 + e_2 + e_3 + e_1)^2 = \begin{vmatrix} e_1e_1 & e_2e_1 & e_3e_1 \\ e_1e_2 & e_2e_2 & e_3e_2 \\ e_1e_3 & e_2e_3 & e_3e_3 \end{vmatrix}$$

Thus, Hindenburg's classes behave as our current sets in set theory, i.e. adding a given element to a set in which that element is already contained yields the same set: $\{a, b, c, \ldots\} = \{a, a, b, c, \ldots\}$. In the works of the combinatorial school, this behavior has been assumed tacitly, and Grassmann not only makes it explicit here, but he also gives a mathematical justification to such a behavior: the idempotent property of addition in the theory of combinations. Then, by applying the definitions given above and the idempotence property of addition, one gets the second classes of variations with and without repetition:

$$\begin{vmatrix} e_1e_1 & e_2e_1 & e_3e_1 \\ e_1e_2 & e_2e_2 & e_3e_2 \\ e_1e_3 & e_2e_3 & e_3e_3 \end{vmatrix} \qquad \begin{vmatrix} - & e_2e_1 & e_3e_1 \\ e_1e_2 & - & e_3e_2 \\ e_1e_3 & e_2e_3 & - \end{vmatrix}$$

and the second classes of combinations with and without repetition:

$$\begin{vmatrix} e_1e_1 & - & - \\ e_1e_2 & e_2e_2 & - \\ e_1e_3 & e_2e_3 & e_3e_3 \end{vmatrix} \qquad \begin{vmatrix} - & - & - \\ e_1e_2 & - & - \\ e_1e_3 & e_2e_3 & - \end{vmatrix}$$

Grassmann's reformulation of the theory of combinations as an algebra with three basic operations is clearly inspired by the work of the combinatorial school, in which the application of the theory of combinations to mathematical analysis depended on interpreting combinations and variations as products, and the corresponding classes as sums of products. However, Grassmann's approach is embedded in the pursuit of mathematical knowledge through the construction of a mathematical language with a combinatorial grammar, a goal that places it in the syntactic current of the nineteenth century.

Grassmann claims that he and his brother Hermann Günther Grassmann (1809–1877) engaged in a joint effort to give the four specific branches of mathematics mentioned above the most complete possible scientific shape, and that Hermann's extension theory is an example of this joint effort (Grassmann 1872b, 14). Thus, the Grassmann brothers shared a general concept of mathematics, although it seems that the theory of combinations had a smaller influence on Hermann's work. In his *Die Ausdehnungslehre*, Hermann uses some elements of the theory of combinations, but this is far from the central place that combinatorics occupied in Robert's symbolic mathematical language (Grassmann 1844, 1862). Thus, Robert's ideas had a rather moderate impact on his brother's work. On the contrary, Robert Grassmann's views on mathematics were decisive in transforming mathematical knowledge into a symbolic language. It is well known that Ernst Schröder (1841–1902) "discovered the analogy between arithmetical and logical connectives" through Robert's *Die Fromenlehre oder Mathematik* (Peckhaus 1994), which led him to deepen his study of mathematics from the perspective of logic. Less well known is the fact that Frege owes a substantial debt to Robert Grassmann. For instance, in Weiner (2020, 52–53) and Petsche et al. (2011), there are slight attempts to link Frege's and Grassmann's thoughts. In the first case, Weiner does not discuss the influence of Grassmann on Frege's logic, but he just points out that Frege thought that ambiguity can be found not only in natural languages, but also in symbolic languages, and, to support his view, he quotes a passage in which Frege criticizes, among others, Grassmann's symbolic language. In the second case, since the book is about Hermann's work, the emphasis is on the link between Frege and Hermann, and a possible link between Frege and Robert is just mentioned quickly and indirectly. However, the passage quoted by Weiner, which can be found in Frege (1993, 112), shows that Frege had carefully studied Robert Grassmann's theory of forms. In his own logic, Frege discarded Grassmann's combinatorial grammar and, certainly, he criticized other aspects of Grassmann's symbolic language, as pointed out by Weiner. However, as seen above, the rule of substitution, one of the most characteristic results of Frege's logic and philosophy of language, was originally formulated by Robert Grassmann. This suggests that Frege's thought was enriched by the work of Grassmann, and, as a consequence, the German combinatorial analysis was a distant precursor of modern logic.

6.2.4 Non-Foundationalist Current

In his *Geschichte der Mathematik*, Johann Heinrich Moritz Poppe (1776–1854) affirms that Hindenburg founded the current combinatorial analysis around 1779, and that many others contributed to its improvement, including Eschenbach, Ernst G. Fischer, Rothe, Töpfer, Burckhardt, Stahl, and Weingärtner (Poppe 1828, 161). This heterogeneous list of authors suggests that Poppe does not want to make a distinction between Hindenburg's combinatorial analysis and the theory of combinations. Even assuming that Poppe was unable to distinguish the German combinatorial analysis from Stahl's and Weingärtner's derivations, it would be really hard to confuse Fischer's theory of dimension symbols with the German combinatorial analysis. Poppe uses "combinatorial analysis" instead of "theory of combinations" deliberately, and thus his list makes sense as a list of mathematicians that contributed to the development of the theory of combinations. In other words, Poppe writes his history of mathematics incorporating the work of the combinatorial school into the most general development of the theory of combinations. A similar historical interpretation was given by Bernhard Friedrich Mönnich (1741–1800) in his *Lehrbuch der Mathematik*, which includes a history of mathematics in an addendum (Mönnich 1801). Mönnich claims that if Fischer affirms that he did not plagiarize Hindenburg's theory, then there was no plagiarism involved at all, and thus Fischer's theory of dimension symbols should be considered as an original contribution to the development of the combinatorial analysis (Mönnich 1801, 551–552). Here again "combinatorial analysis" means "theory of combinations". In mathematics, this assimilation (or reduction) of the German combinatorial analysis to the theory of combinations began, as seen in Chap. 5, with Stahl's and Weingärtner's textbooks, and this tendency kept increasing during the nineteenth century. A large number of textbooks in the nineteenth century shows this tendency in their pages, which is characterized by the absence of any attempt of founding mathematics, or a part of mathematics, on combinatorics. Instead of presenting combinatorics as a foundation of some area of mathematics, these textbooks integrates the theory of combinations into mathematics as just another field. In these manuals, it can be observed that, instead of studying the German combinatorial analysis by its own interest, some elementary results of the combinatorial school have been extracted from their original context (i.e., the German combinatorial analysis) and placed in the context of general mathematics. Thus, these textbooks show a process of conceptual appropriation in which the incorporation of some results of the combinatorial school into mathematics is completely dissociated from the development of the German combinatorial analysis. This is a historical process of assimilation of relevant results and, at the same time, of oblivion of the particular goals, interests, concepts, etc. of a specific research project.

Most of the documents that will be studied in this section are textbooks published in German-speaking countries, whose aim was to supply students of secondary schools with the necessary texts for their mathematical formation. In his analysis of the institutional organization of mathematics teaching, Lorey writes (Lorey 1916, 29):

> However, the combinatorial school did not have the influence on the organization of university studies that Jacobi gained later in Königsberg. On the contrary, in the teaching of secondary schools, its effects can still be felt up to recent times, inasmuch as permutations, variations, and combinations formed a field of study frequently addressed in mathematical teaching, and more precisely in the way Hindenburg addressed those things that he introduced, i.e. with no relation to the concept of group.[7]

---

[7] Auf die Organisation des Hochschulstudiums hat die kombinatorische Schule aber bei weitem nicht den Einfluß gehabt, den Jacobi später von Königsberg aus gewonnen hat. Im Unterricht der höheren Schulen kann man aber ihre Nachwirkung noch bis in die letzte Zeit spüren, insofern Permutationen, Variationen, Kombinationen ein vielfach behandeltes Gebiet im mathematischen Unterricht bildeten,

On the basis of the textbooks that will be analyzed in this section, it can be said that the influence of the combinatorial school on the teaching of secondary schools occurred mainly through the non-foundationalist current. However, as will be seen below, the second part of Lorey's remark is wrong because, as far as textbooks are concerned, the way in which the theory of combinations is addressed in most of those textbooks substantially differs from the way in which Hindenburg formulated his theory of combinations. It is worth mentioning that the non-foundationalist current represents the branch of the combinatorial current of thought of the nineteenth century that is more closely related to mathematics teaching. It is well known that the organization of German academic institutions experienced significant changes during the first half of the nineteenth century, and, naturally, most of the following textbooks belong to this movement of reorganizing mathematics education. There are excellent studies focused on social, cultural, and institutional aspects of mathematics education in the Germany of the nineteenth century, such as those of Jahnke, Lorey, and Schubring, that can be consulted in order to better understand the general social and academic context in which these textbooks emerged (Jahnke 1990; Lorey 1916; Schubring 2010, 2012). Here we are interested in the conceptual changes in the treatment of the theory of combinations presented in those textbooks in order to understand the differences between the approach to combinatorics carried out in the non-foundationalist current and the German combinatorial analysis.

Of course, the prevalence of combinatorics in secondary schools does not exclude the teaching of this discipline in universities. For instance, according to Lorey, the *Grundriss* of Bernhard Friedrich Thibaut (1775–1832) can give us an idea of the content of Thibaut's lectures at the University of Göttingen (Lorey 1916, 26). Although Lorey does not specify to which *Grundriss* of Thibaut he is referring, and although the *Grundriss der reinen Mathematik* contains no section of combinatorics, Thibaut emphasizes the importance of the theory of combinations both in his *Grundriss der reinen Mathematik* and in his *Grundriss der allgemeinen Arithmetik oder Analysis*. Thus, it is highly likely that Thibaut included some topics of combinatorics in his lectures. Thibaut spent his entire professional life as a teacher of mathematics at the University of Göttingen. It seems that he excelled as a university teacher, captivating his audience with his mastery of the subject and the lyric beauty of his language (in a letter to Pfaff, Christian Ludwig Gerling (1788–1864) tells that Thibaut spoke like Goethe wrote (Dunnington 1937, 323; Lorey 1916, 26, footnote 1)). An enthusiastic description of Thibaut's talents as a lecturer was given by Adolph Tellkampf (1798–1869) (Dunnington 1937, 319 ff.; Tellkampf 1841), who became a mathematics teacher at a Gymnasium in Hamm in 1824 and later "a major spokesman of German mathematics teachers" (Schubring 2010, 115). Given his enthusiasm for the pedagogical capacities of Thibaut, probably Tellkampf considered Thibaut as a role model for mathematics teaching. Thibaut's impact on students was not

---

und zwar in der Art, wie Hindenburg diese von ihm eingeführten Dinge behandelte, d.h. also ohne Beziehung zum Gruppenbegriff.

restricted to future teachers of mathematics, but it also reached future major figures of German culture. For instance, Arthur Schopenhauer (1788–1860) attended the lectures on mathematics by Thibaut, and in a draft of his *Die Welt als Wille und Vorstellung*, he refers to Thibaut's *Grundriss der reinen Mathematik* to exemplify the progress in the "substitution of perceptual evidence for the logical method of proof" (Schopenhauer 2016, 626, note 55). Thus, it can be said that Thibaut's views on mathematics mattered deeply to his contemporaries.

At the very beginning of the nineteenth century, the non-foundationalist current seems to be pretty close to the combinatorial school's project. In 1801, Thibaut published his *Grundriss der reinen Mathematik* in which he claims that, in a system of mathematical sciences, the theory of combinations should be presented at the beginning because of its basic concepts and principles, but practical and pedagogical reasons impose, sometimes, the adoption of a different organization (Thibaut 1801, iii). This could suggest that Thibaut embraced a foundationalist position concerning the theory of combinations, but his laconic remark cannot be taken as evidence of such a fact. In his *Grundriss der reinen Mathematik*, Thibaut does not present the theory of combinations at the beginning, but neither does he cover it in the middle or at the end of his book: the theory of combinations has been completely omitted. Further information on the matter can be found in his *Grundriss der allgemeinen Arithmetik oder Analysis*, in which an even stronger remark stresses that combinatorics is "one of the most important foundations of mathematical analysis" (Thibaut 1809, iii). It should be noted that this stronger remark already indicates that Thibaut does not hold a foundationalist position, since the theory of combinations is not, for him, *the* foundation of mathematical analysis, but just a more elementary field compared to others. He points out that it would be interesting to investigate the limits of the theory of combinations as an independent field, that is to say, to investigate which mathematical principles belong to the theory of combinations strictly, and thus to precisely determine the areas of mathematical analysis that depend on combinatorics, but this investigation, Thibaut adds, cannot be carried out in his textbook, although his textbook gives a tacit answer to these questions (Thibaut 1809, v). The first chapter of his book is devoted to basic arithmetic, mostly to the exposition of natural numbers and the four elementary arithmetical operations. The second chapter of his book addresses the theory of combinations. In the first chapter, no combinatorial concepts have been used, and therefore Thibaut tacitly informs us that basic arithmetic does not rest on the theory of combinations. This confirms the previous observation that, for Thibaut, the theory of combinations is not the foundation of mathematical analysis.

Since the German combinatorial analysis was, as seen in previous chapters, a research project that aimed to found mathematical analysis on the theory of combinations, Thibaut's non-foundationalist position should be enough to distinguish his work from the combinatorial school's project. Indeed, Thibaut explicitly distances himself from the German combinatorial analysis. He says that his 1809 textbook gathers the most recent results in the theory of combinations, but, on the contrary, "the terminology and notations of Hindenburg's school were hardly used" in his presentation of combinatorics (Thibaut

1809, v). This claim brings out that, on the one hand, Thibaut does not consider himself to be a member of Hindenburg's school, and, on the other hand, that he conceives his own work as a formulation of the theory of combinations different from the German combinatorial analysis. In fact, a very striking feature of Thibaut's textbook is the almost complete absence of mathematical formulae: mathematical procedures are described in German and the scarce symbolic formulations of mathematical results are usually relegated to footnotes. This is in stark contrast to the superabundance of mathematical symbols in the texts of the combinatorial school, and goes beyond a matter of style to a question of conceptual demarcation of mathematics. As seen in Chap. 3, the development of mathematical notations was an aim of the combinatorial school because of the belief that the progress of mathematics depends, in part, on the development of appropriate mathematical formulae, which implies that natural languages by themselves cannot produce the most perfect expression of mathematics. This feature also separates Thibaut from the syntactic current of the nineteenth century, even more drastically in some cases. For instance, Robert Grassmann claimed that the real progress of the theory of forms lies in its symbolic formulae and a textbook written in words instead of formulae exhibits its lack of scientificity (Grassmann 1872b, 11). Thus, Thibaut's work inaugurated a new way to deal with the German combinatorial analysis, reducing it to the theory of combinations and eliminating several of its most remarkable characteristics. In his 1809 textbook, Thibaut presents the theory of combinations in a very elementary way: he merely explains the notions of "combination", "permutation", and "variation". Hindenburg's method of combinatorial involutions is not discussed, Hindenburg's method of combinatorial tables is not really discussed, Hindenburg's abstract algebra of involutions is not discussed, Kramp's theory of generalized factorials is not discussed, and so on. Therefore, the theory of combinations developed by the combinatorial school has been severely simplified by Thibaut. Concerning the delicate issue of priority between the binomial and multinomial theorems, Thibaut leaned toward the binomial theorem, from which he deduced the multinomial theorem for rational exponents (Thibaut 1809, 196 ff.). In sum, Thibaut presents an elementary theory of combinations, in which it is difficult to recognize the German combinatorial analysis, places this elementary theory in the general framework of mathematical analysis, without pretensions of foundationalism, and conveys the idea that the binomial theorem is more fundamental than the multinomial theorem. A second edition of his 1809 textbook appeared in 1830, and Thibaut's views on the theory of combinations were certainly decisive in the way in which "the combinatorial analysis" was incorporated into German mathematics in the nineteenth century (Thibaut 1830). Probably, the non-foundationalist current had the highest number of supporters among the four branches of the combinatorial current of the nineteenth century because of Thibaut's influence.

Surprisingly, two manuals authored by former members of the combinatorial school can be included in the non-foundationalist current. The first appeared in 1808 under the title *Élémens d'arithmétique universelle* (Kramp 1808). Kramp called "universal arithmetic" what was normally identified as algebra, i.e. the theory of equations, although his book also deals with some topics of Gauss's modular arithmetic and some questions of derivation. The first sections on permutations and combinations appear around the middle of the book, and the previous sections, devoted to algebraic operations, equations, and modular arithmetic, are not based on combinatorial principles. Thus, the theory of combinations occupies a rather modest place in Kramp's textbook, as a mathematical discipline among others. Moreover, Kramp does not offer a complete account of the theory of combinations developed by the combinatorial school: he explains the notions of "combination" and "permutation", then discusses his own theory of generalized factorials, and that is all. The second is a textbook of Prasse published in 1813 under the title *Institutiones analyticae* (von Prasse 1813). Although Prasse covers more topics of the theory of combinations he had contributed to develop, most of the chapters in his *Institutiones* are not based on the theory of combinations; for instance, the first five chapters on algebra, the chapters on differential and integral calculus, which occupy more than half of the book, are treated independently of any combinatorial principle. Furthermore, Prasse reiterates his opinion that the binomial theorem is more fundamental than the multinomial theorem by proving again these theorems according to the deductive order he established in his 1803 proof of the binomial theorem for rational exponents (see Sect. 5.1.3) (von Prasse 1813, 28 ff., 144 ff.). In 1803, Prasse did not give an explicit demonstration of the binomial theorem for positive integer exponents, which is the first step in his deductive order, assuming that elementary proofs were available at the time. In his *Institutiones*, Prasse proved the binomial theorem for positive integer exponents by induction and included the proof in a chapter on algebra (von Prasse 1813, 28 ff.). Thus, instead of using an elementary combinatorial proof, Prasse decided to show that the binomial theorem for integer exponents belongs to the field of algebra, and this field is independent, at least in his *Institutiones*, of the theory of combinations. As in the case of Rothe, who published a two volumes textbook deprived of any combinatorial notion (see Sect. 6.2.2), Kramp's and Prasse's manuals send the message that the German combinatorial analysis provides no answer to the question about the nature of mathematical analysis. Concerning Kramp's and Prasse's manuals, the message is that the theory of combinations is just another discipline in the field of mathematical analysis.

Another influential textbook on arithmetic and algebra came from the pen of Hirsch in 1804. It contains a brief section on combinatorics, focused on the elementary notions of "combination", "permutation", and "variation", which are treated in a very introductory way. Then, it follows an elementary section on the binomial theorem and its corollary, the multinomial theorem. Hirsch points out that Stahl's and Weingärtner's textbooks can be consulted for a more comprehensive study of the theory of combinations. Probably, Hirsch's reduction of the German combinatorial analysis to the theory of combinations was motivated by his reading of Stahl's and Weingärtner's works. In any case, Hirsch presents the theory of combinations as a discipline that provides some methods for solving certain problems of analysis, but its relevance in the theoretical organization of mathematical analysis seems rather slight. Hirsch's appropriation of combinatorial methods developed by the combinatorial school, but without subscribing to the combinatorial school's project, can also be seen in his *Sammlung von Aufgaben aus der Theorie der algebraischen*

Gleichungen, which contains no section on combinatorics but in which Hirsch applies, in particular cases, combinatorial techniques to some problems of analysis (Hirsch 1809). His 1804 textbook was reprinted twenty times between 1804 and 1890, and translated into English and French, although the section on combinatorics was omitted in the French version (Hirsch 1832). In fact, there are two English translations of Hirsch's book, a British version and an American version (Hirsch 1827, 1831). In this latter, the translator claims that Hirsch "is well known as one of the ablest mathematicians in Europe, and, perhaps, as the best teacher of our time." (Hirsch 1831, v) Certainly, Hirsch's views played a decisive role in the way in which the German combinatorial analysis was perceived in the nineteenth century.

As seen in Sect. 6.1, Ohm learned the methods of the combinatorial school from Rothe during his formation at the University of Erlangen, and developed original research in the field of the German combinatorial analysis, although his research never had any impact on German mathematics. Despite his early interest in the German combinatorial analysis, Ohm did not follow this path in his subsequent works. He was convinced, however, that mathematics should take the form of a system, but such a system was different from that proposed by the combinatorial school. His major mathematical work consisted in constructing this system of mathematics, and the first part of the first volume devoted to this task appeared in 1822 (Ohm 1822a). From the beginning, Ohm made it clear that the most fundamental concept in mathematical analysis is the concept of "number". As a consequence, mathematical analysis should be built on this concept, and he called "number theory" (*Zahlenlehre*) the most general branch of analysis. This branch was supposed to be even more general than the theory of magnitudes (*Größenlehre*), since "number" was, for him, a more general concept that "magnitude", and thus the theory of magnitudes would be subordinated to number theory. In the first part of the first volume of his system of mathematics, Ohm presents the first elements of his number theory, from the discussion of the concept of "number" to the theory of equations, and then he begins the exposition of the general theory of magnitudes. The second part of the first volume, also published in 1822, resumes the exposition of his number theory (Ohm 1822b). It contains three brief chapters devoted to combinatorial topics: a chapter on Kramp's theory of generalized factorials, a chapter on combinations, variations, and permutations, and a chapter on the binomial and multinomial theorems for positive integer exponents. Ohm uses the notations invented by his former teacher Rothe for the binomial coefficients, and the multinomial theorem is just a corollary (Ohm 1822b, 48 ff.). But, more importantly, since everything in mathematical analysis depends on the basic concept of "number", all combinatorial notions are derivative in Ohm's system of mathematics. A second volume in several parts deals with differential and integral calculus, whose first part appeared in 1829 (Ohm 1829a). Hence, the theory of combinations has been assimilated as a part of mathematical analysis and it coexists with other parts without being a foundation for any of them. A second edition of the second part of the first volume contains a more complete account of the theory of combinations, but the conclusion is the same: combinatorial notions are derivative in Ohm's system of mathematics (Ohm 1829b). According to Ohm, mathematics should rest on pure mathematical principles and all intrusion of metaphysics should be strictly excluded from mathematical systems (Ohm 1822b, Vorwort). Perhaps this is why he decided to search a non-combinatotial mathematical basis for his system; combinatorial principles had been, after all, more or less at the center of some major philosophical reconstructions of mathematics, such as those of Fries and Krause.

The texts examined so far in this section have a special relationship with the combinatorial school. Some of them were authored by former members of this school, or by a student of a former member of this school, while those of Thibaut and Hirsch were the first to reinterpret the German combinatorial analysis from the point of view of the non-foundationalist current and were, perhaps, the most influential textbooks of this current. There is another textbook that has a special relationship with the combinatorial school, due to the fact that its subject matter seems to be very close to the German combinatorial analysis. It was entitled *Der polynomische Lehrsatz und leichte Anwendungen desselben* by Heinrich Wilhelm Brandes (1777–1834) (Brandes 1820). In the preface, Brandes points out that his textbook is intended to provide an introduction to higher analysis, particularly it is intended to develop students's abstract reasoning skills necessary for the successful study and comprehension of differential calculus. This does not mean that differential calculus is based on the multinomial theorem, but that the theory of combinations on which the multinomial theorem rests can help students to better understand mathematical reasoning in higher analysis. Thus, for Brandes, the theory of combinations belongs to the field of mathematical analysis, but it is not the foundation of this field. Therefore, his textbook falls within the non-foundationalist current of the nineteenth century. It is worth noting that Brandes stresses the fact that Thibaut's work was a valuable source of information on combinatorics and on mathematical analysis in general. Brandes's textbook is composed of three parts. In the first part, he explains the notions of "combination", "permutation", and "variation" at a rather elementary level. The second part opens with a discussion of multiplication of polynomials, but then it focuses on the binomial theorem which is proved first for positive integer exponents, then applied to the multinomial theorem for the same type of exponents, later Brandes establishes the binomial theorem for rational exponents and applies it to the multinomial theorem in order to justify it for rational exponents. The third part presents applications of these theorems to some problems of analysis. First, the multinomial theorem is used to solve the question of reversion of series. Then, the binomial theorem is applied to exponential and logarithmic functions. Brandes also calculates the power series expansion of cosine and sine by means of the binomial theorem. It seems that the title of his book is a little misleading, a better description of the content would be "The binomial theorem and some simple applications". Thus, Brandes's textbook conveys the idea that the theory of combinations is just a part of mathematical analysis and that, in spite of its title, the binomial theorem is more important than its corollary, the multinomial theorem.

The non-foundationalist current continued throughout the nineteenth century. Every decade witnessed the arrival of new manuals on mathematical analysis that followed the same pattern with respect to the theory of combinations. They contain a brief section on combinatorics, which deal with the notions of "combination", "permutation", and "variation", and they also contain a brief section devoted to the binomial theorem and, if the multinomial theorem is treated, it is considered as a consequence of the binomial theorem. In general, the subject is presented at a very introductory level and almost no traces remain of the German combinatorial analysis. These textbooks are linked to the combinatorial school because they make reference either to a former member of this school or to an author of the combinatorial current of thought studied in this chapter, or, when there is no reference, the link can be established indirectly by the use of the characteristic mathematical notations of this school or of the authors of the combinatorial current of thought of the nineteenth century. All these manuals have in common the assimilation of a small part of the German combinatorial analysis as ordinary results of general mathematics, that is to say, those results are no longer considered part of the specific research project of the combinatorial school, but they are considered part of a common mathematical knowledge and are placed in this common mathematical knowledge without any kind of privilege or prerogative.

In the 1810s, besides the authors analyzed above, other representatives of the non-foundationalist current have been identified. In 1817, a textbook of Langsdorf follows the pattern described above: it contains a small section on combinatorics, whose results are presented without using the mathematical notations of the combinatorial school, and, in other sections, the binomial theorem is applied to the calculation of some power series expansions (Langsdorf 1817). Langsdorf affirms that Thibaut's manual on arithmetic and analysis is one of the most important source for this discipline. Perhaps he took the decision of including the theory of combinations in his textbook because of Thibaut, since he admits having found no opportunity to use the German combinatorial analysis in his professional activities (Langsdorf 1817, 11). For further information on the German combinatorial analysis, Langsdorf refers the reader to Stahl's and Weingärtner's textbooks, and he even mentions Gruson in this context, which is rather misleading given that, as seen in Sect. 6.2.2, Gruson held a position opposite to that of Hindenburg. In the same year, Andreas Neubig (1780–?) makes reference to Stahl and Weingärtner, as well as to Hindenburg, in a textbook on mathematical analysis which contains a section on combinatorics (Neubig 1817). The binomial theorem for rational exponents is proved, however, without using combinatorial methods. Some years earlier, in 1812, Johann Baptist Weigl (1783–1852) published a textbook on mathematical analysis which makes no reference to any author of the combinatorial current of thought or of the combinatorial school and in which no mathematical notation related to this current or to this school is used, but it contains a section on combinatorics and a section on the binomial theorem (Weigl 1812). However, in a second edition, Weigl includes an example of combinatorial involution and uses some notations of the combinatorial current of thought (Weigl 1823). Thus, it is sure that his textbook is related to the combinatorial school and belongs to the non-foundationalist current.

Weigl's textbook shows that there can be manuals belonging to the non-foundationalist current that make no reference to and use no mathematical symbols of the combinatorial school or of the combinatorial current of thought. However, for the sake of prudence, such manuals should not be included in the non-foundationalist current; to include them, it is important to make sure that there exists a link between them and the combinatorial school, even if it is an indirect link through the authors of the combinatorial current of thought. Some examples of these kind of textbooks are those of Johann von Gott Bundschue (1784–1851), Friedrich Kries (1768–1849), Maurus Magold (1761–1837), Johann Friedrich Raupach (1775–1819), Friedrich Wilhelm Daniel Snell (1761–1827), Gottfried Christian Vogel (1795–1836), and Friedrich Benedikt Wilhelm von Hermann (1795–1868) (Bundschue 1817; Kries 1826; Magold 1830; Raupach 1815; Snell 1804; Vogel 1826; von Hermann 1845). Even if these books cannot be included in the non-foundationalist current, they can be considered as examples of the incorporation of the theory of combinations into the field of mathematical analysis, and this incorporation was certainly a consequence of the influence of the combinatorial school on nineteenth century German mathematics.

In 1817, Justus Günther Grassmann, who was a student of Klügel, published the first volume of his *Raumlehre* and the second appeared in 1824 (Grassmann 1817, 1824); both are indirectly related to the German combinatorial analysis. These books aim to find a solution to the pedagogical problem of teaching geometry to young learners and of training teachers not only in the art of transmitting some theoretical content of geometry, but also in the more complex and challenging art of awakening a love for geometry in the minds of their pupils. The first volume of the *Raumlehre* was addressed to teachers who were in charge of children aged around 7 or 8 years old (Grassmann 1817, XV), whereas, according to its title, the second volume was addressed to teachers of Gymnasia and *Volksschulen*. Their pedagogical aim expresses a need to fill the didactic gap left by the reorganization of the German education system, which was trying to leave behind an educational model oriented toward classical studies (Greek and Latin) in order to turn to a more scientific education, where mathematics will occupy a more relevant place. Under such circumstances, professional mathematicians, like Grassmann, saw the need for decisive action to revitalize the poor scientific preparation of teachers. Those actions included the development of pedagogical guidelines as those presented by Grassmann in his *Raumlehre*. Grassmann's pedagogical ideas about geometry result from his general philosophical conception of mathematics. Some elements of his philosophical conception were given in the two volumes of the *Raumlehre*, but there is no systematic presentation of it (Grassmann 1817, X ff., 1824, VIII ff.). It is clear that the notions of "synthesis" and "construction", as well as the theory of combinations, play an important role in Grassmann's philosophical conception, but no firm conclusion about this issue can be reached on the basis of the *Raumlehre*. On the contrary, the pedagogical importance of the theory of combinations for geometry teaching is clearly established in the *Raumlehre*. Grassmann formulated a new mathematical theory named "geometric theory of combinations" (*geometrische Combinationslehre*). The key idea of this theory consisted in approaching the study of geometry through the notion of "direction", interpreting direction as a combinatorial property of geometric objects. According to Grassmann, geometric theory of combinations is a constructive science in the sense that all its procedures obey a synthesis of the intellect that allows to "see" in the mind the production of a geometric object. On the other hand, according to Grassmann, the geometry presented by Euclid in his *Elements* is not constructive in this sense, but it follows an axiomatic method. The pedagogical strength of geometric theory of combinations lies in its constructive features, which are better suited to children because of the visual aid required at a young age, whereas the comprehension of the logical structure of Euclid's *Elements* involves the intellectual abilities of adulthood. Besides its pedagogical advantages, geometric theory of combinations was conceived by Grassmann as being a real science by its own right, not just as being a didactic substitute for Euclid's *Elements* in elementary geometry teaching. Indeed, Grassmann was convinced that his combinatorial theory of geometry could mean a real breakthrough in science, and he decided to apply it to crystallonomy and thought that his theory provided the necessary link between mathematics and natural sciences (Grassmann 1829). The influence of Grassmann's approach to crystallography has been analyzed in (Scholz 1989, 1996).

Given his education background and the growing presence of combinatorial methods in German mathematics, it is not surprising that Grassamann considered combinatorics as an important part of this scientific discipline. A more complete account of Grassmann's philosophical conception of mathematics was furnished in *Ueber den Begriff und Umfang der reinen Zahlenlehre*, a text written in the context of his pedagogical programmatic endeavors, which was recently translated into English (Grassmann 1827, 2011). In what follows, we will make reference to the English version which is more easily accessible than the original source. Grassmann's general conception of mathematics rests on a philosophical notion of "synthesis". Grassmann claims that mathematics possesses a particular kind of synthesis, different from that of logic (Grassmann 2011, 457). It has to be said that Grassmann's idea of a logical synthesis is, at the very least, questionable. In accordance with the logic of his time, Grassmann considered that a proposition consists of a subject and a predicate, and what he called logical synthesis consists in attributing "to the proposition a truth or validity if the predicate, as a consequence of its assertion, must really be conjoined to the subject, or not" (Grassmann 2011, 457). Thus, Grassmann called "logical synthesis" what was known to everyone under the name of "judgment", and, in his paper, there is no clue as to the difference between logical synthesis and judgment. His notion of "mathematical synthesis" is not much clearer (Grassmann 2011, 457):

> Now the mathematical synthesis attributes no truth in this sense, and precisely thus it is distinguished from a synthetic proposition. If for example we take that synthesis whereby a number results $(1+1)$, we do indeed find here the same constituents, subject, predicate, copula (except that here there is no essential distinction between subject and predicate), but nothing can be said here about the objective validity of this synthesis—that is, whether one unit really amounts to another—the conjunction can take place unconditionally, and the concept created thereby, the product of this synthesis, is the number two. But if one wishes to say that the proposition is actually this: $1 + 1$ is 2, then on the other hand one would with justice be

reminded that here one has no longer introduced a concept complete in itself, but has simply labeled it by a word.

There are multiple problems in this explanation. When Grassmann talks about synthesis or construction (he did not distinguish between the two notions), his ideas seem to be inspired by a philosophical trend. His terminology inevitably brings to mind Kant's critical philosophy, although no philosopher or philosophical current is mentioned in the text. It should be recalled that, according to Kant, the result of the synthesis performed by the synthetic unity of apperception is, in few words, a judgment. As a consequence, from Kant's point of view, it is not possible to say that "1 + 1" involves any kind of synthesis, for "1 + 1" is not a judgment; "1 + 1 is 2" is a judgment in which the synthetic unity of apperception has performed a synthesis. Thus, Grassmann holds a philosophical position different from that of Kant, but his position seems to arise from the confusion between name (linguistic sign), concept, and object. When he claims that "1 + 1" consists of subject, predicate, and copula "except that here there is no essential distinction between subject and predicate", it should be understood that a mathematician produces in his mind the representation "1 + 1" as being a subject and a predicate at once, that is to say, the representation "1 + 1" depends on an intellectual intuition (in Radu (2000), Radu suggests that Grassmann's belief in intellectual intuition follows a philosophical current whose origins are in Schelling's philosophy). Therefore, in this case "1 + 1" is an object produced by the mathematician's intellectual intuition. However, in the last sentence of the quotation given above, "1 + 1" is treated as a concept and as a name (Grassmann cannot see that "1 + 1" can be a name in itself independently of its appearance in "1 + 1 is 2"). Besides this confusion between name, concept, and object, Grassmann does not seem to be aware of the ontological and epistemological difficulties raised by his belief in intellectual intuition. For example, since intellectual intuition entails the actual production of the object, Grassmann cannot talk of *the* number 2 but just of a number 2 produced by a certain mathematician, and a number 2 produced by another mathematician, and so on; and, in Grassmann's exposition, it is not clear how a science of the number 2 could be achieved, i.e. how to justify the fact that all these mathematicians are discussing the same subject. This obscure notion of mathematical synthesis is used by Grassmann to define mathematics: "*Mathematics is the science of the synthesis according to outer relations, that is as equal or unequal*" (Grassmann 2011, 458). The terms "equal" and "unequal" refer to the fact whether a mathematical expression should be considered with regard to its content or to its form respectively.

If synthesis is performed as equal, one gets magnitude; on the contrary, if synthesis is performed as unequal, one gets combination. On the other hand, Grassmann does not reject the traditional conception of mathematics according to which mathematics can be divided into the science of continuous magnitude and the science of discrete magnitude. However, he points out that such a conception offers an incomplete description of the entire field of knowledge with which mathematics is concerned, inasmuch as it leaves out all the important aspects covered by synthesis as unequal, that is to say, it leaves out the theory of combinations (Grassmann 2011, 459). This is why Grassmann holds that that traditional conception cannot be taken as an adequate definition of mathematics. An adequate definition of mathematics should take into account the theory of combinations, which is, along with geometry and arithmetic, one of the three basic disciplines from which all mathematical knowledge emerges. Indeed, those three mathematical disciplines are like the primary colors for Grassmann: they can be mixed to produce new mathematical knowledge, but they are independent of one another. In particular, Grassmann emphasizes that the theory of combinations is not dependent on arithmetic: "combination theory stands on an equal footing with arithmetic" (Grassmann 2011, 460). In a footnote, this remark is complemented by the following interesting comment (Grassmann 2011, 460, footnote 3):

> Combination theory is still in its infancy, as if in arithmetic one had proceeded no further than addition. Sad to say, it was immediately made the handmaid of analysis; this early servitude has impeded its growth and development, and even now the times are so inclined against it that only by this servitude can its existence be delayed. But the day will come when the stepchild will appear in her undimmed beauty, and be acknowledged, and while one asked nothing of her, assigned her no duties, she will in her harmless presence cast her rays upon all the sciences.

Grassmann was neither the first, nor the only one to have stated that the theory of combinations was still in its infancy. As seen in Sect. 6.2.1, in 1823 Spehr claimed that both the remote past of the theory of combinations, with Lull, Kircher, Jacob Bernoulli, Leibniz, Lambert, Euler, and its recent past, with Hindenburg, Rothe, Kramp, Töpfer, Prasse, Eschenbach, and a "few more" (i.e. the combinatorial school), are characterized by the lack of a scientific, rigorous presentation; and in 1824 Ernst G. Fischer regretted that the Euclid of combinatorics had not yet been born, making clear allusion to Hindenburg. This intellectual attitude stands in contrast to the more common narrative that depicted Hindenburg as the unrivaled benefactor of combinatorics, as the one who led this discipline to its most advanced stage. Spehr's attitude was due to the radicalization of the foundationalist posture concerning the theory of combinations and to his belief that combinatorial methods had been expressed in an unscientific way in the past, while Fischer's attitude was tinged with his rivalry and animosity with the combinatorial school. Evidently, this kind of attitude can also be understood as an indication that times are changing. Grassmann's statement quoted above can be considered as an example of those changes. His claim that combinatorics was in its infancy is conditional on his belief that this discipline was reduced to servitude by mathematicians in order to satisfy the needs of mathematical analysis. The important point here is to understand which mathematicians dared to do such a thing. As seen in Sect. 6.2.1, there are philosophical conceptions of mathematics, as that of Fries, in which combinatorics is conceived as the foundation of the whole of mathematics. Certainly, a servant, a slave could not be at the same time the most important part of mathematics. More generally, none of the mathematicians belonging to the foundationalist current could be accused of enslaving the theory of combinations, since even if their work were mainly focused on mathematical analysis, the theory of combinations is not conceived as a tool but as the theoretical basis of analysis. This last observation applies to the German combinatorial analysis too because it was a foundationalist project, and one might add that it was not strictly confined to mathematical analysis since, as seen in Chap. 4, it was also applied, for example, to cryptography or, more importantly, the theory of combinations was even conceived as the foundation of arithmetic by the combinatorial school. The mathematicians belonging to the syntactic current cannot be charged with slavery either because their aim was to develop a scientific language for mathematics, whose grammatical rules were supposed to obey combinatorial patterns, and thus the theory of combinations was called to eventually impregnate all the areas of mathematics with its syntactic rules. On the contrary, in the countercurrent of algebraic purity, the use of combinatorics in mathematical analysis is tolerated for practical purposes, but combinatorics is absolutely irrelevant from a theoretical point of view. In this case, it can be said that the theory of combinations has been treated like a handmaid of analysis. This is also true of the non-foundationalist current, where the theory of combinations is seen as an appendage of mathematical analysis. Thus, Grassmann's claim quoted above reflects a historical situation in which the ideas of the combinatorial school have become blurred and have given way to different approaches to the theory of combinations. Grassmann's claim is a fair reproach when it comes to the non-foundationalist current and the countercurrent of algebraic purity. Nevertheless, Grassmann does not take the next step of conceptualizing the theory of combinations as the foundation of mathematics (or of mathematical analysis), even though he accords it a more relevant place in mathematics. He embraced then a non-foundationalist position with regard to combinatorics. Despite his fiery discourse in favor of a further development of the theory of combinations and despite his own application of this theory to geometry and crystallography, he never made a contribution to improve the pure theory of combinations, as he called combinatorics as an independent discipline. Grassmann's proposals on the theory of combinations were not strong enough to overpower the force and impetus of the non-foundationalist current that, as will be seen in the rest of this section, continued to treat combinatorics as an appendage of mathematical analysis, with the exception of Diesterweg who adopted Grassmann's combinatorial geometry, and of his son Hermann whose philosophical ideas about mathematics evolved from those of his father.

In 1820, Friedrich Adolph Wilhelm Diesterweg (1790–1866) followed Grassmann's idea of applying the theory of combinations to geometry and presented his own *Geometrische Combinationslehre*, an elementary textbook in which Diesterweg acknowledges his debt to Grassmann, and a reprint appeared in 1839 (Diesterweg 1820, 1839). It should be pointed out that neither Grassmann nor Diesterweg used the combinatorial tools developed by the combinatorial school, and thus the influence of this school on their works was rather theoretical, i.e. they endorsed a general idea according to which the theory of combinations serves to systematize mathematical knowledge, even for pedagogical purposes. While Grassmann's and Diesterweg's textbooks depart radically from the combinatorial school's project, they exemplify the kind of new mathematical approaches that emerged from the German combinatorial analysis.

In the decade of the 1820s, there is an increased number of texts related to the non-foundationalist current. Following Kramp's terminology, Peter Nikolaus Caspar Egen (1793–1849) agrees that arithmetic and algebra can be referred to as "universal arithmetic", which is the term he chose for the title of one of his textbooks on mathematical analysis (Egen 1820a, vii, footnote; 1820b). Egen points out that Hirsch's 1804 textbook is the main reference of his work. For the main sources of the theory of combinations, Egen endorses Hirsch's claim that Stahl's and Weingärtner's textbooks are invaluable for learning combinatorial methods, and he adds some writings of Hindenburg, Rothe, and Töpfer. His presentation of the theory of combinations is elementary. Several editions of Egen's textbook appeared during the nineteenth century. In 1822, Franz Adrian Köcher (1786–1846) published his book *Die Combinationslehre und ihre Anwendung auf die Analysis* (Köcher 1822). It is aimed at Gymnasium students and is designed to guide them to higher levels of mathematical analysis. This is a good example of the deep penetration of the combinatorial thinking into the German-speaking educational system. A new edition of a book by Daniel Christian Ludolph Lehmus (1780–1860) appeared the same year, which was originally published in 1816 and in which the only trail left of the German combinatorial analysis is the use of Rothe's notation for binomial coefficients (see Sect. 6.1) (Lehmus 1822). In 1824, Johann Albert Eytelwein (1764–1848) published his *Grundlehren der höhern Analysis* in two volumes, a comprehensive work on mathematical analysis (Eytelwein 1824a, 1824b). Unlike most of the textbooks we have discussed so far, that of Eytelwein offers a detailed account of the methods developed by the combinatorial school, and Eytelwein possesses extensive knowledge of the relevant literature: he makes reference to writings of Hindenburg, Eschenbach, Rothe, Töpfer, Weingärtner, Lorenz, Brandes, Ohm, among others. There is no doubt that the theory of combinations occupies a much more prominent place in Eytelwein's textbook than in any of the other manuals belonging to the non-foundationalist current of the nineteenth century, including those of Prasse and Kramp, but even so Eytelwein's textbook follows the pattern described above: it contains a section on combinatorics, a section on the binomial theorem, the multinomial theorem is a consequence of the binomial theorem, and the theory of combinations is not the foundation of the whole of mathematical analysis. Indeed, despite his extensive knowledge and genuine interest in the German combinatorial analysis and related topics, it seems that either Eytelwein was not aware of the existence of Schweins's combinatorial proof of the multinomial theorem for rational exponents, which finally established the independence of this theorem with respect to the binomial theorem, as seen in Sect. 6.2.1, or that he decided to ignore Schweins's proof and consciously chose to turn aside from the original objectives of the combinatorial school. In either case, the historical result is the same, namely, Eytelwein's work cannot be considered as a continuation of the combinatorial school's project, since it reduces the German combinatorial analysis to the theory of combinations and inserts this theory into mathematical analysis without even trying to give mathematical analysis the form of the combinatorial system of mathematics outlined by Hindenburg and his school. It should be noted that Eytelwein explicitly replaces Hindenburg's Gothic notation for binomial coefficients with Rothe's elegant notation, perhaps because of the animadversion on Gothic letters among scholars. Eytelwein's excellent textbook belongs, thus, to the non-foundationalist current of the nineteenth century. In 1825, Joseph Salomon (1793–1856) discussed the binomial theorem in a textbook on arithmetic, which contains no section on combinatorics, but Salomon makes reference to Hirsch's 1804 manual and even to Rothe's 1804 textbook on arithmetic, and there is some little influence in mathematical notations (Salomon 1825). In fact, Salomon innovates by proposing the following notation for binomial coefficients (Salomon 1825, 250):

$$\overset{n}{\underset{p}{C}} = \frac{n(n-1)\cdots(n-p+1)}{1\cdot 2\cdots p},$$

which is a clear historical antecedent of our present notation C_n^p. In the same year, Ephraim Salomon Unger (1789–1870) includes two examples of combinatorial involutions in the section of combinatorics of one of his textbooks on mathematical analysis; the inclusion of Hindenburg's combinatorial involutions is really rare in the writings of the non-foundationalist current (Unger 1825). He also discusses the binomial theorem and deduces the multinomial theorem from it, and, as Salomon, he proposes a mathematical notation for binomial coefficients close to ours:

$$\overset{p}{\underset{(1,2,\ldots,n)}{C}} = \frac{n(n-1)\cdots(n-(p-1))}{1\cdot 2\cdots p}.$$

In 1829, Adolph Tellkampf (1798–1869) published his *Vorschule der Mathematik*, which is divided into two parts, arithmetic and geometry (Tellkampf 1829). The first part follows the pattern described above: it contains an elementary section on combinatorics, a section on the binomial theorem, and the multinomial theorem is presented as a consequence of the binomial theorem. Among the works consulted for arithmetic, Tellkampf mentions Thibaut's 1809 textbook, Kramp's *Élémens d'arithmétique universelle*, Egen's textbook discussed above, and even Spehr's *Lehrbegriff der reinen Combinationslehre*. However, Tellkampf decided not to follow Spehr's foundationalist approach. Several editions of Tellkampf's textbook appeared during the nineteenth century.

   More manuals of the non-foundationalist current appeared in the decade of the 1830s. A second edition of Salomon's *Lehrbuch der Arithmetik und Algebra* was published in 1831, whose first edition was issued in 1821 (Salomon 1821, 1831). Unlike his 1825 textbook, this manual includes a section on the theory of combinations. In the introduction, Salomon affirms that mathematics consists of two parts, arithmetic and geometry, and it is interesting to note that he does not consider the theory of combinations as a fundamental division of mathematics. The section on combinatorics was reworked in the second edition. In particular, one example of combinatorial involution was included and Salomon adopted Ettingshausen's combinatorial notations, making explicit reference to Ettingshausen's *Combinatorische Analysis*. On the contrary, Salomon did not adopt Ettingshausen's foundationalist position, even ignoring Ettingshausen's proof of the multinomial theorem for rational exponents, which was carried out without using the binomial theorem. In his textbook, Salomon justifies the multinomial theorem by means of the binomial theorem. Thus, it is clear that Salomon was not interested in further developing the German combinatorial analysis, but he just took some elements of this theory and found a place for them in his exposition of arithmetic and algebra. This free use of the combinatorial school's results can also be seen in one of his workbooks (Salomon 1834). Something similar can be said about Kulik, who also makes reference to Ettingshausen's textbook in the introduction of his *Lehrbuch der höheren Analysis* (Kulik 1831). Although this textbook is focused on differential and integral calculus and does not contain a section on combinatorics, Kulik discusses the binomial theorem in the introduction, using Ettingshausen's notations for binomial coefficients, and applies this theorem to the justification of the multinomial theorem. As Salomon, Kulik is not interested in developing the German combinatorial analysis. In 1833, Johann Schön (1771–1839) points out in a textbook on mathematics that the theory of magnitudes is composed of the theory of equations, the theory of series, the theory of combinations, and infinitesimal calculus (Schön 1833). The section of combinatorics is clearly inspired by the combinatiorial school's work: Schön includes three examples of combinatorial involutions and uses some notations of the school. In an earlier textbook on the theory of magnitudes, Schön had already implemented such a conception of mathematics, but the section on combinatorics of this earlier textbook is not written with the notations of the school and includes no example of combinatorial involutions (Schön 1825). Other examples of manuals belonging to the non-foundationalist current are those of J. J. Caspari, Heinrich Gustav Doerk (1806–?), Johann August Grunert (1797–1872), Gustav Adolph Jahn (1804–1857), Johann Friedrich Kroll (1795–?), and Franz Minsinger, in which a slight influence of the combinatorial current's notations can be perceived in the section devoted to combinatorics (Caspari 1836; Doerk 1839; Grunert 1832, 1835; Jahn 1839; Kroll 1839; Minsinger 1832). Jahn's manual deals with the theory of probability instead of with mathematical analysis, but the theory of combinations is presented as a useful tool, not as a foundation, just like in all the other textbooks. In his *Lehrbuch der Arithmetik und Algebra*, Ludwig Öttinger (1797–1869) claimed that, putting aside geometry, mathematics can be divided into arithmetic, algebra (theory of equations), and analysis (theory of functions) (Öttinger 1837b). He thought that arithmetic possesses three elementary operations: adding, subtracting, and combining. Thus, for him, the theory of combinations forms part of arithmetic. In a textbook on the theory of combinations, Öttinger argued that this theory was still under development and had not yet reached the stage of a mature scientific discipline, partly because it faced opposition from a wide sector of the mathematical community, and he hoped his manual would help to better understanding this discipline (Öttinger 1837a). According to Öttinger, among those who have contributed to this subject are Hindenburg, Rothe, Weingärtner, Hirsch, and Spehr. However, Öttinger makes no distinction among the different approaches of these authors, nor does he note that his own view is different from that of Hindenburg's school, since, for him, the theory of combinations is just a small part of mathematical analysis. This reduction of different approaches to a single unproblematic theory of combinations is not unique to Öttinger, but it is a common feature of the non-foundationalist current, as pointed out at the beginning of this section. This same reduction can be found in the *Lehrbuch der allgemeinen Arithmetik* of Johann Heinrich Traugott Müller (1797–1862), who made no distinction among the approaches of Crelle, Eytelwein, Ernst G. Fischer, Hindenburg, Hirsch, Klügel, Kramp, Kries, Ohm, Öttinger, Rothe, Schweins, and Thibaut, on which, according to him, his own account of the theory of combinations and analysis is based (Müller 1838).

In the next decade, Kulik reiterated his non-foundationalist position in a second edition of his textbook on analysis (Kulik 1843). As Salomon and Kulik, Straßnitzki showed no interest in developing the German combinatorial analysis and he simply ignored, too, Ettingshausen's proof of the multinomial theorem for rational exponents, preferring to approach, in one of his textbooks, universal arithmetic from the point of view that has been explained in this section (von Straßnitzki 1844). Indeed, Straßnitzki does not even mention the multinomial theorem in his textbook. In the same vein, Wunder reworked, in 1844, the first edition of one of his textbooks on pure mathematics, originally published in 1823, and he adopted Ettingshausen's notation for binomial coefficients, but refused to follow Ettingshausen's conception of mathematical analysis (Wunder 1844). As Straßnitzki, Wunder does not deal with the multinomial theorem. In 1844, Johann Carl Hermann Ludowieg (1795–?) published a textbook that can be classified together with the manuals of the non-foundationalist current, since it follows the pattern (Ludowieg 1844). Indeed, an earlier textbook of Ludowieg on arithmetic and algebra contains no section on combinatorics, although Hirsch and Thibaut are among his references, which suggests that, for him, the theory of combinations is not essential for understanding mathematical analysis (Ludowieg 1835). This detachment of the theory of combinations can be observed in manuals in which no particular position is assumed with respect to this theory. For instance, in one of his textbooks on mathematical analysis, Oskar Schlömilch (1823–1901) uses Rothe's notation for binomial coefficients, without assuming any particular position with respect to the theory of combinations (Schlömilch 1845, 130 ff.). Indeed, he does not talk about this theory at all and he does not mention Rothe. He also uses Rothe's notation for binomial coefficients in a textbook devoted to differential and integral calculus, and again he mentions neither Rothe nor the theory of combinations (Schlömilch 1847, 58). Here the use of this mathematical notation has been normalized, that is to say, it is not longer considered as belonging to a particular school or to a particular current, but it is just a standard mathematical notation.

In the 1850s, there is still persistent use of Ettingshausen's notation for the binomial coefficients while avoiding a combinatorial foundationalist position, as can be seen in Reuschle's textbook on arithmetic and algebra, which includes an elementary section on combinatorics, discusses the binomial theorem, and does not address the multinomial theorem (Reuschle 1850). With clear influences from the combinatorial school, Heinrich Borchert Lübsen (1801–1864) belongs to the non-foundationalist current, since his

Lehrbuch der Analysis presents the theory of combinations as a subsidiary part of mathematical analysis (Lübsen 1853). In 1854, Öttinger published a book on Kramp's theory of generalized factorials (Öttinger 1854). The aim of his book is to improve this theory and to integrate it into mathematical analysis. Thus, Öttinger's book can be seen as an attempt to absorb another part of the German combinatorial analysis into the existing mathematical knowledge.

In the 1860s, as in the following decades of the nineteenth century, there was a decrease in the number of manuals associated with the non-foundationalist current, excluding all the reeditions of several textbooks discussed above. A textbook on arithmetic and algebra by Joseph Helmes (1810–1883) includes two examples of combinatorial involutions in its section of combinatorics (Helmes 1862). It also contains a section on the binomial theorem. The *Prinzipien der Arithmetik* of Friedrich Grelle (1835–1878) opens with a chapter devoted to the binomial theorem in which some elements of combinatorics are explained (Grelle 1863).

Hermann Grassmann's *Lehrbuch der Arithmetik*, appeared in 1861, deserves a special mention (Grassmann 1861). As is widely known, this is a seminal book in the history of arithmetic (Ferreirós Domínguez 2005; von Plato 2017, 40 ff.; Wang 1957). It contains what could be seen as the first recursive definitions for the basic operations of addition and multiplication, and mathematical induction is systematically used to prove arithmetic properties. Hermann Hankel (1839–1873) picked up on Grassmann's approach to arithmetic in his *Theorie der complexen Zahlensysteme*, and Schröder in his *Lehrbuch der Arithmetik und Algebra* (Hankel 1867, 1 ff.; Schröder 1873, 5 ff.). In his *Was sind und was sollen die Zahlen?*, Richard Dedekind (1831–1916) refers to Schröder's *Lehrbuch* as one of his main sources, and Giuseppe Peano (1858–1932) acknowledges his debt to Grassmann's *Lehrbuch* in his *Arithmetices principia* (Dedekind 1932, 335, footnote; Peano 1889, v). Although there is a debate among scholars about whether Grassmann held an axiomatic position regarding arithmetic (Radu 2011), his *Lehrbuch* has been fairly considered as an antecedent of the axiomatization of arithmetic that prevails today. On the other hand, Grassmann's *Lehrbuch* contains a brief section on the theory of combinations. Excluding the symbol "!" used by Kramp to designate the factorial function, there is no trace of the theory of combinations developed by the combinatorial school, but there is no doubt that Grassmann's knowledge of this theory was linked, to some extent, to this school through the teachings of his father Justus. However, combinatorics plays no role in the development of arithmetic presented by Grassmann in his *Lehrbuch*. This suggests that the theory of combinations does not assume a foundational role in mathematics, but it is essential to look at Grassmann's philosophical ideas to better understand this question.

In the Introduction to his 1844 *Ausdehnungslehre*, Hermann Grassmann presented his philosophical views behind the mathematics he developed, and they were omitted in the new edition of 1862 for fear that mathematicians would be frightened away by the prospect of a philosophical approach to mathematics (Grassmann 1844, 1862). Nothing could be more untrue: the extension theory of Grassmann was not only elaborated with mathematical rigor and precision but also, as is well known, laid the groundwork for linear algebra, giving form to the concept of vector space. However, in the Introduction to the 1844 *Ausdehnungslehre*, Grassmann reworked his father's philosophical views on mathematics in order to explain the motivations of his own work and to place his extension theory within the whole body of mathematical knowledge. As in the case of his father, Grassmann's general philosophical conception of mathematics is based on the two couples of notions "equal–unequal" and "discrete–continuous", but his father's notion of "synthesis" has been excluded (or, at least, it is not explicitly mentioned in the text). In fact, Grassmann does not define mathematics as his father did by means of the notion of "synthesis"; on the contrary, for him, "pure mathematics is the theory of forms" (*ist reine Mathematik Formenlehre*) (Grassmann 1844, xx). According to him, science can be divided into real sciences and formal sciences. Real sciences are concerned with the external world, whereas formal sciences have no relation at all with the external world. Thus, the theory of forms deals with objects produced by the mind exclusively. Grassmann claims that space is a concept linked to the external world, and, as a consequence, geometry does not belong to pure mathematics. Following his father's ideas, Grassmann argues that, in the domain of discrete magnitudes, objects can be conceptualized either in relation to the notion of "equal" or in relation to the notion of "unequal". In the first case, one gets number theory, in which "number is the algebraic discrete form" (*Zahl ist die algebraisch diskrete Form*), while in the second case one gets the theory of combinations, in which "combination is the combinatorial discrete form" (*Kombination ist die kombinatorisch diskrete Form*) (Grassmann 1844, xxiv). In the domain of continuous magnitudes, by assuming the perspective of the equal, one gets the theory of functions, which consists of differential and integral calculus according to Grassmann, and here the algebraic continuous form is called "intensive magnitude" (*intensive Grösse*); on the contrary, by assuming the perspective of the unequal, one gets the extension theory of Grassmann in which the combinatorial continuous form is called "extensive magnitude" (*extensive Grösse*) (Grassmann 1844, xxiv). Pure mathematics is then composed of these four branches: number theory, theory of combinations, theory of functions, and theory of extension. None of them is the foundation of mathematics. Indeed, as his brother Roger, Hermann Grassmann was convinced that there exists a general mathematical theory whose laws were supposed to be common to the four branches of pure mathematics, namely, the general theory of forms whose laws are thus the foundation of mathematics (Grassmann 1844, xxvi). As a consequence, Grassmann did not hold a foundationalist position with respect to the theory of combinations, although he assigned it a much more important role than the very secondary one that it played in most of the textbooks of the non-foundationalist current.

In 1872, Theodor Wittstein (1816–1894) continues the tradition of the non-foundationalist current by including combinatorics in mathematical analysis, or, more precisely, by including a section on elementary combinatorics and a section on the binomial theorem in his *Lehrbuch der Elementar-Mathematik* (Wittstein 1872). He uses Rothe's notation for binomial coefficients in his exposition of the binomial theorem.

In 1880, Reidt wrote the chapter on arithmetic and algebra of the *Handbuch der Mathematik* edited by Schlömilch, Richard Heger (1846–1919) and himself (Reidt 1880). In his chapter, Reidt includes a section on combinatorics and the binomial theorem, and he uses Ettingshausen's notation for binomial coefficients. However, his text does not seem to be animated by any polemical aims about the nature of mathematical analysis. By this time, the debate about the nature of analysis had ceased to be concerned with the theory of combinations. Reidt simply transmits what has become a matter of fact in his educational tradition, namely, the theory of combinations is a subsidiary part of mathematical analysis.

In an exercise book published in 1893, in which Öttinger and Schlömilch are mentioned, Hans Staudacher is no longer concerned with the combinatorial school's conceptions and challenges (Staudacher 1893). His book on the theory of combinations is confined to reproducing what has become a standard material on the subject.

*

In sum, there are four main branches of the combinatorial current of thought in the nineteenth century. Some of these branches are incompatible with each other. For instance, the foundationalist current is incompatible with the non-foundationalist current from a theoretical point of view, and they are incompatible with the countercurrent of algebraic purity. In itself, this fact shows that the combinatorial current cannot be confused with a single unitary research project, whatever this project could be. In particular, it cannot be confused with the German combinatorial analysis, which was the research project of the combinatorial school. These four branches emerged from the theoretical discrepancies that appeared around the combinatorial school at the end of the eighteenth century, discrepancies that have been analyzed in Chap. 5. Therefore, the combinatorial current of thought in the nineteenth century can be considered as a historical consequence of the dissolution of the German combinatorial analysis. On the other hand, the incompatibilities are not restricted to different branches. In a same given branch, there are incompatible positions. For instance, Fries's psychological foundationalism is incompatible with the objective, technical foundationalism of Spehr. This means that the combinatorial current of thought in the nineteenth century is characterized by the coexistence of an impressive amount of personal conceptions and personal projects about mathematical analysis. Therefore, there is no collaborative work among the authors of the combinatorial current of thought. One of the main features of the combinatorial school was the collaborative work of its members in order to achieve a common objective. Hence, the combinatorial current of thought of the nineteenth century cannot be considered as a prolongation of the combinatorial school.

6.3   Vicissitudes Abroad

The German combinatorial analysis was a local phenomenon. It never took root in non-German-speaking countries. Hindenburg was well aware of this problem and never could explain it. From time to time, the ritornello returned in the works of those interested in the German combinatorial analysis to recall that this was an important theory for mathematical analysis, but that it had been neglected by scholars. This final section gives a brief overview of the vicissitudes of the German combinatorial analysis in foreign countries.

In France, the German combinatorial analysis encountered an adverse situation. In Chap. 5, it has been described how Bürmann had to face the hostility of the French Academy of Sciences against his innovative mathematical notations. In fact, not only Bürmann's project of improving algebraic notations was regarded with disbelief by French mathematicians, but also that of the combinatorial school. In 1805, Lalande pointed out that the notation system of the combinatorial school, particularly its Gothic letters, was the main cause for the confinement of the German combinatorial analysis in German-speaking countries (de Lalande 1805). Although he added that this was not a good reason, but a prejudice, the fact is that neither he nor anyone else in France adopted the combinatorial school's project of creating new mathematical notations. In fact, avoiding the creation of mathematical notations appeared to be, sometimes, a question of scientific merit. For instance, concerning some mathematical results, Servois claims that (Servois 1814, 50):

> I was able to deduce from these formulae, without using any new mathematical notation, those main formulae that were based until now on *combinatorial analysis* or on *derivation calculus*.[8]

In this passage, Servois is convinced that avoiding the creation of new mathematical notations is a desirable feature of mathematical research. Besides, this passage also shows the interest of French mathematicians in finding mathematical alternatives to the combinatorial analysis.

The German combinatorial analysis was never cultivated in France, except for Kramp who was a member of the combinatorial school. Arbogast created his calculus of derivations which deals with the same kind of problems as Hindenburg's theory does and which was in open competition with it. But mostly, the ideas of the combinatorial school were relegated to short marginal comments in which they are criticized and oversimplified, or they are mentioned in passing but only to discard them. For instance, Lacroix makes some comments on the German combinatorial analysis without presenting any concrete result or method belonging to this theory, and, on this basis, he criticizes the work of "the German geometers" who spend their time and effort studying polynomial expansions (Lacroix 1810, xxviii ff.). In his comment, Lacroix also offers a very

---

[8] J'y suis parvenu à déduire de ces formules, sans avoir besoin de recourir à aucune notation nouvelle, les formules principales fondées jusqu'ici sur l'*analise combinatoire* ou sur le *calcul des dérivations*.

misleading explanation of the historical origins of the German combinatorial analysis. Very much along the same lines, Joseph François Français (1768–1810) affirmed that he had discovered a way to completely improve Hindenburg's combinatorial analysis, but, instead of explaining his discovery, he preferred to present another method which was "more simple" and "more analytic" (Français 1815, 79). In fact, Français mentions the German combinatorial analysis tangentially because it is a historical antecedent of Arbogast's calculus of derivations, on which his study is focused.

At the beginning of the nineteenth century, Lacroix wrote a letter to Français which could be related to the German combinatorial analysis, and which has been interpreted by Schubring as a representative example of the negative opinion that "can be regarded as typical of the Parisian inner circle" against the work of the combinatorial school (Schubring 1996a, 371). This letter has been reproduced by Grattan-Guinness in Lacroix (1990). The letter is not dated and it was not sent because Français died, but Grattan-Guinness estimates that it was written shortly before Français's death, thus around October 1810. Schubring hypothesizes that the letter was written in response to a request by Français "for the reasons why he [Lacroix] and other Paris mathematicians neglected the research results achieved by the combinatorial school" (Schubring 1996a, 371). However, this hypothesis does not agree with the motivations given by Lacroix at the beginning of the letter (Lacroix 1990, 1325):

> Please do not think that this response to some points of your interesting letters aims to raise controversy over the views that you hold or to pass an absolute judgment on the calculus of derivations. [...] After all, we may both be in the wrong, you for having cultivated this branch of analysis too much, and I for having stayed far away from its procedures.[9]

Thus, the discussion topic between Lacroix and Français is the calculus of derivations created by Arbogast, not the German combinatorial analysis, and throughout the letter Lacroix explains his reasons for rejecting this theory, constantly and explicitly mentioning the name "calculus of derivations". Lacroix summarizes his criticism against the calculus of derivations in the following passage (quoted by Schubring in his paper) (Lacroix 1990, 1326; Schubring 1996a, 371–372):

> Pure analysis and pure geometry are without a doubt very beautiful speculations in themselves, highly propitious to the exercise of the mind, and they can provide an occasion for developing quite a bit of sagacity; but I must admit that I have never been able to ascribe great value to those advantages when they are in my opinion the only reason to study those sciences. I have always believed that there were ways of exercising one's reason, and especially of

---

[9] En répondant, Monsieur, à quelques uns des articles de vos intéressantes lettres, ne croyez pas que j'aye pour but d'établir une controverse sur les points que vous avancez, ni de porter un jugement absolu sur le calcul des dérivations. [...] Nous pouvons peut-être tous deux nous tromper à cet égard, vous pour avoir trop cultivé cette branche de l'analyse, et moi pour être demeuré trop étranger à ses procédés.

favoring the activity of one's mind, much more satisfying than combinations of exhausting results, which, when carried too far, isolate you more and more from the rest of mankind. Besides common applications, besides the "reasoned" exposition of great methods that is offered by the philosophy of science and that shows the path traversed by the human mind in the research into the properties of magnitude, it would seem to me that the science of calculation would be nothing but a sort of chess game if it were unable to offer the key of several phenomena whose laws would remain otherwise unreachable. I evaluates, thus, each analytic discovery by the hope it offers for the progress of physical-mathematical sciences.[10]

According to Lacroix's letter, the calculus of derivations developed by Arbogast and cultivated by Français is nothing but a sterile game of chess that does not contribute at all to the prosperity and progress of science. As pointed out by Schubring, Lacroix's severe criticism could be a consequence of the dominant perspective in France that privileged "applied" mathematics over "empty" theoretical abstractions (Schubring 1996a, 372). In any case, Lacroix states his conclusion in the last two sentences of the latter as follows (also quoted by Schubring) (Lacroix 1990, 1329; Schubring 1996a, 372):

> Such are, without a doubt, the reasons that prevented French geometers from taking an interest in the inquiries of German analysts and in those of Arbogast. "It would be necessary to envelop them rather than expand them", claimed Lagrange [...].[11]

Despite the fact that Lacroix has been talking about derivations calculus throughout the entire letter, he inexplicably infers conclusions for the inquires of "German analysts". Although it is impossible to read minds to confirm who are those German analysts, it is almost certain that "German analysts" means the combinatorial school. Needless to say, the fact of drawing a conclusion about a given theory by analyzing another different theory is colossal nonsense. But the unserious and careless attitude of Lacroix shows his utter contempt and ignorance of the theory developed by the combinatorial school. As pointed out in different places of this book, Lacroix's opinions about the German

---

[10] L'analyse et la géométrie pures sont sans doute en elles-mêmes de très-belles spéculations, très propres à exercer l'esprit, et peuvent offrir l'occasion de développer beaucoup de sagacité ; mais j'avoue que je n'ai jamais pu attacher un grand prix à ces avantages lorsque je les ai considérés comme l'unique objet de l'étude de ces sciences. J'ai toujours cru qu'il y avait des manières d'exercer sa raison, et surtout d'alimenter l'activité de son esprit, beaucoup plus satisfaisantes que des combinaisons de résultats fatigants, qui, lorsqu'elles sont poussées très loin, vous isolent de plus en plus du reste des hommes. Après les applications usuelles, après l'exposition « raisonnée » des grandes méthodes, qui fait connaître la philosophie de la science et montre la route que suit l'esprit humain dans la recherche des propriétés de la grandeur, la science du calcul ne me paraîtrait plus qu'une sorte de jeu d'échecs si elle n'offrait pas la clef de beaucoup de phénomènes dont les lois seraient inaccessibles sans son recours. J'examine donc toute découverte analytique, relativement aux espérances qu'elle peut donner pour l'avancement des sciences physico-mathématiques.

[11] Tels sont, sans doute, les motifs qui ont empêché les géomètres français de prendre intérêt aux recherches des analystes allemands et à celles d'Arbogast. « Il faudrait plus les envelopper que développer » disait La Grange [...].

combinatorial analysis are very often theoretically wrong and historically inaccurate, but they are symptomatic of the disdain this theory received in foreign countries.

There are other texts in which the combinatorial analysis is mentioned, but without making reference to the combinatorial school, and thus it is difficult to say whether the author refers to the German combinatorial analysis or to the theory of combinations. For instance, in the context of discussion of the binomial theorem and its application to the calculation of the power series expansion of the exponential function, Janot de Stainville (1783–1828) shows that this power series expansion can be deduced from the principles of the combinatorial analysis, but there is no element in his book to decide whether he is talking about the German combinatorial analysis or about the theory of combinations (de Stainville 1815, 346 ff.). The same kind of ambiguity can be found in some texts of Antoine Augustin Cournot (1801–1877) (Cournot 1841a, 74, 1841b, 472, 486).

There is an important example of a French mathematician influenced by the German combinatorial current that did not acknowledge this influence. Augustin Louis Cauchy (1789–1857) used Rothe's notation for binomial coefficients (see Sect. 6.1) on at least two of his works, but he did not say a word about Rothe or the German combinatorial current of thought (Cauchy 1833, 9, 1841, 389). As seen in Sect. 6.2.4, Rothe's notation was adopted by several authors of the German combinatorial current of thought of the nineteenth century, so if Cauchy did not take it directly from Rothe, he could take it from other author. This is not a minor issue, since there is other mathematical result that links Cauchy to Rothe: the q-binomial theorem. As seen in Sect. 6.2.2, Rothe published in 1811 a version of the q-binomial theorem. In 1843, Cauchy published three papers related to this theorem (Cauchy 1843b, 1843c, 1843a). In fact, in the relevant literature, Cauchy, Rothe, and others are supposed to have discovered the q-binomial theorem independently of each other (Johnson 2020; Mc Laughlin 2018). The use of Rothe's notation and Cauchy's silence about it cast serious doubt on Cauchy's independent discovery of the q-binomial theorem, since it was not impossible for him to know about Rothe's q-binomial theorem, and it is not impossible that, later, he did not feel the need to mention it, just as he did not feel the need to mention the origin of his notation for binomial coefficients.

The German combinatorial analysis was introduced in England in 1818 by Peter Nicholson (1765–1844) through his *Essays on the combinatorial analysis* (Nicholson 1818). As far as we know, this is the only book on the German combinatorial analysis published in a foreign land and written in a foreign language. The approach of the book is more practical than theoretical, placing relatively more emphasis on examples than on theorems or conceptual discussions. While Nicholson appears to have a genuine interest in the subject, his ignorance of German is a real handicap to him. Nicholson tells that he had a friend help him translate some papers of Hindenburg. As a consequence, the range of topics addressed in his book is rather small, and, in several places, he moves away from the methods of the combinatorial school, although he claims that he did so on purpose, in order to improve Hindenburg's methods. As almost every foreign scholar, Nicholson criticizes Hindenburg's mathematical notations and chooses not to use them. Thus, what the reader gets from the book is an incomplete and distorted picture of the German combinatorial analysis. A more elementary account of Nicholson's combinatorial analysis was included in another anthology 3 years later, in which Hindenburg is not mentioned (Nicholson 1821).

Nicholson's combinatorial analysis did not become popular in England. Indeed, in itself, the German combinatorial analysis never was studied and developed by British mathematicians other than Nicholson. On the other hand, the combinatorial school's ideas seem to have had a considerable impact on some major mathematicians. As seen at the end of Sect. 5.1.1, Peacock's principle of the permanence of equivalent forms was inspired by Klügel's principle of similitude, although Peacock did not say it explicitly. For the study of the German combinatorial analysis, Peacock recommended Eytelwein's textbook on mathematical analysis discussed in Sect. 6.2.4 (Peacock 1834, 288, footnote), but he never elaborated on the extent of influence of the German combinatorial analysis on his own work. Unfortunately, given the lack of explicit acknowledgment of their debt to the combinatorial school, the discussion of the influence of the combinatorial school's ideas on British mathematicians becomes a matter of interpretation. In any event, there are, for instance, punctual references to the German combinatorial analysis in some works of Augustus De Morgan (1806–1871) and Charles Babbage (1791–1871). De Morgan includes Kramp's generalized factorials in his book on differential and integral calculus but without mentioning Kramp; instead, he makes reference to Nicholson's book on combinatorial analysis in which Kramp's generalized factorials are treated (De Morgan 1842, 254). De Morgan states that (De Morgan 1842, 335):

> The *combinatorial analysis* mainly consists in the analysis of complicated developments by means of a priori consideration and collection of the different combinations of terms which can enter the coefficients. The first theorem of the kind which the student usually meets with is the well known development of $(1 + x)^n$, when n is a whole number [...].

But he makes no reference to any author in particular. In any case, if he is talking about the German combinatorial analysis, then reducing it to the binomial theorem for positive integer exponents is not only misleading, but a very poor oversimplification. For his part, Babbage thought that analytic operations could be mechanized by using one of three options: differential and integral calculus, Hindenburg's combinatorial analysis, or Arbogast's derivation calculus (Babbage 1864, 137):

> Each of these systems professes to expand any function according to any laws. Theoretically each method may be admitted to be perfect; but practically the time and attention required are, in the greater number of cases, more than the human mind is able to bestow. Consequently, upon several highly interesting questions relative to the Lunar theory, some of the ablest and most indefatigable of existing analysts are at variance.
> The Analytical Engine is capable of executing the laws prescribed by each of these methods. At one period I examined the Combinatorial Analysis, and also took some pains to ascertain from several of my German friends, who had had far more experience of it than myself, whether it could be used with greater facility than the Differential system. They seemed to think that it was more readily applicable to all the usual wants of analysis.

I have myself worked with the system of Arbogast, and if I were to decide from my own limited use of the three methods, I should, for the purposes of the Analytical Engine, prefer the Calcul des Derivations.

Despite Babbage's preference for Arbogast's calculus, it is interesting to ask for the extent of the influence of Hindenburg on Babbage concerning the idea of constructing a mechanical device to perform analytic operations. It should be recalled that Hindenburg considered mathematical tables to be the principal mode of implementing analytic operations in mechanical devices. Thus, Hindenburg's combinatorial analysis was, in principle, machinable by conception. Did this idea guide Babbage's conception of the Analytical Engine? Since Babbage did not elaborate on his discovery and assimilation of the German combinatorial analysis, this will remain a mystery.

In Italy, the difficulties in accessing primary sources on the German combinatorial analysis were described by Nicola Trudi (1811–1884) in 1877 (Trudi 1879). Trudi tells that he was interested in knowing Hindenburg's method of involutions for integer partitions, but for a long time he could not fulfill his wish, since Hindenburg's works were not available in any Italian library. Finally, he managed to get a copy of Hindenburg's scientific journal *Sammlung* from Germany. This shows that the difficulty in accessing the combinatorial school's writings was a real obstacle to the dissemination of the German combinatorial analysis in Italy. In his article, Trudi did not further develop Hindenburg's method of involutions, but he used Kramp's notation for factorials and Rothe's notation for binomial coefficients, although he thought, erroneously, that Cauchy was the inventor of Rothe's notation. Besides the poor dissemination of the works of Hindenburg's school, some Italian mathematicians were not convinced of the scientific character of the German combinatorial analysis. For instance, Guillaume Libri (1803–1869) claimed that Hindenburg's formula for calculating the general coefficient of the multinomial expansion was not a real mathematical formula, but a "didactic rule", since it depends on the calculation of combinatorial tables (Libri 1827, 3). Naturally, the aim of his article was to provide mathematical analysis with an actual mathematical formula for the multinomial expansion. His article was reprinted in Crelle's *Journal für die reine und angewandte Mathematik* 4 years later (Libri 1831). In a philosophical essay, Francesco Bertinaria (1816–1892) held that the theory of combinations can also be called "theory of series" (Bertinaria 1846, 349). Although he does not talk about the combinatorial school, his identification of both theories seems to be a remote consequence of the views of this school.

Some news about the German combinatorial analysis reached Spain in 1827. José Mariano Vallejo y Ortega (1779–1846) included an addendum in the second edition of his *Compendio de matématicas puras y mistas* in which he informed his readers about some "new" mathematical disciplines appeared in other European countries, including the German combinatorial analysis (Vallejo y Ortega 1827). The aim of the addendum is not to develop the mathematical details of those disciplines, but to describe them informally. It seems clear that Vallejo y Ortega had not read the original works of the combinatorial school, apart from some books of Kramp, since his news consisted in translating what Français said about the German combinatorial analysis in his 1815 article to which we have made reference above. This was an unfortunate decision because what Français says comes down to the personal remark that Arbogast's derivation calculus is better than Hindenburg's combinatorial analysis. This not very flattering image marked the German combinatorial analysis's entry into Spain. Because of this or because of other obstacles, the Spanish mathematical community never showed any interest in the combinatorial school's ideas. Late in the nineteenth century, Francisco Giner de los Ríos (1839–1915) contributed to the introduction of Krause's philosophy of mathematics in Spain by translating and publishing, alongside some philosophical essays of his own, several writings of Krause, including an article in which Krause discusses the German combinatorial analysis and its place in mathematics (Krause 1876). Thus, as part of the Spanish Krausism, some ideas of the combinatorial school broke into philosophy in Madrid. And finally, way over there in the Iberian Peninsula, the trail of the German combinatorial analysis goes cold in the meanders of Spanish philosophy.

Epilogue

7

The previous pages narrate the story of the German combinatorial analysis, a particular mathematical theory born in Leipzig toward the end of the eighteenth century. It had a trouble life, marked by conceptual changes and theoretical battles, as well as by stigma and rejection, and it died young, isolated by the European mathematical communities. The aim of this research was to understand the main episodes of its conceptual life and to delineate the traits of the theoretical environments in which it was born, grew up, and died. Our interest in the human beings and institutions related to this theory was conditional on their relevance to improving our understanding of its development. While this is not a history about human beings or their institutions, this research has allowed to reach some conclusions about the combinatorial school, a particular group of scholars who tried for a time to promote the growth of the German combinatorial analysis. Before moving on to the German combinatorial analysis, let us take a look at those conclusions.

Carl Friedrich Hindenburg was fully committed to the theory he invented, even if his research was paced by long periods of inactivity. His commitment encouraged him to establish the aims of his research project, to identify the areas with which the project was concerned, and to formulate a new mathematical method; later, his commitment led him to reshape the aims of his project, to arouse the interest of other mathematicians in his project, to reformulate his mathematical method, and to find the theoretical foundation of his project. But more importantly, his commitment convinced him that the study of mathematics could be systematized by recourse to the notion of "combination", or more generally to the theory of combinations. With the exception of two or three papers, his entire work depends on that notion and he refused to use other mathematical methods, such as differential ones, because he thought that they were unnecessary given the combinatorial nature of mathematics. His general vision of mathematics and his refusal of other mathematical fields made him a "combinatorialist". For some of his contemporaries, he was not just a combinatorialist, but the most important of them all, more than the

E. Noble, *The Rise and Fall of the German Combinatorial Analysis*, Frontiers in the History of Science, https://doi.org/10.1007/978-3-030-93820-8_7

members of the Bernoulli family, more than Euler, more than Moivre, more than Leibniz. Moreover, some of his contemporaries considered him one of the greatest mathematicians of all time, even greater than Descartes, even greater than Newton. Over time this kind of opinions ended up losing much of their verve, to the point that a radical shift in academic opinion occurred in the second half of the nineteenth century and the first half of the twentieth, when historians like Bell or Cajori openly mocked Hindenburg for his lack of mathematical talent, or like Lorey who called Hindenburg a narrow-minded man because, according to Lorey, Hindenburg was blinded by his beliefs in combinatorics and was then unable to appreciate the quality and scientific importance of other mathematical theories. The historical analysis of Hindenburg's work allows to rectify such unfounded opinions. Certainly, Hindenburg was not one of the greatest mathematicians of all time, but there is no doubt that he was a talented mathematician. For instance, he solved a mathematical problem that had been opened for almost a century, namely, that of finding a non-recursive formula for multinomial expansions. It is also false that he was narrow-minded. Although it is true that he ceaselessly advocated his ideas for grounding mathematical analysis on the theory of combinations, he displayed on several occasions a remarkable openness of spirit. As chief editor of several mathematical journals, he had the power to censor any articles before they were published in his journals or to simply refuse them, but he never did such a thing. On the contrary, he included in his journals articles by Tetens and Pasquich, for instance, which not only contained no combinatorial methods, but they pointed to the opposite direction of which Hindenburg was heading. He also published Bürmann's works knowing full well that the success of Bürmann's theories would entail the suppression of his own. He made these editorial decisions for the sake of science, whose progress was more important for him than his personal views on mathematics. His spirit of tolerance can be seen as well in his enthusiastic reception of Stahl's and Weingärtner's manuals, even though they did not exactly reflect his mathematical theory. In other words, he was not blinded by an irrational obsession and had the good sense to open a space for the diffusion of mathematical theories that challenged his own. Certainly, he argued in favor of his theory, but his arguments were not based on impossible dreams, or on unreachable ideals, or on personal illusions: they were based on the fact that his theory provided effective mathematical methods to solve problems in different areas of mathematics. It should not be forgotten that rational discussion of ideas plays an important part for the advancement of science. Thus, Hindenburg was a talented mathematician and a scholar more concerned about the progress of science than about the imposition of his ideas. He was also a careful mathematics teacher, who succeeded in transmitting his theoretical interests in this science to his students.

A research on the German combinatorial analysis inevitably raises a delicate histori-ographical issue regarding the historiographical category of "combinatorial school". To avoid confusion in historiographical discussions, it is necessary to clarify the content of that category. Contrary to common opinion, the combinatorial school was not founded by Hindenburg. Furthermore, it was not founded by anyone. The formation of the combinatorial school resulted from a complex historical process in which a small group of former students of Hindenburg joined him in order to denounce the plagiarism by Ernst G. Fischer and to prevent him from being deprived of the credit he deserved. The combinatorial school emerged fortuitously around 1793 as a means of preventing the usurpation of Hindenburg's ideas by Fischer. There is quite a difference between the willful act of founding a research team and the circumstantial, sporadic formation of a team. In the second case, there is no formal structure for the team and its organization and cohesion are very fragile. In fact, while the number of members of the combinatorial school increased, its structure and organization did not improve. From a historical point of view, this informs us that the notion of "research team" was rather obscure at the time. But there is no doubt that they were a research team. At the time, several scholars referred to that group as "Hindenburg's school", meaning that different mathematicians had joined forces to develop Hindenburg's ideas. Thus, the research team focused on Hindenburg's project existed and was visible to many scientists at the time. This research team, Hindenburg's school, disappeared somewhere between 1803 and 1811, as attested by the isolated research on the German combinatorial analysis of the beginning of the nineteenth century. This isolated research makes evident that Hindenburg's school does not exist anymore, since its former members ignored the results of that isolated research. Here we have to make a historiographical decision: either we decide to equate "combinatorial school" with "Hindenburg's school" or to keep them apart as two different notions. We have decided to equate them since it seemed to be the most reasonable choice. Effectively, the notion of "combinatorial school" contains the idea of "mathematicians working on Hindenburg's combinatorial theories", which is also the main idea contained in the notion of "Hindenburg's school". Therefore, the combinatorial school emerged around 1793 and disappeared between 1803 and 1811.

On the other hand, an intellectual movement inspired by the combinatorial school started at the beginning of the nineteenth century and lasted till the last decade of the same century. At the very beginning, that is to say during the first decade of the nineteenth century, this movement coexisted with the combinatorial school or with what was left of this school. The historiographical differentiation between this movement and the combinatorial school rests on two main facts. First, the quasi-totality of the representatives of the movement was not interested in developing the German combinatorial analysis (i.e. the theoretical ideas of Hindenburg's team). Second, it is clear that the representatives of the movement did not consider themselves as members of the combinatorial school: for instance, Thibaut pointed out explicitly that his work was scarcely based on the work of Hindenburg's school, or Spehr plainly emphasized that his work on the theory of combinations was different from that of the combinatorial school. Concerning the first point, the influence of the combinatorial school is reflected in the belief, shared by the representatives of the movement, that the theory of combinations should play some role in mathematical analysis or in mathematics in general. However the theory of combinations used by this movement is not the theory developed by the combinatorial school. Some traces of the combinatorial school's theory remain in the movement, but they are mostly reduced to the use of some mathematical notations invented by that school. Additionally, other than the belief that the theory of combinations should play a role in mathematics, the movement did not hold a common doctrine or follow a specific "teacher". This particular point suggests that this movement was not a "school", which is another good reason for restraining the use of the term "combinatorial school" to Hindenburg's team. Another good reason is that, as pointed out above, some of the representatives of the movement explicitly claimed that they did not belong to Hindenburg's school, and thus it would be really strange to distort this fact in order to fit (and perpetuate) some misleading historiographical models. Therefore, this was an intellectual movement, called in this book "combinatorial current of thought of the nineteenth century", whose origins were in the ideas of the combinatorial school, but whose aim was neither to further develop those ideas, nor to follow the same path of the combinatorial school. Let us examine now the German combinatorial analysis.

It is possible to say that the development of mathematical analysis during the eighteenth century was associated with the binomial theorem. This association was not confined to a specific place or European nation, but it was a common feature of European mathematics. This phenomenon is particularly perceptible in England and in German-speaking countries, where the binomial theorem was promptly used as a powerful tool of analysis. Eventually, this theorem went from being a powerful tool to being one of the major theoretical elements of mathematical analysis. At the time, mathematical analysis was roughly divided into finite analysis and analysis of the infinite, where the first was concerned with topics of arithmetic and algebra, while the second was mainly conceived, by mathematicians like Euler, as the theory of series, and there was a sharp debate among scholars on whether or not differential calculus should be considered as a legitimate part of infinite analysis. The binomial theorem got involved in the debate because, on the one hand, sometimes it was justified by means of differential techniques and, on the other hand, it became for some mathematicians like Euler the most important theorem of analysis in a foundational sense. The question, then, was to know whether this theorem belonged to finite analysis or to infinite analysis (via differential calculus). If this theorem could be proved on the basis of algebraic methods, then it would belong to finite analysis and it would unify this branch with infinite analysis (reduced in this case to the theory of series), leaving out differential calculus as a set of tolerated techniques or as an applied science. The multinomial theorem was mostly treated as a subsidiary topic in the research carried out around the binomial theorem.

On the other hand, in the eighteenth century, the need for a robust and reliable database of calculations became more and more evident to scientists working in different fields of knowledge, from physics to statistics. Such a database consisted in what is known as "mathematical tables". These tables were of great utility for preventing repetitive calculations in science. In this sense, the interest of mathematical tables was rather practical than theoretical, and the emphasis on the practical side was stressed by the parallel activity of constructing or designing mechanical devices to perform the tabulation of tables.

The brilliant idea of Hindenburg's strategy to prove the multinomial theorem for rational exponents in 1778, which was his first major work on mathematical analysis, was to conceptualize mathematical tables as a theoretical instrument and to assume that this instrument was a valid means of justifying a theorem of pure mathematics. This is an important conceptual transformation regarding the way in which mathematical tables were conceived at the time, and this conceptual transformation allowed Hindenburg to retain the power of calculation intrinsic to mathematical tables, while at the same time providing a theoretical justification for the theorem. The result of his strategy was the solution of the old problem of finding a non-recursive formula for the multinomial expansion, a mathematical problem that had been in the open since Moivre published his paper on the multinomial theorem in 1697. Hindenburg gained some notoriety for this achievement, but more importantly he realized that his method of combinatorial tables could become a general theoretical method for mathematical analysis. So, in 1779, he generalized his method and applied it to the calculation of the power series expansion of several functions. His method of powers, as he called it in 1779, turned out to be a very versatile approach to mathematical analysis, to the point that he convinced himself that his method of combinatorial tables was the key to unify the branch of finite analysis with the branch of infinite analysis, removing at the same time the dubious methods of differential calculus from infinite analysis. In this sense, his method of combinatorial tables, discovered in the context of his research on the multinomial theorem, seemed to offer an answer to the foundational problem related to the binomial theorem discussed above. This laid the ground for the formulation, in 1781, of a foundational research program for developing mathematical analysis (or, more precisely, the theory of series) on the basis of the theory of combinations. The 1781 research program of Hindenburg was the first stage in the historical evolution of the German combinatorial analysis. In this stage, the German combinatorial analysis can be seen as an algebraization program of analysis inasmuch as one of its main objectives is to avoid the use of questionable techniques based on the obscure notion of "infinite", such as those of differential calculus. Thus, the first stage of Hindenburg's program belongs to the general trend of algebraization of analysis that prevailed in the second half of the eighteenth century. However, it should be noted that there is nothing to indicate that Hindenburg was influenced by other programs of algebraization of analysis, whether that of Lagrange or of anyone else. The available historical evidence suggests that Hindenburg's program was the result of his own work on mathematics: first he invented a mathematical method to solve a specific problem (that of finding a non-recursive formula for multinomial expansions), then he generalized the method to other problems of analysis, and finally he wondered whether this method could serve to unify mathematical analysis, i.e. he formulated his research program.

Instead of recapitulating the events that led to the second and last stage in the historical evolution of the German combinatorial analysis, which can be easily consulted in the previous pages of this book, we want to focus now on answering the question of which features specifically characterize the theory of combinations developed by the combinatorial school. To our knowledge, this question has never been addressed by historians, and an adequate historical understanding of what was the German combinatorial analysis requires a clear answer to that question. Regarding the first stage of evolution of the German combinatorial analysis, this question has no answer. Back then, the distinctiveness of Hindenburg's combinatorial method resided in the use of partitions of numbers as a means of calculating the coefficients in the multinomial expansion. However, although his method was certainly innovative, partitions of numbers had been studied by other mathematicians, like Euler, in the context of combination theory and the theory of series. In other words, partition of numbers was a topic included in the theory of combinations that preceded the work of Hindenburg. To avoid confusion, we will call "classic theory of combinations" the theory that existed before Hindenburg. Classic theory of combinations included the definition of "combination", "permutation", "variation"; the study of methods for enumerating combinations, permutations, variations; formulae for calculating the number of combinations, permutations, variations with and without repetition; different topics on partition of numbers. Thus, in the first stage of the German combinatorial analysis, Hindenburg did not develop a new theory of combinations or a new combinatorial procedure, he simply applied some elements of classic theory of combinations to mathematical analysis in an innovative manner.

The German combinatorial analysis was completely reshaped between 1794 and 1796, when Hindenburg decided to study again, and more carefully, Moivre's paper on the multinomial theorem. The outcome of Hindenburg's reexamination constitutes a remarkable improvement to his previous work, and it was the second and last stage in the historical evolution of the German combinatorial analysis. In this stage, the German combinatorial analysis became a theory of combinatorial involutions. A combinatorial involution was a mathematical object that could be represented through a mathematical table, but it was not a mathematical table. This mathematical object was characterized by a group of algorithmic rules, i.e. the object was defined by precise algorithmic rules. The theory of combinatorial involutions sketched by Hindenburg was a sort of abstract algebra in which combinatorial involutions behaved in a similar way to sets in set theory and in which Hindenburg tried to define algebraic operations, such as addition of involutions and multiplication, and he also assumed the existence of the empty involution, that played the role of neutral element for addition of involutions. They were called "combinatorial involutions" because the algorithmic rules that defined them were considered to be combinatorial rules. Thus, the answer to the question of the previous paragraph is that the theory of combinations developed by the combinatorial school was the theory of involutions. This theory was completely new and different from classic theory of combinations. Although, unfortunately, it was never further developed by Hindenburg or anyone else, this theory was an incipient, but clear, attempt to approach mathematics by means of the study of abstract structures. To clarify this point, let us consider an example. In the nineteenth century, some mathematicians started to explore the idea of defining natural numbers as a specific kind of sets because a set was considered to be a more general mathematical object on the basis of which arithmetic and other areas of mathematics could be rebuilt. Thus, instead of dealing with a multiplicity of mathematical objects, set theory offered the possibility of redefining different mathematical objects by means of one single entity (called "set") and some rules in order to give some structure to this entity. Essentially, the idea is that all mathematical objects are sets but it is possible to distinguish, for instance, a natural number from other object because of its structure. In the case of the German combinatorial analysis, if someone had asked Hindenburg in 1794 what is a natural number, Hindenburg would have told him that a natural number is a combinatorial involution whose structure was defined by some algorithmic rules, as he actually wrote it when studying in 1794 the structure of number systems by means of the notion of "involution". Similarly, since a power series can be characterized by its coefficients and its coefficients can be expressed by combinatorial involutions, a power series is nothing else but a sum of combinatorial involutions, which is well defined in the abstract algebra of involutions. This suggests that arithmetic and the theory of series could be reduced, in principle, to the theory of combinatorial involutions, that is to say, to the particular theory of combinations developed by the combinatorial school. The German combinatorial analysis went from being a program of algebraization of analysis in its first stage to being an incipient foundational theory on general mathematical structures in its last stage of evolution. This is why Hindenburg was indeed the greatest combinatorialist of all time, because he was transforming the theory of combinations into an abstract algebra that opened up the possibility of founding different mathematical theories on general mathematical structures. We must insist, however, that Hindenburg's abstract algebra of involutions was not further developed and remained in embryonic stage. On the other hand, it is worth recalling that two central aspects of nineteenth and twentieth century mathematics are their focus on the study of general abstract structures and, in the twentieth century, the study of algorithms in computational mathematics. Algorithms and general mathematical structures are the core of Hindenburg's theory of combinatorial involutions. Additionally, the algorithms used by Hindenburg served to construct mathematical tables, which could be translated into mechanical devices, according to Hindenburg's early work on mathematical tables. In this light, the German combinatorial analysis, in its second stage of historical evolution, turns out to be a very promising theory. Thus, Felix Klein was wrong in claiming that the ideas of the combinatorial school were rather "a ramification of old scientific trends" than the "beginning of a new scientific development". Quite the opposite, in its last stage of evolution, the German combinatorial analysis pointed in the direction that mathematics will take in the two following centuries.

Sadly, the promising abstract algebra of involutions was doomed to oblivion because the combinatorial school, including Hindenburg, never turned to look at it again, and it went unnoticed by other mathematicians who did not join this school, including the mathematicians belonging to the combinatorial current of thought of the nineteenth century. Perhaps German mathematicians were distracted by the new theories that emerged around the German combinatorial analysis, so that they failed to grasp the deep implications of the abstract algebra of involutions. For instance, Hindenburg himself wasted a lot of time pointing out the similarities and differences between the German combinatorial analysis and the calculus of derivations of Arbogast and arguing in favor of his ideas, instead of concentrating his efforts on the purpose of moving his unfinished theory forward. On the other hand, it is true that the German combinatorial analysis was forced to face all those theories that were formulated during the last years of the eighteenth century and that wanted to take its place on the mathematical scene, posing a threat to its survival. In particular, it is possible to identify three major threats. First, there are new theories whose aim is to algebraize mathematical analysis and can be seen as belonging to the general trend of algeraization of analysis that took place during the second half of the eighteenth century, theories like those of Pasquich and Gruson, but also the ideas of Tetens. In this case, the threat consists in the belief that combinatorial concepts and methods are elements foreign to mathematical analysis, that is to say, they are not analytic concepts and methods, and thus they cannot provide mathematical analysis with a theoretical justification. Ironically, those theories were inspired by the work of the combinatorial school but aimed to eliminate combinatorics from that work. Needless to say, the elimination of combinatorics from the work of the combinatorial school entails the elimination of the German combinatorial analysis from mathematics. The threat was even more pronounced because of the fact that Hindenburg edited the works of Tetens and Pasquich, and published them in his scientific journals. Apparently, this fact continues to have effects contrary to the interests of the German combinatorial analysis in the present, where some historians, like Pradier and Phili, surprisingly and erroneously suggest that Tetens and Pasquich were members of the combinatorial school with no other apparent reason than Hindenburg published their works in his scientific journals (it is true that Phili's claim about Pasquich remains a little ambiguous). Second, Bürmann's theory of combinatorial characteristic openly defied the German combinatorial analysis. Hindenburg and Bürmann engaged in a discussion in which each of them argued in favor of his own theory, without reaching agreement, and Hindenburg ended up claiming that the future of their theories would be decided by the advancement of science itself. Despite the intellectual confrontation between Hindenburg and Bürmann, as well as the deep differences between the German combinatorial analysis and Bürmann's combinatorial characteristics, Bürmann's ideas were mixed with those of the combinatorial school. In this case, it is possible to observe a similar phenomenon to that observed in the case of Tetens and Pasquich: the fact of publishing Bürmann's works in Hindenburg's journals and anthologies nourished the false narrative that Bürmann was a member of the combinatorial school, even though he was in reality an adversary. Third, the publication of Stahl's and Weingärtner's manuals on Hindenburg's theory drove a wedge between the German combinatorial analysis and the way in which the theory of combinations should be perceived. According to the German combinatorial analysis in its second stage of historical evolution, the theory of combinations should be conceived as a self-founding science and as the theoretical solution to unify the field of mathematical analysis, that is to say, to rebuild finite analysis and infinite analysis on one single ground, including arithmetic, algebra, and the theory of series. On the contrary, Stahl and Weingärtner presented the theory of combinations as a bunch of methods that could be successfully applied to the theory of series, and, as a consequence, the theoretical relevance of the theory of combinations disappears in this approach. This conceptual difference between the German combinatorial analysis and the approach of Stahl and Weingärtner will bring about serious consequences for the future of the German combinatorial analysis.

On the other hand, the internal dispute about the theoretical priority with respect to the binomial and the multinomial theorems among the members of the combinatorial school at the beginning of the nineteenth century must also be taken into consideration in the history of the German combinatorial analysis. This discrepancy of opinion and the three points mentioned in the previous paragraph undermined the unity of the research program of the combinatorial school. As a consequence, the unity of the combinatorial school fell apart as well. Other reasons that explain the disband of this research team are the lack of Hindenburg's leadership in the first decade of the nineteenth century, the reorientation of the intellectual interests of its members, and a failure to renew the team's members. After the dissolution of the combinatorial school, there are still a very few research studies on the German combinatorial analysis (we were only able to detect two: a dissertation of Ohm elaborating on the ideas of his teacher Rothe, and a work of Rothe whose content was in fact developed before the dissolution of the combinatorial school but it was published quite late, almost 20 years late), but they are isolated research works that were ignored both by the former members of the combinatorial school and by the rest of German mathematicians. While the life of the combinatorial school was fading along with that of the German combinatorial analysis, a strong intellectual movement arose in German-speaking countries to some extent inspired by the combinatorial school's ideas. This movement is actually a combinatorial current of thought that spans the entire nineteenth century and that is composed of four main branches: a foundationalist current, a countercurrent of algebraic purity, a syntactic current, and a non-foundationalist current. As the combinatorial school, the foundationalist current bet on the theory of combinations as a good candidate for justifying mathematical knowledge, although the meaning of combinatorics was in general profoundly different from that of the combinatorial school. The branch of algebraic purity followed the path opened by Tetens and Pasquich; the syntactic current followed that opened by the mixture of Hindenburg's and Bürmann's theories; and the non-foundationalist current traveled in the direction indicated by Stahl and Weingärtner. Thus, the combinatorial current of thought of the nineteenth century was a consequence of the dissolution of the research program of the combinatorial school (i.e. the German combinatorial analysis) and of the rise of the three new approaches mentioned in the previous paragraph.

The four branches of the combinatorial current of thought have three main elements in common. First, none of them deals with the theory of combinations developed by the combinatorial school, that is to say, with the abstract algebra of involutions. There are very few examples of documents that present in some detail the notion of "combinatorial involution" (for example, the textbooks of Lorenz, Struve, or Prasse's 1813 textbook), but the study of the theory of involutions has been omitted. There are also some very few cases of documents that include one or two examples, no more, of a mathematical table in the shape of the tables used by the combinatorial school for involutions, but these examples do not aim to present the theory of involutions, which is absent from the documents, they are presented just as a means of enumerating a partition of a given number. In the vast majority of documents, combinatorial involutions are not even mentioned. Since the German combinatorial analysis had become a theory of involutions and this theory was overlooked by the four branches of the combinatorial current of thought, it is not possible to say that the German combinatorial analysis underwent a transformation in the texts of the combinatorial current of thought of the nineteenth century, no matter what "transformation" may be. On the contrary, what we have here is the isolation and abandonment of a scientific theory. In fact, the quasi-totality of the texts of the combinatorial current of thought deals with what was called above the classic theory of combinations, which was not an invention of the combinatorial school. The influence of this school on the combinatorial current was exerted either conceptually or through the interest in a specific item studied by the combinatorial school. For instance, this school influenced the foundationalist current conceptually through the idea that the theory of combinations could provide mathematical analysis with a theoretical foundation, even though each mathematician of this branch conceptualized both the theory of combinations and the idea of "foundation" differently; the branch of algebraic purity was influenced through the interest in specific items of study, such as Tetens's interest in Hindenburg's formula for multinomial expansions or Crelle's interest in Kramp's theory of general factorials; the syntactic branch was conceptually influenced through the idea that mathematics needed its own language, and it was also influenced through the interest in the symbols invented by the combinatorial school; the non-foundationalist current was influenced through the idea that the theory of combinations could be advantageously used in mathematical analysis.

Second, the four branches of the combinatorial current of thought contain strong philosophical positions. This means that the work of the combinatorial school crossed the frontiers of mathematics and reached the lands of philosophy. In the foundationalist current, the main problem was to understand what "foundation" means, which, in itself, is a philosophical problem. In particular, a philosophical response to this problem, like that of Fries, gave rise to a philosophical approach of mathematics, called "philosophical mathematics" at the time, which consisted in systematizing mathematical knowledge on the basis of ontological and epistemological concepts. The branch of algebraic purity was indeed animated by the philosophical idea that the nature of the knowledge of mathematical analysis resides in its analytic procedures, and thus combinatorial methods cannot justify the knowledge produced in mathematical analysis. For its part, the syntactic current shows other examples of "philosophical mathematics", namely, the system of mathematics of Krause, but also the project of Robert Grassmann. In particular, Grassmann succeeded in applying a philosophical conception of mathematics to the actual development of mathematics. His incipient philosophy of language allowed him to reformulate the theory of combinations in terms of a logical system, which is an immediate and direct antecedent for the logical research of Schröder and Frege. Thus, his work bridges the historical gap between the German combinatorial analysis and the development of mathematical logic.

In the non-foundationalist current, the works of his father Justus and his brother Hermann provide more examples of philosophical conceptions of mathematics. On the basis of his philosophical conception of mathematics, Justus Grassmann believed that the theory of combinations (via combinatorial geometry) could bridge the theoretical gap between mathematics and natural sciences. And reworking the philosophical views of his father, Hermann Grassmann was convinced that his extension theory had a combinatorial nature. It is curious to note that the study of combinatorics carried out by the Grassmann brothers led them to an approach of mathematics in which the emphasis is on abstract structures (logical structures for Robert and "vector spaces" for Hermann), just as Hindenburg's study of combinatorics led him to the approach of an abstract algebra of involutions, but there is no evidence that the abstract approach of the Grassmann bothers depended on that of Hindenburg directly.

Third, the four branches of the combinatorial current of thought are strongly connected with mathematics teaching. The majority of the documents produced by the combinatorial current are indeed textbooks addressed to students either of secondary schools or of universities. Certainly, this phenomenon is attached to the historical and cultural circumstances of education in German-speaking countries at the time. In particular, mathematics was gaining a more important place in a number of college and secondary school programs because of the reformation of education that wanted to leave behind the model based on classical studies. The combinatorial approach to the theory of series was seen by several mathematics teachers as one much easier to handle than other based on differential calculus, especially in secondary schools. However, the combinatorial approach is not supposed to eliminate differential calculus from mathematics; on the contrary, it is conceived as an introduction to the subject and an intellectual preparation for the more challenging concepts of differential calculus. It is possible to find claims in this sense in the textbooks of the combinatorial current of thought (for instance, in his 1820 textbook, Brandes claims that his book was intended to assist students in the learning process of higher analysis, particularly regarding differential calculus). On the other hand, there are cases in which the combinatorial approach and that based on differential calculus coexist in the same textbook, in either a single volume or several separate volumes. In these cases, there is no priority, whether conceptual or pedagogical, between the two approaches. Therefore, we should exercise caution when reaching conclusions concerning the nature of mathematics on the basis of this pedagogical tendency. For instance, a conclusion according to which the combinatorial approach included in these textbooks points in the direction of the algebraization of analysis would be inappropriate because many of these textbooks also include a differential approach or, at least, do not intend to eliminate it from mathematics. It should also be noted that some texts of the combinatorial current are genuine research projects, such as the texts of the Grassmann brothers or the texts of philosophers like Fries and Krause.

Some years before his death, Hindenburg regretted that his theory, the German combinatorial analysis, was misunderstood. The history of the German combinatorial analysis shows that he had every reason to regret it. His theory was ridiculed, trivialized, distorted, stolen, rejected. However, the German combinatorial analysis was an interesting theory. It is an early example of a mathematical theory focused on the study of general mathematical structures and on the implementation of theoretical results in powerful tools of calculation, even on their implementation in mechanical devices. The ideas of the combinatorial school also prompted the development of German mathematics through their influence on the combinatorial current of thought of the nineteenth century. From a historical point of view, this school indirectly contributed to the development of important fields of mathematics, as mathematical logic and even linear algebra. We hope that this study will contribute, in its turn, to finally give the German combinatorial analysis its right place in the history of mathematics.

References

Aepinus, F. U. T. (1763). Demonstratio generalis theorematis Neutoniani de binomio ad potentiam indefinitam elevando. *Novi Commentarii Academiae Scientiarum Imperialis Petropolitanae, 8*(1760–1761), 27–29, 169–180.

Andrews, G. E. (2007). Euler's "De partitio numerorum". *Bulletin (New Series) of the American Mathematical Society, 44*(4), 561–573.

Anonym (1798a). Séance du 1 Floréal an 6. In *Procès-verbaux des séances de l'Académie tenues depuis la fondation de l'Institut jusqu'au mois d'août 1835*, Vol. 1, an IV–VII (1795–1799), pp. 377–378. Hendaye: Imprimerie de l'Observatoire d'Abbadia, 1910.

Anonym (1798b). Séance du 21 Prairial an 6. In *Procès-verbaux des séances de l'Académie tenues depuis la fondation de l'Institut jusqu'au mois d'août 1835*, Vol. 1, an IV–VII (1795–1799), pp. 403–407. Hendaye: Imprimerie de l'Observatoire d'Abbadia, 1910.

Anonym (1798c). Séance du 6 Nivôse an VII. In *Procès-verbaux des séances de l'Académie tenues depuis la fondation de l'Institut jusqu'au mois d'août 1835*, Vol. 1, an IV–VII (1795–1799), pp. 504–506. Hendaye: Imprimerie de l'Observatoire d'Abbadia, 1910.

Anonym (1801). Mémoires que la classe a jugés dignes d'être imprimés dans le volume des Savans étrangers. In *Mémoires de l'Institut National des sciences et arts. Sciences mathématiques et physiques*, Vol. 3, pp. 52–53. Paris: Imprimeur de l'Institut National.

Anonym (1807). Nekrolog. Georg Friedrich von Tempelhoff. *Intelligenzblatt der allgemeine Literatur-Zeitung, 67*, 537–542.

Anonym (1820). Combination. In *Allgemeine deutsche Real-Encyclopädie für die gebildeten Stände (Conversations-Lexicon)* (Vol. 2). Leipzig: F. A. Brockhaus.

Anonym (1833). Nekrolog. Conrad Diedrich Martin Stahl. *Intelligenzblatt der jenaischen allgemeine Literatur-Zeitung, 11*, 84.

Apian, P. (1527). *Eyn newe unnd wolgegründte Underweysung aller Kauffmansz Rechnung in dreyen Büchern: mit schünen Regeln unnd Fragstücken begriffen.* Ingolstadt: Georgium Apianum.

Arbogast, L. F. A. (1800). *Du calcul des dérivations.* Strasbourg: Imprimerie de Levrault.

Arneth, A. (1852). *Geschichte der reinen Mathematik in ihrer Beziehung zur Geschichte der Entwickelung des menschlichen Geistes.* Stuttgart: Franckh'sche Verlagshandlung.

Ayyangar, A. A. K. (1925–1926). The mathematics of Āryabhaṭa. *Quarterly Journal of the Mythic Society, 16*, 158–179.

Babbage, C. (1864). *Passages from the life of a philosopher.* London: Longman, Green, Longman, Roberts, & Green.

Bartels, J. C. M. (1833). *Vorlesungen über mathematische Analysis mit Anwendungen auf Geometrie, Mechanik und Wahrscheinlichkeitslehre. Erster Band.* Dorpat: Gedruckt bei J. C. Schünmann Universitäts-Buchdrucker.

© The Author(s), under exclusive license to Springer Nature Switzerland AG 2022
E. Noble, *The Rise and Fall of the German Combinatorial Analysis*, Frontiers in the History of Science, https://doi.org/10.1007/978-3-030-93820-8

Bauer, F. L. (2000). *Entzifferte Geheimnisse: Methoden und Maximen der Kryptologie*. Berlin: Springer-Verlag.

Baumgarten, A. G. (1766). *Metaphysik*. Halle: Carl Hermann Hemmerde.

Baur, S. (1816). *Neues historisch-biographisch-literarisches Handwörterbuch, von der Schöpfung der Welt bis zum Schlusse des Jahres 1810* (Vol. 6). Ulm.

Bayes, T. (1736). *An introduction to the doctrine of fluxions, and defence of the mathematicians against the objections of the autor of the* Analyst, *so far as they are designed to affect their general methods of reasoning*. London: Printed for J. Noon.

Beiser, F. C. (2014). *The genesis of Neo-Kantianism, 1796–1880*. New York: Oxford University Press.

Bell, E. T. (1940). *The development of mathematics*. New York: McGraw-Hill.

Bell, J. (2010). A summary of Euler's work on the pentagonal number theorem. *Archive for History of Exact Sciences, 64*(3), 301–373.

Berkeley, G. (1734). *The analyst; or, a discourse addressed to an infidel mathematician*. London: Printed for J. Tonson.

Berkeley, G. (1735). *A defence of free-thinking in mathematics: In answer to a pamphlet of Philalethes Cantabrigiensis, intituled, Geometry no friend to infidelity, or a defence of Sir Isaac Newton, and the British mathematicians. Also an appendix concerning Mr. Walton's Vindication of the principles of fluxions*. London: Printed for J. Tonson.

Bernoulli, J. (1713). *Ars conjectandi, opus posthumum*. Basileae: impensis Thurnisiorum.

Bernoulli, J. (1744). Attollere infinitinomium ad potestatem indefinitam. In G. Cramer (Ed.), *Opera* (Vol. 2, pp. 993–998). Genevae: sumptibus haeredum Cramer et fratrum Philibert.

Bernoulli, J. (1975). Aus den meditationes von Jakob Bernoulli. In *Die Werke von Jakob Bernoulli* (Vol. 3). Basel: Birkhäuser Verlag.

Bernoulli, J. (1993). *Die Werke von Jakob Bernoulli* (Vol. 4). Basel: Birkhäuser Verlag.

Bernoulli, J. III (1783). Extrait de la correspondance de M. Bernoulli. *Nouveaux Mémoires de l'Académie Royale des Sciences et Belles-Lettres de Berlin*, 31–35.

Bernoulli, J. (1695). Epistola XI. Bernoullii ad Leibnitium. In Bernoulli, J. et Leibniz, G. W., *Virorum celeberr. Got. Gul. Leibnitii et Johan. Bernoullii commercium philosophicum et mathematicum* (Vol. 1, pp. 52–64). Lausannae et Genevae: Sumt. Marci.Michaelis Bousquet & Socior, 1745.

Bernoulli, J. (1742). Perfectio regula suae editae in libro gall. *Analyse des infiniment petits*, art. 163. pro determinando valore fractionis cujus numerator & denominator certo casu evanescunt. In Gabriel Cramer (Ed.), *Opera omnia, tam antea sparsim edita, quam hactenus inedita* (Vol. 1). Lausannae et Genevae: sumptibus M. Bousquet et sociorum.

Bernoulli, J. (1988). *Der Briefwechsel von Johann Bernoulli. Der Briefwechsel mit Pierre Varignon. Erster Teil, 1692–1702* (Vol. 2). Basel/Boston/Berlin: Birkhäuser Verlag.

Bernoulli, N. (1709). *Dissertatio inauguralis mathematico-juridica de usu artis conjectandi in jure*. Basilae: typis Johannis Conradi.

Bertinaria, F. (1846). Concetto della filosofia e delle scienze inchiuse nel dominio di essa. *Antologia italiana. Giornale di scienze, lettere ed arti, I*, 332–359.

Bertrand, J. (1864). *Traité de calcul différentiel et de calcul intégral* (Vol. 1). Paris: Gauthier-Villars.

Beskiba, J. (1847). *Lehrbuch der Arithmetik für Real-Schulen*. Wien: Im Verlage der k. k. Schulbücher-Verschleiß-Administration bei St. Anna in der Johannis-Gasse.

Bhāskara (1150). *Līlāvatī*, dans H. T. Colebrooke (éd.), *Algebra, with arithmetic and mensuration, from the sanscrit of Brahmegupta and Bháscara*. London: John Murray, 1817.

Blay, M. (1986). Deux moments de la critique du calcul infinitésimal: Michel Rolle et George Berkeley. *Revue d'histoire des sciences, 39*(3), 223–253.

Bolzano, B. (1816). *Der binomische Lehrsatz, und als Folgerung aus ihm der polynomische, und die Reihen, die zur Berechnung der Logarithmen und Exponentialgrößen dienen.* Prag: C. W. Enderschen Buchhandlung.

Book Review (1790). Kleine Schriften. Mathematik. *Allgemeine Literatur-Zeitung, 2*(125), 279.

Book Review (1794). Mathematik. *Allgemeine Literatur-Zeitung, 3*(283), 545.

Book Review (1795). Anfangsgründe der Analysis endlicher Grössen. Abgefaßt von A. G. Kästner. *Archiv der reinen und angewandten Mathematik, 1,* 233–235.

Book Review (1796a). Mathematik. *Allgemeine Literatur-Zeitung, 4*(381), 585–589.

Book Review (1796b). Mathematik. *Allgemeine Literatur-Zeitung, 380,* 577–583.

Book Review (1797). Mathematics. *The Analytical Review, XXV,* 100.

Borz, G. H. (1769). *De rationibus regularum, quas calculus differentialis in constituendis punctis curvarum multiplicibus, et subtangentibus in iis ad haec puncta ducendis offeret.* Lipsiae: ex officina Langenhemia.

Boscovich, R. J. (1747). Metodo di alzare un'infinitinomio a qualunque potenza. *Giornale de'letterati Decembre, (articolo XXXI),* 393–404.

Boscovich, R. J. (1748a). Parte prima delle reflessioni sul metodo di alzare un infinitinomio a qualunque potenza. *Giornale de'letterati Gennaio (articolo III),* 12–27.

Boscovich, R. J. (1748b). Parte seconda delle riflessioni sul metodo di alzare un infinitinomio a qualunque potenza. *Giornale de'letterati Marzo*(articolo XII), 84–99.

Bottazzini, U. (1989). Lagrange et le problème de Kepler. *Revue d'histoire des Sciences, 42*(1–2), 27–42.

Bourbaki, N. (1984). *Éléments d'histoire des mathématiques.* Paris: Masson.

Boyer, C. B. (1950). Cardan and the Pascal triangle. *The American Mathematical Monthly, 57*(6), 387–390.

Boyer, C. B. (1959). *The history of the calculus and its conceptual development.* New York: Dover Publications.

Brandes, H. W. (1820). *Vorbereitungen zur höheren Analysis. Der polynomische Lehrsatz und leichte Anwendungen desselben zum ersten Unterricht für Anfänger dargestellt.* Leipzig: bei Johann Ambrosius Barth.

Brauen, F. (1982). Athanasius Kircher (1602–1680). *Journal of the History of Ideas, 43*(1), 129–134.

Broch, K. (1967). The theory of risk. *Journal of the Royal Statistical Society. Series B (Methodological), 29*(3), 432–467.

Brockliss, L. (2003). Science, the universities, and other public spaces: Teaching science in Europe and the Americas. In P. Roy (Ed.), *The Cambridge history of science. Eighteenth-century science* (pp. 44–86). New York: Cambridge University Press.

Bullynck, M. (2006). *Vom Zeitalter der Formalen Wissenschaften. Anleitung zur Verarbeitung von Erkenntnissen anno 1800, vermittelst einer parllelen Geschichte.* Ph.D. thesis, Ghent University.

Bullynck, M. (2009a). Decimal periods and their tables: a German research topic (1765–1801). *Historia Mathematica, 36*(2), 137–160.

Bullynck, M. (2009b). Modular arithmetic before C. F. Gauss: systematizations and discussions on remainder problems in 18th-century Germany. *Historia Mathematica, 36*(1), 48–72.

Bullynck, M. (2010). Factor tables 1657–1817, with notes on the birth of number theory. *Revue d'histoire des mathématiques, 16*(2), 141–224.

Bundschue, J. v. G. (1817). *Lehrbuch der Arithmetik um Gebrauche in den Schulen und zum Selbstunterrichte. Vierter Theil.* Kempten: gedruckt und im Verlag bei Dannheimer.

Burckhardt, J. K. (1794). *Methodus combinatorio-analytica evolvendis fractionum continuarum valoribus maxime idonea.* Lipsiae: ex officina Klaubarthia.

Burckhardt, J. K. (1799). Anwendung der combinatorischen Analytik zur Bestimmung der trigonometrischen Linien der einzelnen Winkel gegeben sind. *Nova acta Academiae electoralis Moguntinae scientiarum utilium quae Erfurti, 1*, 293–316.

Bürmann, H. (1798a). Aus zween Briefen von Herrn Bürmann (17. Aug. u. 15 Sept. 1798). *Archiv der reinen und angewandten Mathematik, 2*(8), 509–510.

Bürmann, H. (1798b). Essai de calcul fonctionnaire aux constantes ad libitum. Archives de l'Académie des sciences de Paris. Institut de France, (manuscrit) pochette de la séance du 21 Prairial an VI.

Bürmann, H. (1798c). Formules de développement, de retour et d'intégration. Fonds ancien de l'École des Ponts ParisTech, (manuscrit) MS.1715.

Bürmann, H. (1798d). Lettre du 20 Vendémiaire an VII. Fonds ancien de l'École des Ponts ParisTech, (manuscrit) MS.1715.

Bürmann, H. (1798e). Lettre du 23 mai 1798. Archives de l'Académie des sciences de Paris. Institut de France, (manuscrit) pochette de la séance du 21 Prairial an VI.

Bürmann, H. (1798f). Lettre du 8 avril 1798. Archives de l'Académie des sciences de Paris. Institut de France, (manuscrit) pochette de la séance du 21 Prairial an VI.

Bürmann, H. (1798g). Lettre à Lalande du 13 juin 1798. Archives de l'Académie des sciences de Paris. Institut de France, (manuscrit) pochette de la séance du 21 Prairial an VI.

Bürmann, H. (1798h). Lettre à Legendre, Thermidor an VI. Archives de l'Académie des sciences de Paris. Institut de France, (manuscrit) pochette de la séance du 21 Prairial an VI.

Bürmann, H. (1798i). Supplément à l'Essai de calcul fonctionnaire aux ad-libitum. Archives de l'Académie des sciences de Paris. Institut de France, (manuscrit) pochette de la séance du 21 Prairial an VI.

Bürmann, H. (1798j). Versuch einer vereinfachten Analysis; ein Auszug eines Auszuges. *Archiv der reinen und angewandten Mathematik, 2*(8), 495–499.

Bürmann, H. (1801a). Développement général aux fonctions arbitraires. In C. F. Hindenburg (Ed.), *Über combinatorische Analysis und Derivations-Calcul* (pp. 29–50). Leipzig: Schwickertschen Verlage, 1803.

Bürmann, H. (1801b). Essai de caractéristique combinatoire ou notation universelle déduite d'élémens simples systématiquement combinés. In C. F. Hindenburg (Ed.), *Über combinatorische Analysis und Derivations-Calcul* (pp. 1–28). Leipzig: Schwickertschen Verlage, 1803.

Bürmann, H. (1801c). Polynome combinatoire. In C. F. Hindenburg (Ed.), *Über combinatorische Analysis und Derivations-Calcul* (pp. 51–130). Leipzig: Schwickertschen Verlage, 1803.

Bürmann, H. (1807a). *Eudoxe, ein neu-occidentalischer Hochgesang der Liebe in 8 Zuschriften*. Mannheim: Schwan und Goetz.

Bürmann, H. (1807b). *Programme de la pangraphie partie fondamentale de la caractéristique syntactique, système de notation universelle, déduit dééléments simples, méthodiquement combinés*. Mannheim: Imprimerie de Kaufmann et Friederich.

Buzengeiger, K. H. I. (1797). Von einigen werkwürdigen Eigenschaften der Binomial-Coefficienten. *Archiv der reinen und angewandten Mathematik, 2*(7), 161–173.

Cajori, F. (1890). History of infinite series. In *The teaching and history of mathematics in the United States* (pp. 361–376). Washington: Gouverment Printing Office.

Cajori, F. (1892). Evolution of criteria of convergence. *Bulletin of the New York Mathematical Society, 2*(1), 1–10.

Cajori, F. (1911). Historical note on the Newton-Raphson method of approximation. *The American Mathematical Monthly, 18*(2), 29–32.

Cajori, F. (1919a). *A history of mathematics* (2nd ed.). New York: Macmillan.

Cajori, F. (1919b). *A history of the conceptions of limits and fluxions in Great Britain from Newton to Woodhouse*. Chicago/London: The Open Court Publishing Company.

Cajori, F. (1929). *A history of mathematical notations* (Vol. 2). Chicago/Illinois: The Open Court Publishing Company.

Cardano, G. (1570). *Opus novum de proportionibus numerorum, motuum, ponderum, sonorum, aliarumque rerum mesurandarum.* Basileae: ex officina Henricpetrina.

Caspari, J. J. (1836). *Ausführliches Lehrbuch der Algebra von den ersten Elementen bis zur Analysis oder der Lehre von den Funktionen für Gymnasien und höhere Lehr-Anstalten.* Coblenz: Bei J. Hölscher.

Cassirer, E. (1907). *Das Erkenntnisproblem in der Philosophie und Wissenschaft der neueren Zeit* (Vol. 2). Berlin: Bruno Cassirer.

Castillon, J. (1742). Read at a meeting of the Royal Society, on may 6, 1742. *Philosophical Transactions, 42,* 91–98.

Cauchy, A. L. (1833). *Résumés analytiques.* Turin: de l'imprimerie royale.

Cauchy, A. L. (1841). Mémoire sur la théorie des intégrales définies singulières. Appliquée généralement à la détermination des intégrales définies, et en particulier à l'évaluation des intégrales eulériennes. In *Exercices d'analyse et de physique mathématique. Tome deuxième* (pp. 358–410). Paris: Bachelier.

Cauchy, A. L. (1843a). Mémoire sur l'application du calcul des résidus au développement des produits composés d'un nombre infini de facteurs. *Comptes rendus hebdomadaires des séances de l'Académie des sciences, 17,* 572–581.

Cauchy, A. L. (1843b). Mémoire sur les fonctions dont plusieurs valeurs sont liées entre elles par une équation linéaire, et sur diverses transformations de produits composés d'un nombre indéfini de facteurs. *Comptes rendus hebdomadaires des séances de l'Académie des sciences, 17,* 523–531.

Cauchy, A. L. (1843c). Second Mémoire sur les fonctions dont plusieurs valeurs sont liées entre elles par une équation linéaire. *Comptes rendus hebdomadaires des séances de l'Académie des sciences, 17,* 567–572.

Cayley, A. (1869). On the binomial theorem, factorials, and derivations. In A. R. Forsyth (Ed.), *The Collected Mathematical Papers of Arthur Cayley* (Vol. VIII, pp. 463–473). Cambridge: Cambridge University Press, 1895.

Chemla, K. (1994). Similarities between Chinese and Arabic mathematical writings: (I) root extraction. *Arabic Sciences and Philosophy, 4,* 207–266.

Cheyne, G. (1703). *Fluxionum methodus inversa; sive quantitatum fluentium leges generaliores.* Londoni: typis J. Matthews.

Clairaut, A. C. (1746). *Éémens d'algèbre.* Paris: chez les frères Guérin, David et Durand.

Coolidge, J. L. (1949). The story of the binomial theorem. *The American Mathematical Monthly, 56*(3), 147–157.

Correia, M. (2002). Categorical propositions and logica inventiva in Leibniz's *Dissertatio de arte combinatoria. Studia Leibnitiana, 34*(2), 232–240.

Costabel, P. (1966). *Pierre Varignon (1654–1722) et la diffusion en France du calcul différentiel et intégral.* Paris: Palais de la découverte.

Coumet, E. (1972). Mersenne: dénombrements, répertoires, numérotations de permutations. *Mathématiques et sciences humaines, 38,* 5–37.

Cournot, A. A. (1841a). *Traité élémentaire de la théorie des fonctions et du calcul infinitésimal. Tome premier.* Paris: Chez L. Hachette.

Cournot, A. A. (1841b). *Traité élémentaire de la théorie des fonctions et du calcul infinitésimal. Tome second.* Paris: Chez L. Hachette.

Couturat, L. (Ed.) (1903). *Opuscules et fragments inédits de Leibniz.* Paris: Félix Alcan.

Craig, J. (1685). *Methodus figurarum lineis rectis & curvis comprehensarum quadraturas determinandi.* Londini: impensis Mosis Pitt.

Cramer, G. (1750). *Introduction à l'analyse des lignes courbes algébriques*. Genève: chez les frères Cramer & Cl. Philibert.

Crelle, A. L. (1823). *Versuch einer allgemeinen Theorie der analytischen Facultäten, nach einer neuen Entwickelungs-Methode; vorbereitet durch einen Versuch einer critischen Untersuchung über die Potenzen, Logarithmen und Exponential-Grössen und begleitet von Bemerkungen und Erörterungen, die Theorie der Winkel-Functionen betreffend*. Berlin: Gedruckt und verlegt bei G. Reimer.

Crelle, A. L. (1825). *Lehrbuch der Arithmetik und Algebra, vorzüglich zum Selbstunterrichte*. Berlin: Gedruckt und verlegt bei G. Reimer.

d'Alembert, J. L. R. (1753). Combinaison. In J. L. R. d'Alembert, & D. Diderot (Eds.), *Encyclopédie, ou dictionnaire raisonné des sciences, des arts et des métiers*. Paris: Briasson.

Datta, B., & Avedhesh N. S. (1962). *History of Hindu mathematics: a source book*. Bombay: Asia Publishing House.

de Fontenelle, B. (1704). Éloge de M. le Marquis de l'Hôpital. *Histoire de l'Académie royale des Sciences*, 125–136.

de Fontenelle, B. (1719). Éloge de M. Rolle. *Histoire de l'Académie royale des Sciences*, 116–124.

de l'Hopital, G. F. A. (1696). *Analyse des infiniment petits, pour l'intelligence des lignes courbes*. Paris: Imprimerie royale.

de Lalande, J. (1805). Différens fragmens d'Analyse combinatoire et de Calcul de dérivation, rassemblés et publiés par le professeur C. F. Hindenbourg (en allemand). *Magasin encyclopédique, ou journal des sciences, des lettres et des arts. Tome I*, 196–198.

de Maimieux, J. (1797). *Pasigraphie, premiers élémens du nouvel art-science d'écrire et d'imprimer en une langue de manière à être lu et entendu dans toute autre langue sans traduction*. Paris: Imprimerie de C. J. Gelé.

de Moivre, A. (1697). A method of raising an infinite multinomial to any given given power, or extracting any given root of the same. *Philosophical Transactions, 19*, 619–625.

de Moivre, A. (1698). A method of extracting the root of an infinite equation. *Philosophical Transactions, 20*, 190–193.

de Moivre, A. (1704). *Animadversiones in D. Georgii Cheynaei tractatum de fluxionum methodo inversa*. Londoni: impensis D. Midwinter et T. Leigh.

de Moivre, A. (1718). *The doctrine of chances, or a method of calculating the probability of events in play*. London: by W. Pearson.

de Moivre, A. (1730). *Miscellanea analytica de seriebus et quadraturis*. Londini: excudebant J. Tonson & J. Watts.

De Morgan, A. (1842). *The differential and integral calculus*. London: Robert Baldwin.

De Morgan, A. (1852). On the early history of infinitesimals in England. *The London, Edinburgh and Dublin Philosophical Magazine and Journal of Science. Fourth Series, 4*(26), 321–330.

De Risi, V. (2007). *Geometry and monadology. Leibniz's analysis situs and philosophy of space*. Basel: Birkhäuser Verlag.

de Stainville, J. (1815). *Mélanges d'analyse algébrique et de géométrie*. Paris: Chez Mme Ve Courcier, Imprimeur-Libraire pour les mathématiques.

Dedekind, R. (1932). Was sind und was sollen die Zahlen? (1888). In *Gesammelte mathematische Werke. Dritter Band* (pp. 335–391). Braunschweig: Druck und Verlag von Friedr. Vieweg & Sohn Akt.-Ges.

Deidier, A. (1740). *Le calcul différentiel et le calcul intégral, expliqués et appliqués à la géométrie*. Paris: Charles-Antoine Jombert.

Desberger, F. E. (1831). *Algebra oder die Elemente der mathematischen Analysis*. München: bei F. G. Franckh.

Desberger, F. E. (1847). *Lehrbuch der Arithmetik* (3rd ed.). München: Druck und Verlag von Georg Franz.

Deslauriers, G., & Dubuc, S. (1996). Le calcul de la racine cubique selon Héron. *Elemente der Mathematik, 51*(1), 28–34.

Dhombres, J. (1986). Quelques aspects de l'histoire des équations fonctionnelles liés à l'évolution du concept de fonction. *Archive for History of Exact Sciences, 36*(2), 91–181.

Dhombres, J., & Pensivy, M. (1988). Esprit de rigueur et présentation mathématique au XVIIIème siècle: le cas d'une démonstration d'Aepinus. *Historia Mathematica, 15*(1), 9–31.

Diesterweg, F. A. W. (1820). *Geometrische Combinationslehre. Zur Beförderung des Elementar-Unterrichts in der Formen- und Größenlehre, nebst einer Sammlung von Aufgaben, zu zweckmäßiger Beschäftigung mehrerer Abtheilungen einer Schulklasse.* Elberfeld: Verlag von R. L. Friderichs.

Diesterweg, F. A. W. (1839). *Geometrische Combinationslehre. Zur Beförderung des Elementar-Unterrichts in der Formen- und Größenlehre, nebst einer Sammlung von Aufgaben, zu zweckmäßiger Beschäftigung mehrerer Abtheilungen einer Schulklasse* (2nd ed.). Elberfeld: Schönian'sche Buchhandlung.

Dikson, L. E. (1919–1923). *History of the theory of numbers, 3 vols.* Washington: The Carnegie Institution of Washington.

Ditton, H. (1706). *An institution of fluxions: containing the first principles, the operations, with some of the uses and applications of that admirable method; according to the scheme perfix'd to his tract of Quadratures, by (its first inventor) the incomparable Sir Isaac Newton.* London: printed by W. Botham.

Djebbar, A. (2005). *L'algèbre arabe: genèse d'un art.* Paris: Vuibert/ADAPT.

Djebbar, A., & Rashed, R. (1981). *L'œuvre algébrique d'al-Khayyām.* Syrienne: University of Aleppo.

Doerk, H. G. (1839). *Lehrbuch der Mathematik für Gymnasien und höhere Bürgerschulen. Erster Band. Lehrbuch der Arithmetik und Algebra.* Elbing: Verlag von Fr. L. Levin.

Dunnington, G. W. (1937). B. F. Thibaut (1775–1832), early master of the art of teaching and popularizing mathematics. *National Mathematics Magazine, 11*(7), 318–323.

Durner, M. (1990). Schellings Begegnung mit den Naturwissenschaften in Leipzig. *Archiv für Geschichte der Philosophie, 72*(2), 220–236.

Dyck, M. (1960). *Novalis and mathematics: a study of Friedrich von Hardenberg's Fragments on mathematics and its relation to magic, music, religion, philosophy, language, and literature.* Chapel Hill: University of North Carolina Press.

Edwards, A. W. F. (1987). *Pascal's Arithmetical Triangle.* London: Charles Griffin and Co.

Egen, P. N. C. (1820a). *Handbuch der allgemeinen Arithmetik. Theil I. Die Buchstabenrechnung.* Berlin: Verlag von Duncker und Humblot.

Egen, P. N. C. (1820b). *Handbuch der allgemeinen Arithmetik. Theil II. Die Algebra.* Berlin: Verlag von Duncker und Humblot.

Ersch, J. S., Gruber, J. G., & Hoffmann, A. G. (Eds.) (1831). *Allgemeine Encyklopädie der Wissenschaften und Künste. Zweite Section. Achter Theil.* Leipzig: F. A. Brockhaus.

Eschenbach, H. C. W. (1785). *Ad fratrem Christian. Gotthold Eschenbach, ordinariam chemiae professionem adeuntem, epistola Hieronymi Christophori Vilelmi Eschenbach, Inest in locum Kaestnerianum de multipli angulorum tangentibus commentatio.* Lipsiae: Litteris Breitkopfiis.

Eschenbach, H. C. W. (1789). *De serierum reversione formulis analytico-combinatoriis exhibita specimen.* Lipsiae: ex Officina Breitkopfia.

Euler, L. (1748). *Introductio in analysin infinitorum.* 2 vols., Lausannae: apud Marcum-Michaelem Bousquet & Socios.

Euler, L. (1755). *Institutiones calculi differentialis cum eius usu in analysi finitorum ac doctrina serierum*. Berolini: Impensis Academiae imperialis scientiarum petropolitanae, ex. off. Michaelis.

Euler, L. (1775). Demonstratio theorematis neutoniani de evolutione potestatum binomii pro casibus quibus exponentes non sunt numeri integri. *Novi Commentarii Academiae Scientiarum Imperialis Petropolitanae, 19*(1774), 103–111.

Euler, L. (1784). De mirabilibus proprietatibus unciarum, quae in evolutione binomii ad potestatem quamcunque evecti occurrunt. *Acta Academiae Scientiarum Imperialis Petropolitanae, 5*, 74–111.

Euler, L. (1785). De insignibus proprietatibus unciarum binomii ad uncias quorumvis polynomiorum extensis. *Acta Academiae Scientiarum Imperialis Petropolitanae, 5*, 76–89.

Euler, L. (1789). Nova demonstratio quod evolutio potestatum binomii neutoniana etiam pro exponentibus fractis valeat. *Nova Acta Academiae Scientiarum Imperialis Petropolitanae, 5*(1787), 52–58. (présenté en 1776).

Euler, L. (1813). De serie maxime memorabili, qua potestas binomialis quaecunque exprimi potest. *Mémoires de l'Académie des sciences de Saint-Pétersbourg, 4*(1811), 75–87. (présenté en 1779).

Eytelwein, J. A. (1824a). *Grundlehren der höhern Analysis. Erster Band*. Berlin: Gedruckt und verlegt bei G. Reimer.

Eytelwein, J. A. (1824b). *Grundlehren der höhern Analysis. Zweiter Band*. Berlin: Gedruckt und verlegt bei G. Reimer.

Fabbianelli, F. (2009). Ein unbekanntes Gutachten von Schelling aus dem Jahre 1804. In J. Stolzenberg, K. Ameriks & F. Rush (Eds.), *Internationales Jahrbuch des Deutschen Idealismus. Romantik (2008)/International Yearbook of German Idealism. Romaniticism (2008)* (pp. 301–310). Berlin: Walter de Gruyter.

Fahland, H. (1869). Die Combinationslehre und der binomische Lehrsatz. In C. W. Osterwald (Ed.), *Jahres–Bericht über das Gymnasium zu Mühlhausen* (pp. 1–19). Mühlhausen: Druck der W. Rode'schen Buchdruckerei–Th. Vorhauer.

Fatio de Duillier, N. (1699). *Lineae brevissimi descensus investigatio geometrica duplex, cui addita est investigatio geometrica solidi rotundi in quod minima fiat resistentia*. Londini: apud J. Taylor.

Felkel, A. (1776a). *Tabula omnium factorum simplicum numerorum per 2, 3, 5 non divisibilium, ab 1 usque 10 000 000*. Wien: ex typographia A. Gheleniana.

Felkel, A. (1776b). *Tafel aller einfachen Factoren der durch 2, 3, 5 nicht theilbaren Zahlen von 1 bis 10 000 000. I. Theil. Enthaltend die Factoren von 1 bis 144000*. Wien: von Ehelenschen.

Ferraro, G. (2007). The foundational aspects of Gauss's work on the hypergeometric, factorial and digamma functions. *Archive for History of Exact Sciences, 61*(5), 457–518.

Ferraro, G., & Panza, M. (2003). Developing into series and returning from series: A note on the foundations of eighteenth-century analysis. *Historia Mathematica, 30*(1), 17–46.

Ferreirós Domínguez, J. (1993). *El nacimiento de la teoría de conjuntos*. Madrid: Ediciones de la Universidad Autónoma de Madrid.

Ferreirós Domínguez, J. (1999). *Labyrinth of thought: A history of set theory and its role in modern mathematics*. Basel: Birkäuser Verlag.

Ferreirós Domínguez, J. (2005). Richard Dedekind (1888) and Giuseppe Peano (1889), booklets on the foundations of arithmetic. In I. Grattan-Guinness (Ed.), *Landmark writings in Western mathematics 1640–1940* (pp. 613–626). Amsterdam: Elsevier.

Fink, K. (1890). *Kurzer Abriss einer Geschichte der Elementar-Mathematik mit Hinweisen auf die sich anschliessenden höheren Gebiete*. Tübingen: H. Laupp'schen Buchhandlung.

Fischer, E. G. (1792). *Theorie der Dimensionszeichen nebst ihrer Anwendung auf verschiedene Materien aus der Analysis endlicher Größen, 2 vols*. Halle: in der Buchhandlung des Waisenhauses.

Fischer, E. G. (1794). *Ueber den Ursprung der Theorie der Dimensionszeichen und ihr Verhältniß gegen die combinatorische Analytik des Herrn Professor Hindenburg*. Halle: Buchhandlung des Waisenhauses.

Fischer, E. G. (1797). Ueber die Wegschaffung der Wurzelgrößen aus den Gleichungen. *Archiv der reinen und angewandten Mathematik, 2*(7–8), 180–195, 426–440.

Fischer, E. G. (1824). *Anmerkungen zu seinem Lehrbuch der Mathematik. Drittes Heft, welches Anmerkungen zu den Ergänzungen der Arithmetik, desgleichen zu der ebenen und sphärischen Trigonometrie enthält*. Berlin und Leipzig: gedruckt und verlegt bei G. C. Nauck.

Fischer, L. J., & Krause, K. C. F. (1812). *Lehrbuch der Combinationlehre und der Arithmetik als Grundlage des Lehrvortrages und des Selbstunterrichtes, nebst einer neuen und fasslichen Darstellung der Lehre vom Unendlichen und Endlichen, und einem Elementarbeweise des binomischen und polynomischen Lehrsatzes*. Dresden: in der Arnoldschen Buchhandlung.

Français, J. F. (1815). Du calcul des dérivations, ramené à ses véritables principes, ou théorie du développement des fonctions, et du retour des suites. *Annales de mathématiques pures et appliquées. Tome sixième*, 61–111.

Frege, G. (1993). Über die wissenschaftliche Berechtigung einer Begriffsschrift. In I. Angelelli (Ed.), *Begriffsschrift und andere Aufsätze* (pp. 106–114). Hildesheim-Zürich-New York: Georg Olms Verlag.

Friedelmeyer, J.-P. (1994). *Le calcul des dérivations d'Arbogast dans le projet d'algébrisation de l'analyse à la fin du XVIIIe siècle*. Number 43 in Cahiers d'Histoire et de Philosophie des Sciences. Paris: Société française d'Histoire des Sciences et des Techniques.

Friedelmeyer, J.-P. (1997). La création des premières revues de mathématiques. *Philosophia Scientiæ, 2*(3), 1–26.

Friederich, J. B. (1831). *Lehrbuch der Arithmetik für die lateinischen Schulen in Bayern*. Nürnberg: Druck und Verlag von Friedrich Campe.

Friedländer, P. (1937). Athanasius Kircher und Leibniz. Ein Beitrag zur Geschichte der Polyhistorie im XVII Jahrhundert. *Atti della Pontificia Accademia Romana di Archeologia. Rendiconti. Serie III, XIII*, 229–247.

Fries, J. F. (1807). *Neue Kritik der Vernunft*. Heidelberg: bey Mohr und Zimmer.

Fries, J. F. (1811). *System der Logik. Ein Handbuch für Lehrer und zum Selbstgebrauch*. Heidelberg: bei Mohr und Zimmer.

Fries, J. F. (1842). *Versuch einer Kritik der Principien der Wahrscheinlichkeitsrechnung*. Braunschweig: Verlag von Friedr. Vieweg u. Sohn.

Fritsch, R. (1979). Ein Lehrer und zwei Schüler: Buzengeiger, v. Staudt und Feuerbach. Biographische Notizen. In H. Sund & M. Timmermann (Eds.), *Auf den Weg gebracht. Idee und Wirklichkeit der Gründung der Universität Konstanz* (pp. 139–160). Konstanz: Universitätsverlag Konstanz Gmbh.

Gerhardt, C. I. (1877). *Geschichte der Mathematik in Deutschland*. München: R. Oldenbourg.

Girlich, H.-J. (2009). Über Wege zu ersten mathematischen Fachzeitschriften in Europa. In I. Kästner (Ed.), *Wissenschaftskommunikation in Europa im 18. und 19. Jahrhundert* (pp. 213–228). Aachen: Shaker Verlag.

Glaisher, J. W. L. (1878). On factor tables, with an account of the mode of formation of the factor table for the fourth million. *Proceedings of the Cambridge Philosophical Society, 3*, 99–138.

Grassmann, H. G (1844). *Die Wissenschaft der extensiven Grösse oder die Ausdehnungslehre, eine neue mathematische Disciplin. Erster Theil. Die lineale Ausdehnungslehre ein neuer Zweig der Mathematik, dargestellt und durch Anwendungen auf die übrigen Zweige der Mathematik, wie auch auf die Statik, Mechanik, die Lehre vom Magnetismus und die Krystallonomie erläutert*. Leipzig: Verlag von Otto Wigand.

Grassmann, H. G. (1861). *Lehrbuch der Arithmetik für höhere Lehranstalten*. Berlin: Verlag von Th. Chr. Fr. Enslin.

Grassmann, H. G. (1862). *Die Ausdehnungslehre. Vollständig und in strenger Form bearbeitet*. Berlin: Verlag von Th. Chr. Fr. Enslin.

Grassmann, H. G (2009). Description of the life of Professor Justus Günther Graßmann of Stettin, 1779 to 1852. In H.-J. Petsche, L. Kannenberg, G. Keßler, & J. Liskowacka (Eds.), *Hermann Graßmann. Roots and traces. Autographs and unknown documents* (pp. 36–44). Basel-Boston-Berlin: Birkhäuser.

Grassmann, J. G (1817). *Raumlehre für Volksschulen. Erster Theil. Ebene räumliche Verbindungslehre*. Berlin: In der Realschulbuchhandlung.

Grassmann, J. G. (1824). *Raumlehre für die untern Klassen der Gymnasien, und für Volksschulen. Zweiter Theil. Ebene räumliche Größenlehre*. Berlin: Bei G. Reimer.

Grassmann, J. G. (1827). *Ueber den Begriff und Umfang der reinen Zahlenlehre* (Program des Königl. und Stadt-Gymnasiums zu Stettin). Stettin.

Grassmann, J. G. (1829). *Zur Mathematik und Naturkunde. Erster Band. Zur physischen Krystallonomie und geometrischen Combinationslehre*. Stettin: bei Friedr. Heinr. Morin.

Grassmann, J. G. (2011). On the concept and extent of pure theory of number (1827). In H.-J. Petsche, A. C. Lewis, J. Liesen, & S. Russ (Eds.), *Hermann Graßmann. From past to future: Graßmann's work in context. Graßmann Bicentennial Conference, September 2009*. Basel: Birkhäuser.

Grassmann, R. (1872a). *Die Bindelehre oder Combinationslehre. Drittes Buch der Formenlehre oder Mathematik*. Stettin: Druck und Verlag von R. Grassmann.

Grassmann, R. (1872b). *Die Formenlehre oder Mathematik*. Stettin: Druck und Verlag von R. Grassmann.

Grassmann, R. (2009). Lebenslauf. In H.-J. Petsche, L. Kannenberg, G. Keßler, & J. Liskowacka (Eds.), *Hermann Graßmann. Roots and traces. Autographs and unknown documents* (pp. 97–105). Basel-Boston-Berlin: Birkhäuser.

Grelle, F. (1863). *Prinzipien der Arithmetik*. Hannover: Carl Rümpler.

Gretschel, C. C. C. (1830). *Die Universität Leipzig in der Vergangenheit und Gegenwart*. Dresden: Hilscher'sche Buchhandlung.

Grunert, J. A. (1832). *Lehrbuch der Mathematik für die obern Classen höherer Lehranstalten. Erster Theil. Lehrbuch der allgemeinen Arithmetik*. Brandenburg: bei J. J. Wiesike.

Grunert, J. A. (1835). *Lehrbuch der Mathematik für die obern Classen höherer Lehranstalten. Erster Theil. Lehrbuch der allgemeinen Arithmetik* (2nd ed.). Brandenburg: bei J. J. Wiesike.

Gruson, J. P. (1802). *Mémoire sur le calcul d'exposition*. Berlin.

Gruson, J. P. (1803). Suite du mémoire sur le calcul d'exposition. *Mémoires de l'Académie royale des Sciences et Belles-Lettres de Berlin*, 157–188.

Guicciardini, N. (1989). *The development of newtonian calculus in Britain 1700–1800*. Cambridge: Cambridge University Press.

Guicciardini, N. (1999). *Reading the* Principia: *the debate on Newton's mathematical methods for natural philosophy from 1687 to 1736*. Cambridge: Cambridge University Press.

Guicciardini, N. (2009). *Isaac Newton on mathematical certainty and method*. Cambridge-Massachusetts: Massachusetts Institute of Technology.

Haberman, S. & Sibbett, T. A. (1995). *History of actuarial science* (Vol. IV). London: W. Pickering.

Hamberger, G. C., & Meusel, J. G. (Eds.) (1797). *Das gelehrte Teutschland, oder Lexikon der jetzt lebenden teutschen Schriftsteller* (Vol. 3). Meyerschen Buchhandlung.

Hankel, H. (1867). *Theorie der complexen Zahlensysteme insbesondere der gemeinen imaginären Zahlen und der Hamilton'schen Quaternionen nebst ihrer geometrischen Darstellung*. Leipzig: Leopold Voss.

Hannah, R. W. (1981). *The Fichtean dynamic of Novalis' poetics*. Bern: P. Lang.

Harris, J. (1702). *A new short treatise of algebra; with the geometrical construction of equations, as far as the fourth power or dimension. Together with a specimen of the nature and algorithm of fluxions*. London: printed by J. M.

Hartmann von Franzens-Huld, M. (1843). *Grundlehren der allgemeinen Arithmetik*. Wien: Verlag und Druck von Joh. Bapt. Wallishausser.

Hauff, J. K. F. (1807). *Lehrbuch der Arithmetik zum Gebrauche auf hohen und niedern Schulen wie zum Selbstunterricht* (2nd ed.). Marburg: in der neuen akademischen Buchhandlung.

Hayes, C. (1704). *A treatise of fluxions: or, an introduction to mathematical philosophy*. London: printed by Edw. Midwinter.

Heath, T. L. (Ed.) (1908). *The thirteen books of Euclid's Elements* (Vol. 1). Cambridge: Cambridge University Press.

Heath, T. L. (1921a). *A history of Greek mathematics*. Oxford: Oxford Clarendon Press.

Heath, T. L. (1921b). Greek mathematics and science. *The Mathematical Gazette, 10*(153), 289–301.

Heidegger, M. (1925). *Prolegomena zur Geschichte des Zeitbegriffs, Gesamtausgabe* (Vol. 20). Frankfurt am Main: Vittorio Klostermann, 1979.

Helmes, J. (1862). *Die Elementar-Mathematik nach den Bedürfnissen des Unterrichts streng wissenschaftlich dargestellt. Erster Band. Die Arithmetik und Algebra*. Hannover: Hahn'sche Hofbuchhandlung.

Hennert, J. F. (1799). Ueber das Ausziehen der Wurzeln aus binomischen Grössen. *Archiv der reinen und angewandten Mathematik, 9*, 50–61.

Hennert, J. F. (1805). *Mathematische Abhandlungen nebst einem Verzeichniss seiner sämmtlichen Schriften*. Leipzig: Gerhard Fleischer.

Heron of Alexandria (1903). Metrica. In H. Schöne (Ed.), *Heronis Alexandrini Opera quae supersunt omnia* (Vol. III). Lipsiae: B. G. Teubner.

Heuser, M.-L. (1996). Geometrical product-exponentiation-evolution. Justus Günther Grassmann and dynamist Naturphilosophie. In G. Schubring (Ed.), *Hermann Günther Graßmann (1809–1877): visionary mathematician, scientist, and neohumanist scholar. Papers from a sesquicentennial conference* (pp. 47–58). Dordrecht: Kluwer Academic Publishers.

Hindenburg, C. F. (1763). *Specimen animadversionum philologico-criticarum in Musaeum praemittit simulque summos in philosophia honores*. Lipsiae: ex officina Langenhemia.

Hindenburg, C. F. (1769). *Animadversiones quibus Xenophontis memorabilium Socratis dictorum*. Lipsiae: apud S. L. Crusium.

Hindenburg, C. F. (1776a). *Beschreibung einer ganz neuen Art, nach einem bekannten Gesetze fortgehende Zahlen, durch Abzählen oder Abmessen bequem und sicher zu finden*. Leipzig: bey Siegfried Lebrecht Crusis.

Hindenburg, C. F. (1776b). Brif. Hindenburg an Lambert, den 3 Aug. 1776. In J. III Bernoulli (Ed.), *Johann Heinrich Lamberts deutscher gelehrter Briefwechsel* (Vol. 5, pp. XX). Berlin, 1785: bey dem Herausgeber.

Hindenburg, C. F. (1778a). *Infinitinomii dignitatum indeterminatarum leges ac formulae*. Gottingae: Litteris Joh. Christ. Dieterich.

Hindenburg, C. F. (1778b). *Methodus nova et facilis serierum infinitarum exhibendi dignitates exponentis indeterminati*. Lipsiae: ex officina Langenhemiana.

Hindenburg, C. F. (1779). *Infinitinomii dignitatum exponentis indeterminati historia leges ac formulae*. Gottingae: litteris Joh. Christ. Dieterich.

Hindenburg, C. F. (1781a). Ein Zusatz hierzu. *Leipziger Magazin zur Naturkunde, Mathematik und Oekonomie, 1*, 459–462.

Hindenburg, C. F. (1781b). *Novi systematis permutationum combinationum ac variationum primae lineae et logisticae serierum formulis analytico-combinatoriis per tabulas exhibendae conspectus et specimena.* Lipsiae: apud S. L. Crusium.

Hindenburg, C. F. (1781c). *Novi systematis permutationum combinationum ac variationum primas lineas et logisticae serierum formulis analytico-combinatoriis per tabulas exhibendae conspectum.* Lipsiae: in officina Breitkopfia.

Hindenburg, C. F. (1781d). Über die Schwürigkeit bey der Lehre von den Parallellinien. Neues System der Parallellinien. *Leipziger Magazin zur Naturkunde, Mathematik und Oekonomie, 1,* 145–168.

Hindenburg, C. F. (1784). Praefatio. In Rüdiger, C . F.: *Specimen analyticum de lineis curvis secundi oridinis in dilucidationem analyseos finitorum kaestnerianae.* Lipsiae: in bibliopolio I. G. Mülleriano.

Hindenburg, C. F. (1785). Über den Schachspieler des Herrn von Kempelen, nebst einer Abbildung und Beschreibung seiner Sprachmaschine. *Leipziger Magazin zur Naturkunde, Mathematik und Oekonomie, 4,* 235–269.

Hindenburg, C. F. (1786). Verbindungsgesetz cyklischer Perioden; Natur und Eingenschaften derselben; ihr Gebrauch in der diophantischen oder unbestimmten Analytik. *Leipziger Magazin für reine und angewandte Mathematik, 1*(3), 281–324.

Hindenburg, C. F. (1793). *Problema solutum maxime universale ad serierum reversionem formulis localibus et combinatorio-analyticis.* Lipsiae: ex officina Klaubarthia.

Hindenburg, C. F. (1794a). Combinatorische Verfahren, zu Bestimmung der Werthe der continuirlichen Brüche, in und außer der Ordnung. *Archiv der reinen und angewandten Mathematik, 1*(1–2), 47–68, 154–195.

Hindenburg, C. F. (1794b). Kritisches Verzeichnis aller bis hieher herausgekommenen, die combinatorische Analytik unmittelbar oder mittelbar betreffenden Schriften. Ein Beytrag zur künstigen Geschichte dieser neuen Wissenschaft. *Archiv der reinen und angewandten Mathematik, 1*(1), 111–119.

Hindenburg, C. F. (1794c). Ueber combinatorische Involutionen und Evolutionen, und ihren Einfluß auf die combinatorische Analytik. *Archiv der reinen und angewandten Mathematik, 1*(1), 13–46.

Hindenburg, C. F. (1794d). Ueber das Umkehrungsproblem des Herrn de la Grange. *Archiv der reinen und angewandten Mathematik, 1*(1), 88–93.

Hindenburg, C. F. (1795a). Allgemeine Darstellung des Polynomialtheorems nach de Moivre und Boscovich, nebst verschiedenen Bemerkungen über die dabey zum Grunde liegenden lexikographischen Involutionen. *Archiv der reinen und angewandten Mathematik, 1*(4), 385–423.

Hindenburg, C. F. (1795b). Fragen eines Ungenannten über die Art durch Gitter geheim zu schreiben, und vorläufige Beantwortung derselben. *Archiv der reinen und angewandten Mathematik, 1*(3), 347–351.

Hindenburg, C. F. (1795c). Mehrere große Mathematiker sind der Erfindung der combinatorischen Involutionen ganz nahe gewesen. *Archiv der reinen und angewandten Mathematik, 1*(3), 319–336.

Hindenburg, C. F. (1795d). *Terminorum ab infinitinomii dignitatibus Coefficientes Moiuraeanos sequi Ordinem Lexicographicum ostenditur.* Lipsiae: ex officina Klaubarthia.

Hindenburg, C. F. (1795e). Vorrede. *Archiv der reinen und angewandten Mathematik, 1.*

Hindenburg, C. F. (1796a). Die Combinationslehre ist eine selbstständige Grundwissenschaft; ihre Verbindung mit der Analysis ist die engste und natürlichste; die unmittelbarste Anwendung derselben zeigt sich bey allgemeinen Produkten- und Potenzenprobleme der Reihen; Vergleichung des von Hrn. Tetens bey diesen Problemen angebrachten Substitutionsverfahren mit der Hindenburgischen Combinationsmethode; Nothwendigkeit einer in die Analysis einzuführenden allgemeinen, größtentheils combinatorischen, Charakteristik. *Sammlung combinatorisch-*

analytischer Abhandlungen. Der polynomische Lehrsatz das wichtigste Theorem der ganzen Analysis nebst einigen verwandten und andern Sätzen, 1, 153–304.

Hindenburg, C. F. (1796b). Ueber Gitter und Gitterschrift, fernere Aeusserung des Ungenannten. Ubersetzung der von ihm mitgetheilten geheimen Gitterschrift. *Archiv der reinen und angewandten Mathematik, 2*(5), 81–99.

Hindenburg, C. F. (1800). Vorbericht. *Sammlung combinatorisch-analytischer Abhandlungen, 2*, V–XXII.

Hindenburg, C. F. (1803a). Anmerkungen zu Bürmann's Essai de caractéristique combinatoire. In C. F. Hindenburg (Ed.), *Über combinatorische Analysis und Derivations-Calcul*. Leipzig: Schwickertschen Verlage.

Hindenburg, C. F. (1803b). Der Derivations-Calcul und die combinatorische Analysis in Beziehung auf einander. In C. F. Hindenburg (Ed.), *Über combinatorische Analysis und Derivations-Calcul* (pp. 167–294). Leipzig: Schwickertschen Verlage.

Hindenburg, C. F. (1992). Von Carl Friedrich Hindenburg, 24 may 1793. In U. Joost & A. Schöne (Eds.), *Georg Christoph Lichtenberg Briefwechsel* (Vol. IV, pp. 96–97). München: C. H. Beck.

Hindenburg, C. F., & Leske, N. G. (Eds.) (1785). *Leipziger Magazin zur Naturkunde, Mathematik und Oekonomie* (Vol. 4). Leipzig.

Hirsch, M. (1804). *Sammlung von Beispielen, Formeln und Aufgaben aus der Buchstabenrechnung und Algebra*. Berlin: Duncker und Humblot.

Hirsch, M. (1809). *Sammlung von Aufgaben aus der Theorie der algebraischen Gleichungen*. Berlin: Bey Duncker und Humblot.

Hirsch, M. (1827). *Collection of examples, formulae, & calculations, on the literal calculus and algebra* (Vol. I). London: Black, Young, and Young.

Hirsch, M. (1831). *A collection of arithmetical and algebraic problems and formulae*. Boston: Carter, Hendee and Babcock.

Hirsch, M. (1832). *Exemples, formules et problèmes du calcul littéral et de l'algèbre*. Berlin: Duncker et Humblot.

Hodgson, J. (1736). *The doctrine of fluxions, founded on Sir Isaac Newton's method, published by himself in his tract upon the quadrature of curves*. London: printed by T. Wood.

Hughes, B. (1989). The arithmetical triangle of Jordanus de Nemore. *Historia Mathematica, 16*(3), 213–223.

Hérigone, P. (1634). Algebra/Algèbre. In *Cursus mathematici/Cours Mathématique* (Vol. 2). Paris: chez l'auteur et chez Henry le Gras.

Jahn, G. A. (1839). *Die Wahrscheinlichkeitsrechnung und ihre Anwendung auf das wissenschaftliche und praktische Leben*. Leipzig: Bei E. B. Schwickert.

Jahnke, H. N. (1987). Motive und Probleme der Arithmetisierung der Mathematik in der ersten Hälfte des 19. Jahrhunderts – Cauchys Analysis in der Sicht des Mathematikers Martin Ohm. *Archive for History of Exact Sciences, 37*(2), 101–182.

Jahnke, H. N. (1990). *Mathematik und Bildung in der Humboldtschen Reform*. Göttingen: Vandenhoeck & Ruprecht.

Jahnke, H. N. (1991). Mathematics and culture: the case of Novalis. *Science in Context, 4*(2), 279–295.

Jahnke, H. N. (1992). A structuralist view of Lagrange's algebraic analysis and the German combinatorial school. In J. Echeverria, A. Ibarra, & T. Mormann (Eds.), *The space of mathematics. Philosophical, epistemological, and historical explorations* (pp. 280–295). Berlin: Walter de Gruyter.

Jahnke, H. N. (1993). Algebraic analysis in Germany, 1780–1840: Some mathematical and philosophical issues. *Historia Mathematica, 20*(3), 265–284.

Jahnke, H. N. (1996). The development of algebraic analysis from Euler to Klein and its impact on school mathematics in the nineteenth century. In R. Calinger (Ed.), *Vita mathematica: Historical research and integration with teaching* (pp. 145–152). Washington: Mathematical Association of America.

Jesseph, D. M. (1993). *Berkeley's philosophy of mathematics*. Chicago: University of Chicago Press.

Johnson, W. P. (2020). *An introduction to* q-analysis. Providence, Rhode Island: American Mathematical Society.

Jones, W. (1706). *Synopsis palmariorum matheseos: Or, a new introduction to the mathematics: Containing the principles of arithmetic & geometry demonstrated, in a short and easie method.* London: printed by J. Matthews.

Jurin, J. (1734). *Geometry no friend to infidelity: or, a defence of Sir Isaac Newton and the British mathematicians, in a letter to the author of the* Analyst. London: printed for T. Cooper.

Jurin, J. (1735). *The minute mathematician: Or, the free-thinker no just-thinker. Set forth in a second letter to the author of the Analyst; containing a defense of Sir Isaac Newton and the British mathematicians, against a late pamphlet, entitled, A defence of free-thinking in mathematicks.* London: printed for T. Cooper.

Jushkevich, A. P. (1976). *Les mathématiques arabes (VIIIe-XVe siècles)*. Paris: J. Vrin.

Jushkevich, A. P. (1983). L. Euler's unpublished manuscript *Calculus Differentialis*. In E. A. Fellmann (Ed.), *Leonhard Euler 1707–1783: Beiträge zu Leben und Werk* (pp. 161–170). Basel: Birkhäuser Verlag.

Karsten, W. J. G. (1760). *Mathesis theoretica elementaris atque sublimior*. Rostochii et Gryphiswaldiae: apud Anton. Ferdin. Röseum.

Keach, W. (1997). Poetry, after 1740. In H. B. Nisbet & C. Rawson (Eds.), *The Cambridge history of literary criticism* (Vol. IV, pp. 117–166). Cambridge: Cambridge University Press.

Keiding, N. (1987). The method of expected number of deaths, 1786–1886–1986. *International Statistical Review/Revue Internationale de Statistique, 55*(1), 1–20.

Kitcher, P. (1973). Fluxions, limits, and infinite littlenesse. A study of Newton's presentation of the calculus. *Isis, 64*(1), 33–49.

Klein, F. (1967). *Vorlesungen über die Entwicklung der Mathematik in 19. Jahrhundert. Teil I*. New York: Chelsea Publishing Company.

Klügel, G. S. (1770). *Analytische Trigonometrie*. Braunschweig: Verlag der Fürstl. Waisenhausbuchhandlung.

Klügel, G. S. (1796). Bemerkungen über den Polynomischen Lehrsatz. *Sammlung combinatorisch-analytischer Abhandlungen. Der polynomische Lehrsatz das wichtigste Theorem der ganzen Analysis nebst einigen verwandten und andern Sätzen, 1*, 48–90.

Klügel, G. S. (1800). Erläuterungen über den Beweis des polynomischen Lehrsatzes von dem Verfasser des Beweises. *Sammlung combinatorisch-analytischer Abhandlungen, 2*, 145–154.

Klügel, G. S. (1803). *Mathematisches Wörterbuch oder Erklärung der Begriffe, Lehrsätze, Aufgaben und Methoden der Mathematik. Erste Abtheilung. Die reine Mathematik. Erster Theil von A bis D*. Leipzig: Schwickertschen Verlage.

Klügel, G. S. (1808). *Mathematisches Wörterbuch oder Erklärung der Begriffe, Lehrsätze, Aufgaben und Methoden der Mathematik. Erste Abtheilung. Die reine Mathematik. Dritter Theil von K bis P*. Leipzig: im Schwickertschen Verlage.

Klügel, G. S., Carl Brandan M., & Johann August G. (1831). *Mathematisches Wörterbuch oder Erklärung der Begriffe, Lehrsätze, Aufgaben und Methoden der Mathematik. Erste Abtheilung. Die reine Mathematik. Fünfter Theil. Zweiter Band. V bis Z*. Leipzig: bey C. V. Schwickert.

Knobloch, E. (1976). *Die mathematischen Studien von G. W. Leibniz zur Kombinatorik*. Wiesbaden: F. Steiner Verlag. (*Studia liebnitiana. Supplementa*, XVI).

Knobloch, E. (2001). Déterminants et élimination chez Leibniz. *Revue d'histoire des sciences, 54*(2), 143–164.

Knobloch, E., & Berlin, W. (1974). The mathematical studies of G. W. Leibniz on combinatorics. *Historia Mathematica, 1*(4), 409–430.

Knorr, W. R. (1975). *The evolution of the Euclidean elements: A study of the theory of incommensurable magnitudes and its significance for early Greek geometry.* Dordrecht-Boston: D. Reidel.

Köcher, F. A. (1822). *Die Combinationslehre und ihre Anwendung auf die Analysis zum Gebrauche für Gymnasien und höhere Anstalten.* Leipzig: Bei Paul Gotthelf Kummer.

Kollerstrom, N. (1992). Thomas Simpson and "Newton's Method of Approximation": An Enduring Myth. *The British Journal for the History of Science, 25*(3), 347–354.

König, A. R. J. (1827). *Lehrbuch der Arithmetik für höhere Bürgerschulen und zum Selbstunterrichte. Erster Theil.* Nürnberg: bei Joh. Leonh. Schrag.

König, A. R. J. (1828). *Lehrbuch der Arithmetik für höhere Bürgerschulen und zum Selbstunterrichte. Zweiter Theil.* Nürnberg: bei Joh. Leonh. Schrag.

Kramp, C. (1796). Coefficient des allgemeinen Gliedes jeder willkührlichen Potenz eines infinitinomiums; Verhalten zwischen Coefficienten der Gleichungen und Summen der Produkte und der Potenzen ihrer Wurzeln; Transformation und Substitution der Reihen durch einander. *Sammlung combinatorisch-analytischer Abhandlungen. Der polynomische Lehrsatz das wichtigste Theorem der ganzen Analysis nebst einigen verwandten und andern Sätzen, 1,* 91–122.

Kramp, C. (1799). *Analyse des réfractions astronomiques et terrestres.* Strasbourg: Philippe Jacques Dannbach.

Kramp, C. (1800). Verschiedene combinatorisch-analytisch bearbeitete Aufgaben. *Sammlung combinatorisch-analytischer Abhandlungen, 2,* 341–352.

Kramp, C. (1808). *Élémens d'arithmétique universelle.* Cologne: de l'imprimerie de Th. F. Thiriart.

Krause, K. C. F. (1804). *Grundlage der Arithmetik. Erster Theil.* Jena und Leipzig: bei Christian Ernst Gabler.

Krause, K. C. F. (1807). Review. *Neue Leipziger Literaturzeitung, 4*(132), 2097–2105.

Krause, K. C. F. (1876). La ciencia de la forma. In F. Giner de los Ríos (Ed.), *Estudios filosóficos y religiosos,* pp. 177–215. Madrid: Librería de Francisco Góngora.

Krause, K. (2003). *Alma Mater Lipsiensis: Geschichte der Universität Leipzig von 1409 bis zur Gegenwart.* Leipzig: Leipziger Universitätsverlag.

Kries, F. (1826). *Lehrbuch der reinen Mathematik* (4th ed.). Jena: Friedrich Frommann.

Kroll, J. F. (1839). *Grundriß der Mathematik für Gymnasien und andere höhere Lehr-Anstalten.* Eisleben: Verlag von Georg Reichardt.

Kulik, J. P. (1831). *Lehrbuch der höheren Analysis.* Prag: in Kommission bei Kronberger und Weber.

Kulik, J. P. (1843). *Lehrbuch der höheren Analysis. Erster Band. Lehrbuch der höheren Arithmetik und Algebra* (2nd ed.). Prag: in Kommission bei Kronberger und Rziwnatz.

Kästner, A. G. (1745). *Demonstratio theorematis binomialis.* Lipsiae: Litteris Breitkopfianis.

Kästner, A. G. (1758). *Theorema binomiale universaliter demonstrat praelectionesque suas indicat.* Gottingae: Litteris Pockwitzii et Barmeieri.

Kästner, A. G. (1759). *Infinitinomii ad potentiam indefinitam elevati formula.* Gottingae: ex officina Schultzia.

Kästner, A. G. (1760). *Anfangsgründe der Analysis endlicher Grössen.* Göttingen: Wittwe Vandenhoeck.

Kästner, A. G. (1761). *Anfangsgründe der Analysis des Unendlichen.* Göttingen: Wittwe Vandenhoeck.

Kästner, A. G. (1771). *Dissertationes mathematicae et physicae.* Altenburg: ex officina Richteria.

Lacroix, S.-F. (1810). *Traité du calcul différentiel et du calcul intégral* (Vol. 1). Paris: chez Courcier.

Lacroix, S.-F. (1819). *Traité du calcul différentiel et du calcul intégral* (Vol. 3). Paris: Courcier.

Lacroix, S.-F. (1990). Réflexions générales sur les méthodes analytiques et particulièrement sur celles des dérivations. In I. Grattan-Guinness (Ed.), *Convolutions in French mathetnatics, 1800–1840. From the calculus and mechanics to mathematical analysis and mathematical physics* (Vol. III, pp. 1325–1329). Basel: Springer.

Lagrange, J.-L. (1770). Nouvelle méthode pour résoudre les équations littérales par le moyen des séries. *Histoire de l'Académie royale des Sciences et Belles-Lettres de Berlin*, 251–326.

Lagrange, J.-L. (1771). Sur le problème de Kepler. *Histoire de l'Académie royale des Sciences et Belles-Lettres de Berlin*, 204–233.

Lagrange, J.-L. (1868). Mémoire sur l'utilité de la méthode de prendre le milieu entre les résultats de plusieurs observations. In J.-A. Serret (Ed.), *Œuvres de Lagrange* (Vol. 2, pp. 173–234). Paris: Gauthier-Villars.

Lagrange, J.-L., & Legendre, A.-M. (1799). Rapport sur deux mémoires d'analyse du professeur Bürmann. *Mémoires de l'Institut National des sciences et arts. Sciences mathématiques et physiques, 2*, 13–17.

Lai, T. (1975). Did Newton renounce infinitesimals? *Historia Mathematica, 2*(2), 127–136.

Lambert, J. H. (1770a). *Beyträge zum Gebrauch der Mathematik und deren Anwendung. Zweyter Theil. Erster Abschnitt.* Berlin: Buchhandlung der Realschule.

Lambert, J. H. (1770b). *Zusätze zu den logarithmischen und trigonometrischen Tabellen.* Berlin: bey Haude und Spener.

Langsdorf, K. C. (1817). *Leichtfaßliche Anleitung zur Analysis endlicher Größen und des Unendlichen und zur höheren Geometrie für Physiker, Architekten, Hydrötekten, Berg- und Salzwerksbeamte, Ingenieurs und Technologen.* Mannheim und Heidelberg: in der Schwan- und Gößischen Buchhandlung.

Laugwitz, D. (1996). *Bernhard Riemann: 1826–1866. Wendepunkt in der Auffassung der Mathematik.* Basel: Birkhäuser Verlag.

Lay-Yong, L. (1980). The Chinese connection between the Pascal triangle and the solution of numerical equations of any degree. *Historia Mathematica, 7*(4), 407–424.

Legendre, A.-M. (1798a). Copie d'une lettre du c. Legendre en date du 6 fructidor an 6. Archives de l'Académie des sciences de Paris. Institut de France, pochette de la séance du 21 Prairial an VI.

Legendre, A.-M. (1798b). Démonstration du théorème de M. Burmane. Fonds ancien de l'École des Ponts ParisTech, (manuscrit) MS.1715.

Legendre, A.-M. (1817). *Exercices de calcul intégral sur divers ordres de transcendantes et sur les quadratures* (Vol. 2). Paris: Vve Courcier.

Lehmus, D. C. L. (1822). *Lehrbuch der Zahlen-Arithmetik, Buchstaben-Rechnung und Algebra. Zum Gebrauch in höheren Schulen und zum Selbststudium eingerichtet. Neue, ganz umgearbeitete Ausgabe.* Leipzig: In der Wienbrack'schen Buchhandlung.

Lefébure de Fourcy, L., Vincent, M., & Ritter, L. F. (1844). *Neun Abhandlungen über eben so wichtige als interessante Gegenstände aus der Algebra und niedern Analysis.* Stuttgart: E. Schweizerbart'sche Verlagshandlung.

Leibniz, G. W. (1666a). *Disputatio arithmetica de complexionibus.* Lipsiae: Literis Spörelianis.

Leibniz, G. W. (1666b). *Dissertatio de arte combinatoria.* Lipsiae: apud J. S. Fickium et J. P. Seaboldum.

Leibniz, G. W. (1676a). Leibniz to Oldenburg, 12 May 1676. In Collins, J., *Commercium epistolicum D. Johannis Collins et aliorum de analysi promota: jussu Societatis regiae in lucem editum* (pp. 45). Londini: typis Pearsonianis, 1712.

Leibniz, G. W. (1676b). Leibniz to Oldenburg, 27 August 1676. In Collins, J., *Commercium epistolicum D. Johannis Collins et aliorum de analysi promota: jussu Societatis regiae in lucem editum* (pp. 58–65). Londini: typis Pearsonianis, 1712.

Leibniz, G. W. (1677). Leibniz an Galloys, 1677. In K. I. Gerhardt (Ed.), *Leibnizens mathematische Schriften* (Vol. 1, pp. 178–181). Berlin: A. Asher & Comp., 1849.

Leibniz, G. W. (1679a). Entwurf der geometrischen Charakteristik. Aus einem Brief an Huygens. In E. Cassirer (Ed.), *Hauptschriften zur Grundlegung der Philosophie* (Vol. 1, pp. 56–61). Hamburg: Felix Meiner Verlag, 1996.

Leibniz, G. W. (1679b). Tentamina de definitione quantitatis. In H. Schepers, M. Schneider, G. Biller, U. Franke, & H. Kliege-Biller (Eds.), *Sämtliche Schriften und Briefe, Sechste Reihe* (Vol. 4, pp. 162–164). Berlin: Akademie Verlag, 1999.

Leibniz, G. W. (1683). Elementa nova matheseos universalis. In H. Schepers, M. Schneider, G. Biller, U. Franke, & H. Kliege-Biller (Eds.), *Sämtliche Schriften und Briefe, Sechste Reihe* (Vol. 4, pp. 513–524). Berlin: Akademie Verlag, 1999.

Leibniz, G. W. (1684, Octobris). Nova methodus pro maximis et minimis, itemque tangentibus, quae nec fractas, nec irrationales quantitates moratur, & singulare pro illis calculi genus. *Acta Eruditorum, Octobris*, 467–473.

Leibniz, G. W. (1687–1696). Definitiones: Ens, possibile, existens. In H. Schepers, M. Schneider, G. Biller, U. Franke, & H. Kliege-Biller (Eds.), *Sämtliche Schriften und Briefe, Sechste Reihe* (Vol. 4, pp. 867–870). Berlin: Akademie Verlag, 1999.

Leibniz, G. W. (1691). Ars combinatoria. *Acta eruditorum*, 63–64.

Leibniz, G. W. (1693). De analysi situs. In K. I. Gerhardt (Ed.), *Leibnizens mathematische Schriften. Zweite Abtheilung* (Vol. I, pp. 178–183). Halle: H. W. Schmidt, 1858.

Leibniz, G. W. (1694, Augusti). Constructio propria problematis de curva isochrona paracentrica. *Acta Eruditorum, Augusti*, 364–375.

Leibniz, G. W. (1695). Epistola X. Leibnitii ad Bernoullium. In *Bernoulli, J. et Leibniz, G. W., Virorum celeberr. Got. Gul. Leibnitii et Johan. Bernoullii commercium philosophicum et mathematicum* (Vol. 1, pp. 46–51). Lausannae et Genevae: Sumt. Marci.Michaelis Bousquet & Socior, 1745.

Leibniz, G. W. (1697). Leibniz à André Morell. In *Sämtliche Schriften und Briefe, Reihe I* (Vol. 14). Berlin: Deutsche Akademie der Wissenschaften.

Leibniz, G. W. (1700). Responsio ad Dn. Nic. Fatii Duillerii imputationes. Accessit nova artis analytica promotio specimine indicata; dum designatione per numeros assumtitios loco literarum, algebra ex combinatoria arte lucem capit. *Acta Eruditorum Maji*, 198–208.

Leibniz, G. W. (1702). Justification du calcul des infinitesimales par celuy de l'algebre ordinaire. In *Leibnizens mathematische Schriften. Erste Abtheilung* (Vol. 4, pp. 104–106). Halle, 1859: Druck und Verlag von H. W. Schmidt.

Leibniz, G. W. (1705). Lettre de Leibniz à Bernoulli, 1705. *Histoire de l'Académie royale des Sciences et Belles-Lettres de Berlin, 1757(1759)*, 475–477.

Leibniz, G. W. (1715). Initia rerum mathematicarum metaphysica. In K. I. Gerhardt (Ed.), *Leibnizens mathematische Schriften. Zweite Abtheilung* (Vol. III, pp. 17–29). Halle: H. W. Schmidt, 1863.

Leibniz, G. W. (1768). Monitum de characteribus algebraicis. In L. Dutens (Ed.), *Gothofredi Guillelmi Leibnitii opera omnia* (Vol. 3, pp. 416–420). Genevae: apud Fratres de Tournes.

Leibniz, G. W. (1859). *Leibnizens mathematische Schriften* (Vol. 4). Halle: Druck und Verlag von H. W. Schmidt.

Leske, N. G. (Ed.) (1779–1780). *Abhandlungen zur Naturgeschichte, Physik und Oekonomie aus den Philosophischen Transaktionen und Sammlungen mit einigen Anmerckungen übersetzt*. Leipzig: Weygand.

L'Huilier, S. A. J. (1795). *Principiorum calculi differentialis et integralis expositio elementaris*. Tubingae: apud Joh. Georg. Cottam.

Libri, G. (1827). Mémoire sur quelques formules générales d'analyse. In *Mémoires de mathématique et de physique. Premier Cahier* (pp. 3–15). Pise: de l'imprimerie di Prosperi.

Libri, G. (1831). Mémoire sur quelques formules générales d'analyse. *Journal für die reine und angewandte Mathematik, 7*(1), 57–67.

Lieber, F. (Ed.) (1830). *Encyclopaedia Americana, a popular dictionary of arts, sciences, literature, history, politics and biography* (Vol. III). Philadelphia: Carey and Lea.

Lorenz, J. F. (1803). *Lehrbegriff der Mathematik. Erster Theil, die gesammte Logistik, oder die Arithmetik, Syntactik, Algebra und Analysis. Erste Abtheilung. Lehrbegriff der gemeinen und allgemeinen Arithmetik*. Magdeburg: Georg Christian Keil.

Lorenz, J. F. (1806). *Lehrbegriff der Mathematik. Erster Theil, die gesammte Logistik, oder die Arithmetik, Syntactik, Algebra und Analysis. Zweyte Abtheilung. Lehrbegriff der Syntactik, oder Combinationslehre*. Magdeburg: Georg Christian Keil.

Lorey, W. (1916). *Das Studium der Mathematik an den deutschen Universitäten seit Anfang des 19. Jahrhunderts*. Leipzig und Berlin: Verlag und druck von B. G. Teubner.

Loria, G. (1888). Il passato e il presente delle principali teorie geometriche. *Memorie della Reale Accademia delle Scienze di Torino. Serie seconda, XXXVIII*, 327–376.

Loria, G. (1933). *Storia delle matematiche. Dall'alba del secolo XVIII al tramonto del secolo XIX* (Vol. 3). Torino: Società Tipografico-Editrice Nazionale.

Lubet, J.-P. (1998). De Lambert à Cauchy: la résolution des équations littérales par le moyen des séries. *Revue d'histoire des mathématiques, 4*(1), 73–129.

Lubet, J.-P. (2010). Calcul symbolique et calcul intégral de Lagrange à Cauchy. *Revue d'histoire des mathématiques, 16*(1), 63–131.

Lübsen, H. B. (1853). *Ausführliches Lehrbuch der Analysis, zum Seblstunterricht mit Rücksicht auf die Zwecke des praktischen Lebens*. Hamburg: Perthes-Besser & Mauke.

Ludowieg, J. C. H. (1835). *Lehrbuch der Arithmetik und der Anfangsgründe der Algebra, für Gymnasien und höhere Lehranstalten* (2nd ed.). Hannover: Im Verlage der Hahnschen Hof-Buchhandlung.

Ludowieg, J. C. H. (1844). *Grundriß der reinen Mathematik, oder Leitfaden für den Unterricht in der gesammten Elementar-Mathematik. Zum Gebrauche für die oberen Classen der Gymnasien und höheren Lehranstalten*. Hannover: Im Verlage der Hahnschen Hof-Buchhandlung.

Maclaurin, C. (1742). *A Treatise of fluxions. In two books*. 2 vols., Edinburgh: Printed by T.W. and T. Ruddimans.

Magold, M. (1830). *Lehrbuch der Arithmetik zum Gebrauche öffentlicher Vorlesungen* (4th ed.). München: in der Anton Weberschen Buchhandlung.

Mancosu, P. (1989). The metaphysics of the calculus: A foundational debate in the Paris Academy of Sciences, 1700–1706. *Historia Mathematica, 16*(3), 224–248.

Margantin, L. (1999). *Système minéralogique et cosmologie chez Novalis ou les plis de la terre*. Paris/Montréal: L'Harmattan.

Martin, T. (1996a). Cournot et les mathématiques. In E. Barbin & M. Caveing (Eds.), *Les philosophes et les mathématiques* (pp. 193–212). Paris: Ellipses.

Martin, T. (1996b). *Probabilités et critique philosophique selon Cournot*. Paris: J. Vrin.

Martin, T. (2005). Cournot, philosophe des probabilités. In T. Martin (Ed.), *Actualité de Cournot* (pp. 51–68). Paris: J. Vrin.

Martzloff, J.-C. (1988). *Histoire des mathématiques chinoises*. Paris: Masson.

Mayer, J. T. (1818). *Vollständiger Lehrbegriff der höhern Analysis. Erster Theil. Die Differenzial-rechnung*. Göttingen: Im Verlage bey Vandenhoek und Ruprecht.

Mc Laughlin, J. (2018). *Topics and methods in q-series*. Singapore: World Scientific.

Mercator, N. (1668). *Logarithmo-technia: sive methodus logarithmos nova, accurata, & facilis*. Londini: impensis Mosis Pitt.

Mersenne, M. (1635). *Harmonicorum libri, in quibus agitur de sonorum natura, causis, & effectibus*. Lutetiae Parisiorum: sumptibus Guillelmi Baudry.

Michel, P.-H. (1958). *Les nombres figurés dans l'arithmétique pythagoricienne.* Paris: Université de Paris.

Minsinger, F. (1832). *Lehrbuch der Arithmetik und Algebra.* Augsburg: Verlag der Karl Kollman'schen, vormals Jos. Wolff'schen, Buchhandlung.

Mittelstrass, J., & Schroeder-Heister, P. (1986). Zeichen, Kalkül, Warscheinlichkeit. Elemente einer Mathesis universalis bei Leibniz. In H. Stachowiak (Ed.), *Pragmatik. Handbuch pragmatischen Denkens* (pp. 392–414). Hamburg: Felix Meiner Verlag.

Molk, J., Pringsheim, A., & Faber, G. (1911). Analyse algébrique. In J. Molk (Ed.), *Encyclopédie des sciences mathématiques pures et appliquées* (Vol. 2, pp. 1–93). Paris: Gauthier-Villars.

Mönnich, B. F. (1801). *Lehrbuch der Mathematik mit Rücksicht auf solche, welche sie erlernen, um sie bei ihren mehr oder weniger damit in Verbindung stehenden Berufsgeschäften zu benutzen. Erster Theil. Zweite Abtheilung* (2nd ed.). Berlin: Bei Gottlieb August Lange.

Montucla, J.-E. (1758). *Histoire des mathématiques, dans laquelle on rend compte de leurs progrès depuis leurs origines jusqu'à nos jours* (Vol. 2). Paris: chez Ch. Ant. Jombert.

Müller, A. (1822). *Dissertatio mathematico-historica de binomii ac polynomii problematis quam pro facultate legendi auctoritate illustris philosophorum ordinis in Academia Ruperto-Carolina rite obtinenda.* Heidelbergae: Typis Joannis Michaëlis Gutmanni, Universitatis typographi.

Müller, J. H. T. (1838). *Lehrbuch der allgemeinen Arithmetik für Gymnasien und Realschulen, nebst vielen Uebungsaufgaben und Excursen.* Halle: Verlag der Buchhandlung des Waisenhauses.

Muller, J. (1736). *A mathematical treatise: Containing a system of conic-sections; with the doctrine of fluxions and fluents, applied to various subjects.* London: printed by T. Gardner.

Naux, C. (1966–1971). *Histoire des logarithmes de Neper à Euler.* 2 vols., Paris: A. Blanchard.

Netto, E. (1908). Kombinatorik, Wahrscheinlichkeitsrechnung, Reihen Imaginäres. In M. Cantor (Ed.), *Vorlesungen über der Geschichte der Mathematik* (Vol. 4, pp. 199–318). Leipzig: Druck und Verlag von R. G. Teubner.

Neubig, A. (1817). *Anfangsgründe der mathematischen Analysis oder der Algebra, Differential- und Integral-Rechnung.* Nürnberg: In der C. H. Zeh'schen Buchhandlung.

Newton, I. (1665a). The binomial theorem invented in Wallisian style. In D. T. Whiteside (Ed.), *The mathematical papers of Isaac Newton* (Vol. I, pp. 104–112). Cambridge: Cambridge University Press, 1967. Document's title by Whiteside.

Newton, I. (1665b). The logarithmic series. In D. T. Whiteside (Ed.), *The mathematical papers of Isaac Newton* (Vol. I, pp. 112–115). Cambridge: Cambridge University Press, 1967. Document's title by Whiteside.

Newton, I. (1666). The october 1666 tract on fluxions. In D. T. Whiteside (Ed.), *The mathematical papers of Isaac Newton* (Vol. I, pp. 400–448). Cambridge: Cambridge University Press, 1967. Document's title by Whiteside.

Newton, I. (1676a). Epistola posterior, 24 October 1676. In Collins, J., *Commercium epistolicum D. Johannis Collins et aliorum de analysi promota: jussu Societatis regiae in lucem editum* (pp. 67–86). Londini: typis Pearsonianis, 1712.

Newton, I. (1676b). Epistola prior, 13 June 1676. In Collins, J., *Commercium epistolicum D. Johannis Collins et aliorum de analysi promota: jussu Societatis regiae in lucem editum* (pp. 49–57). Londini: typis Pearsonianis, 1712.

Newton, I. (1687). *Philosophiae naturalis principia mathematica.* Londini: jussu Societatis regiae ac typis Josephi Streater.

Newton, I. (1704). Tractatus de Quadratura Curvarum. In *Opticks: Or, a treatise of the reflexions, refractions, inflexions and colours of light. Also two treatises of the species and magnitude of curvilinear figures* (pp. 164–211). London: Printed for S. Smith and B. Walford.

Newton, I. (1711). De analysi per aequationes numero terminorum infinitas. In W. Jones (Ed.), *Analysis per quantitatum series, fluxiones, ac differentias: Cum enumeratione linearum tertii ordinis* (pp. 1–21). Londini: ex Officina Pearsoniana.

Newton, I. (1713). *Philosophiae naturalis principia mathematica* (2nd ed.). Cantabrigiae.

Newton, I. (1736). *The method of fluxions and infinite series; with its applications to the geometry of curve-lines.* London: Henry Woodfall.

Newton, I. (1745). *Sir Isaac Newton's two treatises of the quadrature of curves, and analysis by equations of an infinite number of terms, explained: Containing the treatises themselves, translasted into English, with a large commentary.* London: John Nourse.

Nicholson, P. (1818). *Essays on the combinatorial analysis.* London: Printed for the author, and published by Longman, Hurst, Rees, Orme, and Brown, and by the author.

Nicholson, P. (1821). *Analytical and arithmetical essays.* London: Printed for the author, published by Davis and Dickson, Mathematical and Philosophical Booksellers.

Nikolantonakis, K., & Yao-Yong, D. (2011). The algorithm of extraction in Greek and Sino-Indian mathematical traditions. In B. S. Yadav & M. Mohan (Eds.), *Ancient Indian leaps into mathematics* (pp. 171–184). New York: Birkhäuser.

Novalis (1993). *Das allgemeine Brouillon: Materialien zu Enzyklopädistik 1789/99.* Hamburg: F. Meiner Verlag.

Ohm, M. (1811). *De elevatione serierum infinitarum secundi ordinis ad potestatem exponentis indeterminati.* Erlangae: typis Adolphi Ernesti Junge.

Ohm, M. (1822a). *Versuch eines vollkommen consequenten Systems der Mathematik. Erster Theil. Lehrbuch der Arithmetik, Algebra und Analysis.* Berlin: Gedruckt und verlegt bei G. Reimer.

Ohm, M. (1822b). *Versuch eines vollkommen consequenten Systems der Mathematik. Zweiter Theil. Lehrbuch der Arithmetik, Algebra und Analysis.* Berlin: Gedruckt und verlegt bei G. Reimer.

Ohm, M. (1829a). *Versuch eines vollkommen consequenten Systems der Mathematik. Dritter Theil. Lehrbuch der höhern Analysis. Erster Theil. Differenzialrechnung enthaltend.* Berlin: bei T. H. Riemann.

Ohm, M. (1829b). *Versuch eines vollkommen consequenten Systems der Mathematik. Zweiter Theil. Lehrbuch der niedern Analysis* (2nd ed.). Berlin: bei T. H. Riemann.

Otte, M. (1989). The ideas of Hermann Grassmann in the context of the mathematical and philosophical tradition since Leibniz. *Historia Mathematica, 16*(1), 1–35.

Öttinger, L. (1837a). *Die Lehre von den Combinationen nach einem neuen Systeme.* Freiburg: Druck und Verlag der Gebrüder Groos.

Öttinger, L. (1837b). *Lehrbuch der reinen Mathematik. Erster Theil. Lehrbuch der Arithmetik und Algebra.* Freiburg: Druck und Verlag der Gebrüder Groos.

Öttinger, L. (1854). *Theorie der analytischen Facultäten nebst ihrer Anwendung auf Analysis, Kreisfunctionen und bestimmte Integrale.* Freiburg im Breisgau: Universitäts-Buchhandlung von J. Diernfellner.

Oughtred, W. (1652). *Clavis mathematicae denuo limata, sive potius fabricata. Cum aliis quibusdam ejusdem commentationibus, quae in sequenti pagina recensentur. Editio tertia auctior & emendatior.* Oxoniae: excudebat Leon Lichfield.

Panza, M. (1992). *La forma della quantità: analisi algebrica e analisi superiore: il problema dell'unità della matematica nel secolo dell'illuminismo, 2 vols. Cahiers d'Histoire et de Philosophie des Sciences,* 38–39.

Panza, M. (1995). Da Wallis à Newton: una via verso il calcolo. Quadrature, serie e rappresentazioni infinite delle quantità e delle forme trascendenti. In M. Panza & C. S. Roero (Eds.), *Geometria, flussioni e differenziali: tradizione e innovazione nella matematica del seicento* (pp. 131–219). Napoli: La città del sole.

Panza, M. (2005). *Newton et les origines de l'analyse: 1664–1666.* Paris: A. Blanchard.

Panza, M. (2007). Euler's Introductio in analysin infinitorum and the program of algebraic analysis: Quantities, functions and numerical partitions. In R. Baker (Ed.), *Euler reconsidered. Tercentenary essays* (pp. 119–166). Heber city, UT: Kendrick press.

Pappas, J. (1996). R. J. Boscovich et l'Académie des sciences de Paris. *Revue d'histoire des sciences, 49*(4), 401–414.

Pascal, B. (1665). *Traité du triangle arithmétique, avec quelques autres petits traitez sur la mesme matière.* Paris: Guillaume Desprez.

Pasquich, J. (1798a). Anfangsgründe einer neuen Exponentialrechnung. *Archiv der reinen und angewandten Mathematik, 2*(8), 385–425.

Pasquich, J. (1798b). Nachricht von einer neuen Rechnung, welche, ganz unabhängig von allen Begriffen des unendlich kleinen, und auf den einfachsten Gründen beruhend, dienen soll, alles das, was bisher nur immer die Differensialrechnung geleistet hat, eben so schnell und leicht zu leisten. *Intelligenzblatt der allgemeinen Literatur-Zeitung, 99,* 832.

Pasquich, J. (1798c). *Unterricht in der mathematischen Analysis und Machinen-Lehre.* Leipzig: in der Weidmannischen Buchhandlung.

Pasquich, J. (1812). *Anfangsgründe der gesammten theoretischen Mathematik. Erster Band. Anfangsgründe der allgemeinen Grössenlehre, und decadischen Arithmetik.* Wien: bei Schaumburg und Compagnie.

Peacock, G. (1834). Report on the recent progress and present state of certain branches of analysis. In *Report of the Third Meeting of the British Association for the Advancement of Science; held at Cambridge in 1833* (pp. 185–352). London: John Murray.

Peano, G. (1889). *Arithmetices principia nova methodo exposita.* Romae–Florentiae: Augustae Taurinorum ediderunt Fratres Bocca.

Peckhaus, V. (1994). Wozu Algebra der Logik? Ernst Schröders Suche nach einer universalen Theorie der Verknüpfungen. *Modern Logic, 4*(4), 357–381.

Peckhaus, V. (1997). *Logik, mathesis universalis und allgemeine Wissenschaft: Leibniz und die Wiederentdeckung der formalen Logik im 19. Jahrhundert.* Berlin: Akademie Verlag.

Peletier, J. (1549). *L'arithmétique de Jacques Peletier du Mans, departie en quatre livres.* Poitiers.

Pensivy, M. (1987–1988). Jalons historiques pour une épistémologie de la série infinie du binôme. *Science et Techniques en perspective, 14,* 1–231.

Petsche, H.-J., A. C. Lewis, J. Liesen, & S. Russ (Eds.) (2011). *Hermann Graßmann. From past to future: Graßmann's work in context. Graßmann Bicentennial Conference, September 2009.* Basel: Birkhäuser.

Pfaff, J. F. (1794a). Ableitung der Localformel für die Reversion der Reihen, aus dem Satze des Herrn de la Grange. *Archiv der reinen und angewandten Mathematik, 1*(1), 85–88.

Pfaff, J. F. (1794b). Analysis einer wichtigen Aufgabe des Herrn de la Grange. *Archiv der reinen und angewandten Mathematik, 1*(1), 81–85.

Pfaff, J. F. (1796a). Bemerkungen über eine besondere Art von Gleichungen, nebst Beyspielen von ihrer Auflösung. *Sammlung combinatorisch-analytischer Abhandlungen. Der polynomische Lehrsatz das wichtigste Theorem der ganzen Analysis nebst einigen verwandten und andern Sätzen, 1,* 144–152.

Pfaff, J. F. (1796b). Sätze über Potenzen und Produkte gewisser Reihen. *Sammlung combinatorisch-analytischer Abhandlungen. Der polynomische Lehrsatz das wichtigste Theorem der ganzen Analysis nebst einigen verwandten und andern Sätzen, 1,* 123–143.

Pfaff, J. F. (1797a). De progressionibus arcuum circularium, quorum tangentes secundum datam legem procedunt. In *Disquisitiones analyticae maxime ad calculum integralem et doctrinam serierum pertinentes* (Vol. 1, pp. 1–132). Helmstadt: apud C. G. Fleckeisen.

Pfaff, J. F. (1797b). Tractatus de reversione serierum, sive de resolutione aequationum per series. In *Disquisitiones analyticae maxime ad calculum integralem et doctrinam serierum pertinentes* (Vol. 1, pp. 227–348). Helmstadt: apud C. G. Fleckeisen.

Phili, C. (1990). La théorie des fonctions analytiques de Lagrange et son influence postérieure en Europe. In *Échanges d'influences scientifiques et techniques entre pays européens de 1780 à 1830. Actes du 114e Congrès national des sociétés savantes (Paris 3–9 avril 1989)* (pp. 105–124). Paris: Éditions du CTHS.

Poppe, J. H. M. (1828). *Geschichte der Mathematik seit der ältesten bis auf die neueste Zeit.* Tübingen: bei C. F. Osiander.

Pradier, P.-C. (1998). *Concepts et mesures du risque en théorie économique. Essai historique et critique.* Ph.D. thesis, ENS-Cachan.

Pradier, P.-C. (2003). L'actuariat au siècle des Lumières: risque et décision économiques et statistiques. *Revue économique, 54*(1), 139–156.

Prakash, S. (1968). *A critical study of Brahmagupta and his works: A most distinguished Indian astronomer and mathematician of the sixth century A.D.* New Delhi: Indian Institute of Astronomical & Sanskrit Research.

Pringsheim, A., & Faber, G. (1909). Algebraische Analysis. In H. Burkhardt & W. Wirtinger (Eds.), *Encyklopädie der mathematischen Wissenschaften mit Einschluss ihrer Anwendungen* (Vol. 3, pp. 1–46). Leipzig: B. G. Teubner.

Pulte, H. (2006). Kant, Fries and the expanding universe of science. In M. Friedman & A. Nordmann (Eds.), *The Kantian legacy in nineteenth-century science* (pp. 101–122). Cambridge-Massachusetts: MIT Press.

Rabouin, D. (Ed.) (2018). *Mathesis universalis: écrits sur la mathématique universelle.* Paris: Librairie philosophique J. Vrin.

Radu, M. (2000). Justus Grassmann's contributions to the foundations of mathematics: Mathematical and philosophical aspects. *Historia Mathematica, 27*(1), 4–35.

Radu, M. (2011). Axiomatics and self-reference. Reflections about Hermann Grassmann's contribution to axiomatics. In H.-J. Petsche, A. C. Lewis, J. Liesen, & S. Russ (Eds.), *Hermann Graßmann. From past to future: Graßmann's work in context. Graßmann Bicentennial Conference, September 2009* (pp. 101–116). Basel: Birkhäuser.

Raju, C. K. (2007). *Cultural foundations of mathematics. The nature of mathematical proof and the transmission of the calculus from India to Europe in the 16th c. CE.* Delhi: Centre for Studies in Civilizations.

Raphson, J. (1690). *Analysis aequationum universalis, seu, ad aequationes algebraicas resolvendas methodus generalis, et expedita: ex nova infinitarum serierum doctrina deducta ac demonstrata.* Londini: Prostant venales apud Abelem Swalle.

Rashed, R. (1972). L'induction mathématique: al-Karajī, as-Samaw'al. *Archive for History of Exact Sciences, 9*(1), 1–21.

Rashed, R. (1984). *Entre arithmétique et algèbre: recherches sur l'histoire des mathématiques arabes.* Paris: Les Belles Lettres.

Raupach, J. F. (1815). *Die Elemente der Algebra und Analysis, nebst ihrer Anwendung auf die Geometrie.* Breslau: bei Wilhelm Gottlieb Korn.

Reidt, F. (1880). Arithmetik und Algebra. In O. Schlömilch, F. Reidt, & R. Heger (Eds.), *Encyklopaedie der Naturwissenschaften. I. Abtheilung. II. Theil: Handbuch der Mathematik* (Vol. 1, pp. 1–166). Breslau: Verlag von Eduard Trewendt.

Reiff, R. A. (1889). *Geschichte der unendlichen Reihen.* Tübingen: H. Laupp'schen Buchhandlung.

Reiss, H. (1994). The "naturalization" of the term "Ästhetik" in eighteenth-century German: Alexander Gottlieb Baumgarten and his impact. *The Modern Language Review, 89*(3), 645–658.

Reuschle, C. G. (1850). *Die Arithmetik in der Hand des Schülers. Ein kurzes und vollständiges Lehrbuch der elementaren Arithmetik und Algebra nach dem neuesten Standpunkt der Wissenschaft in der altklassischen Form abgefaßt.* Stuttgart: E. Schweizerbart'sche Verlagshandlung und Durckerei.

Reyneau, C. (1708). *Analyse démontrée, ou la méthode de résoudre les problèmes des mathématiques, et d'apprendre facilement ces sciences.* 2 vols., Paris: chez Jacque Quillau.

Rigaud, S. P. (1838). *Historical essay on the publication of Sir Isaac Newton's Principia.* Oxford: Oxford University Press.

Robins, B. (1735). *A discourse concerning the nature and certainty of Sir Isaac Newton's methods of fluxions: and of prime and ultimate ratios.* London: printed by W. Innys and R. Manby.

Robins, B. (1736). *Remarks on the remainder of the considerations relating to fluxions, &c. that was published by Philalethes Cantabrigiensis in the Republick of Letters for the last Month. To which is added by Dr. Pemberton a postscript occasioned by a passage in the said considerations.* London: printed by W. Innys and R. Manby.

Rolle, M. (1696). Extrait d'une letre de R. L. au sujet de l'algebre. *Le journal des sçavans, XXI,* 244–249.

Rolle, M. (1702). Règles et remarques, pour le problème general des tangentes. *Le journal des sçavans, XVI,* 239–254.

Rolle, M. (1703). Du nouveau système de l'infini. *Mémoires de mathématique et de physique de l'Académie royale des sciences, Académie royale des sciences,* 312–336.

Rome, A. (Ed.) (1936). *Commentaires de Pappus et de Théon d'Alexandrie sur l'Almageste,* Volume II. Roma: Biblioteca apostolica vaticana.

Rothe, H. A. (1793). *Formulae de serierum reversione demonstratio universalis signis localibus combinatorio-analyticorum vicariis exhibita.* Lipsiae: Litteris Sommeriis.

Rothe, H. A. (1794). Lokalformeln für höhere Differenziale von Potenzen und ihren Producten, wo alle höhere Differenziale beybehalten werden. *Archiv der reinen und angewandten Mathematik, 1*(2), 228–232.

Rothe, H. A. (1795). Lokal- und combinatorisch-analytische Formeln für höhere Differenziale. *Archiv der reinen und angewandten Mathematik, 1*(4), 431–449.

Rothe, H. A. (1796). *Theorema binomiale ex simplicissimis analyseos finitorum fontibus universaliter demonstratum.* Lipsiae: ex officina Sommeria.

Rothe, H. A. (1804a). *De divisione peripheriae circuli in XVII et XIII partes aequales disquisitio analytica.* Erlangae: typis Kunstmannianis.

Rothe, H. A. (1804b). *Handbuch der reinen Mathematik. Systematisches Lehrbuch der Arithmetik. Erster Theil.* Leipzig: bei Johann Ambrosius Barth.

Rothe, H. A. (1811). *Handbuch der reinen Mathematik. Systematisches Lehrbuch der Arithmetik. Zweyter Theil.* Leipzig: bei Johann Ambrosius Barth.

Rothe, H. A. (1820). *Theorie der combinatorischen Integrale erfunden, dargestellt, und mit mehrern Anwendungen auf die Analysis versehen.* Nürnberg: bei Riegel und Wiessner.

Rothe, H. A., & Ohm, M. (1814). *Solutio problematis summe memorabilis atque generalis ad diuisionem polygonorum per diagonales spectantis.* Erlangae.

Salomon, J. (1821). *Lehrbuch der Arithmetik und Algebra zum öffentlichen Gebrauche und Selbstunterrichte. Erste Abtheilung.* Wien und Triest: im Verlage der Geistinger'schen Buchhandlung.

Salomon, J. (1825). *Versuch eines gemeinfaßlichen Unterrichtes in der Arithmetik. Ein Handbuch für Alle, welche im praktischen Leben Gebrauch von der Mathematik machen wollen.* Wien: gedruckt und im Verlage bey Carl Gerold.

Salomon, J. (1831). *Lehrbuch der Arithmetik und Algebra* (2nd ed.). Wien: Gedruckt und im Verlage bey Carl Gerold.

Salomon, J. (1834). *Sammlung von Formeln, Aufgaben und Beyspielen aus der Arithmetik und Algebra, nebst vier Tafeln über die Vergleichung der vorzüglichsten Maße, Gewichte und Münzen mit den österreichischen und französischen.* Wien: Gedruckt und im Verlage bey Carl Gerold.

Saurin, J. (1702). Réponse à l'écrit de M. Rolle de l'Ac. R. des Sc. inseré dans le journal du 13 avril 1702 sous le titre de Regles et remarques pour le problème general de tangentes. *Le journal des sçavans, XXXIII*, 519–534.

Schaffer, J. F. (1824). *Vollständiger Lehrbegriff der höhern, auf Combination der Größen gegründeten, Analysis, und der höhern phoronomischen Geometrie.* Oldenburg: in der Schulze'schen Buchhandlung.

Schefer, O. (2005). *Résonance du romantisme.* Bruxelles: La lettre volée.

Scherk, H. F. (1825). *Mathematische Abhandlungen.* Berlin: Gedruckt und verlegt bei G. Reimer.

Scheubel, J. (1545). *De Numeris et diversis rationibus seu regulis computationum opusculum.* Lipsiae: Blum.

Schlömilch, O. (1845). *Handbuch der mathematischen Analysis. Erster Theil. Algebraische Analysis.* Jena: Friedrich Frommann.

Schlömilch, O. (1847). *Handbuch der Differenzial- und Integralrechnung. Erster Theil.* Greifswald: Verlag von Ferd. Otte.

Schmeißer, F. (1817). *Lehrbuch der reinen Mathesis zu einem zum Selbstfinden leitenden Vortrage derselben nach Platonischer Weise in Gymnasien. Erster Theil. Lehrbuch der Arithmetik.* Berlin: Realschulbuchhandlung.

Schneider, I. (1968). Der Mathematiker Abraham de Moivre (1667–1754). *Archive for History of Exact Sciences, 5*(3–4), 177–317.

Scholz, E. (1989). The rise of symmetry concepts in the atomistic and dynamistic schools of crystallography, 1815–1830. *Revue d'histoire des sciences, 42*(1–2), 109–122.

Scholz, E. (1996). The influence of Justus Grassmann's crystallographic works on Hermann Grassmann. In G. Schubring (Ed.), *Hermann Günther Graßmann (1809–1877): visionary mathematician, scientist and neohumanist scholar* (pp. 37–45). Dordrecht: Kluwer.

Schön, J. (1825). *Lehrbuch der niedern, reinen, allgemeinen Größenlehre, oder der Buchstabenrechnung und Algebra, zum Behufe öffentlicher Vorlesungen und des Selbstunterrichtes.* Würzburg: im Verlage der Stahel'schen Buchhandlung.

Schön, J. (1833). *Kurzer Lehrbegriff der höhern Mathematik, oder Lehrbuch der höhern Analysis und höhern Geometrie, auf dem Grunde der niedern Mathematik und zum Behufe öffentlicher Vorlesungen und des Selbstunterrichtes.* Sulzbach: in der J. E. v. Seidel'schen Buchhandlung.

Schoof, C. L. (1858). *Arithmetik und Algebra für höhere Lehranstalten und zum Selbstunterricht. Drittes Heft.* Hannover: Hahn'sche Hofbuchhandlung.

Schopenhauer, A. (2016). *The world as will and presentation* (Vol. 1). New York: Routledge.

Schröder, E. (1873). *Lehrbuch der Arithmetik und Algebra für Lehrer und Studirende. Erster Band. Die sieben algebraischen Operationen.* Leipzig: Druck und Verlag von B. G. Teubner.

Schubring, G. (1990). Les échanges entre les mathématiciens français et allemands sur la rigueur dans le concept d'arithmétique et d'analyse. In *Échanges d'influences scientifiques et techniques entre pays européens de 1780 à 1830. Actes du 114e Congrès national des sociétés savantes (Paris 3–9 avril 1989)* (pp. 89–104). Paris: Éditions du CTHS.

Schubring, G. (1996a). Changing cultural and epistemological views on mathematics and different intitutional contexts in nineteenth-century Europe. In C. Goldstein, J. Gray, & J. Ritter (Eds.), *L'Europe mathématique. Histoires, mythes, identités* (pp. 361–388). Paris: Éditions de la Maison des sciences de l'homme.

Schubring, G. (1996b). The cooperation between Hermann and Robert Grassmann on the foundations of mathematics. In G. Schubring (Ed.), *Hermann Günther Graßmann (1809–1877):*

Visionary mathematician, scientist, and neohumanist scholar. Papers from a sesquicentennial conference (pp. 59–70). Dordrecht: Kluwer Academic Publishers.

Schubring, G. (2005). *Conflicts between generalization, rigor, and intuition. Number concepts underlying the development of analysis in 17–19th century France and Germany.* New York: Springer.

Schubring, G. (2007). Documents on the mathematical education of Edmund Külp (1800–1862), the mathematics teacher of Georg Cantor. *ZDM–The International Journal on Mathematics Education, 30*(1), 107–118.

Schubring, G. (2010). How to relate regional history to general patterns of history? The case of mathematics teaching in Westphalia. *Bolema, 23*(35), 101–122.

Schubring, G. (2012). Antagonisms between German states regarding the status of mathematics teaching during the 19th century: processes of reconciling them. *ZDM Mathematics Education, 44*, 525–535.

Schultz, J. (1803). *Sehr leichte und kurze Entwickelung einiger der wichtigsten mathematischen Theorien.* Königsberg: bey Friedrich Nicolovius.

Schweins, F. F. (1820). *Analysis.* Heidelberg: Auf Kosten des Verfassers und in Commission bei Mohr und Winter.

Schweins, F. F. (1825). *Theorie der Differenzen und Differentiale, der gedoppelten Verbindungen, der Producte mit Versetzungen, der Reihen, der wiederholenden Functionen, der allgemeinsten Facultäten und der fortlaufenden Brüche.* Heidelberg: Verlag der Universitäts-Buchhandlung von C. F. Winter.

Sebestik, J. (1992). *Logique et mathématique chez Bernard Bolzano.* Paris: J. Vrin.

Segner, J. A. (1758). Elementa analyseos finitorum. In *Cursus Mathematici* (Vol. 2). Halae Magdeburgicae: prostat in Officina Rengeriana.

Segner, J. A. (1761). Elementorum analyseos infinitorum. Pars I. In *Cursus Mathematici* (Vol. 3). Halae Magdeburgicae: prostat in Officina Rengeriana.

Segner, J. A. (1779). Demostratio universalis theorematis binomialis Newtoni. *Nouveaux mémoires de l'Académie royale des sciences et des belles lettres de Berlin, 1777*, 37–41.

Séguin, P. (2005). La recherche d'un fondement absolu des mathématiques par l'Ecole combinatoire de C. F. Hindenburg (1741–1808). *Philosophia Scientiae. Cahier spécial, 5*, 61–79.

Séguin, P. (2006). Ars combinatoria universalis: un rêve poético-mathématique de Novalis et C. F. Hindenburg. In L. Dahan-Gaida (Ed.), *Conversations entre la littérature, les arts et les sciences* (pp. 59–72). Besançon: Presses universitaires de Franche-Comté.

Semler, C. A. (1809). Ueber die combinatorische Methode. *Neues allgemeines Intelligenzblatt für Literatur und Kunst* (1, 5, 15, 20, 25, 32, 40, 41), 1–12, 65–78, 226–232, 305–316, 385–390, 509–517, 625–640, 641–653.

Semler, C. A. (1811). *Versuch über die combinatorische Methode, ein Beytrag zur angewandten Logik und allgemeinen Methodik.* Dresden: In der Waltherschen Hofbuchhandlung.

Semler, C. A. (1822). *Versuch über die combinatorische Methode, ein Beytrag zur angewandten Logik und allgemeinen Methodik* (2nd ed.). Dresden: In der Waltherschen Hofbuchhandlung.

Serres, M. (1982). *Le système de Leibniz et ses modèles mathématiques: étoiles, schémas, points.* Paris: Presses universitaires de France.

Serret, J.-A. (Ed.) (1882). *Œuvres de Lagrange.* Paris: Gauthier-Villars.

Servois, F.-J. (1814). *Essai sur un nouveau mode d'exposition des principes du calcul différentiel.* Nismes: Imprimerie de P. Blachier-Belle.

Simpson, T. (1750). *The doctrine and application of fluxions.* London: John Nourse.

Snell, F. W. D. (1804). *Anfangsgründe der Arithmetik und Algebra.* Giesen: bey Tasche und Müller.

Spehr, F. W. (1823). *De quantitate fluente tractatus, in quo explicantur fundamenta calculi differentialis nonnisi ex notione fluentis deducta. Accedit theorematis infinitinomialis indeterminati exponentis, sine ponendo theoremate binomiali demonstratio universalis.* Brunsvigae: Friderici Viewegii.

Spehr, F. W. (1824). *Vollständiger Lehrbegriff der reinen Combinationslehre mit Anwendungen derselben auf Analysis und Wahrscheinlichkeitsrechnung.* Braunschweig: Im Kunst- und geographischen Büreau.

Spehr, F. W. (1826). *Neue Principien des Fluentencalculus, enthaltend die Grundsätze der Differential- und Variationsrechnung unabhängig von der gewöhnlichen Fluxionsmethode, von den Begriffen des unendlich Kleinen oder der verschwindenden Größen, von der Methode der Grenzen und der Functionenlehre, zugleich als Lehrbuch dieser Wissenschaft dargestellt, und mit Anwendungen auf analytische Geometrie und höhere Mechanik verbunden. Erster Theil.* Braunschweig: bei G. C. E. Meyer.

Spehr, F. W. (1840). *Vollständiger Lehrbegriff der reinen Combinationslehre mit Anwendungen derselben auf Analysis und Wahrscheinlichkeitsrechnung* (2nd ed.). Braunschweig: Verlag von Eduard Leibrock.

Srinivasiengar, C. N. (1967). *The history of ancient Indian mathematics.* Calcutta: World Press.

Stahl, C. D. M. (1800). *Grundriss der Combinationslehre nebst Anwendung derselben auf die Analysis.* Jena und Leipzig: bei Christian Ernst Gabler.

Stahl, C. D. M. (1801). *Einleitung in das Studium der Combinationslehre nebst einem Anhange über die Involutionen und deren Anwendung auf die continuirlichen Brüche.* Jena und Leipzig: Christian Enst Gabler.

Staudacher, H. (1893). *Lehrbuch der Kombinatorik. Ausführliche Darstellung der Lehre von den kombinatorischen Operationen (Permutieren, Kombinieren, Variieren).* Stuttgart: Verlag von Julius Maier.

Stewart, J. (1745). Sir Isaac Newton's *Analysis by equations of an infinite number of terms,* explained. In J. Stewart (Ed.), *Sir Isaac Newton's two treatises of the quadrature of curves, and analysis by equations of an infinite number of terms, explained: Containing the treatises themselves, translasted into english, with a large commentary* (pp. 344–479). London: John Nourse.

Stifel, M. (1544). *Arithmetica integra.* Norimbergae: apud Johan Petreium.

Struve, J. (1808). *Handbuch der Mathematik für angehende Studirende und zum Selbstunterrichte. Erster Theil. Arithmetik.* Altona: bey Johann Friedrich Hammerich.

Struve, J. (1809). *Handbuch der Mathematik für angehende Studirende und zum Selbstunterrichte. Zweyter Theil. Syntaktik, oder Combinationslehre.* Altona: bey Johann Friedrich Hammerich.

Sylla, E. D. (2006). Introduction. In E. D. Sylla (Ed.), *The art of conjecturing: Together with "Letter to a friend on sets in court tennis".* Baltimore: The Johns Hopkins University Press.

Tartaglia, N. (1556). *General trattato di numeri, et misure.* 2 vols., Vinegia: Curtio Troiano dei Navo.

Tellkampf, A. (1829). *Vorschule der Mathematik.* Berlin: bei August Räcker.

Tellkampf, A. (1841). Erinnerungen an B. F. Thibaut. *Hallische Jahrbücher für deutsche Wissenschaft und Kunst* (74–76), 295–296, 299–300, 303–304.

Tetens, J. N. (1763). Methodus inveniendi curvas, maximum vel minimum efficientes, universaliter, et ex analyticis principiis demonstrata. *Nova Acta Eruditorum,* 502–515.

Tetens, J. N. (1785). *Einleitung zur Berechnung der Leibrenten und Anwartschaften die vom Leben und Tode einer oder mehrerer Personen abhangen mit Tabellen zum practischen Gebrauch.* Leipzig: Weidmanns Erben und Reich.

Tetens, J. N. (1786). *Einleitung zur Berechnung der Leibrenten und Anwartschaften. Zweyter Theil.* Leipzig: Weidmanns Erben und Reich.

Tetens, J. N. (1796). Formula Polynomiorum. Eine allgemeine Formel für die Potenzen mehrtheiliger Größen. *Sammlung combinatorisch-analytischer Abhandlungen. Der polynomische Lehrsatz das wichtigste Theorem der ganzen Analysis nebst einigen verwandten und andern Sätzen, 1*, 1–47.

Thibaut, B. F. (1801). *Grundriss der reinen Mathematik zum Gebrauch bey academischen Vorlesungen.* Göttingen: bey Philipp Georg Schröder.

Thibaut, B. F. (1809). *Grundriss der allgemeinen Arithmetik oder Analysis zum Gebrauch bey academischen Vorlesungen.* Göttingen: Bey Heinrich Dieterich.

Thibaut, B. F. (1830). *Grundriss der allgemeinen Arithmetik oder Analysis zum Gebrauch bey academischen Vorlesungen* (2nd ed.). Göttingen: in der Dieterichschen Buchhandlung.

Thiel, C. (1995). Nicht aufs Gerathewohl und aus Neuerungssucht: die Begriffsschrift 1879 und 1893. In I. Max & W. Stelzner (Eds.), *Logik und Mathematik. Frege-Kolloquium Jena 1993* (pp. 20–37). Berlin: Walter de Gruyter.

Trudi, N. (1879). Intorno ad alcuni punti di analisi dipendenti dalla partizione dei numeri. *Atti della Reale accademia delle scienze fisiche e matematiche (Napoli), VIII,* 1–88. Letta nell'Adunanza del di 8 dicembre 1877.

Turnbull, H. W. (Ed.) (1960). *The correspondence of Isaac Newton* (Vol. II). Cambridge: Cambridge University Press.

Töpfer, H. A. (1793). *Combinatorische Analytik und Theorie der Dimensionszeichen, in Parallele gestellt.* Leipzig: bey Siegfried Lebrecht Crusius.

Unger, E. S. (1825). *Handbuch der mathematischen Analysis zum Gebrauch für Alle, die diese Wissenschaft zu erlernen und anzuwenden wünschen. Zweiter Band. Die Lehre von den Gleichungen, Funktionen und Reihen, und ihre Anwendung.* Erfurt und Gotha: in der Hennings'schen Buchhandlung.

Vallejo y Ortega, J. M. (1827). *Compendio de matématicas puras y mistas* (2 ed., Vol. II). Madrid: imprenta que fué de García.

Vandermonde, A.-T. (1772). Mémoire sur des irrationnelles de différens ordres avec une application au cercle. *Histoire de l'Académie royale des Sciences,* 489–498.

Varignon, P. (1700). Séance du 11 août 1700. *Registres des Procès-Verbaux de l'Académie royale des Sciences, 19,* f. 311r–317v.

Vogel, G. C. (1826). *Lehre der Arithmetik. Lehre der Buchstabenrechnung, Algebra, arithmetischen Reihen, Potenzen und Wurzeln dann Permutationen, Combinationen und Variationen.* Kulmbach: zu haben bei dem Verfasser.

von Ettingshausen, A. (1826). *Die combinatorische Analysis als Vorbereitungslehre zum Studium der theoretischen höhern Mathematik.* Wien: Druck und Verlag von J. B. Wallishausser.

von Goethe, J. W. (1798). Ernennung des Conrad Dietrich Martin Stahl zum a.o. Professor in Jena. In H. Dahl (Ed.), *Goethes amtliche Schriften. Veröffentlichung des Staatsarchivs Weimar 2. Goethes Tätigkeit in Geheimen Consilium, die Schriften der Jahre 1788–1819* (Vol. 2, pp. 583–585). Weimar: Hermann Böhlaus Nachfolger, 1970.

von Hermann, F. B. W. (1845). *Lehrbuch der Arithmetik und Algebra zum Gebrauch in Schulen und beim Selbstunterricht* (2nd ed.). Nürnberg: bei Riegel und Wießner.

von Liliencron, R. F., & von Wegele, F. X. (Eds.) (1875–1912). *Allgemeine deutsche Biographie.* 56 vols. Leipzig: Duncker & Humblot.

von Plato, J. (2017). *The great formal machinery works. Theories of deduction and computation at the origins of the digital age.* Princeton: Princeton University Press.

von Prasse, M. (1796). *Usus logarithmorum infinitinomii in theoria aequationum.* Lipsiae: Christ. Theoph. Rabenhorst.

von Prasse, M. (1799). *De reticulis cryptographicis.* Lipsiae: impressit Carolus Tauchnitz.

von Prasse, M. (1803a). *Functiones logarithmicae et trigonometricae in series infinitas solutae.* Lipsiae.

von Prasse, M. (1803b). *Theorematis binomialis demonstratio elementaris*. Lipsiae: Universitate Literarum Lipsiensi.

von Prasse, M. (1804). *Commentationes mathematicae*. Lipsiae: Christ. Theoph. Rabenhorst.

von Prasse, M. (1813). *Institutiones analyticae*. Lipsiae: apud auctorem et in bibliopolio kühniano.

von Sommer, F. (1823). *System der topisch-arithmetischen Combinations-Lehre, und der allgemeinen Auflösung aller Gleichungen*. Braunschweig: bei G. C. E. Meyer.

von Straßnitzki, L. C. S. (1844). *Handbuch der besondern und allgemeinen Arithmetik für Praktiker, zunächst für das Gelbststudium gemeinverständlich abgefaßt*. Wien: Gedruckt und im Verlage bei Carl Gerold.

von Tempelhoff, G. F. (1769). *Anfans-Gründe der Analysis endlicher Größen*. Berlin: bey Arnold Wever.

von Tempelhoff, G. F. (1770). *Anfangsgründe der Analysis des Unendlichen. Erster Theil, welcher die Differential-Rechnung enthält*. Berlin und Stralsund: Gottlieb August Lange.

Wallis, J. (1685). *A treatise of algebra, both historical and practical. Shewing, the original, progress, and advancement thereof, from time to time; and by what steps it hath attained to the heighth at which now it is. With some additional treatises*. London: John Playford.

Wallis, J. (1693). *De algebra tractatus; historicus & practicus. Operum Mathematicorum Volumen alterum*. Oxoniae: E Theatro Sheldoniano.

Wang, H. (1957). The axiomatization of arithmetic. *The Journal of Symbolic Logic, 22*(2), 145–158.

Watkins, E. (2005). *Kant and the metaphysics of causality*. New York: Cambridge University Press.

Weber, G. (1869). *Algebra. Selbstbelehrung. Beispielen und Aufgaben*. Stuttgart: Verlag der J. B. Metzler'schen Buchhandlung.

Weigl, J. B. (1812). *Lehrbuch der Arithmetik und Algebra zum öffentlichen Gebrauche und Selbstunterrichte*. Sulzbach: im Verlage der J. E. Seidelschen Kunst- und Buchhandlung.

Weigl, J. B. (1823). *Lehrbuch der Arithmetik und Algebra zum öffentlichen Gebrauche und Selbstunterrichte* (2nd ed.). Sulzbach: in des Kommerzienraths J. E. von Seidel Kunst- und Buchhandlung.

Weiner, J. (2020). *Taking Frege at his word*. Oxford-New York: Oxford University Press.

Weingärtner, J. C. (1800). *Lehrbuch der combinatorischen Analysis nach der Theorie des Herrn Professor Hindenburg* (Vol. 1). Leipzig: Gerhard Fleischer.

Weingärtner, J. C. (1801). *Lehrbuch der combinatorischen Analysis nach der Theorie des Herrn Professor Hindenburg* (Volume 2). Leipzig: Gerhard Fleischer.

Weingärtner, J. C. (1831). *Ueber die Bezeichnung in der combinatorischen Analysis*. Erfurt: Verlag von F. W. Otto.

Westfall, R. S. (1980). *Never at rest: a biography of Isaac Newton*. Cambridge: Cambridge University Press.

Whiteside, D. T. (1961). Patterns of mathematical thought in the later seventeenth century. *Archive for History of Exact Sciences, 1*(3), 179–388.

Whiteside, D. T. (1966). Newton's marvellous year: 1666 and all that. *Notes and Records of the Royal Society of London, 21*(1), 32–41.

Whiteside, D. T. (Ed.) (1967). *The mathematical papers of Isaac Newton, 1664–1666* (Vol. I). Cambridge: Cambridge University Press.

Whiteside, D. T. (Ed.) (1969). *The mathematical papers of Isaac Newton, 1670–1673* (Vol. III). Cambridge: Cambridge University Press.

Whiteside, D. T. (Ed.) (1976). *The mathematical papers of Isaac Newton, 1691–1695* (Vol. VII). Cambridge: Cambridge University Press.

Wittstein, T. (1872). *Lehrbuch der Elementar-Mathematik. Dritter Band. Erste Abtheilung. Analysis*. Hannover: Hahn'sche Hofbuchhandlung.

Wolff, C. (1705). Methodus serierum infinitarum. In *Meletemata mathematico-philosophica cum erudito orbe literarum commercio communicata* (pp. 290–319). Halae Magdeburgicae: in Bibliopoleo Rengeriano, 1755.

Wolff, C. (1710). *Der Anfangs-Gründe aller mathematischen Wissenschaften. Letzter Theil.* Halle im Magdeburgischen: Rengerischer Buchhandlung.

Wolff, C. (1713). *Elementa matheseos universae.* Halae Magdeburgicae: prostat in Officina Rengeriana.

Wolff, C. (1715). Meditatio de similitudine figuram praesertim curvilinearum & constructione lunularum cyclico-parabolicarum similium datamque inter se rationem habentium. In *Meletemata mathematico-philosophica* (pp. 72–77). Magdeburgicae: in Bibliopoleo Rengeriano, 1755.

Wolff, C. (1732). *Elementa matheseos universae* (Vol. 1). Genevae: apud Marcum-Michaelem Bousquet & Socios.

Wolff, C. (1747). *Vernünftige Gedanken von Gott, der Welt und der Seele des Menschen, auch allen Dingen überhaupt.* Halle: Rengerischen Buchandl.

Wunder, C. G. (1826). *Ueber Kombinationen des zweiten Grades oder Kombinationen von Kombinationen.* Wittenberg: Gedruckt in der Rübenerschen Buchdruckerei.

Wunder, C. G. (1844). *Versuch einer heuristischen Entwickelung der Grundlehren der reinen Mathematik zum Gebrauche bei dem Unterrichte auf Gelehrtenschulen* (2nd ed.). Leipzig: Bei E. B. Schwickert.

Yadegari, M. (1980). The binomial theorem: a widespread concept in medieval Islamic mathematics. *Historia Mathematica, 7*(4), 401–406.

Ypma, T. J. (1995). Historical development of the Newton-Raphson method. *SIAM Review, 37*(4), 531–551.

Zhmud, L. (1989). Pythagoras as a mathematician. *Historia Mathematica, 16*(3), 249–268.

Printed in the United States
by Baker & Taylor Publisher Services